Fluid Balance, Hydration, and Athletic Performance

Fluid Balance, Hydration, and Athletic Performance

Edited by
Flavia Meyer
Zbigniew Szygula
Boguslaw Wilk

CRC Press
Taylor & Francis Group
Boca Raton London New York

CRC Press is an imprint of the
Taylor & Francis Group, an **informa** business

CRC Press
Taylor & Francis Group
6000 Broken Sound Parkway NW, Suite 300
Boca Raton, FL 33487-2742

First issued in hardback 2019
First issued in paperback 2021

© 2016 by Taylor & Francis Group, LLC
CRC Press is an imprint of Taylor & Francis Group, an Informa business

No claim to original U.S. Government works

ISBN 13: 978-1-03-209815-9 (pbk)
ISBN-13: 978-1-4822-2328-6 (hbk)

Visit the Taylor & Francis Web site at
http://www.taylorandfrancis.com

and the CRC Press Web site at
http://www.crcpress.com

Dedication

In memory of Professor Oded Bar-Or, MD (1937–2005), with thanks for his mentorship and valuable scientific guidance that has been leading us throughout our professional lives

Flavia Meyer
Zbigniew Szygula
Boguslaw Wilk

Contents

SECTION I   The Fundamentals

SECTION II   Effects of Fluid Imbalance on Body Functions and Performance

SECTION III   *Special Populations*

SECTION IV   Recommendations

Preface

This book presents a comprehensive review of aspects relating to body fluid balance, rehydration, sport, and physical exercise. The content is scientifically supported, practical, and suitably written for a range of audiences, including academics (professors and students) and sports and health professionals (coaches, physical educators, nutritionists, and physicians), as well as athletes and individuals involved in physical activities.

Compared to other books previously published in this area, *Fluid Balance, Hydration, and Athletic Performance* does not limit body hydration issues to the average or elite adult athlete; it also addresses aspects relevant to a range of individuals of different ages (adolescents and master athletes) competing in various sports. In recognition of the growing number of individuals with specific medical conditions who have been exercising more and even participating in competitive sports, separate chapters on prevalent diseases or medical conditions associated with risks of body fluid homeostasis are also presented. To achieve such a complete and qualified publication, the book is written by top experts and professionals experienced in their respective research areas.

The book presents the basics of fluid balance and provides updates on controversial fluid intake–related issues such as hyponatremia, optimal recovery, intermittent sports, and perceptual responses.

We hope that readers will find this book useful and easy to apply to their respective theoretical and/or practical needs.

Flavia Meyer
Universidade Federal do Rio Grande do Sul
Porto Alegre, Rio Grande do Sul, Brazil

Zbigniew Szygula
University School of Physical Education
Kraków, Poland

Boguslaw Wilk
McMaster Children's Hospital and McMaster University
Hamilton, Ontario, Canada

Acknowledgments

We thank Dr. Ira Wolinsky for the invitation and encouragement to edit this book. This task would have been impossible without the positive response of so many well-known scientists who accepted our invitation and devoted several hours to share their expertise in the respective chapters of this book. We thank the authors of all chapters, as well as to all researchers whose cited works provided the scientific basis for the content of this book. We are also grateful for all reviewer comments, as well as suggestions from our friends and colleagues. These enabled us to prepare the concept, structure, and content of this book.

Editors

Dr. Flavia Meyer, MD, PhD, is currently a professor at the School of Physical Education, Federal University of Rio Grande do Sul (UFRGS), Brazil. She is also a faculty member for the Human Movement Graduate Program at UFRGS, where she has supervised 7 PhD and over 25 master's degree students.

Dr. Meyer received her medical doctor degree from UFRGS, which was followed by a two-year residency in pediatrics at Hospital de Clinicas de Porto Alegre, UFRGS. As well, Dr. Meyer completed a specialization course in sports medicine at UFRGS under the supervision of Professor Eduardo H. de Rose, who encouraged her to proceed with her studies in the area of pediatric sports medicine. Dr. Meyer was then accepted to the PhD medical sciences program at McMaster University in Canada to work under the supervision of Professor Oded Bar-Or at the Children's Exercise and Nutrition Centre.

Dr. Meyer's doctoral thesis involved a series of original studies on fluid balance, thermoregulation, and hydration in exercising youth. Dr. Meyer has researched and published in the field of pediatric exercise and nutrition science, and more specifically in hydration, thermoregulation, obesity, and chronic diseases.

Dr. Zbigniew Szygula, MD, PhD, is currently a professor and the head of the Department of Sports Medicine and Human Nutrition at the University of Physical Education in Kraków, Poland.

Dr. Szygula received his medical degree from the Medical Academy in Kraków, Poland, as well as medical specializations in general rehabilitation and sports medicine. He also received his MSc and PhD degrees in exercise sciences from the University of Physical Education in Kraków, Poland. Dr. Szygula's area of expertise is in exercise physiology and sports medicine, with an emphasis on sport's hematology, hypoxia, and thermoregulation. He has published extensively and has served as a reviewer in various international scientific journals. He was the founding partner and main editor of the journal *Medicina Sportiva* for 18 years. He is a member of the Polish Society of Sports Medicine, and he served as a member of the Board of the Society for many years. He is also a member of the Polish Scientific Society of Physical Culture,

Polish Society for Mountain Medicine and Rescue, and the International Society of Exercise and Immunology.

Dr. Szygula actively participates in several international conferences and has also organized international congresses in Poland. He is a sports physician for many athletes in Kraków.

Dr. Boguslaw "Bogdan" Wilk, PhD, FACSM, is currently a research associate with the Child Health and Exercise Medicine Program at the Children's Exercise and Nutrition Centre, Department of Pediatrics, McMaster University, Hamilton, Ontario, Canada.

Dr. Wilk received his MSc and PhD degrees in exercise sciences from the Academy of Physical Education (University School of Physical Education) in Kraków, Poland, where he worked as an assistant professor, until he moved to Canada in 1991. Upon arriving in Canada, Dr. Wilk secured a position as a research associate with the late Dr. Oded Bar-Or at the Children's Exercise and Nutrition Centre, McMaster University.

In 2003, Dr. Wilk began working as a clinical exercise physiologist at the Children's Exercise and Nutrition Centre, McMaster Children's Hospital in Hamilton, Ontario. He continued in this position until 2015, when he commenced his current role as a research associate with Dr. Brian Timmons.

Dr. Wilk's main research interest areas are, fluid and electrolyte replenishment in the exercising child and adult, and thermal regulation and tolerance of exercise in high and low environmental temperatures in health and disease. Throughout Dr. Wilk's research and clinical roles, he has continued publishing in a variety of scientific journals as well as serving as a referee and a member of international scientific editorial boards.

Dr. Wilk was previously a member of several scientific organizations in Poland and North America, and he is currently a fellow of the American College of Sports Medicine.

Contributors

Luis Fernando Aragón-Vargas
Human Movement Science Research
    Center
University of Costa Rica
San Jose, Costa Rica

Lindsay B. Baker
Gatorade Sports Science Institute
Barrington, Illinois

Kelly A. Barnes
Gatorade Sports Science Institute
Barrington, Illinois

Michael F. Bergeron
Youth Sports of the Americas
Lemak Sports Medicine
Birmingham, Alabama

Louise M. Burke
Sports Nutrition
Australian Institute of Sport
Belconnen, Australian Capital Territory,
    Australia

and

Sports Nutrition
Mary MacKillop Centre for Health
    Research
Australian Catholic University
Fitzroy, Victoria, Australia

François Carré
Service Explorations Fonctionnelles
Unité de Biologie et Médecine du Sport,
    CHU
Pontchaillou, Université Rennes
Rennes, France

Douglas J. Casa
Department of Kinesiology
and
Korey Stringer Institute
University of Connecticut
Storrs, Connecticut

Philippe Connes
CRIS EA 647 Laboratory
Vascular Biology and Red Blood Cell
University Claude Bernard Lyon
Villeurbanne, France

and

Institut Universitaire de France
and
Laboratory of Excellence in Red Blood
    Cell (LABEX GR-Ex)
PRES Sorbonne
Paris, France

Brent C. Creighton
Department of Kinesiology
University of Connecticut
Storrs, Connecticut

Scott L. Davis
Department of Applied Physiology &
    Wellness
Southern Methodist University
Dallas, Texas

Jeremy Fransen
College of Nursing
Loyola University
Chicago, Illinois

Victoria Goosey-Tolfrey
The Peter Harrison Centre for
    Disability Sport
School of Sport, Exercise and Health
    Sciences
Loughborough University
Loughborough, United Kingdom

Terri Graham-Paulson
The Peter Harrison Centre for
    Disability Sport
School of Sport, Exercise and Health
    Sciences
Loughborough University
Loughborough, United Kingdom

Emily Haymes
Department of Nutrition, Food and
    Exercise Sciences
Florida State University
Tallahassee, Florida

Craig Horswill
Department of Kinesiology and
    Nutrition
University of Illinois
Chicago, Illinois

Daniel A. Judelson
Korey Stringer Institute
University of Connecticut
Storrs, Connecticut

Stavros A. Kavouras
Department of Health, Human
    Performance and Recreation
University of Arkansas
Fayetteville, Arkansas

G. Patrick Lambert
Department of Exercise Science and
    Pre-Health Professions
Creighton University
Omaha, Nebraska

Ronald J. Maughan
School of Sport, Exercise and Health
    Sciences
Loughborough University
Loughborough, United Kingdom

Ricardo Mora-Rodriguez
Center of Socio-Sanitary Studies and
    Sports Sciences Faculty
Universidad of Castilla-La Mancha
Ciudad Real, Spain

Juan F. Ortega
Center of Socio-Sanitary Studies and
    Sports Sciences Faculty
Universidad of Castilla-La Mancha
Ciudad Real, Spain

Dennis H. Passe
Scout Consulting LLC
Hebron, Illinois

Dilip R. Patel
School of Medicine
Western Michigan University
Kalamazoo, Michigan

Thomas Paulson
The Peter Harrison Centre for
    Disability Sport
School of Sport, Exercise and Health
    Sciences
Loughborough University
Loughborough, United Kingdom

J. Luke Pryor
Department of Kinesiology
University of Connecticut
Storrs, Connecticut

Vimal M.S. Raj
Pediatric Nephrology
Children's Hospital of Illinois College
  of Medicine
Peoria, Illinois

Michael C. Riddell
School of Kinesiology & Health
  Science
York University
Toronto, Ontario, Canada

Anita M. Rivera-Brown
Department of Physical Medicine,
  Rehabilitation and Sports Medicine
and
Department of Physiology and
  Biophysics
University of Puerto Rico
San Juan, Puerto Rico

Paulo L. Sehl
School of Physical Education
Federal University of Rio Grande
  do Sul
Porto Alegre, Brazil

Manabu Shibasaki
Faculty of Human Life and
  Environment
Nara Women's University
Nara, Japan

Susan M. Shirreffs
School of Sport, Exercise and Health
  Sciences
Loughborough University
Loughborough, United Kingdom

John R. Stofan
Gatorade Sports Science Institute
Barrington, Illinois

Jane E. Yardley
Augustana Campus
University of Alberta
Camrose, Alberta, Canada

Section I

The Fundamentals

1 Body Water
Balance, Turnover, Regulation, and Evaluation

Craig Horswill and Jeremy Fransen

CONTENTS

1.1   INTRODUCTION

Undeniably, water, the nutrient comprising the greatest percentage of mass in the human body, has the most dramatic impact on function and structure when its balance is upset. The human body is resilient to some change. However, during physical activities, such as athletics, the effects of water deficit can be magnified to that point where physical and mental functions are diminished. Additionally, depending on the extent of water imbalance, the safety and well-being can be at risk.

Specific functions that water serves in the body include providing the medium for metabolic reactions, serving as a *substrate* or a reactant for specific reactions, transporting nutrients and waste products to specific locations, contributing in several ways to thermoregulation, and contributing to lubrication. For the athlete, transport of nutrients and waste and thermoregulation are especially critical roles in the body. The athlete's ability to sustain performance and avoid fatigue is in many ways attributed to the maintenance of homeostasis with substrate availability, transport of metabolites, and transfer of heat from the muscle, all of which require adequate blood flow at some point during the performance. Blood flow is highly dependent on having adequate blood volume, which hydration status will impact.

The objective of this chapter is to provide an overview of hydration of the athlete's body. Our review will include defining total body water (TBW) and its components; defining water balance and turnover; identifying the factors that influence water balance particularly during exercise; describing mechanisms of body water regulation to ensure balance is maintained; and identifying and describing methods of assessing hydration status and the state of fluid balance. In the spirit of this textbook, care will be taken to provide examples and data that are most relevant to the athlete.

1.2   TBW AND ITS COMPONENTS

The cumulative water volume in the body is called *total body water*. TBW comprises about 50%–65% of body weight (Mack and Nadel 2011; Schoeller 2005). In the reference adult male of 70 kg and 15% body fat, his body water volume is approximately 42 L. TBW is dependent on age, gender, and body composition. TBW is highest in infants (~70%) and decreases over the lifespan. Much of the age-associated decline in TBW is due to the increase in body fat, which decreases the percentage of TBW. Likewise, women usually have a higher body fat percentage than men, and therefore, TBW average about 50% of total body weight. Because body mass is highly variable based on the fat mass, and fat mass is essentially anhydrous, the TBW as a percentage of body mass is a poor indicator of the hydration status. In Section 1.5 of this chapter, we will return to the topic of hydration assessment.

TBW is compartmentalized into extracellular fluid (ECF) space and intracellular fluid (ICF) space. The ECF is further divided into the blood plasma found in the vascular space, and the interstitial fluid, which baths the cells but is outside of the vascular space. One other general ECF compartment that can be identified but will not be discussed further in this chapter is the transcellular space. This includes fluids in the cerebrospinal space, intraocular space, synovial space, and peritoneal and pericardial spaces. The total combined transcellular fluid amounts to approximately 1–2 L.

1.2.1 EXTRACELLULAR FLUID

The fluid outside the cell is called the *extracellular fluid*. The ECF is divided between the interstitial fluid, which is approximately 11 L, and the blood plasma, which equals about 3 L (Mack and Nadel 2011; Schoeller 2005). Therefore, for the average 70 kg man, the ECF totals approximately 14 L. The plasma is the fluid of the blood that does not contain cells. The plasma is in continuous exchange with the interstitial fluids through pores in the capillary membrane. Therefore, with the exception of higher concentrations of proteins in the plasma, the plasma and interstitial fluids have about the same composition. The ECF contains a high concentration of chloride and sodium ions with low concentrations of potassium, calcium, phosphate, and magnesium.

1.2.2 INTRACELLULAR FLUID

The fluid inside the cells is called the *intracellular fluid*. The ICF is about 28 L or approximately 40% of total body weight of the average person (Mack and Nadel 2011; Schoeller 2005). In contrast to the ECF, the ICF contains high concentrations of proteins, potassium, and phosphate and low concentrations of sodium and chloride.

1.2.3 PLASMA VOLUME

The plasma volume is the extracellular component of the blood volume. Blood plasma is nearly 95% water that contains dissolved proteins, clotting factors, electrolytes, glucose, hormones, and small amounts of oxygen and carbon dioxide. Besides plasma volume, the blood volume is also composed of ICF most of which is contained in the red blood cells. In the resting, normally hydrate state, roughly 60% of the blood is plasma with the remaining 40% red blood cells in humans. For the average adult, the blood volume equals about 7% of body weight or approximately 5 L (Astrand and Rodahl 1977).

1.3 WATER BALANCE AND TURNOVER

1.3.1 BALANCE AND EUHYDRATION

TBW is constantly changing with gains and losses of fluid even in the sedentary, weight-stable adult. The volume is regulated through a complex system of exchange of fluids, solutes, and ions within the compartments in the body and also influenced

TABLE 1.1
Routes and Rates of Daily Fluid Loss in Sedentary Human

Route of Loss	Volume (L/24 h)
Insensible from skin and breath (water vapor)	700 ml
Feces	100 ml
Urine	0.5–20
Sweat	0.1–7.0

Source: Horswill, C. A., *Int. J. Sport Nutr.*, 8, 175–95, 1998.

by the environment. A human is described as euhydrated or in a state of euhydration when the volume of water is that which supports good health and normal function. The assessment of hydration status will be addressed later in this chapter, but suffice to say, a single accurate and precise indicator of euhydration is yet to be identified (Armstrong 2007). To maintain the proverbial *well-hydrated* state, variable daily fluid intake or gain must match the daily losses. Water gain originates from ingested beverages and foods (~90%) and from metabolism during the oxidation of carbohydrate and lipids within the cell (~10%). Losses occur through a number of routes including insensible loss from skin, water vapor in the breath, urine production, fecal loss, and perspiration or sweating (see Table 1.1 for summary).

1.3.2 WATER TURNOVER

Closely related to water balance is daily water turnover or the volume of body water that is exchanged in 24 h. In the sedentary adult male, daily water turnover is approximately 3.6 L, and in the sedentary adult female, it is about 3 L (Raman et al. 2004). The amount is lower in female mainly due to the smaller body mass and pool of body water. Using these data, the Institute of Medicine (IOM) has identified daily water needs to be 3.7 L/24 h for the sedentary adult male and 2.7 L/24 h for the sedentary female (Institute of Medicine 2004). Body water balance is the mathematical difference between the sum of all intakes and sum of all losses in 24 h. A positive balance indicates a water surplus and a negative balance indicates a water deficit.

1.3.3 PROCESS OF CHANGE AND RESULTANT STATES OF HYDRATION

When fluid intake does not match fluid loss, a water deficit occurs. The process leading to a deficit is defined as *dehydration*. A human who undergoes dehydration and remains in that state is said to be in a steady-state of *hypohydration*. This can occur voluntarily such as in athletes that must make a weight class for competition or involuntarily in athletes who are at altitude or in a desert climate and may not realize the greater body water loss that has occurred. The process of fluid replacement to correct the deficit defines *rehydration*. Complete rehydration would restore euhydration. In the case of excess fluid intake that creates a water surplus, the new state produced is one of *hyperhydration*. Hyperhydration might be accomplished transiently by forced ingestion of excess fluids; however, to maintain a state of hyper-hydration would require the addition of osmolytes that generate an osmotic pressure for fluid retention. Strategies to help achieve complete rehydration post exercise or hyperhydration before exercise in the heat are provided in Chapter 18 of this book.

Dehydration can occur in several ways, and the specific processes discussed here are the ones having relevance to the athlete. The most common type is referred to as *hypovolemic hypertonic dehydration,* which has also been described as *hypernatremia dehydration.* High sweat rates during training or competition typically promote hypovolemic hypertonic dehydration. Because the sweat gland draws on the plasma water to produce sweat, plasma volume decreases (hypovolemic). The sweat gland reabsorbs electrolytes, and thus, water is lost disproportionately to the ions in the vascular space. Consequently, the remaining sodium, chloride, and other ions are concentrated resulting in greater tonicity of the blood (hypertonic). Because sodium is the most prominent ion in the blood, sodium concentrations will rise above normal producing hypernatremia, which is an increase in plasma sodium concentration. Although ECF is the frontline source of water being lost, the higher tonicity or osmolality of the blood will draw fluid from the ICF. Consequently, both compartments may experience a decrease in water volume. Hypertonicity, or hyperosmolality, plays a critical role in water regulation, which is discussed in Section 1.4 of this chapter.

Less probable but still possible, dehydration can result from a loss of sodium chloride and partial rehydration that dilutes blood sodium; this state is termed *isotonic hypovolemic dehydration.* This type of dehydration may occur with diuretic use such as in athletes who make weight or use diuretics to mask the use of banned substances. Certain diuretics work by increased ion excretion by the kidney; expulsion of sodium, potassium, or other minerals draws water too. Consequently, water and the electrolytes are lost proportionately to give the appearance of the blood not being dehydrated. Besides the ill effects of dehydration, the large potassium losses might put individuals at risk of cardiac arrhythmia.

Finally, *hyponatremia dehydration* has been reported, though infrequently, at medical tents for endurance events conducted in the heat. Tremendous losses of sodium in the sweat as a consequence of high sweat sodium concentrations and high sweat rates for an extended period are risk factors for hyponatremia. In addition, consumption of fluids that partially rehydrate but fail to replace any sodium contributes to dilution of the already compromised blood sodium content. This phenomenon has been reported during training in the heat by the American football players (Horswill et al. 2009) and in distance events (Laird and Johnson 2012).

1.3.4 FACTORS CONTRIBUTING TO WATER IMBALANCE IN ATHLETES

Hydration status can be compromised by factors that include behavior of drinking, access to fluids, physical activity including mode, intensity, and duration of the effort; the macro- and micro-environment. Macro-environment consists of the external surroundings, and micro-environment is determined by clothing such as athletic uniform or protective gear. For example, as an extreme case, an American football player wearing full gear lost ~7.7 kg while training for 2 h in a hot humid environment (author's observations while conducting—the study by Horswill et al. 2009).

Likewise, despite minimal uniform, a triathlete in the World Championship in Hawaii competes in the heat for ~8.5–15 h depending on his or her level of ability and risks severe dehydration and electrolyte depletion with high sweat rates and inadequate intake along the race course.

1.3.4.1 Environment

The external environment influences body water balance through air temperature, humidity, atmospheric pressure (i.e., altitude), microgravity, and hyperbaric conditions (i.e., deep water diving). Several of the factors may be present for additive effects. For example, mountaineering at a moderate to high terrestrial elevations (>2300 m) along with cold temperatures and dry air can increase water loss through the respiratory passages. Most often, the environmental factor that can impair TBW balance is high ambient temperature. When combined with a high relative humidity, heat storage in the body can increase resulting in an increase in sweat rate to cool the body and thus a decrease in TBW unless fluid loss is replaced with fluid consumption. In microgravity, body fluids shift to the head and torso resulting in a diuresis effect, increasing urine output thus decreasing TBW. At the same time, weightlessness can cause nausea and vomiting, which would compound fluid losses. Effects in specific environments will be briefly discussed in the next paragraphs.

1.3.4.1.1 Altitude

Acute and chronic exposure to altitude can impact water turnover through several physiologic responses and adaptations. Upon ascent to altitude, the immediate compensatory response to the low partial pressure of oxygen (PO_2) is an increase in pulmonary ventilation (V_E). Increases in V_E facilitate respiratory water loss due to the respiratory passages heating and humidifying the air breathed into the trachea. Exposure to moderate-to-high altitude further facilitates respiratory water loss because air at high terrestrial altitude is often cold and dry, which accentuates the gradient between lung moisture and the environment. With the elevated V_E, bicarbonate production is increased, which is excreted by kidneys. However, the expulsion of the bicarbonate anion and renal clearance rate increases water excretion, which is a diuresis that promotes dehydration with a reduction in blood plasma and TBW. With sustained exposure to elevation, the body undergoes adaptations, and through the stimulation of thirst, fluids are regained and help to eventually restore plasma volume and TBW.

1.3.4.1.2 Microgravity

Spaceflight promotes a negative body water balance as a result of several different physiological stressors during the take-off: hyper-gravity g-forces, the microgravity environment itself, and the hydration status of the individual along with psychological stress. TBW, ECF volume, and plasma volume have been measured on several NASA flight missions. After the Skylab missions of the 1970s, TBW decreased by an average of 1.7% at landing (Leach and Rambaut 1977). ECF volume was 2% lower than pre-flight levels following the Skylab mission and decreased by 10% after 24 h of spaceflight on the shuttle missions. ECF volume returned to normal values upon landing (Huntoon et al. 1998). Plasma volume has been shown to decrease by up to 17% within 24 h of spaceflight and then seems to stabilize at 10%–15% of pre-flight values after 60 days (Huntoon et al. 1998). In addition, the length of stay in microgravity can have an impact on the physiologic adaptations that take place. Without gravity, fluids from an interstitial space return to the vascular space and shift to the upper body. This increases venous return, atrial stretching, and consequently the release of atrial natriuretic peptide, which increases urine output. The decreased plasma volume and blood volume has consequences on systemic cardiovascular function. Over time, a decreased blood volume can decrease cardiac size. Orthostatic intolerance or the ability to maintain normal blood pressure and reduced blood flow to the brain while standing results from the decrease in blood volume and cardiac size. Upon returning to a gravitational field when landing, individuals are more susceptible to dizziness and fainting.

To combat these negative consequences, different fluid intake strategies have been suggested by NASA. Astronauts usually consume a fluid beverage containing moderate amounts of sodium to expand plasma volume prior to reentry to mitigate orthostatic intolerance (Buckley 2006).

1.3.4.1.3 Underwater

Impact of the underwater environment has been examined using head-out immersion in thermoneutral water. This causes a translocation of fluid from the limbs and increases intrathoracic blood volume (Epstein 1992). The transient fluid shifts increase venous return, atrial stretch, and compensatory diuresis. The shift in transcapillary fluid from the extravascular compartment into the blood results in an initial expansion of plasma volume, which returns to normal following diuresis and increased urine output. The fluid shift is from the intracellular compartment, and the kidney maintains this balance by increasing water loss to prevent a large increase in plasma volume and cardiac output (Krasney 1996; Miki et al. 1986). Plasma volume can initially increase by approximately 7% during immersion, and as immersion depth increases, there is a concomitant increase in tissue pressure, capillary pressure, and central venous pressure. While there are no changes in plasma osmolality, the hemodilution results in a decline in plasma oncotic pressure. Interstitial fluid remains constant suggesting the shift in fluids come from the ICF compartment (Pendergast and Lundgren 2009). Similar to the astronauts returning to a gravitational field, the diver returning from underwater will experience the challenges of orthostatic hypotension due to hypohydration and compromised cardiac output.

1.3.4.2 Heat and Humidity

When the human body is exposed to higher than normal ambient temperatures (i.e., above 28°C), skin vessels vasodilate to increase blood flow and facilitate heat loss via conduction and convection. If high ambient temperatures and/or humidity exist, the sweat glands are activated because heat dissipation may not be adequate merely by conduction, convection, and/or radiation. Heat is then lost via evaporation of the sweat from the surface of the skin. Evaporation of water via sweat production on the surface of the skin provides approximately 600 kcal, or 2400 kJ of heat transfer per liter of sweat. The relative humidity also plays a large role in the ability to transfer heat from evaporative sweat. In high humidity, the evaporation of sweat is decreased in part to higher water vapor pressure independent of the air temperature. For example, lower ambient temperatures with high humidity (30°C; 70% RH; ~3 kPa) will have a higher water vapor pressure and decrease evaporative heat loss compared to a higher ambient temperature and lower relative humidity (40°C; 20% RH; ~1.5 kPa). In addition to skin, respiratory heat exchange is responsible for conductive, convective, and evaporative heat loss. The mechanisms are described later in this chapter, but briefly, water conservation engaged during thermal stress is dependent on individual sweat rate and the ability of the kidneys to conserve water and electrolytes. With acute exposure to high ambient temperatures, urine output may be reduced to half of normal, from approximately 22 to 10 ml/h (Lee 1964). With progressive dehydration, urine output continues to decrease with minimal excretion for removal of waste as the kidney attempts to conserve body water. Regardless, though, if fluid loss continues with sweating, heat production continues due to exercise, and fluids are not ingested to replace, the safety and function of the individual will eventually be compromised.

1.3.4.3 Exercise and Heat Acclimation

Independent of environmental effects, exercise increases metabolic heat production, which raises core temperature and signals the sweating response. Exercise-induced sweating is then dependent on heat production, which is affected by the mode, duration, and intensity of the physical activity (Sawka and Wenger 1988). In addition, sweat capacity can vary considerably between individuals due to the number of sweat glands, activation of the sweat glands and their efficiency (Burke and Hawley 1997; Maughan 1991). The ability of an individual to sweat decreases with age due to the declining function of the sweat glands (Inoue 1996). Sweat production can exceed several liters per hour and may reach as high as 6–7 L in a day when workers are exposed to a hot environment. There have been reports of sweat loss totaling 12 L within a 24-h time period (Leithead and Lind 1964). As a result of exercise with chronic heat exposure, that is, 90–100 min of exercise a day for 10–14 days, heat acclimatization will elevate human sweat production at a given ambient temperature by about 50%. The composition of the sweat becomes more dilute during acclimatization, as the sweat gland increases electrolyte reabsorption and conserves the ions (Armstrong and Maresh 1991; Kirby and Convertino 1986). Prolonged sweating at a higher sweat rate promotes greater water loss and increases the chances of an athlete developing chronic dehydration with training for several days and particularly if multiple training sessions are held each day in a hot environment.

1.3.4.4 Illness and Travel

Illness can affect hydration of athletes as a result of overtraining, exposure to viral or other pathogens, and during traveling for competition. Illness accompanying diarrhea, vomiting, and fever that results in sweating will promote water loss. Inability to hold down fluids or stop diarrhea also compromises nutritional status beyond hydration and includes deficits of electrolytes such as sodium, potassium, and chloride. The consequences of illness can be hyponatremia and hypokalemia with decreases in ECF volume and plasma volume. As described above, hyponatremia can cause severe swelling of the brain. If the hyponatremia occurs slowly over time, the brain responds by transporting electrolytes and solutes from the cells to the ECF compartment thus attenuating the osmotic flow of water into the cells mitigating swelling. If hyponatremia is severe with a rapid onset, osmotic injury of the neurons and demyelination occur.

Independent of illness, travel, particularly air travel, can disrupt fluid balance. Confined to a cabinet of relatively dry air that circulates rapidly, insensible fluid loss from the skin is accelerated. Some estimates put fluid loss during a 3-to-4 h flight to that similar to losses during 8 h of sleep. While the risk may be low during flight, trans-Atlantic or transpacific air travel that induces dehydration may increase the risk of deep vein thrombosis particularly with limited movement (Chee and Watson 2005).

1.3.4.5 Other Factors

Cultural events, religion expression, and sports-specific behaviors that are based on misconceptions can influence water turnover, balance, and potentially alter hydration state. Fasting for religious reasons such as Ramadan will potentially reduce fluid intake and be quantified as reduced fluid turnover during the month of observation. It is clear that fasting promoted dehydration (Leiper et al. 2003) and that in athletes observing Ramadan, the potential for acute dehydration during training or competition is great but the variability does not seem to adversely affect sweating for thermoregulation (Shirreffs and Maughan 2008). Food and fluid restriction to make weight for competition in combative sports reduces fluid turnover and promotes hypohydration over a period of days according to one case control (Horswill 2009). Finally, misconceptions such as avoiding water consumption for a period of time in hopes of *oxidizing more body fat* also reduce fluid ingestion and promote hypohydration that could adversely affect subsequent training and athletic competition.

1.4 REGULATION OF FLUID BALANCE

1.4.1 Mechanisms

Given the importance of water and its functions in the body, regulation to ensure preservation and restoration of an adequate amount and distribution within the body would seem to be essential. The body has two general mechanisms by which to do so: that which is driven by osmoreceptors and that driven by baroreceptors. Thorough reviews have been provided by Mack and Nadel (2011), Wade and Freund (1990), and Zambraski (1990). In brief and simplified terms, chemical and physical changes are detected by the vascular compartment of the body and lead to hormonal signals that result in physiological or behavioral responses to conserve or replenish the vascular water space, thereby restoring TBW and in particular the ECF compartment, which includes the plasma volume. Additional details of each mechanism are presented in the following Sections 1.4.1.1 through 1.4.1.3.

1.4.1.1 Osmoreceptor Mechanisms and Responses

As water is lost from the body in the form of sweat during exercise, the remaining ECF becomes more concentrated with osmolytes such as sodium. Sweat is hypotonic relative to the blood thereby increasing the particle content in the body fluid.

The increased osmolality of the blood, which is a component of the ECF compartment, is detected by osmo- or chemoreceptors located in the anterior hypothalamus. This stimulates the release of arginine vasopressin (AVP), also known as antidiuretic hormone, from the posterior hypothalamus. The release of AVP stimulates the kidney to decrease free water clearance as a means to conserve body water. Urine production decreases and osmolyte clearance, namely sodium retention, increases. AVP also stimulates thirst, which will induce the drinking behavior to help restore body water via exogenous sources in contrast to the other responses merely conserving existing body water.

1.4.1.2 Baroreceptor Mechanisms and Responses

Besides increasing osmolality, the reduction in water volume of the vascular space decreases the blood pressure. This is sensed by baro- or pressure-receptors located in the cardiopulmonary system. In response, aldosterone is secreted by the adrenal cortex and renin from nephrons. Renin release also stimulates angiotensin secretion, which indirectly can stimulate aldosterone release. This hormone system stimulates a sodium appetite leading to increased sodium intake that could promote water retention. The hormones also stimulate the nephrons of the kidney to reabsorb more sodium which will simultaneously draw more water for reabsorption and conservation. Renin–angiotensin also causes vasoconstriction to help conserve the central blood volume to maintain cardiac output when the body is functioning during a fluid deficit. The end result of these actions is an increase in extracellular sodium; an increase in arterial pressure, which would help restore blood pressure to normal; and an increase in glomerular filtration rate.

1.4.1.3 Other Mechanisms of Conserving Fluid Balance

Other responses that occur during dehydration and exercise include an elevation of plasma catecholamines, epinephrine, and norepinephrine and an increase in the autonomic nervous system activity. These responses are mechanisms that help to maintain central blood volume at a time when it may be decreasing due to sweat loss and vasodilation in the active skeletal muscle. However, they do not seem to be directly involved in body water conservation or replacement from exogenous sources.

1.4.2 Limitations

During physical exertion such as training or competition, the mechanisms previously described are intact. However, because AVP doesn't initiate a thirst response until plasma osmolality reaches a critical level, 1%–2% reduction in body weight can occur before an athlete or exerciser has the desire to drink. This phenomenon has been described as *involuntary dehydration*, as in an unintended outcome (Greenleaf 1966; Greenleaf and Sargent 1965). The outcome is fairly consistent among children and young and older adults but may differ between men and women (Baker et al. 2005; O'Neal et al. 2012; Passe et al. 2007; Wilk and Bar-Or 1996). Males seem to develop dehydration, whereas females are more likely to consume adequate or excess volumes (Baker et al. 2005; Wilk and Bar-Or 1996; Wilk et al. 2007). The reason is unknown with speculation being that males, the hunters in ancient civilizations, have the capability to override the drive to drink in pursuit of food or avoidance of predators, while females that could possibly be carrying offspring are more assured of staying well hydrated for the safety of the fetus.

Quite surprisingly, limitations of these mechanisms appear to exist in the opposite direction; in some cases, the biological system appears to allow overhydration during exercise and has received increasing scrutiny in research and clinical settings. Acute overhydration is a relatively infrequent event during prolonged athletic competition such as marathons or triathlons; however, the result can be and has been deadly. Excess water results in the dilution of plasma sodium. If the concentration drops below 136 mmol/L, hyponatremia is indicated (Montain et al. 2001). The relative decrease in sodium reduces the osmotic effect for maintaining fluid in the extracellular space. Consequently, fluids shift into cells (ICF), to an extent that tissues such as the brainstem become impaired if not damaged, and ventilation and brain function are disrupted. The exact mechanism for hyponatremia is unclear. In most individuals, the excess fluid load would promote free water clearance with the excretion of dilute urine, but for whatever reason, this does not happen in victims of hyponatremia. Syndrome of inappropriate antidiuretic secretion, that is, excess secretion of AVP at a time when it is not needed to conserve fluid, has been implicated. Epidemiological-type studies at endurance events indicate the prevalence of symptomatic hyponatremia may be 0.1%–4% of participants with percentages being higher, between 3% and 27% of those athletes who report to the medical tent (Almond et al. 2005; Montain et al. 2001). The risk of hyponatremia is further discussed in Chapter 2.

1.5 EVALUATION OF BODY HYDRATION STATUS/BALANCE

1.5.1 RESEARCH METHODS

1.5.1.1 Total Body Water

To assess TBW, the criterion method is to use a tracer that equilibrates in the body water space. The extent of dilution, once the tracer has been given enough time to equilibrate, establishes the volume of water—the TBW—required to achieve that dilution. The tracer used is a heavy form of water, either a stable isotope such as deuterium oxide (D_2O) or a very small amount of a radioactive isotope, tritium oxide. Alternatively, a tracer with labeled oxygen, that is, ^{18}O-water, can be used. A sample of body fluid (D_2O or ^{18}O-water) or breath sample that is in equilibration with the body water (^{18}O-water) is collected before dosing, to establish background levels and then 3 or 4 h post ingestion when the tracer has had adequate time to equilibrate in the body water pool (Schoeller et al. 1980).

TBW, though, does not establish the hydration status of an individual; it is merely a measure of the absolute water volume, which is related to the size of the individual. Because of a fairly fixed ratio with the fat-free tissue of the body, fat-free mass (FFM) can be established once TBW has been measured. Table 1.2 lists the generally accepted ratio across the life span. The general pattern is a decrease in the hydration status of the fat-free body due to the rate of increase in solids (protein and mineral) surpassing the rate of increase in water with maturation. In contrast to what

TABLE 1.2
Ratio of Total Body Water to Fat-Free Mass (TBW/FFM) in Healthy Females (F) and Males (M)

Phase of Life	TBW/FFM Ratio
Infant	0.806
Child, 10 years	F: 0.765; M: 0.746
Young adult, 21–30 years	0.73 with range of 0.69–0.77
Elderly adults, ~70 years	0.725–0.733

Source: Schoeller, D. A., Hydrometry. In S. B. Heymsfield, T. G. Lohman, Z. Wang, and S. B. Going (Eds.), *Human Body Composition,* 2nd ed., Human Kinetics, Champaign, IL, 2005, pp. 35–50; Bossingham, M. J. et al., *Am. J. Clin. Nutr.,* 81, 1342–50, 2005; Wells, J. C. K. et al., *Am. J. Clin. Nutr.,* 69, 904–12, 1999.

is thought for elderly, hydration of the fat-free body may not differ that much from the commonly cited value of 0.738 (Brozek et al. 1963).

Therefore, to determine the hydration status, an independent assessment of the FFM is needed along with the TBW water assessment. Densitometry, whole body potassium, or DEXA can be used to provide FFM. The ratio of TBW from the tracer method to FFM from the independent method will reveal the hydration status of the individual and can be contrasted to the expected norms for the appropriate reference group.

Quite surprising, the ratio has been found to fluctuate in college-age athletes who are presumed to be biologically mature, and the discrepancy has generated a debate about what is normal for athletes. Researchers at the University of Georgia have reported that the TBW to FFM ratio may be higher or lower depending on gender and the type of athlete being assessed (Prior et al. 2001). Table 1.3 summarizes mean values for the athletes studied.

The fluctuation could be due to a variety of factors including genetics, diet, acute and chronic state of the subjects, and specific training. Athletes that *survive* youth and adolescent sport leagues to go on to participate at the collegiate level likely

TABLE 1.3
Hydration Status of Athletes Based on Total Body Water (TBW) and Fat-Free Mass (FFM)

Athlete	TBW/FFM Ratio
Male football players	0.744
Male swimmer	0.752
Female swimmer	0.750
Female gymnast	0.690

Source: Prior, B. M. et al., *J. Appl. Physiol.,* 90, 1520–31, 2001.

have a genetic predisposition that gives them the phenotype required to succeed in their sport. The natural selection process (minimal success, injury, etc.) may weed out those athletes that lack the ideal phenotype and composition. Diet, particularly a high-carbohydrate diet, could promote hyperhydration in the skeletal muscle. Several decades ago, Olsson and Saltin (1970) reported an increase in TBW due to carbohydrate loading. Dietary supplements such as creatine may act as an osmolyte to promote water retention within skeletal muscle fibers. Creatine in fact has been shown to elevate TBW (Powers et al. 2003). If athletes are acutely but unknowingly dehydrated at the time of their assessment, the water to FFM ratio is expected to be lower due to less dilution of the tracer or an inflated estimate of body density if FFM was determined by hydrostatic weighing (Girandola et al. 1977). The subjects from which the data in Table 1.3 were derived were reportedly euhydrated based on an average urine specific gravity of 1.020 ± 0.009. As an example of the effect of chronic state, heat acclimation results in expanded body water possibly due to sodium retention and an increase in plasma albumin both of which would contribute to greater retention of water in extracellular compartment. Finally, the type of training the subjects had done over the course of their sports career, more impact training for example in football or gymnastics, might alert the bone mass, which would change the denominator, that is, FFM and thereby alter the ratio.

Besides using a tracer to measure TBW in a static state, the same tracer can be used to assess fluid turnover, the dynamic state of water flux as defined early in this chapter. While this has relevance in research settings, it would not be practical or cost effective in field settings. The observed rates of water turnover in athletes are summarized in Section 1.3.

1.5.1.2 Extracellular Fluid and Intracellular Fluid

As an approximate index, ICF comprises 67% of the TBW and ECF fluid accounts for the remaining third (Schoeller 2005). ECF is composed of interstitial fluid and the plasma volume. Similar to the methodology used for TBW, administering a tracer that equilibrates only in the ECF fluid provides a means to quantifying the ECF volume via the dilution of the tracer. The difference between TBW and ECF provides the ICF volume. Among the tracers used to measure ECF, sodium bromide is effective. Bromide being a halogen behaves similarly to chloride by equilibrating primarily in the extracellular space. Its dilution, 2–3 h after a specific dose is ingested, indicates the ECF volume with appropriate corrections for tracer loss. Currently, a tracer to directly assess ICF volume is not available. If one were, it would require invasive sampling such as a biopsy or biopsies of tissue to examine and quantify the dilution in the intracellular space.

1.5.1.3 Blood Volume and Plasma Volume

Blood volume is quantified with a tracer such as Evan's blue dye, radioactively labeled albumin, or carbon monoxide. The principles of the method rely on (1) a marker that is introduced into the vascular space; (2) the marker does not to leave that space, or if losses occur, they can be estimated or quantified; and (3) the dilution of the marker after a relatively short equilibration period is established by the volume of blood required to achieve the dilution.

The use of a tracer for the direct measure of blood volume is infrequent due to required technology, invasiveness and risks to the subjects, and time required to make analytical quantifications. More common is the measurement of relative plasma volume, or the change in plasma volume, known as the Dill and Costill method (1974). Unlike the tracer methods that provide an absolute volume, the Dill–Costill approach assumes starting with 100% of the blood volume in a rested, euhydrated state; acute changes can be quantified by using changes in hemoglobin and hematocrit as a result of exercise, dehydration, rehydration, or posture changes.

1.5.2 CLINICAL METHODS

The choice of method to categorize hydration status depends on the situation and environment. The choice made also depends on efficiency for the number of assessments, invasiveness, costs and resources, and the consequences of false negatives and false positives if categorization is in error. The criteria used to estimate the category for various methods are summarized in Table 1.4. A brief description of the methods is provided in the subsequent section.

1.5.2.1 Blood and Urine Parameters

In research and clinical settings, plasma osmolality is often considered the gold standard for verifying or determining the hydration status. Measurements are done using osmometers relying on freezing point depression or vapor pressure (change in the dew point temperature for pure water as a result of a change in vapor pressure of the sample). Osmolality of human plasma can range from 275 to 300 mOsm/kg (Scott et al. 1999) with normal values between 260 and 280 mOsm/kg for a euhydrated state.

As a proxy for blood osmolality, urine samples can be obtained noninvasively and assessed for osmolality. Again, freezing point depression or the change in vapor

TABLE 1.4
Summary of Criteria of Body Fluid Samples Used to Estimate Hydration Status

Criteria	Euhydrated	Dehydrated (>−2% Mass)
Plasma osmolality (mOsm/kg)	260–280	>290
Hemoglobin (g/dL)		
Males, 15–44 years	11.7–17.3	Increase versus baseline
Females, 15–44 years	11.7–15.5	Increase versus baseline
Hematocrit (%)		
Males, 15–44 years	35–48	Increase versus baseline
Females, 15–44 years	34–44	Increase versus baseline
Urine osmolality (mOsm/kg)	200–400	>900
Urine-specific gravity	1.001–1.020	>1.020

Source: Burtis, C. A. and E. R. Ashwood, *Tietz Textbook of Clinical Chemistry*, 3rd ed. Philadelphia, PA: Saunders, pp. 1817–26, 1999; Horswill, C. A. and L. M. Janas. *Am. J. Lifestyle Med.* 5, 304–15, 2011; Kavouras, S. A. *Curr. Opin. Clin. Nutr. Metab. Care.* 5,519–24, 2002; Kenefick, R. W. and S. N. Cheuvront. *Nutr. Rev.* 70(Suppl. 2), S137–42, 2012; Oppliger, R. A. and C. Bartok. *Sports Med.* 32, 959–71, 2002.

pressure method could be used to make the assessment. Urine osmolality values less than 600 mOsm/kg are generally regarded as indicating euhydration, whereas values at or above 900 mOsm/kg are considered to indicate hypohydration (Shirreffs and Maughan 1998).

1.5.3 FIELD AND NONINVASIVE ASSESSMENT

For the purpose of athletics and tracking hydration status during training and competition, field methods are most practical as long as they are reliable and accurate. Field methods and those not requiring relatively long and invasive procedures are also of great value when coaches or sports medicine personnel are evaluating a large group, for example, team of athletes, and need to be efficient and effective with time and limited exposure to the athletes. It should be noted in the case of predictions, there is always a prediction error, technically the standard error of estimate from the regression analyses. This means applying a prediction equation developed from a sample of people will have an error when applying the prediction to an individual.

1.5.3.1 Bioelectrical Impedance

Conduction of an electrical current through the body has been used to track changes in the hydration status of individuals. The principle of the method is that the ability of the body to conduct an electrical current is dependent on conductivity and resistance of individual tissues in the body, and the water and electrolyte content of the tissues determined, in large part, the tissues ability to conduct electricity. Having established a baseline measure of a person in a euhydrated state, a subsequent measure after an abrupt change in hydration status should be detectable and quantifiable. However, the original impedance technology on the market was not consistent in detecting acute changes in TBW as a result of exercise-induced acute weight loss.

1.5.3.2 Urine Characteristics

Collecting urine to assess it as a proxy of internal body fluids has also been used to predict the hydration status. The primary application of urine markers has been for discrete categorization: a person as either euhydrated or hypohydrated, and possibly *predicted to be inadequately hydrated* would be a more accurate description than hypohydrated. Urine osmolality was described earlier as a clinical measure of hydration status. Given that osmometers are expensive and rarely found outside of clinical or research laboratories, osmolality has little use in the field for a rapid and real-time assessment. Instead, urine-specific gravity and/or urine color is commonly used. The correlation between urine osmolality and urine specific gravity is highly significant (Stover et al. 2006).

Urine density or more commonly urine-specific gravity, the measure of the mass-to-volume ratio standardized to water, increases when a person is not drinking adequately or the body is in a state of energy conservation such as during exercising. (Exercise induces hemo-concentration and volume reduction independent of sweat loss, and the kidneys respond by decreasing urine product to conserve the blood and body water volume.) The urine subsequently passed is more concentrated—has a high particle content and mass relative to the volume; this is detected as an increase in specific gravity. Specific gravity is quantified by using a refractometer or a reagent strip that uses color change to evaluate the urine as a proxy for hydration status. Sports medicine experts and professional societies have agreed that 1.020 is the cut-off for decision making: urine having a specific gravity of less than or equal to 1.020 indicates euhydration, while urine-specific gravity in excess indicates inadequate hydration (Casa et al. 2000). The potential error with urine specific gravity is that the change in urine characteristics can lag behind the internal state of the body (Popowski et al. 2001; Ryan et al. 1998).

Urine electrical conductivity can also be used to predict osmolality. Shirreffs and Maughan (1998) demonstrated consistency between the two methods: testing athletes in a euhydrated state and again in a dehydration state (~2% body weight reduction).

The appearance of the urine will increase in intensity, darken in shade of color, and the urine becomes more translucent than transparent. Color charts have been developed as a simplified method for athletes (Casa et al. 2000). While practical and simple, color may be oversimplified, influenced by diet independent of hydration status, and varies depending upon environmental conditions (e.g., color appearance depends on the type of overhead lighting, dilution in toilet water, etc.).

Despite the limitations, urine-specific gravity continues to be used commonly as a practical assessment and has been demonstrated as being highly sensitive and specific when changes in TBW are assessed using the isotope methods previously described (Bartok et al. 2004). The debate over its validity continues. A recent review argues against using a single void at any given time of the day (spot urine concentration) to establish the hydration status (Cheuvront et al. 2015). However, given the required application for weight-class athletes who frequently practice dehydration to achieve a lighter weight for competition, the barrier of passing the urine test reduces the severity of the dehydration that otherwise puts the participants in peril.

1.5.3.3   Change in Body Mass

Acute change in body weight consequent to fluid lost as sweat during exercise, diuresis, fasting, or diarrhea and/or vomiting due to illness is a very practical and often applied means to tracking relative hydration status. The assumption during exercise is that the majority of the change in body mass is due to fluid lost as sweat. In the case of athletes who exercise for prolonged periods, the accuracy of this method has been debated (Baker et al. 2009; Tam et al. 2011; Maughan et al. 2007). In the case of exercise-induced weight change in warm or hot environments, the majority of mass lost is that of water and minor amounts due to substrate oxidation. In cooler environments, in which sweating is curtailed due to more efficient heat exchange with a greater heat gradient between the active body and environment, substrate oxidation may play a more prominent component of the reduction in mass, possibly approaching 25% of the weight change if respiratory water loss is also included (Ly et al. 2013).

1.6   SUMMARY AND CONCLUSION

TBW makes up at least 50% of the body mass and closer to 65% in the lean athlete. It is distributed within various compartments within the body and serves critical functions. The plasma volume is of particular importance because of its role in cardiovascular function and signaling the responses that lead to water conservation or water replacement. Numerous factors impact the turnover of the body water and the potential to alter the hydration state of the individual. Various methods of assessment can help guide research and application in the field to better understand water flux and develop strategies to restore euhydration in athletes, respectively.

REFERENCES

Almond, C. S., A. Y. Shin, E. B. Fortescue et al. 2005. Hyponatremia among runners in the Boston Marathon. *N. Engl. J. Med.* 352: 1550–6.

Armstrong, L. E. 2007. Assessing hydration status: The elusive gold standard. *J. Am. Coll. Nutr.* 26: 575S–84S.

Armstrong, L. E., and C. M. Maresh. 1991. The induction and decay of heat acclimatization in trained athletes. *Sports Med.* 12: 302–12.

Astrand, P.-O., and K. Rodahl. 1977. *Textbook of Work Physiology*. New York: McGraw-Hill, p. 134.

Baker, L. B., J. A. Lang, and W. L. Kenney. 2009. Change in body mass accurately and reliably predicts change in body water after endurance exercise. *Eur. J. Appl. Physiol.* 105: 959–67.

Baker, L. B., T. A. Munce, and W. L. Kenney. 2005. Sex differences in voluntary fluid intake by older adults during exercise. *Med. Sci. Sports Exerc.* 37: 789–96.

Bartok, C., D. A. Schoeller, J. C. Sullivan, R. R. Clark, and G. L. Landry. 2004. Hydration testing in collegiate wrestlers undergoing hypertonic dehydration. *Med. Sci. Sports Exerc.* 36: 510–17.

Bossingham, M. J., N. S. Carnell, and W. W. Campbell. 2005. Water balance, hydration status, and fat-free mass hydration in younger and older adults. *Am. J. Clin. Nutr.* 81: 1342–50.

Brozek, J., F. Grande, J. T. Anderson, and A. Keys. 1963. Densitometric analysis of body composition: Revisions of some quantitative assumptions. *Ann. N. Y. Acad. Sci.* 110: 113–40.

Buckley, J. C. 2006. *Space Physiology*. New York: Oxford University Press, pp. 140–4.

Burke, L. M., and J. A. Hawley. 1997. Fluid balance in team sports. Guidelines for optimal practices. *Sports Med.* 24: 38–54.

Burtis, C. A., and E. R. Ashwood. 1999. *Tietz Textbook of Clinical Chemistry*, 3rd ed. Philadelphia, PA: Saunders, pp. 1817–26.

Casa, D. J., L. E. Armstrong, S. K. Hillman et al. 2000. National athletic trainers' association position statement: Fluid replacement for athletes. *J. Athl. Train.* 35: 212–24.

Chee, Y. L., and H. G. Watson. 2005. Air travel and thrombosis. *Br. J. Haematol.* 130: 671–80.

Cheuvront, S. N., R. W. Kenefick, and E. J. Zambraski. 2015. Spot urine concentrations should not be used for hydration assessment: A methodology review. *Int. J. Sport Nutr. Exerc. Metab.* 25: 293–7.

Dill, D. B., and D. L. Costill. 1974. Calculation of percentage changes in volumes of blood, plasma, and red cells in dehydration. *J. Appl. Physiol.* 37: 247–8.

Epstein, M. 1992. Renal effects of head-out water immersion in humans: A 15-year update. *Physiol. Rev.* 72: 563–621.

Girandola, R. N., R. A. Wiswell, and G. Romero. 1977. Body composition changes resulting from fluid ingestion and dehydration. *Res. Q.* 48: 299–303.

Greenleaf, J. E. 1966. Some observations on the effects of heat, exercise and hypohydration upon involuntary hypohydration in man. *Int. J. Biometeorol.* 10: 71–6.

Greenleaf, J. E., and F. Sargent 2nd. 1965. Voluntary dehydration in man. *J. Appl. Physiol.* 20: 719–24.

Horswill, C. A. 1998. Effective fluid replacement. *Int. J. Sport Nutr.* 8: 175–95.

Horswill, C. A. 2009. Making weight and cutting weight. In R. Kordi, N. Maffulli, R. Wroble, and W. A. Wallace (Eds.), *Sports Medicine and Science in Combat Sports*. London: Springer, pp. 21–40.

Horswill, C. A., and L. M. Janas. 2011. Hydration and health. *Am. J. Lifestyle Med.* 5: 304–15.

Horswill, C. A., J. R. Stofan, M. Lacambra, T. A. Toriscelli, E. R. Eichner, and R. Murray 2009. Sodium balance during U. S. football training in the heat: Cramp-prone vs. reference players. *Int. J. Sports Med.* 30: 789–94.

Huntoon, C. S. L., A. L. Grigoriev, and Y. V. Natochin. 1998. *Fluid and Electrolyte Regulation in Spaceflight*. Science and Technology Series 94. San Diego, CA: Univelt Publishers, pp. 16–17.

Inoue, Y. 1996. Longitudinal effects of age on heat-activated sweat gland density and output in healthy active older men. *Eur. J. Appl. Physiol.* 74: 72–7.

Institute of Medicine. 2004. *Dietary Reference Intakes for Water, Potassium, Sodium, Chloride, and Sulfate*. Washington, DC: The National Academies.

Kavouras, S. A. 2002. Assessing hydration status. *Curr. Opin. Clin. Nutr. Metab. Care.* 5: 519–24.

Kenefick, R. W., and S. N. Cheuvront. 2012. Hydration for recreational sport and physical activity. *Nutr. Rev.* 70(Suppl. 2): S137–42.

Kirby, C. R., and V. A. Convertino. 1986. Plasma aldosterone and sweat sodium concentrations after exercise and heat acclimation. *J. Appl. Physiol.* 61: 967–70.

Krasney, J. A. 1996. Physiological responses to head-out water immersion: Animal studies. In M. J. Fregly and C. M. Blatteis (Eds.), *Handbook of Physiology. Environmental Physiology*. Am. Physiol. Soc. NY: Oxford University Press, Section 4, Vol. II, Chapter 38, pp. 855–88.

Laird, R. H., and D. Johnson. 2012. The medical perspective of the Kona Ironman Triathlon. *Sports Med. Arthrosc.* 20: 239.

Leach, C. S., and P. C. Rambaut. 1977. Biomedical responses of the Skylab crewman: An overview. In R. S. Johnston and L. F. Dietlein (Eds.), *Biomedical Results from Skylab*. NASA SP 3-77.Washington, DC: NASA, pp. 204–16.

Lee, D. K. 1964. Terrestrial animals in dry heat: Man in the desert. In D. B. Dill (Ed.), *Handbook of Physiology. Adaptation to Environment*. Am. Physiol. Soc. Washington, DC, Section 4, Chapter 35, pp. 551–82.

Leiper, J. B., A. M. Molla, and A. M. Molla. 2003. Effects on health of fluid restriction during fasting in Ramadan. *Eur. J. Clin. Nutr.* 57(Suppl. 2): S30–8.

Leithead, C. S., and A. R. Lind. 1964. *Heat Stress and Heat Disorders*. London: Cassell.

Ly, N., K. Hamstra-Wright, and C. Horswill. 2013. Does respiratory water loss and substrate oxidation contribute substantially to body mass reduction during 90 min of training? *Paper Presented at the Sports Cardiovascular and Wellness Nutritionists' Annual Meeting*. Chicago, IL.

Mack, G. W., and E. R. Nadel. 2011. Body fluid balance during heat stress in humans. *Compr. Physiol.* 10: 187–214.

Maughan, R. J. 1991. Fluid and electrolyte loss and replacement in exercise. *J. Sports Sci.* 9(Spec No): 117–42.

Maughan, R. J., S. M. Shirreffs, and J. B. Leiper. 2007. Errors in the estimation of hydration status from changes in body mass. *J. Sports Sci.* 25: 797–804.

Miki, K., G. Hajduczok, S. K. Hong, and J. A. Krasney. 1986. Plasma volume changes during head-out water immersion in the conscious dog. *Am. J. Physiol.* 251: R284–9.

Montain, S. J., M. N. Sawka, and C. B. Wenger. 2001. Hyponatremia associated with exercise: Risk factors and pathogenesis. *Exerc. Sport Sci. Rev.* 29: 113–17.

Olsson, K. E., and B. Saltin. 1970. Variation in total body water with muscle glycogen changes in man. *Acta Physiol. Scand.* 80: 11–18.

O'Neal, E. K., S. P. Poulos, and P. A. Bishop. 2012. Hydration profile and influence of beverage contents on fluid intake by women during outdoor recreational walking. *Eur. J. Appl. Physiol.* 112: 3971–82.

Oppliger, R. A., and C. Bartok. 2002. Hydration testing of athletes. *Sports Med.* 32: 959–71.

Passe, D., M. Horn, J. Stofan, C. Horswill, and R. Murray. 2007. Voluntary dehydration in runners despite favorable conditions for fluid intake. *Int. J. Sport Nutr. Exerc. Metab.* 17: 284–95.

Pendergast, D. R., and C. E. G. Lundgren. 2009. The underwater environment: Cardiopulmonary, thermal, and energetic demands. *J. Appl. Physiol.* 106: 276–83.

Popowski, L. A., R. A. Oppliger, G. P. Lambert, R. F. Johnson, A. K. Johnson, and C. V. Gisolfi. 2001. Blood and urinary measures of hydration status during progressive acute dehydration. *Med. Sci. Sports Exerc.* 33: 747–53.

Powers, M. E., B. L. Arnold, A. L. Weltman et al. 2003. Creatine supplementation increases total body water without altering fluid distribution. *J. Athl. Train.* 38: 44–50.

Prior, B. M., C. M. Modlesky, E. M. Evans et al. 2001. Muscularity and the density of the fat-free mass in athletes. *J. Appl. Physiol.* 90: 1520–31.

Raman, A., D. A. Schoeller, A. F. Subar, R. P. Troiano, A. Schatzkin, T. Harris, D. Bauer, S. A. Bingham, J. E. Everhart, A. B. Newman, F. A. Tylavsky. 2004. Water turnover in 458 American adults 40–79 yr of age. *Am. J. Physiol. Renal Physiol.* 286(2): F394–401.

Ryan, A. J., G. P. Lambert, X. Shi, R. T. Chang, R. W. Summers, and C. V. Gisolfi. 1998. Effect of hypohydration on gastric emptying and intestinal absorption during exercise. *J. Appl. Physiol.* 84: 1581–8.

Sawka, M. N., and C. B. Wenger. 1988. Physiologic responses to acute heat stress. In K. B. Pandolf, M. N. Sawka, and R. R. Gonzalez (Eds.), *Human Performance Physiology and Environmental Medicine at Terrestrial Extremes.* Carmel, IN: Cooper Publishing Group, pp. 114–21.

Schoeller, D. A. 2005. Hydrometry. In S. B. Heymsfield, T. G. Lohman, Z. Wang, and S. B. Going (Eds.), *Human Body Composition,* 2nd ed. Champaign, IL: Human Kinetics, pp. 35–50.

Schoeller, D. A., E. van Santen, D. W. Peterson, W. Dietz, J. Jaspan, and P. D. Klein. 1980. Total body water measurement in humans with [18]O and 2H labeled water. *Am. J. Clin. Nutr.* 33: 2686–93.

Scott, M. G., J. W. Heusel, V. A. LeGrys, and O. Siggaard-Anderson. 1999. Electrolytes and blood gases. In C. A. Burtis, and E. R. Ashwood (Eds.), *Tietz Textbook of Clinical Chemistry,* 3rd ed. Philadelphia, PA: Saunders, p. 1066.

Shirreffs, S. M., and R. J. Maughan. 1998. Urine osmolality and conductivity as indices of hydration status in athletes in the heat. *Med. Sci. Sports Exerc.* 30: 1598–602.

Shirreffs, S. M., and R. J. Maughan. 2008. Water and salt balance in young male football players in training during the holy month of Ramadan. *J. Sports Sci.* 26(Suppl.): S47–54.

Stover, E. A., H. J. Petrie, D. Passe, C. A. Horswill, B. Murray, and R. Wildman. 2006. Urine specific gravity in exercisers prior to physical training. *Appl. Physiol. Nutr. Metab.* 31: 320–7.

Tam, N., H. W. Nolte, and T. D. Noakes. 2011. Changes in total body water content during running races of 21.1 km and 56 km in athletes drinking ad libitum. *Clin. J. Sport Med.* 21: 218–25.

Wade, C. E., and B. J. Freund. 1990. Hormonal control of blood volume during and following exercise. In C. V. Gisolfi and D. R. Lamb (Eds.), *Perspectives in Exercise Science and Sports Medicine Volume 3. Fluid Homeostasis during Exercise.* Carmel, IN: Benchmark Press, pp. 207–46.

Wells, J. C. K., N. J. Fuller, O. Dewit, M. S. Fewtrell, M. Elia, and T. J. Cole. 1999. Four-component model of body composition in children: Density and hydration of fat-free mass and comparison with simpler models. *Am. J. Clin. Nutr.* 69: 904–12.

Wilk, B., and O. Bar-Or. 1996. Effect of drink flavor and NaCl on voluntary drinking and hydration in boys exercising in the heat. *J. Appl. Physiol.* 80: 1112–17.

Wilk, B., A. M. Rivera-Brown, and O. Bar-Or. 2007. Voluntary drinking and hydration in non-acclimatized girls exercising in the heat. *Eur. J Appl. Physiol.* 101: 727–34.

Zambraski, E. J. 1990. Renal regulation of fluid homeostasis during exercise. In C. V. Gisolfi, and D. R. Lamb (Eds.), *Perspectives in Exercise Science and Sports Medicine Volume 3. Fluid Homeostasis during Exercise.* Carmel, IN: Benchmark Press, pp. 247–80.

2 Sodium Balance during Exercise and Hyponatremia

Stavros A. Kavouras

CONTENTS

2.1 INTRODUCTION

Dietary sodium intake is vital for cellular homeostasis and physiological function. Sodium is the most abundant cation of the extracellular space and one of the primary regulators of the extracellular fluid volume (Ball 2013). Sodium balance influences plasma volume and atrial pressure, playing a key role in the regulation of cardiac output and stroke volume both at rest and during exercise. Sodium is also involved in the acid–base regulation, neural signal transmission, muscular contraction, and metabolism (Farquhar et al. 2015; Montain, Sawka, and Wenger 2001). Dietary sodium is readily absorbed from the digestive track and travels freely in the blood (Fordtran, Rector, and Carter 1968). The renal system is the main regulator of sodium. Excessive intake of sodium can lead to natriuresis (urinary sodium excretion) in healthy individuals (Mack and Nadel 1996). Dietary sodium intake has drawn a lot of media attentions since excessive sodium intake has been linked to hypertension. Even though the minimum daily amount of sodium required to maintain homeostasis is very low (<500 mg), most Americans consume more than 3,000 mg. The U.S National Academy of Sciences Food and Nutrition Board has established adequate intake (AI: recommended average daily intake) and tolerable upper level (UL: the individuals' maximum level of daily nutrient intake that is likely to pose no adverse effects) of 1.5 and 2.3 g/day, respectively (Institute of Medicine US Panel on Dietary Reference 2005). Salt is ~40% sodium by weight. So, 1 g of salt consists of 400 mg of sodium. AI and UL for sodium refer mainly to inactive individuals, while physically active individuals and athletes who lose large volume of sweat regularly would require much greater dietary sodium intake. For instance, during a 90-min soccer training session, sodium losses could range from 600 mg to 3.12 g (Maughan et al. 2010). During exercise, an average athlete's sweat contains around 40 mmol or 920 mg of sodium per liter of sweat. During intense exercise in the heat, sweat volume could exceed 2 L/h (American College of Sports et al. 2007). So it is not unreasonable for an athlete to have well over 2–3,000 mg of sodium lost from sweat alone, which is almost twice that of recommended intake (Rehrer 2001; Sallis 2008).

2.2 DEFINITION OF HYPONATREMIA

Hyponatremia is clinically defined by plasma or serum sodium concentration lower than 135 mmol/L (Montain, Sawka, and Wenger 2001). Due to the fact that hyponatremia between 134 and 130 mmol/L is usually asymptomatic, it has been also described as *mild hyponatremia* (Sallis 2008). Even though hyponatremia develops more often in clinical conditions, not related to exercise, for the purpose of this chapter we will focus on the exercise-associated hyponatremia (EAH) (Hew-Butler et al. 2008). The first line of symptoms is non-specific and they can develop in the absence of EAH (bloating, nausea, vomiting, headache bloating, and puffiness). During more severe EAH, the symptoms are associated with cerebral edema encephalopathy causing changes in mental status like confusion or disorientation and even seizures or delirium. When EAH cell swelling is developed in the lungs (pulmonary edema), breathing becomes short and difficult (Ayus, Varon, and Arieff 2000). As symptoms progress, coma or even death can happen. The intensity of the symptoms is related to how rapid hyponatremia has developed, as well as the level of serum sodium. There are two types of EAH: (1) dilutional hyponatremia and (2) hyponatremia due to mixed sodium and water loss (Armstrong 1999). The most common type of EAH is the first one when hyponatremia develops as a response to increased total body water relative to total sodium content in the circulation. In the dilutional hyponatremia, there is a concomitant increase in the body weight.

2.3 INCIDENCE OF HYPONATREMIA

Several studies have reported that endurance athletes that participate in events lasting more than 3 h can develop asymptomatic hyponatremia. The incidence of clinically significant hyponatremia is 0.1%–0.3%, but the actual mortality rate is very low, although exact numbers are unknown (Carter 2008). The first case of EAH was reported in South Africa during the Comrades Marathon by Dancaster and Whereat (1971). Later on, Noakes et al. (1985) reported four cases of water intoxication (hyponatremia) in athletes participating in endurance exercise. One of the largest cross-sectional studies was published in the *New England Journal of Medicine* based on the 2002 Boston Marathon (Almond et al. 2005). The data indicated that most of the hyponatremic runners were either asymptomatic or mildly symptomatic, even though 13% of the subjects has serum sodium <135 mmol/L while 0.6% has <120 mmol/L. In a study from the New Zealand Ironman triathlon, the incidence of hyponatremia defined as plasma sodium <135 mmol/L was 18% (Speedy, Noakes, and Schneider 2001). In a different study, based on the Houston marathon, the prevalence of hyponatremia (<135 mmol/L) was 0.4%, while the incidence of severe hyponatremia (<120 mmol/L) was 0.04% (Hew et al. 2003). Noakes and colleagues (2005) compiled data from 8 different endurance races and reported hyponatremia (<130 mmol/L) in 1.4% of the runners. They also concluded that EAS occurred in athletes that drank in excess while running and that retained fluid due to an inadequate suppression of the antidiuretic hormone (ADH).

2.4 PREDISPOSING RISK FACTORS

2.4.1 EXCESSIVE DRINKING

Over-consumption of hypotonic drinks (mainly water) during exercise above fluid losses (urine and sweat) leading to weight gain is the main cause of hyponatremia. This type of hyponatremia has been described as *dilutional hyponatremia*. ADH or vasopressin is the main hormone that regulates water balance. Based on the regulation of vasopressin, someone would expect that over-drinking would stimulate diuresis in order to maintain homeostasis (Robertson 2011; Robertson, Shelton, and Athar 1976). Over-drinking decreases plasma osmolality, which in turn will rapidly decrease ADH leading to large urinary excretion. Interestingly, case studies have indicated that during hyponatremia, even though plasma osmolality and sodium are low, ADH remains high. Based on this observation, the term *inappropriate antidiuretic hormone secretion* has been used, and it has been linked in the development of EAH (Siegel et al. 2007; Speedy et al. 2001). A potential explanation is that some of the non-osmotic stimuli of ADH include nausea and exercise. Figure 2.1 demonstrates that gaining weight during exercise is a strong predictor of hyponatremia (Noakes et al. 2005). Also the study from the 2002 Boston marathon with 488 runners indicated that those who gained 3–4.9 kg during the race had 30% and 70% greater risk of developing severe (<130 mmol/L) or mild hyponatremia (<135 mmol/L) (Almond et al. 2005). However, runners who gain 3–4.9 kg in a marathon (42 km), for example, will have to drink about 715–1,167 mL above their overall body fluid losses (sweat + urine) every 10 km.

2.4.2 DRINK COMPOSITION

The composition of the drinks can also play an important role on the development of EAH. Several studies have shown that sodium-free fluid ingestion during prolonged

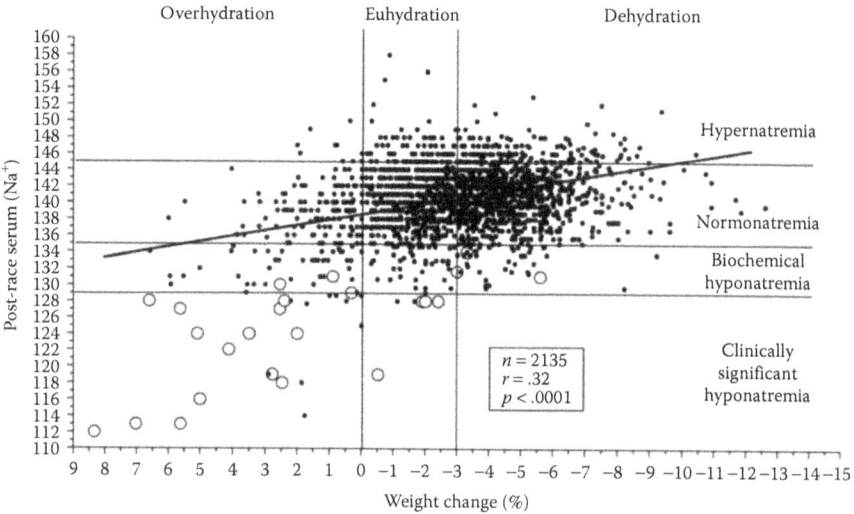

FIGURE 2.1 The relationship between the serum [Na⁺] after racing and the weight change during exercise in the 2,135 athletes in this study is a linear function with a negative slope ($p < .0001$). Asymptomatic athletes are denoted by the closed circle, and athletes with symptoms considered compatible with EAH are denoted by the open circle o. (Data from Noakes, T. D. et al., Three independent biological mechanisms cause exercise-associated hyponatremia: Evidence from 2,135 weighed competitive athletic performances, *Proc. Natl. Acad. Sci. USA.*, 102, 18550–5. Copyright 2005 National Academy of Sciences, U.S.A.)

exercise in the heat will decrease plasma sodium, since the volume is replaced but not the solutes lost in the sweat (Anastasiou et al. 2009; Takamata et al. 1994; Vrijens and Rehrer 1999). Therefore, administration of sodium drinks could prevent the development or delay the development of hyponatremia (Anastasiou et al. 2009).

2.4.3 Sweat Rate and Sodium Content

EAH can develop at the same time with dehydration. In Figure 2.1, there are several data point depicted with the open circles at the right part of the graph indicating water deficit. In this case, EAH is a result of mixed sodium and water deficit. Even though sodium sweat concentration in the general healthy population is around 25–45 mmol/L, studies have indicated that some athletes tend to have extremely high levels of sodium excretion (~100 mmol/L) in the absence of any pathologies (i.e., cystic fibrosis). These subjects described as *salty sweaters* have been identified as a high-risk group for the development of hyponatremia (Eichner 2008; Maughan et al. 2010; Shirreffs 2010). Heavy salt stains on the clothes are an indication of high sweat sodium content (see Figure 2.2). Also athletes with very high sweat rate (heavy sweaters) have been identified as high risk for the development of hyponatremia (Buono, Ball, and Kolkhorst 2007). The combination of a high volume and sodium content sweat could further increase the risk of the development of EAH. In an effort to

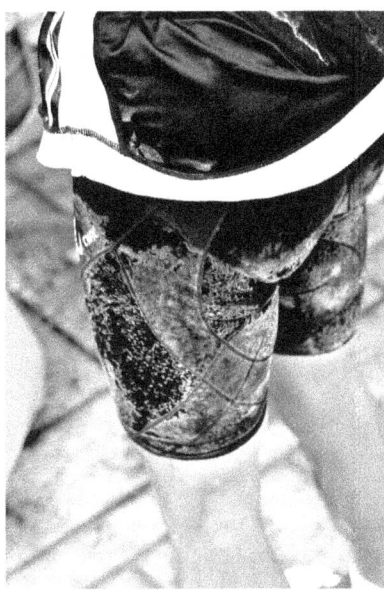

FIGURE 2.2 Heavy salt stains during exercise could indicate high sweat sodium content. The picture was taken during the 2010 Sparthathlon ultra-marathon in the 90-km checkpoint.

quantitatively understand the etiology of hyponatremia, Montain and his colleagues (2006) developed a mathematical model to predict the effect of different drinking volumes on plasma sodium levels when body mass, body composition, running speed, weather conditions, and sweat sodium concentration were varied. The results indicated that the combination of fluid overloading and salt depletion in those who secrete salty sweat is a critical risk factor for hyponatremia.

2.4.4 BODY SIZE

Smaller body size (or body mass) has also been considering a predisposing risk factor for hyponatremia mainly due to the overall lower volume of body water. Dilutional hyponatremia is easier to develop in a smaller athlete with less body water volume than in a bigger one. Several studies have associated females with a greater risk for developing hyponatremia (Almond et al. 2005). However, the risk is probably linked mainly to the fact that women usually have smaller body mass and consequently less body water content.

2.4.5 ENVIRONMENTAL TEMPERATURE

Cooler weather has also been linked to the development of hyponatremia. Lower environmental temperature leads to less sweating and lower risk of dehydration. In other words, it is more likely that fluid intake could be greater than fluid loss in cooler than in warmer climates (Hew-Butler et al. 2008). Of course, the role of cooler temperature as a predisposing factor is mainly for the dilutional type of hyponatremia.

2.4.6 EXERCISE DURATION

Long duration of the exercise is also a critical risk factor. The longer the exercise event, the greater the risk for EAH. The study by Almond et al. (2005) suggested that specifically for marathon races, finishing time greater than 4 h is a predisposing factor for hyponatremia. Also looking at the incidence of hyponatremia during prolonged events, the higher percentages have been reported in ultra-endurance events like Ironman triathlon or ultra-marathons. The duration might also be related to the lower absolute exercise intensity that generates lower metabolic heat, and lower need for heat dissipation via sweating.

2.4.7 DIET

Low dietary sodium intake could also be a serious risk in the development of EAH. Even though low sodium intake might be a healthy habit for inactive and hypertensive adults, it may not always be a good practice for an endurance athlete (Armstrong 1999). Athletes with low dietary sodium intake might start a prolonged race in a lower than normal serum sodium concentration, accelerating then the development of hyponatremia during the exercise. A case study has identified low normal pre-exercise serum sodium as a predisposing factor for EAH (Armstrong et al. 1993).

2.5 PREVENTION OF HYPONATREMIA

2.5.1 HYDRATION STRATEGY

Since over-drinking is the most common cause for the development of hyponatremia, knowing the fluid needs during exercise is of critical importance. It is not difficult for an athlete to assess his/her sweat rate and then estimate the volume of fluid needed for a given exercise intensity, duration, and environmental conditions. Unfortunately, the inter-subject variation of sweat rate is so large that no one-size-fits-all advice can be given (American College of Sports et al. 2007). However, if the individual hydration protocol has not been developed, someone can try to guide drinking based on thirst. Even though drinking to thirst can lead to dehydration and impair exercise performance especially in the heat, the risk of hyponatremia is significantly reduced.

2.5.2 DIETARY SODIUM

Increased dietary sodium the days prior to an endurance event can prevent starting a race with low blood sodium concentration. More importantly, moderate sodium intake during exercise in the form of an electrolyte solution can delay or prevent the development of hyponatremia. The American College of Sports Medicine recommends the consumption of beverages with sodium (20–50 mmol/L) and/or small amounts of salted snacks or sodium-containing foods at meals (American College of Sports et al. 2007). Also the ingestion of sodium during exercise can help by stimulating thirst and retaining the consumed fluids.

2.5.3 Acclimatization

It has been suggested for more than 50 years that heat acclimatization not only enhances exercise performance and thermoregulation but also spares sodium by decreasing sodium content in sweat (Taylor 1986). This will lead to lower sodium losses and lower risk for developing hyponatremia. Additionally, heat-acclimatization has a beneficial effect on exercise performance by enhancing cardiovascular function and decreasing heat strain (Allan and Wilson 1971; Eichner 2009; Montain 2008).

2.6 CONCLUSION

In summary, sodium balance is of critical importance during endurance exercise. EAH, even though relatively a rare condition, can lead to serious health issues and even death. Drinking higher volumes than sweating losses during exercise can lead to dilutional hyponatremia. Following an individualized hydration protocol based on predetermined sweating can prevent over-drinking. Also for events lasting longer than 3 h, sodium intake during exercise can prevent serum sodium drop and development of hyponatremia.

REFERENCES

Allan, J. R., and C. G. Wilson. 1971. Influence of acclimatization on sweat sodium concentration. *J. Appl. Physiol.* 30(5): 708–12.

Almond, C. S., A. Y. Shin, E. B. Fortescue, R. C. Mannix, D. Wypij, B. A. Binstadt, C. N. Duncan, D. P. Olson, A. E. Salerno, J. W. Newburger, and D. S. Greenes. 2005. Hyponatremia among runners in the Boston Marathon. *N. Engl. J. Med.* 352(15): 1550–6.

American College of Sports Medicine, M. N. Sawka, L. M. Burke, E. R. Eichner, R. J. Maughan, S. J. Montain, and N. S. Stachenfeld. 2007. American College of Sports Medicine position stand: Exercise and fluid replacement. *Med. Sci. Sports Exerc.* 39(2): 377–90.

Anastasiou, C. A., S. A. Kavouras, G. Arnaoutis, A. Gioxari, M. Kollia, E. Botoula, and L. S. Sidossis. 2009. Sodium replacement and plasma sodium drop during exercise in the heat when fluid intake matches fluid loss. *J. Athl. Train.* 44(2): 117–23.

Armstrong, L. E. (Ed.). 1999. Exertional hyponatremia. In *Performance in Extreme Environments*. Champaign, IL: Human Kinetics, pp. 103–35.

Armstrong, L. E., W. C. Curtis, R. W. Hubbard, R. P. Francesconi, R. Moore, and E. W. Askew. 1993. Symptomatic hyponatremia during prolonged exercise in heat. *Med. Sci. Sports Exerc.* 25(5): 543–9.

Ayus, J. C., J. Varon, and A. I. Arieff. 2000. Hyponatremia, cerebral edema, and noncardiogenic pulmonary edema in marathon runners. *Ann. Intern. Med.* 132(9): 711–14.

Ball, S. G. 2013. How I approach hyponatraemia. *Clin. Med.* 13(3): 291–5.

Buono, M. J., K. D. Ball, and F. W. Kolkhorst. 2007. Sodium ion concentration vs. sweat rate relationship in humans. *J. Appl. Physiol. (1985)* 103(3): 990–4.

Carter, R. 2008. Exertional heat illness and hyponatremia: An epidemiological prospective. *Curr. Sports Med. Rep.* 7(4): S20–7.

Dancaster, C. P., and S. J. Whereat. 1971. Fluid and electrolyte balance during the comrades marathon. *S. Afr. Med. J.* 45(6): 147–50.

Eichner, E. R. 2008. Genetic and other determinants of sweat sodium. *Curr. Sports Med. Rep.* 7(4): S36–40.

Eichner, E. R. 2009. Six paths to hyponatremia. *Curr. Sports Med. Rep.* 8(6): 280–1.

Farquhar, W. B., D. G. Edwards, C. T. Jurkovitz, and W. S. Weintraub. 2015. Dietary sodium and health: More than just blood pressure. *J. Am. Coll. Cardiol.* 65(10): 1042–50.

Fordtran, J. S., F. C. Rector, Jr., and N. W. Carter. 1968. The mechanisms of sodium absorption in the human small intestine. *J. Clin. Invest.* 47(4): 884–900.

Hew, T. D., J. N. Chorley, J. C. Cianca, and J. G. Divine. 2003. The incidence, risk factors, and clinical manifestations of hyponatremia in marathon runners. *Clin. J. Sport Med.* 13(1): 41–7.

Hew-Butler T., M. H. Rosner, S. Fowkes-Godek, J. P. Dugas, M. D. Hoffman, D. P. Lewis, R. J. Maughan, K. C. Miller, S. J. Montain, N. J. Rehrer, W. O. Roberts, I. R. Rogers, A. J. Siegel, K. J. Stuempfle, J. M. Winger, and J. G. Verbalis. 2015. Statement of the third international exercise-associated hyponatremia consensus development conference, Carlsbad, CA. *Clin. J. Sport Med.* 25: 303–20.

Institute of Medicine US Panel on Dietary Reference. 2005. *Dietary Reference Intakes for Water, Potassium, Sodium, Chloride, and Sulfate.* Washington, DC: The National Academies Press.

Mack, G. W., and E. R. Nadel. 1996. Body fluid balance during heat stress in humans. In M. J. Fregly and C. M. Blatteis (Eds.), *Environmental Physiology.* New York: Oxford University Press, pp. 187–214.

Maughan, R. J., S. M. Shirreffs, K. T. Ozgunen, S. S. Kurdak, G. Ersoz, M. S. Binnet, and J. Dvorak. 2010. Living, training and playing in the heat: Challenges to the football player and strategies for coping with environmental extremes. *Scand. J. Med. Sci. Sports* 20(Suppl. 3): 117–24.

Montain, S. J. 2008. Strategies to prevent hyponatremia during prolonged exercise. *Curr. Sports Med. Rep.* 7(4): S28–35.

Montain, S. J., S. N. Cheuvront, and M. N. Sawka. 2006. Exercise associated hyponatraemia: Quantitative analysis to understand the aetiology. *Br. J. Sports Med.* 40(2): 98–105; discussion 98–105.

Montain, S. J., M. N. Sawka, and C. B. Wenger. 2001. Hyponatremia associated with exercise: Risk factors and pathogenesis. *Exerc. Sport Sci. Rev.* 29(3): 113–17.

Noakes, T. D., N. Goodwin, B. L. Rayner, T. Branken, and R. K. Taylor. 1985. Water intoxication: A possible complication during endurance exercise. *Med. Sci. Sports Exerc.* 17(3): 370–5.

Noakes, T. D., K. Sharwood, D. Speedy, T. Hew, S. Reid, J. Dugas, C. Almond, P. Wharam, and L. Weschler. 2005. Three independent biological mechanisms cause exercise-associated hyponatremia: Evidence from 2,135 weighed competitive athletic performances. *Proc. Natl. Acad. Sci. U. S. A.* 102(51): 18550–5.

Rehrer, N. J. 2001. Fluid and electrolyte balance in ultra-endurance sport. *Sports Med.* 31(10): 701–15.

Robertson, G. L. 2011. Vaptans for the treatment of hyponatremia. *Nat. Rev. Endocrinol.* 7(3): 151–61.

Robertson, G. L., R. L. Shelton, and S. Athar. 1976. The osmoregulation of vasopressin. *Kidney Int.* 10(1): 25–37.

Sallis, R. E. 2008. Fluid balance and dysnatremias in Athletes. *Curr. Sports Med. Rep.* 7(4): S14–19.

Shirreffs, S. M. 2010. Hydration: Special issues for playing football in warm and hot environments. *Scand. J. Med. Sci. Sports* 20(Suppl. 3): 90–4.

Siegel, A. J., J. G. Verbalis, S. Clement, J. H. Mendelson, N. K. Mello, M. Adner, T. Shirey, J. Glowacki, E. Lee-Lewandrowski, and K. B. Lewandrowski. 2007. Hyponatremia in marathon runners due to inappropriate arginine vasopressin secretion. *Am. J. Med.* 120(5): 461.e11–17.

Speedy, D. B., T. D. Noakes, N. E. Kimber, I. R. Rogers, J. M. Thompson, D. R. Boswell, J. J. Ross, R. G. Campbell, P. G. Gallagher, and J. A. Kuttner. 2001. Fluid balance during and after an ironman triathlon. *Clin. J. Sport Med.* 11(1): 44–50.

Speedy, D. B., T. D. Noakes, and C. Schneider. 2001. Exercise-associated hyponatremia: A review. *Emerg. Med. (Fremantle)* 13(1): 17–27.

Takamata, A., G. W. Mack, C. M. Gillen, and E. R. Nadel. 1994. Sodium appetite, thirst, and body fluid regulation in humans during rehydration without sodium replacement. *Am. J. Physiol.* 266(5 Pt 2): R1493–502.

Taylor, N. A. 1986. Eccrine sweat glands. Adaptations to physical training and heat acclimation. *Sports Med.* 3(6): 387–97.

Vrijens, D. M., and N. J. Rehrer. 1999. Sodium-free fluid ingestion decreases plasma sodium during exercise in the heat. *J. Appl. Physiol. (1985)* 86(6): 1847–51.

3 Human Perspiration and Cutaneous Circulation

Manabu Shibasaki and Scott L. Davis

CONTENTS

3.1 INTRODUCTION

Dynamic exercise induces metabolic heat production in exercising muscle, which subsequently elevates body temperature. The thermoregulatory system dissipates this heat in order to maintain body temperature within narrow limits (Figure 3.1). The importance of these mechanisms is exemplified in the calculation that if heat were not liberated from skin, internal temperature would reach an upper *safe* limit within 10 min of moderate exercise. Even well-trained individuals are unwilling to continue exercising when internal body temperature exceeds 40°C. Many factors modulate thermoregulatory mechanisms during prolonged exercise. Increases in temperature outside of this narrow range not only impact exercise performance but can be life threatening. These complications may be amplified for individuals with an impaired thermoregulatory system. The primary focus of this chapter is to examine recent advances in the understanding of thermoregulatory responses influenced by exercise-related factors, focusing on human perspiration and cutaneous circulation. This chapter will also discuss the influence of physiological (including body fluid status) and health conditions on these thermoregulatory responses.

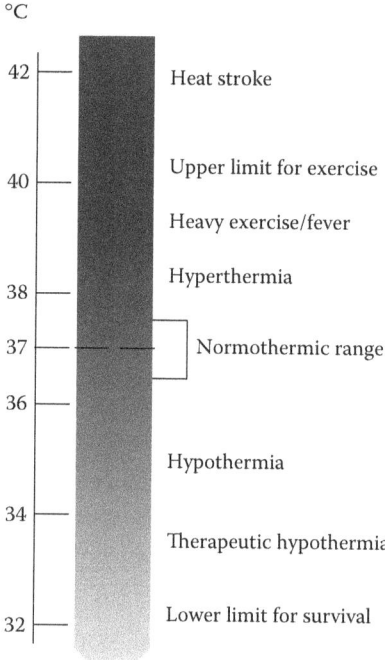

FIGURE 3.1   Regulatory range of internal body temperature. Under resting conditions, internal body temperature is regulated in a narrow normothermic range (±0.5°C) around 37°C. Heat dissipation is critical for humans during exercise especially in a hot environment. Elevations in skin blood flow and sweating are the primary heat exchange mechanisms in humans during exercise.

Increased body temperature, either through active or passive thermal stress, is transferred to the environment via non-evaporative and evaporative heat loss mechanisms. Non-evaporative heat loss (i.e., dry heat loss) mechanisms include conduction, convection, and radiation, and in humans, they are the predominant mechanisms of heat loss at rest. The effectiveness of these mechanisms is dependent on the temperature gradient between the environment and the skin. Skin temperature is primary modulated by increasing/decreasing the blood flow to the skin, which allows the transfer of heat from the internal core of the body to the ambient environment. The evaporative heat loss mechanisms include increased ventilation (i.e., panting), saliva spreading, and perspiration. While evaporative heat loss can occur through panting or saliva spreading in other mammals and birds (Robertshaw 2006), sweating coupled with increased cutaneous vasodilation is the predominant heat dissipation mechanism in humans (White 2006).

Dynamic exercise rapidly increases the rate of metabolic heat production requiring immediate and effective heat dissipation. The balance between heat production and heat dissipation determines the evaporative requirement needed by an individual. While greater rate of metabolic heat production requires increased sweating capacity, sweat production is limited. Profuse sweating can be problematic in the maintenance of fluid balance as the regulation of body fluid conflicts with this evaporative heat loss. In addition, heat dissipation, either non-evaporative or evaporative, is also affected by environmental conditions. Under conditions where environmental temperature is greater than skin temperature, sweating becomes more critical for heat dissipation in humans. However, evaporative heat loss in exceedingly humid environments at a given temperature is less effective, increasing the reliance on non-evaporative heat loss via cutaneous vasodilation. Conversely, while dynamic exercise requires blood flow distribution toward active muscle, increased core temperatures generated by active muscles also demand increased blood flow to the skin for heat dissipation, creating competing influences within the cardiovascular system. Taking these factors into account, we will detail the physiological mechanisms of sweating and skin blood flow during exercise.

Disease and injury can also lead to impairments in thermoregulation which can impact the quality of life by limiting the ability to tolerate physical activity as well as bringing about additional risk for heat-related illness. These impairments are magnified when physical activity is performed under extreme environmental conditions (high ambient temperature and high relative humidity). Disease and injury can affect the neural control of thermoregulation as well as the physical disruption of heat-dissipating structures in the skin (i.e., cutaneous blood vessels and sweat glands). Understanding the physiological and physical characteristics of thermoregulatory dysfunction in disease and injury has led to the development of interventions and strategies that enable individuals with disease or injury to not only perform routine activities of daily living but also reap the benefits of exercise.

3.1.1 NEURAL CONTROL OF SWEATING AND SKIN BLOOD FLOW

The thermoregulatory center is located within the pre-optic hypothalamic regions of the brain, integrating thermal information detected in the central nervous system coupled with information transmitted from thermoreceptors of the skin and core (Aronsohn and Sachs 1885; Kahn 1904; Moorhouse 1911; Ott 1877). The identification of neural pathways responsible for sweating or cutaneous active vasodilation in humans is difficult and as such the exact neurological pathways are not entirely understood. Based on evidence from both animal studies and human anatomical data (Kuno 1956; Low 2004; Nakamura et al. 2004; Sato et al. 1989), the neural pathway from the brain to thermoregulatory organs is thought to be the following: efferent signals from the pre-optic area in the hypothalamus travel via the tegmentum of the pons and the medullary raphe regions to the intermediolateral cell column of the spinal cord. In the spinal cord, neurons emerge from the ventral horn, pass through the white ramus communicans, and then synapse in the sympathetic ganglia. Postganglionic non-myelinated C-fibers (i.e., skin sympathetic nerves) pass through the gray ramus communicans, combine with peripheral nerves, and travel to sweat glands and cutaneous vessels, with these nerve fibers *entwined around* the periglandular tissue of the target organs (Uno 1977).

Sympathetic nerves of the skin have a role in controlling both sweating and skin blood flow. Sympathetic neural innervation of thermoregulatory organs is complex. Sweating is governed by a sympathetic cholinergic sudomotor system, whereas skin blood flow is governed by both a sympathetic adrenergic vasoconstrictor system and a separate sympathetic non-adrenergic vasodilator system. Activation of cholinergic nerves increases both sweating and cutaneous vasodilation as confirmed by blocking neurotransmitter release from cholinergic nerves with botulin toxin. However, it remains unknown whether sweat glands and cutaneous vessels have shared innervation by cholinergic nerves or are separately innervated given different onset times and opposite effector responses observed especially during exercise (Johnson and Proppe 1996).

The primary neurotransmitter for adrenergic nerves and cholinergic nerves is noradrenaline and acetylcholine, respectively. In addition to each primary neurotransmitter, co-transmitters released from sympathetic nerves contribute to thermoregulatory responses. Neuropeptide Y, a co-transmitter from adrenergic nerves, also contributes to cutaneous active vasoconstriction during cold exposure (Stephens et al. 2004). While cutaneous active vasodilation receives neural control by cholinergic nerves, it is also modulated by multiple other cofactors including nitric oxide (NO), vasoactive intestinal peptide, prostaglandins, and substance P (Charkoudian 2010; Holowatz and Kenney 2010; Johnson and Kellogg 2010; Kellogg 2006). Currently, research groups are further elucidating the mechanisms responsible for regulating sweating and skin blood flow.

3.1.2 EVALUATION OF THERMOREGULATORY RESPONSES

Nielsen (1938) indicated that internal temperature increases at the start of dynamic exercise and, as exercise continues, is maintained at a higher value relative to pre-exercise. Nielsen and Nielsen (1965) demonstrated that the magnitude of elevated internal temperature due to exercise was proportional to work intensities at varying environmental conditions (5°C–30°C). Sweat rate during steady-state exercise at each level of work closely followed the rise in internal temperature. Thus, the rate of evaporative heat loss is determined by the requirement for evaporative heat loss (Gagnon and Kenny 2012). In an environment that permits adequate evaporation, the rate of evaporative heat loss is determined by the evaporation required for heat balance until this requirement exceeds the individual's maximal sweating capacity (Gagnon, Jay, and Kenny 2013).

Thermoregulatory responses including sweating, cutaneous vasodilation, and vasoconstriction as well as non-shivering thermogenesis in brown adipose tissues are primary controlled by thermal factors integrated in the pre-optic area of the hypothalamus (Nakamura and Morrison 2008, 2010). In the late 1960s, Benzinger (1969) was the first to describe this dynamic relationship between body temperature and thermoregulatory responses at various environmental temperatures. From these studies, thermoregulatory responses can be described by expressing increases in sweating and cutaneous vasodilation as a function of mean body temperature calculated using a weighted summation of internal and skin temperatures (Nadel et al. 1971; Saltin, Gagge, and Stolwijk 1970). The changes in thermoregulatory responses can be observed at a given level of mean body temperature and increase proportionally with increases in the mean body temperature. In effect, these responses can be characterized by the temperature thresholds corresponding with the onset of sweating and/or cutaneous vasodilation and the sensitivities for sweating and/or cutaneous vasodilation—the relationship between the change in internal temperature and the change in response

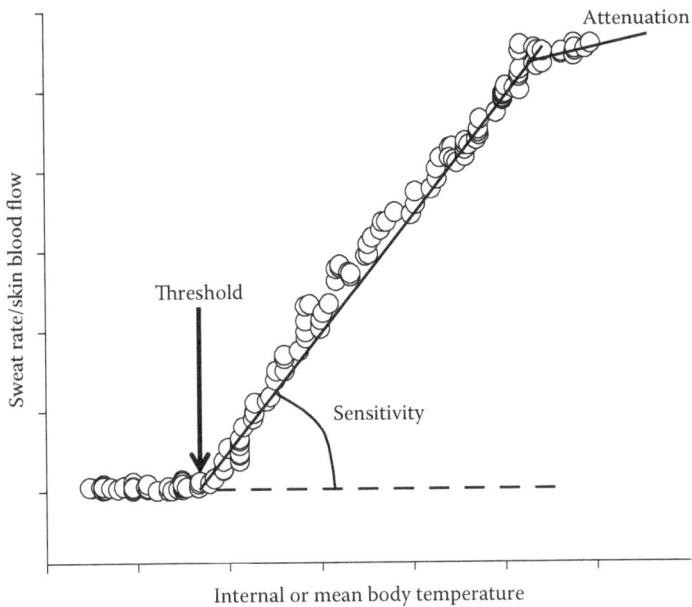

FIGURE 3.2 Increases in body temperature and sweating/skin blood flow. Temperature threshold for sweating or cutaneous vasodilation is defined as the internal or mean body temperature at which sweating or skin blood flow increases above baseline. Sweating and skin blood flow increase proportionally to increases in internal or mean body temperature up to a given temperature level where attenuation of the relationship occurs. The change in gain or slope of the regression line of the sweating or skin blood flow responses between the temperature threshold and the temperature attenuation is termed *sensitivity*.

(Figure 3.2). During dynamic exercise, the rate of metabolic heat production increases immediately, creating a mismatch with heat dissipation. Therefore, thresholds for the onset of sweating and/or cutaneous vasodilation are shifted rightward to a higher mean body temperature relative to passive heat stress. Other factors that influence heat loss thresholds and sensitivities include hydration, baroreceptor loading, circadian rhythm, menstrual cycle, training/detraining, heat acclimation, and aging.

3.2 FACTORS MODIFYING THE CONTROL OF SWEATING AND SKIN BLOOD FLOW

Dynamic exercise instantly and rapidly generates heat in the active muscles. Sweating occurs immediately at the onset of dynamic exercise in warm environmental conditions prior to any measurable elevation of internal temperature, suggesting that non-thermal factors associated with exercise contribute to the increase in sweating (van Beaumont and Bullard 1963). In contrast, skin blood flow decreases regardless of thermal conditions during dynamic exercise (Christensen, Nielsen, and Hannisdahl 1942; Johnson and Park 1982; Taylor et al. 1984). Kellogg et al. (1991) demonstrated that cutaneous vasoconstriction at the onset of dynamic exercise is caused by increases in vasoconstrictor neural activity. The primary finding from these studies is that non-thermal factors associated with the onset of dynamic exercise modulate thermoregulatory responses. Similar conclusions were made from studies in which subjects cycled on an ergometer with workload changing in a sinusoidal manner as thermoregulatory responses followed the sinusoidal pattern of exercise rather than changes in internal or skin temperatures (Yamazaki 2002; Yamazaki, Sone, and Ikegami 1994). Moreover, as exercise continues, the cardiovascular and body fluid system intervene thermoregulatory responses to maintain these regulatory systems. The objective of this portion of the chapter is to summarize the primary non-thermal modifiers of sweat rate and skin blood flow during exercise.

3.2.1 Exercise-Originated Factors

Johansson (1895) postulated that two separate and distinct neural mechanisms control cardiovascular responses during exercise. One mechanism dominating cardiovascular responses *irradiates* impulses from the motor cortex within the central nervous system. Krogh and Lindhard (1913) termed this central mechanism as *cortical irradiation* and later defined it as central command (Goodwin, McCloskey, and Mitchell 1972). The other mechanism, the exercise pressor reflex, originates from the stimulation of afferent nerve endings within the skeletal muscle and is engaged during muscle contraction (Alam and Smirk 1937). The exercise pressor reflex can be evoked by mechanically sensitive afferent nerves (mechanoreceptors) and/or metabolically sensitive afferent nerves (metaboreceptors) (Mitchell 1990).

There are a number of experimental approaches to identify and dissect the role of central command and peripheral afferent receptors during exercise. Active load-less cycling can be used to examine the effect of central command. However, changes in thermoregulatory responses during load-less cycling are quite small (Kondo et al. 1997). Partial neuromuscular blockade can also be used to clearly show the contribution of central command (Iwamoto et al. 1987; Leonard et al. 1985; Mitchell 1985). The effect of central command on thermoregulatory responses has also been tested during isometric handgrip exercise with administration of a partial neuromuscular blocking agent (Shibasaki et al. 2003, 2005). In these studies, subjects were unable to maintain the desired force of contraction following administration of the neuromuscular blocking agent, but continued to attempt the exercise throughout a 2-min period. Despite force production being close to zero, heart rate and blood pressure were increased during the exercise period. Sweat rate was increased but cutaneous vascular conductance (CVC), a measurement of skin blood flow, was reduced during this period of exercise suggesting that both sweating and skin blood flow can be modulated by factors associated with central command. However, the change in CVC was only observed when subjects were heat stressed. Crandall et al. (1995, 1998) observed the reduction in CVC at a site where adrenergic vasoconstriction was abolished (via bretylium treatment) and at an adjacent untreated site, suggesting that this observed reduction in CVC occurred through the withdrawal of active vasodilator nerve activity. On the other hand, Kellogg et al. (1991) demonstrated that the cutaneous vasoconstriction at the onset of dynamic exercise is due to enhanced active vasoconstrictor tone regardless of thermal status. Thus, cutaneous vasoconstriction occurs through differences in neural control depending on the type of exercise (static or dynamic). Similarly, the effect of central command on cutaneous vasoconstriction is different during dynamic exercise relative to isometric exercise. Friedman et al. (1991) evaluated the contribution of central command by causing a reduction in cutaneous vasoconstrictor responses at the onset of dynamic exercise in normothermic subjects following curare administration. In spite of augmented central command by curare during low intensity of dynamic exercise, typical reductions in CVC at the onset of dynamic exercise were not observed suggesting that cutaneous vasoconstriction at the onset of dynamic exercise is not due to the activation of central command. Taken together, central command enhances sweating and cutaneous vasoconstriction due to withdrawal of active vasodilation rather than enhanced active vasoconstriction.

Post-exercise ischemia via cuff occlusion or positive pressure in active muscles can be used to stimulate muscle metaboreceptors. Alam and Smirk (1937) showed that blood pressure increased during dynamic exercise and remained elevated following occlusion of limb circulation. Upon release of the occlusion, blood pressure returned to pre-exercise levels. Their observations led to numerous ongoing studies investigating the role of muscle metaboreceptors in modulating blood pressure during exercise. Using this perturbation, the possible role of metaboreceptors in modulating sweating responses during exercise was determined. Sweat rate increased during isometric handgrip exercise and remained elevated during post-exercise ischemia; sweat rate then returned to pre-exercise levels with the release of the cuff (Crandall, Stephens, and Johnson 1998; Kondo et al. 1999; Shibasaki, Kondo, and Crandall 2001). These observations suggest that metaboreceptor activation is capable of modulating sweat rate during isometric handgrip exercise. However, blood pressure also increased during isometric exercise and remained elevated throughout post-exercise ischemia, raising the issue of whether increases in sweating during isometric exercise and subsequent post-exercise ischemia occurred through loading of baroreceptors rather than metaboreceptor activation. To remove the influence of baroreceptor loading, elevated blood pressures during post-exercise ischemia period were restored to pre-exercise levels via intravenous infusion of sodium nitroprusside. Despite the decrease in blood pressure, sweat rate remained elevated throughout the ischemic period similar to when blood pressure was elevated (Shibasaki, Kondo, and Crandall 2001). A further follow-up study showed similar sweating responses during handgrip exercise and post-exercise ischemia while undergoing lower body negative pressure (LBNP) to decrease exercise blood pressure (Binder et al. 2012). Taken together, these studies indicate that the elevation in sweat rate during isometric exercise and post-exercise ischemia occurs through activation of metaboreceptors independent of changes in blood pressure.

CVC during isometric handgrip exercise does not change in normothermia, but decreases in a warm condition. This decrease occurs through the withdrawal of active cutaneous vasodilation (Crandall, Stephens, and Johnson 1998; Crandall et al. 1995). Crandall et al. (1998) found that CVC remained diminished during post-exercise ischemia and then returned to pre-exercise levels upon release of ischemia. They concluded that the observed reduction in CVC during isometric exercise while heat stressed could be mediated by muscle metaboreceptor stimulation. Applying lower body positive/negative pressure, Binder et al. (2012) found that baroreflexes also contribute to metaboreceptor-induced cutaneous vasoconstriction. This baroreceptor involvement was only observed during hyperthermia but not in normothermia. However, McCord and Minson (2005) demonstrated that the decrease in CVC during isometric handgrip exercise is associated with the initial level of skin blood flow prior to the onset of exercise as CVC initially increased by local heating was reduced by exercise and thus suggested that the observed cutaneous vasoconstriction during isometric handgrip exercise in heat stressed subjects involved non-neural mechanism. Using an axillary nerve blockade, Shibasaki et al. (2009) later demonstrated that neural and non-neural mechanisms contribute to the reduction in CVC during isometric handgrip exercise during heat stress. Taken together, the modulation of CVC due to muscle metaboreflex occurred only when subjects were heat stressed but the cutaneous vasoconstriction partially involves non-neural mechanisms.

Passive limb movement can be used to selectively stimulate muscle mechanoreceptors. Kondo et al. (1997) observed an increase in sweat rate during a 2-min bout of passive limb movement in mildly heated subjects with previously stable sweat rates. However, the increase in sweat rate during this procedure was quite small causing difficulties to ascertain whether this change in sweat rate was due to mechanoreceptor stimulation or due to spontaneous fluctuations of sweat rate (i.e., not attributed to mechanoreceptor stimulation). Carter et al. (2002) performed load-less cycling (i.e., 0 W) and non-pedaling recovery conditions following moderate exercise in the upright position. Sweat rate and CVC during load-less cycling (i.e., 0 W) were greater relative to when the subjects did no pedaling during recovery. In a follow-up study, Wilson et al. (2004) performed a similar protocol with subjects in the supine position to minimize baroreceptor unloading associated with recovery in the upright position. Similar to previous findings, sweat rate remained elevated during unloaded pedaling relative to when subjects did no pedaling. In contrast, no differences in CVC were observed between active cycling and non-pedaling. Given that active cycling includes both the activation of central command and stimulation of muscle mechanoreceptors, they concluded that either central command or mechanoreceptor stimulation modulated sweat rate during the recovery from exercise. Using a tandem cycle ergometer, the role of mechanoreceptors on sweat rate and CVC after moderate exercise was further investigated while subjects remained seated while a second person on the tandem cycle ergometer cycled for the subject during recovery (Journeay et al. 2004; Shibasaki et al. 2004). Thus, muscle mechanoreceptors were stimulated passively by the movement of the second individual, thereby removing central command. This muscle mechanoreceptor stimulation during passive cycling modulated sweat rate, with no effect on CVC. Taken together, these findings suggest that stimulation of muscle mechanoreceptors is capable of modulating sweat rate but not CVC.

3.2.2 BAROREFLEXES

The baroreflex control of the cutaneous vasculature is widely recognized (reviewed by Johnson 1986, 2010), but the effect of baroreflexes on sweating is not fully elucidated. Baroreceptors (cardiopulmonary, aortic, and carotid) are crucial for the regulation of blood pressure primarily through neural control of heart rate and vascular resistance. Loading or unloading baroreceptors via postural changes (standing or supine), simulated hemorrhage (LBNP), neck collar pressure/suction, transfusions, and/or pharmacological agents have been utilized to experimentally examine the function of baroreceptors.

Johnson and Park (1981) compared internal temperature thresholds for cutaneous vasodilation during the following conditions: supine passive heating, upright passive heating, supine exercise, and upright exercise. Both upright posture and exercise contributed to an elevation in the internal temperature threshold for cutaneous vasodilation. Moreover, reductions of blood volume and LBNP delay the onset of cutaneous vasodilation during supine dynamic exercise (Mack, Cordero, and Peters 2001; Mack, Nose, and Nadel 1988; Nishiyasu, Shi, and Mack 1991). Conversely, the effect of baroreceptor unloading on sweating was inconsistent. Johnson and Park (1981) reported no differences in internal temperature threshold for sweating during exercise in upright position or slightly lower thresholds relative to the supine positioning. In contrast, Mack et al. (2001) reported elevated internal temperature thresholds for sweating during exercise in conjunction with application of LBNP.

Similar discrepancies have been observed during passive heating studies. Dodt et al. (1995) applied either LBNP or 30° head-up tilt to unload baroreceptors during passive heating, and concluded that this baroreceptor unloading could modulate skin sympathetic nerve activity (SSNA) and sweating in moderately warm subjects. However, others have found differing responses (Vissing, Scherrer, and Victor 1994; Wilson, Cui, and Crandall 2005). Vissing et al. (1994) did not observe changes in SSNA and sweating during LBNP. Wilson et al. (2005) performed multiple 30° head-up tilts during whole-body heating. These groups concluded that baroreceptor unloading did not modulate both sweating and SSNA.

The neck chamber technique is useful for selective stimulation of carotid baroreceptors. Crandall et al. (1996) did not observe changes in CVC during carotid baroreceptor unloading using 500 ms pulses of neck collar pressure by each cardiac cycle over 3 min, regardless of thermal conditions. However, when performed by open-loop carotid baroreceptor stimulation (i.e., 4 repeated 5 s pulse trials of neck pressure or neck suction separated by at least 45 s) during whole-body heating, carotid baroreceptor unloading and loading caused cutaneous vasoconstriction and vasodilation, respectively (Keller et al. 2006). This technique allows for an increased mean carotid sinus pressure equivalent to the chamber pressure achieved for each trial.

Pharmacological agents have been used to change blood pressure. As aforementioned, Shibasaki et al. (2001) performed a systemic infusion of sodium nitroprusside during post-isometric handgrip exercise. Isometric handgrip exercise elevated both blood pressure and sweat rate. Following ischemia, blood pressure and sweat rate remained elevated due to metaboreflex stimulation. Although blood pressure was restored to pre-exercise levels by systemic infusion of sodium nitroprusside, sweat rate did not change, suggesting that the elevation in blood pressure accompanying metaboreceptor stimulation during post-exercise ischemia (i.e., unloading of baroreceptors) is not a mechanism by which sweat rate is elevated during isometric exercise. Similarly, Wilson et al. (2001) used bolus and steady-state infusions of pharmacological agents (nitroprusside and phenylephrine) to acutely and chronically perturb baroreceptors, respectively, in normothermic and heat stressed humans. Despite changes in blood pressure, neither SSNA nor sweat rate was significantly affected. Pharmacologically induced decreases in blood pressure may cause greater unloading of arterial baroreceptors relative to cardiopulmonary baroreceptors, and therefore, arterial baroreceptor stimulation may not contribute to the modulation of sweat rate.

Since cutaneous circulation is under the control of baroreflexes, cutaneous vascular responses are modulated by baroreceptor loading/unloading. While still unclear, it appears that baroreflexes are not capable of modulating sweating.

3.2.3 Body Fluid Regulation—Hydration Status and Plasma Osmolality

Cardiovascular drift (maintenance of cardiac output by an increase in heart rate despite decreases in stroke volume) occurs after ~10 min of moderate to high intensity exercise (Coyle and Gonzalez-Alonso 2001). The magnitude of cardiovascular drift is greater in hot environments relative to thermo-neutral environments (Lafrenz et al. 2008). However, cardiovascular drift during prolonged exercise can be restored by body fluid replacements (Hamilton et al. 1991). Conversely, prolonged exercise can decrease cardiac output due to body water loss (sweating) and redistribution of blood volume (cutaneous vasodilation). Dill et al. (1933) observed that when a man and a dog walked 32 km in a hot environment with water intake ad libitum, the dog maintained its weight but the man lost ~3 kg of his body weight. Since man lost both salt and water via sweating, he drank only enough water to maintain a constant osmolality of extracellular fluid, whereas the dog drank enough water to maintain body weight as only water was lost through respiratory evaporation (i.e., panting). In humans, prolonged exercise in hot environments induces water deficits due to profuse sweating, and this fluid deficit lowers both intracellular and extracellular volume leading to plasma hyperosmolality and hypovolemia.

Plasma osmolality is tightly regulated within a small range with very small increases evoking physiological responses, such as vasopressin secretion and increased thirst. Fortney et al. (1984) performed exercise trials under both iso-osmotic and hyperosmotic conditions with plasma volume maintained. During the hyperosmotic exercise trial, internal temperature thresholds for both sweating and forearm blood flow were elevated relative to the responses during exercise under iso-osmotic conditions. The slopes for sweating and cutaneous vasodilation as a function of the elevation in internal temperature were not affected by increased plasma osmolality. These results were confirmed during passive heating (via lower leg immersion in warm water) in which the subjects received an infusion of 0.9% (iso-osmotic) or 3% (hyperosmotic) saline (Takamata et al. 1995, 1997). In a follow-up study, Shibasaki et al. (2009) identified that the delay in the elevation of skin blood flow was due primarily to a delay in the onset of cutaneous active vasodilator activity, as opposed to increased vasoconstrictor activity. In addition, they found that plasma hyperosmolality did not modulate cutaneous vasoconstrictor system during cold stress. These studies changed plasma osmolality without reduction in plasma volume. Ito et al. (2005) and Lynn et al. (2012) demonstrated an additive effect of plasma hyperosmolality and baroreceptor unloading on CVC but not sweating.

Thus, an increased plasma osmolality independent of plasma volume impairs sudo-motor and active vasodilator system while not changing cutaneous vasoconstrictor system. Impairments in the cutaneous vasodilator system due to hyperosmolality are modulated by baroreflex stimulation while sweating responses are unaffected.

Hypovolemia, in conjunction with an increase in sweat rate and redistribution of blood to the cutaneous vasculature for thermoregulation, results in a reduc-tion of cardiac filling pressure. Skin blood flow increases proportionally to the elevation of internal temperature but the increase in CVC is substantially attenu-ated (Brengelmann et al. 1977; Fortney et al. 1981a; Kellogg et al. 1993; Nadel, Fortney, and Wenger 1980; Nose et al. 1990) when internal temperature reaches ~38°C (Figure 3.2). This attenuation in the rise in CVC when internal temperature is substantially elevated is a feedback mechanism to maintain cardiac filling pres-sure. Both saline infusion and application of negative-pressure breathing following internal temperature elevation (~0.5°C) nullify the attenuation of forearm blood flow and CVC above an internal temperature of ~38°C (Nagashima et al. 1998; Nose et al. 1990). These observations, coupled with the finding that the attenuation in the rise in CVC is controlled by the active vasodilator system (Kellogg et al. 1993), suggest that inhibition of cutaneous vasodilator responses during passive heating and exer-cise may preserve central venous pressure perhaps through the aforementioned baro-reflex mechanisms in hypovolemic subjects. Kamijo et al. (2011) classified SSNA into two components (synchronized with or without cardiac cycle) and suggested that hypovolemic suppression of cutaneous vasodilation was due to the reduction in the SSNA component synchronized with the cardiac cycle. The SSNA component without cardiac synchronization was due to the sweating response and was unaf-fected by hypovolemia in this study. Hypovolemia induced by diuretic drugs (i.e., no change in plasma osmolality) reduced both the peak cutaneous vasodilation and the slope of the relationship between the change in internal temperature and the change in cutaneous vasodilation (thermosensitivity). These drugs also shifted the internal temperature threshold for vasodilation (Fortney et al. 1981a,b; Kamijo et al. 2005; Nadel, Fortney, and Wenger 1980), but had no effect on the control of sweating (Fortney et al. 1981b; Kamijo et al. 2005, 2011).

Body fluid is regulated by two systems, volume regulation and plasma osmoregu-lation, both of which are not affected in the early phase of exercise. However, pro-longed exercise induces a profuse sweating and/or shift of blood volume resulting in dehydration (hypovolemic hyperosmolality), which reduces thermoregulatory cuta-neous vasodilation and sweating. Increased plasma osmolality impairs both cutane-ous vasodilation and sweating responses, while reduced plasma volume only impairs cutaneous vasodilation via baroreflex mechanisms but not sweating.

3.3 IMPAIRMENTS IN SWEATING AND SKIN BLOOD FLOW

Internal body temperature can be maintained in a narrow range if thermoregulatory heat dissipation is fully functional during exercise (Figure 3.1). However, if heat generated during exercise is not liberated from skin, internal temperature would immediately reach the upper *safe* limit (Kenney and Johnson 1992). Certain disease conditions as well as bodily injury can have a detrimental impact on thermoregulation.

Diseases, including multiple sclerosis (MS) and diabetes mellitus (DM), can alter neural pathways within the central nervous system and the peripheral nervous system, respectively, leading to impairments in sweating and skin blood flow. Physical injuries can cause central (spinal cord injury [SCI]) and peripheral (burns) disruptions in the control of sweating and skin blood flow.

3.3.1 MULTIPLE SCLEROSIS

MS is a disabling progressive neurological disorder affecting more than 2.3 million people worldwide (National Multiple Sclerosis Society 2014). The pathophysiology of MS results in a demyelination and ultimately loss of axons and disorganization of normal tissue architecture within the CNS by autoimmune injury responses (Frohman, Racke, and Raine 2006). *Demyelination* is associated with corresponding changes in axonal physiology including a loss of saltatory properties of electrical conduction, a reduction in conduction velocity, and a predisposition to conduction block. Autonomic dysfunction involving the thermoregulatory system is commonly observed in individuals with MS (Haensch and Jorg 2006). The majority of individuals with MS experience transient and temporary worsening of clinical signs and neurological symptoms (Uhthoff's phenomenon) in response to a number of factors, the most prominent of which are increased ambient/core body temperature and exercise (Uhthoff 1889). It is estimated that 60% to 80% of the MS population experience transient and temporary worsening of clinical signs and neurological symptoms as a result of passive heat exposure, exercise (increase in metabolism), or a combination of both (exercise-heat stress). Typically, deficits caused by increases in temperature are reversible by removing heat stressors and allowing subsequent cooling (Davis et al. 2008).

Compounding temperature-related nerve conduction problems, individuals with MS may have impaired neural control of autonomic function (Huitinga et al. 2001). Areas of the sympathetic nervous system that are responsible for controlling thermoregulatory function are susceptible to disease-related pathology in individuals with MS (Andersen and Nordenbo 1997). Lesions affecting conduction within the CNS of individuals with MS impair thermoregulatory effector responses to eccrine sweat glands (Cartlidge 1972; Noronha, Vas, and Aziz 1968; Vas 1969). Abnormal regions of sweating can be identified by using sweat-induced color changes of quinizarin powder placed on the skin of individuals followed by exposure to a heat stress (Fealey and Sato 2008; Kuno 1956). Pharmacological interventions such as intradermal injections, iontophoresis, and intradermal microdialysis can evaluate the function of eccrine sweat glands as well as the cutaneous vasculature independent of the CNS (Davis et al. 2005). Diminished sweat function has been quantified in individuals with MS and is due to reduced sweat output per gland rather than reduced sweat gland recruitment (Davis et al. 2005). Davis et al. (2010) have documented reduced sweating responses in individuals with MS during indirect whole-body heating (increasing core temperature ~1°C) using water-perfused suits. Observed changes in sweating from these areas were due exclusively to neurally mediated responses, not due to factors associated with local heating as the areas where sweating was assessed were not in contact with the water-perfused suit used to heat the subjects. To address the potential for these observed decreases in sweat function in MS being due to detraining or de-adaption, Davis et al. (2005) trained individuals with MS with aerobic exercise for 15 weeks. In healthy individuals, similar exercise training improves sweat output per gland. Despite 15 weeks of training, individuals with MS neither improved sweat gland recruitment nor sweat output per gland. Adaptive thermoregulatory responses following exercise training typically observed in healthy individuals are not seen in individuals with MS, suggesting impaired CNS control of sudomotor function in individuals with MS. Taken together, it is likely that impaired sweating responses may have been due to impairments in neural control of sudomotor pathways and/or neural-induced changes in eccrine sweat glands (Davis et al. 2010). Interestingly, larger increases in cutaneous vasodilation to whole-body heating were observed in individuals with MS compared to healthy controls suggesting neural control of skin blood flow is intact and may compensate for impairments in sweating in an attempt to adequately dissipate heat (Davis et al. 2010).

3.3.2 Diabetes Mellitus

DM is a metabolic disease characterized by the inability to control blood glucose leading to hyperglycemia. According to the World Health Organization, 347 million people worldwide suffer from diabetes (World Health Organization 2014). The two most common classifications of diabetes are Type 1 (insulin-dependent DM) and Type 2 (non-insulin-dependent DM). Type 1 (T1DM) is due to a deficiency of insulin production triggered by autoimmune destruction of insulin producing ß-cells within the pancreas, leading to the inability of cells to uptake glucose. Type 2 (T2DM) results from the impaired signaling and action of insulin by cells, leading to insulin resistance. Over time, uncontrolled hyperglycemia can lead to serious long-term health complications but are not limited to cardiovascular disease, nerve damage, and kidney damage.

Two hallmark complications of diabetes, microvascular damage and peripheral autonomic neuropathy, can have the potential to profoundly impact cutaneous vasodilation and sweating. Due to the potential for peripheral microvascular and nerve damage, cutaneous vasodilator responses have been examined during direct local heating of the skin. Individuals with T1DM (Khan et al. 2000; Stansberry et al. 1997; Wilson, Jennings, and Belch 1992) and T2DM (Fredriksson et al. 2010; Schmiedel, Schroeter, and Harvey 2007; Stansberry et al. 1997; Strom et al. 2010) exhibit reductions in cutaneous vasodilation during local heating compared to healthy controls. Lower absolute maximum vasodilation during local heating has also been observed in individuals with T2DM compared to healthy controls (Sokolnicki et al. 2007; Wick et al. 2006).

Reflex cutaneous vasodilator responses are also attenuated during indirect whole-body heating (increasing core temperature ~1°C) using a water-perfused suit in individuals with T2DM (Sokolnicki et al. 2009). In addition to lower skin blood flow assessed at sites that were not directly exposed to the heating stimulus, the core temperature at which increases in cutaneous vasodilation are observed is higher during whole-body heating suggesting delayed onset of active vasodilation (Wick et al. 2006). Interestingly, reflex cutaneous vasoconstrictor responses to rapid whole-body cooling are preserved in individuals with T2DM (Strom et al. 2011) suggesting the disease only affects cutaneous vasodilation.

Attenuated vasodilator responses observed in individuals with diabetes during local and whole-body heating could be due, in part, to alterations in NO release and/or impaired vascular responsiveness to NO. Despite observed differences in maximal vasodilation in individuals with T2DM, the local sensory nerve mechanisms (Strom et al. 2010) and the relative contribution of NO (Sokolnicki et al. 2007) to the biphasic vasodilation response to local heating of the skin were similar between individuals with T2DM and healthy controls. Peripheral pharmacological interventions (i.e., iontophoresis and intradermal microdialysis) have also been used to further assess cutaneous vascular responses in individuals with diabetes. Endothelial-dependent cutaneous vasodilation induced by administration of acetylcholine (or agonists such as pilocarpine) is reduced in both T1DM (Gomes, Matheus, and Tibirica 2008; Katz et al. 2001; Khan et al. 2000) and T2DM (Arora et al. 1998; Beer et al. 2008; Berghoff et al. 2006; Schmiedel, Schroeter, and Harvey 2007). Endothelial-independent cutaneous vasodilation also appears to be impaired following iontophoretic administration of sodium nitroprusside, a NO donor, in individuals with T1DM (Katz et al. 2001; Khan et al. 2000) and T2DM (Arora et al. 1998; Beer et al. 2008; Schmiedel, Schroeter, and Harvey 2007). Taken together, these findings suggest attenuated vasodilator responsiveness in individuals with diabetes observed during local heating and whole-body heating could be in part due to altered NO responsiveness in cutaneous vessels from diabetic-induced microvascular damage.

When exposed to a whole-body passive heat stress, individuals with T1DM (Fealey, Low, and Thomas 1989) and T2DM (Fealey, Low, and Thomas 1989; Petrofsky 2012) demonstrate decreased sweating. Decreased sweating responses have also been observed in individuals with T2DM during passive exposure to even low ambient air temperatures (Petrofsky and Lee 2005). Impairments in sweat responses have also been observed following acetylcholine iontophoresis in individuals with T1DM (Hoeldtke et al. 2001; Kennedy and Navarro 1989; Kennedy et al. 1984; Kihara, Opfer-Gehrking, and Low 1993) and T2DM (Kihara, Opfer-Gehrking, and Low 1993). Greater levels of peripheral autonomic neuropathy were associated with greater impairments in sweating in individuals with diabetes (Fealey, Low, and Thomas 1989). Impaired sweating responses are likely due to autonomic neuropathy leading to decreased number of active sweat glands and decreased sweat evaporation.

Few studies have examined sweat function in individuals with diabetes during exercise. The previously described findings would suggest that thermoregulatory function would likely be impaired during exercise in individuals with diabetes. However, Stapleton and colleagues (Stapleton et al. 2013) demonstrated no impairments in the ability to dissipate heat in habitually active individuals with T1DM exercising in a whole-body direct calorimeter. Skin blood flow and sweating responses were comparable to healthy controls during light to moderate exercise. However, attenuated sweating in individuals with T1DM is unmasked during exercise of higher intensity (Carter et al. 2014). Older individuals with T2DM (>age 55) also showed impaired heat loss during moderate exercise in a whole-body direct calorimeter (Kenny et al. 2013). These findings suggest that thermoregulatory impairments may become more problematic in individuals with diabetes during higher thermal loads (increased temperature and humidity) and lower fitness levels as wells as increased diabetic complications and progression including neuropathy and poor glucose control (Stapleton et al. 2013).

3.3.3 Spinal Cord Injury

SCI is an insult to the spinal cord resulting in temporary or permanent changes in neural functions. Injury usually results from trauma to the spinal cord causing fractures and compression of vertebrae leading to bone fragments crushing or tearing axons within the spinal cord. An SCI can damage a few, many, or almost all of the axons in an injured area resulting in often devastating neurologic deficits and disability below the injury site. Injury to the cervical spine typically results in neural deficits and disability to both upper and lower extremities (tetraplegia). Injuries at or below the thoracic spinal cord usually involve only the lower extremities (paraplegia). As such, individuals with paraplegia have greater afferent and efferent neural processing compared to tetraplegic counterparts. As expected, SCI results in disturbances in physiological function including thermoregulation. The magnitude of thermoregulatory dysfunction is proportional to the level of the SCI with tetraplegics showing greater impairments compared to paraplegics.

Impairments in the neural control of cutaneous vasodilation are observed in individuals with SCI during both passive heating and exercise. No changes in forearm blood flow, used as an index of skin blood flow, were observed in paraplegic individuals during passive heating of the lower extremities (i.e., insensate skin) (Freund et al. 1984). Only small changes in forearm blood flow were detected during indirect whole-body (both sensate and insensate skin) heating (increasing core temperature ~1°C) using water-perfused suits. Cutaneous dilator responses are also impaired in skin located below the level of SCI during direct local heating of the skin (Nicotra, Asahina, and Mathias 2004). During exercise, cutaneous vasodilation increased approximately fourfold in able-bodied individuals with no observed changes in SCI paraplegics while performing arm crank exercise (Theisen et al. 2000). Similarly, SCI paraplegics experienced much lower CVC than able-bodied individuals during arm cranking at 50% of maximal power output (Theisen et al. 2001). Impairments in cutaneous vasodilation below the level of the SCI during thermal stresses are consistent with disruption of sympathetic vasomotor control of the cutaneous vasculature (Wallin and Stjernberg 1984).

Reduced sweating or the complete absence of sweating responses are commonly observed in the insensate skin of individuals with SCI during both passive heating (Freund et al. 1984; Huckaba et al. 1976; Tam et al. 1978; Totel 1974; Totel et al. 1971) and exercise (Castle et al. 2013; Gass, Gass, and Pitetti 2002; Price and Campbell 1997, 1999a,b). While these sweating impairments are likely due to the interruption of sudomotor neurons controlling sweat gland function, disruption of neural pathways in SCI may also lead to decreased afferent feedback to the hypothalamus that can stimulate sweating during exercise in able-bodied individuals as previously described. In addition, altered sudomotor innervation of sweat glands in

SCI can lead to peripheral changes at the gland itself. Sweat glands below the lesion are less sensitive to cholinergic activation (peripheral administration of acetylcholine via iontophoresis) (Yaggie, Niemi, and Buono 2002).

Due to the impairments in cutaneous vasodilation and sweating, individuals with SCI commonly demonstrate greater increases in skin temperature during heat exposure in insensate skin compared to sensate skin (Price 2006). This is more problematic in tetraplegics as they have a greater surface area of insensate skin leading to a greater increase in skin temperature and subsequent rise in core temperature during heat exposure and exercise compared to paraplegics. To compensate for the loss of thermoregulatory mechanisms in insensate skin below the level of the spinal injury, improvements in sweat function are observed in sensate skin above the spinal injury in both tetraplegics and paraplegics (Huckaba et al. 1976; Petrofsky 1992; Price and Campbell 1997). As such, paraplegics demonstrate similar core temperature responses during exercise in cool or thermoneutral environments compared to able-bodied individuals despite impairments in the neural control of cutaneous vasodilation and sweating (Dawson, Bridle, and Lockwood 1994; Hopman, Oeseburg, and Binkhorst 1993; Price and Campbell 1999a,b). However, impairments in thermoregulatory function become more problematic with exercise in warmer environments (Price 2006).

3.3.4 BURNS AND SKIN GRAFTING

Every year ~1.4 million people in the United States sustain burns. With severe skin burns, typically a majority of the entire dermal layer, which contains the blood vessels and sweat glands necessary for thermoregulation, is excised. This excised area is then covered with donor skin (i.e., skin graft) harvested from non-injured regions of the body. Approximately 90% of skin grafts are split-thickness grafts in which all of the epidermis and a portion of the dermis are removed from a donor site and grafted to an injured site.

Coupled with a lack of sweat glands, grafted areas will be unable to dissipate heat without neural reinnervation and functional revascularization. Individuals with healed burns over 40% of their body surface area have higher core temperatures during a thermal challenge relative to non-burned individuals (Ben-Simchon et al. 1981; McGibbon et al. 1973; Roskind et al. 1978; Shapiro et al. 1982). Cutaneous vascular responses were investigated from both grafted and adjacent uninjured skin (Davis et al. 2007b). Davis et al. found that cutaneous vasodilation at the grafted site was greatly attenuated during indirect whole-body heating (i.e., heating the patient but not the area where skin blood flow is assessed), and this deficit persisted upward to 4+ years post-grafting surgery (Davis et al. 2007b, 2009). Interestingly, cutaneous vasoconstriction assessed during indirect whole-body cooling (i.e., cooling the patient but not the area where skin blood flow is assessed) was similar between the grafted and control sites. This normal vasoconstrictor capacity in grafted skin indicates the presence of functional adrenergic nerves, functional α-adrenergic and related co-transmitter receptors on the cutaneous vasculature, and normal smooth muscle responses to neuronal stimulation (Davis et al. 2007b, 2009). Taken together, grafted skin has normal reinnervation of the cutaneous vasoconstrictor limb but compromised reinnervation of the active vasodilator limb.

Skin grafting did not adversely affect cutaneous vasodilator responses to sustained whole limb local heating, although the responses were extremely variable between subjects (Freund et al. 1981). Davis et al. (2007b, 2009) monitored skin blood flow from both grafted and adjacent non-injured skin under normothermic conditions (i.e., local temperature of 34°C) and throughout 30 min of local heating at 42°C. The magnitude of cutaneous vasodilation to this local heating protocol was significantly less at grafted sites relative to the adjacent control sites, regardless of the maturity of the graft (Davis et al. 2007b, 2009). Although local heating induced vasodilation is significantly impaired in grafted skin, the magnitude of vasodilation to this stimulus was profoundly greater than the dilation that occurred during whole-body heating, when the assessed sites were not directly exposed to the heating stimulus. Thus, grafted skin retains, to a greater extent, the capacity to dilate to a local heating stimulus relative to a whole-body heating stimulus. A rationale for this large difference in cutaneous vasodilation in grafted tissue between these heating stimuli is likely related to the differing mechanisms by which the skin dilates upon increases in local versus core temperatures (Charkoudian 2010).

Attenuated vasodilator responses of grafted skin during local and whole-body heating could be due, in part, to alterations in NO release and/or impaired vascular responsiveness to NO. Davis et al. (2007a, 2009) sought to identify whether skin grafting impairs cutaneous vasodilation by perfusing acetylcholine (endothelial dependent) and sodium nitroprusside (endothelial independent) through intradermal microdialysis probes placed in grafted and adjacent uninjured skin. The maximal increase in skin blood flow due to acetylcholine administration was significantly lower at grafted sites, regardless of graft maturity (Davis et al. 2007a, 2009). The effective drug concentration resulting in 50% of the maximal vasodilator response (EC50) dose response curves from grafted sites was significantly greater relative to uninjured sites, indicative of a rightward shift of the curve representative of a decreased sensitivity of the vasodilator response to acetylcholine (Davis et al. 2007a, 2009). Conversely, maximal cutaneous vasodilator responsiveness, as well as the EC50 of the dose–response curves, to sodium nitroprusside was not different between uninjured and grafted sites (Davis et al. 2007a). Taking the findings from these microdialysis studies, attenuated vasodilator responsiveness during local heating and whole-body heating in grafted skin could be in part due to altered NO release from endothelial sources (Kellogg, Johnson, and Kosiba 1989; Kellogg et al. 1998; Shastry et al. 1998, 2000). However, attenuated NO release alone is unlikely to entirely account for decreases in cutaneous vasodilation in grafted skin during the whole-body heat stress.

An absence of sweating in split-thickness grafts has been previously reported (Conway 1939; McGibbon et al. 1973; McGregor 1950; Ponten 1960). The reported lack of sweating in split-thickness skin grafts was thought to be due to a combination of the initial injury damaging sweat glands at the recipient site and the harvested skin being absent of deeper dermal structures including secretory coils of sweat glands (Ablove and Howell 1997; Ponten 1960). Davis and colleagues (2007b, 2009) evaluated centrally driven sudomotor responses in grafted skin during whole-body heating and demonstrated minimal or no sweating in grafted skin 5–9 months post-graft surgery compared to adjacent uninjured skin. Sweat rate was significantly reduced at the grafted site and the magnitude of this attenuation was not different when compared between graft maturities. These data support the hypothesis that sweating responses to a whole-body heat stress remain disrupted as the graft matures.

It is unknown whether reduced sweating responses were due to absence of functional sweat glands or due to disrupted innervation of the sweat gland. To address this question, sweating responses to a peripheral stimulus (exogenous administration of acetylcholine) were evaluated in a dose-dependent manner (Shibasaki and Crandall 2001). Sweat rate via capacitance hygrometry was measured directly above microdialysis membranes perfused with increasing doses of acetylcholine in grafted and adjacent uninjured grafted skin (Davis et al. 2007a, 2009). Sweating responsiveness was shown to be impaired in skin grafts, likely due to a reduction or absence of functional sweat glands in the grafted tissue, and that sweating responsiveness showed little evidence of normalizing with graft healing (Davis et al. 2007a, 2009).

3.4 CONCLUSION

Heat dissipation is critical for temperature homeostasis, specifically during exposure to high ambient temperatures, during exercise, or the combination of both. Dynamic exercise, even in a thermoneutral environment, increases metabolic heat production instantly leading to elevations in body temperature. The importance of these thermoregulatory mechanisms is exemplified in the calculation that if heat were not liberated from skin, internal body temperature would reach a critical limit with only 10 min of moderate exercise. This chapter summarized a number of factors that modulate sweat function and/or cutaneous blood flow, as well as impairments in thermoregulatory function due to disease conditions (i.e., MS and DM) and injury (i.e., spinal cord and burns). We would like to emphasize that identifying and understanding factors that modify thermoregulatory function has provided critical knowledge to improve the safety and well-being of healthy individuals during exposure to extreme temperatures and exercise. Further investigation into mechanisms that impair thermoregulatory function can lead to the development of therapeutic interventions and strategies so clinical populations can safely tolerate heat exposure and exercise.

REFERENCES

Ablove, R. H., R. M. Howell. 1997. The physiology and technique of skin grafting. *Hand Clinics* 13 (2):163–73.

Alam, M., F. H. Smirk. 1937. Observations in man upon a blood pressure raising reflex arising from the voluntary muscles. *J. Physiol. (Lond)* 89:372–83.

Andersen, E. B., A. M. Nordenbo. 1997. Sympathetic vasoconstrictor responses in multiple sclerosis with thermo-regulatory dysfunction. *Clin. Auton. Res.* 7 (1):13–16.

Aronsohn, E., J. Sachs. 1885. Die beziehungen des gehirns zur korperwairme und zum fieber. *Pflugers Arch.* 37:232–300.

Arora, S., P. Smakowski, R. G. Frykberg, L. R. Simeone, R. Freeman, F. W. LoGerfo, A. Veves. 1998. Differences in foot and forearm skin microcirculation in diabetic patients with and without neuropathy. *Diabetes Care* 21 (8):1339–44.

Beer, S., F. Feihl, J. Ruiz, I. Juhan-Vague, M. F. Aillaud, S. G. Wetzel, L. Liaudet, R. C. Gaillard, B. Waeber. 2008. Comparison of skin microvascular reactivity with hemostatic markers of endothelial dysfunction and damage in type 2 diabetes. *Vasc. Health Risk Manag.* 4 (6):1449–58.

Ben-Simchon, C., H. Tsur, G. Keren, Y. Epstein, Y. Shapiro. 1981. Heat tolerance in patients with extensive healed burns. *Plast. Reconstr. Surg.* 67 (4):499–504.

Benzinger, T. H. 1969. Heat regulation: Homeostasis of central temperature in man. *Physiol. Rev.* 49:671–759.

Berghoff, M., S. Kilo, M. J. Hilz, R. Freeman. 2006. Differential impairment of the sudomotor and nociceptor axon-reflex in diabetic peripheral neuropathy. *Muscle Nerve* 33 (4):494–9.

Binder, K., A. G. Lynn, D. Gagnon, N. Kondo, G. P. Kenny. 2012. Hyperthermia modifies muscle metaboreceptor and baroreceptor modulation of heat loss in humans. *Am. J. Physiol. Regul. Integr. Comp. Physiol.* 302 (4):R417–23.

Brengelmann, G. L., J. M. Johnson, L. Hermansen, L. B. Rowell. 1977. Altered control of skin blood flow during exercise at high internal temperature. *J. Appl. Physiol.* 43 (5):790–94.

Carter, M. R., R. McGinn, J. Barrera-Ramirez, R. J. Sigal, G. P. Kenny. 2014. Impairments in local heat loss in type 1 diabetes during exercise in the heat. *Med. Sci. Sports Exerc.* 46 (12):2224–33.

Carter, R., 3rd, T. E. Wilson, D. E. Watenpaugh, M. L. Smith, C. G. Crandall. 2002. Effects of mode of exercise recovery on thermoregulatory and cardiovascular responses. *J. Appl. Physiol.* 93 (6):1918–24.

Cartlidge, N. E. 1972. Autonomic function in multiple sclerosis. *Brain* 95 (4):661–64.

Castle, P. C., B. P. Kularatne, J. Brewer, A. R. Mauger, R. A. Austen, J. A. Tuttle, N. Sculthorpe, R. W. Mackenzie, N. S. Maxwell, A. D. Webborn. 2013. Partial heat acclimation of athletes with spinal cord lesion. *Eur. J. Appl. Physiol.* 113 (1):109–15.

Charkoudian, N. 2010. Mechanisms and modifiers of reflex induced cutaneous vasodilation and vasoconstriction in humans. *J. Appl. Physiol.* 109 (4):1221–8.

Christensen, E. H., M. Nielsen, B. Hannisdahl. 1942. Investigations of the circulation in the skin at the beginning of muscular work. *Acta Physiol. Scand.* 4:162–70.

Conway, H. 1939. Sweating function in transplanted skin. *Gynecol. Obstet.* 69:756–61.

Coyle, E. F., J. Gonzalez-Alonso. 2001. Cardiovascular drift during prolonged exercise: New perspectives. *Exerc. Sport Sci. Rev.* 29 (2):88–92.

Crandall, C. G., J. M. Johnson, W. A. Kosiba, D. L. Kellogg, Jr. 1996. Baroreceptor control of the cutaneous active vasodilator system. *J. Appl. Physiol.* 81 (5):2192–8.

Crandall, C. G., J. Musick, J. P. Hatch, D. L. Kellogg, Jr., J. M. Johnson. 1995. Cutaneous vascular and sudomotor responses to isometric exercise in humans. *J. Appl. Physiol.* 79 (6):1946–50.

Crandall, C. G., D. P. Stephens, J. M. Johnson. 1998. Muscle metaboreceptor modulation of cutaneous active vasodilation. *Med. Sci. Sports Exerc.* 30 (4):490–96.

Davis, S. L., T. C. Frohman, C. G. Crandall, M. J. Brown, D. A. Mills, P. D. Kramer, O. Stuve, E. M. Frohman. 2008. Modeling Uhthoff's phenomenon in MS patients with internuclear ophthalmoparesis. *Neurology* 70 (13 Pt 2):1098–106.

Davis, S. L., M. A. Korkmas, C. G. Crandall, E. M. Frohman. 2010. Impaired sweating in multiple sclerosis leads to increased reliance on skin blood flow for heat dissipation. *FASEB J* 24:991, 925.

Davis, S. L., M. Shibasaki, D. A. Low, J. Cui, D. M. Keller, G. F. Purdue, J. L. Hunt, B. D. Arnoldo, K. J. Kowalske, C. G. Crandall. 2007a. Skin grafting impairs postsynaptic cutaneous vasodilator and sweating responses. *J. Burn Care Res.* 28 (3):435–41.

Davis, S. L., M. Shibasaki, D. A. Low, J. Cui, D. M. Keller, G. F. Purdue, J. L. Hunt, T. B. Arnoldo, K. J. Kowalske, C. G. Crandall. 2007b. Impaired cutaneous vasodilation and sweating in grafted skin during whole-body heating. *J. Burn Care Res.* 28 (3):427–34.

Davis, S. L., M. Shibasaki, D. A. Low, J. Cui, D. M. Keller, J. E. Wingo, G. F. Purdue, J. L. Hunt, B. D. Arnoldo, K. J. Kowalske, C. G. Crandall. 2009. Sustained impairments in cutaneous vasodilation and sweating in grafted skin following long-term recovery. *J. Burn Care Res.* 30 (4):675–85.

Davis, S. L., T. E. Wilson, J. M. Vener, C. G. Crandall, J. H. Petajan, A. T. White. 2005. Pilocarpine-induced sweat gland function in individuals with multiple sclerosis. *J. Appl. Physiol.* 98 (5):1740–4.

Dawson, B., J. Bridle, R. J. Lockwood. 1994. Thermoregulation of paraplegic and able bodied men during prolonged exercise in hot and cool climates. *Paraplegia* 32 (12):860–70.

Dill, D. B., A. V. Bock, H. T. Edwards. 1933. Mechanisms for dissipating heat in man and dog. *Am. J. Physiol.* 104:36–43.

Dodt, C., T. Gunnarsson, M. Elam, T. Karlsson, B. G. Wallin. 1995. Central blood volume influences sympathetic sudomotor nerve traffic in warm humans. *Acta Physiol. Scand.* 155 (1):41–51.

Fealey, R. D., P. A. Low, J. E. Thomas. 1989. Thermoregulatory sweating abnormalities in diabetes mellitus. *Mayo Clin. Proc.* 64 (6):617–28.

Fealey, R. D., K. Sato. 2008. Disorders of the eccrine sweat glands and sweating. In *Fitzpatrick's Dermatology in General Medicine*, edited by K. Wolff, L. A. Goldsmith, S. I. Katz, B. A. Gilchrest, A. S. Paller and D. J. Leffell, 215–243. Columbus, OH: The McGraw-Hill Companies.

Fortney, S. M., E. R. Nadel, C. B. Wenger, J. R. Bove. 1981a. Effect of acute alterations of blood volume on circulatory performance in humans. *J. Appl. Physiol.* 50 (2):292–8.

Fortney, S. M., E. R. Nadel, C. B. Wenger, J. R. Bove. 1981b. Effect of blood volume on sweating rate and body fluids in exercising humans. *J. Appl. Physiol.* 51 (6):1594–600.

Fortney, S. M., C. B. Wenger, J. R. Bove, E. R. Nadel. 1984. Effect of hyperosmolality on control of blood flow and sweating. *J. Appl. Physiol.* 57 (6):1688–95.

Fredriksson, I., M. Larsson, F. H. Nystrom, T. Lanne, C. J. Ostgren, T. Stromberg. 2010. Reduced arteriovenous shunting capacity after local heating and redistribution of base-line skin blood flow in type 2 diabetes assessed with velocity-resolved quantitative laser Doppler flowmetry. *Diabetes* 59 (7):1578–84.

Freund, P. R., G. L. Brengelmann, L. B. Rowell, L. Engrav, D. M. Heimbach. 1981. Vasomotor control in healed grafted skin in humans. *J. Appl. Physiol.* 51 (1):168–71.

Freund, P. R., G. L. Brengelmann, L. B. Rowell, E. Halar. 1984. Attenuated skin blood flow response to hyperthermia in paraplegic men. *J. Appl. Physiol.* 56 (4):1104–9.

Friedman, D. B., J. M. Johnson, J. H. Mitchell, N. H. Secher. 1991. Neural control of the forearm cutaneous vasoconstrictor response to dynamic exercise. *J. Appl. Physiol.* 71 (5):1892–6.

Frohman, E. M., M. K. Racke, C. S. Raine. 2006. Multiple sclerosis—The plaque and its pathogenesis. *N. Engl. J. Med.* 354 (9):942–55.

Gagnon, D., O. Jay, G. P. Kenny. 2013. The evaporative requirement for heat balance determines whole-body sweat rate during exercise under conditions permitting full evaporation. *J. Physiol. (Lond)* 591 (Pt 11):2925–35.

Gagnon, D., G. P. Kenny. 2012. Sex differences in thermoeffector responses during exercise at fixed requirements for heat loss. *J. Appl. Physiol.* 113 (5):746–57.

Gass, E. M., G. C. Gass, K. Pitetti. 2002. Thermoregulatory responses to exercise and warm water immersion in physically trained men with tetraplegia. *Spinal Cord.* 40 (9):474–80.

Gomes, M. B., A. S. Matheus, E. Tibirica. 2008. Evaluation of microvascular endothelial function in patients with type 1 diabetes using laser-Doppler perfusion monitoring: Which method to choose? *Microvasc. Res.* 76 (2):132–3.

Goodwin, G. M., D. I. McCloskey, J. H. Mitchell. 1972. Cardiovascular and respiratory responses to changes in central command during isometric exercise at constant muscle tension. *J. Physiol. (Lond)* 226 (1):173–90.

Haensch, C. A., J. Jorg. 2006. Autonomic dysfunction in multiple sclerosis. *J. Neurol.* 253(Suppl. 1):I3–9.

Hamilton, M. T., J. Gonzalez-Alonso, S. J. Montain, E. F. Coyle. 1991. Fluid replacement and glucose infusion during exercise prevent cardiovascular drift. *J. Appl. Physiol.* 71 (3):871–7.

Hoeldtke, R. D., K. D. Bryner, G. G. Horvath, R. W. Phares, L. F. Broy, G. R. Hobbs. 2001. Redistribution of sudomotor responses is an early sign of sympathetic dysfunction in type 1 diabetes. *Diabetes* 50 (2):436–43.

Holowatz, L. A., W. L. Kenney. 2010. Peripheral mechanisms of thermoregulatory control of skin blood flow in aged humans. *J. Appl. Physiol.* 109 (5):1538–44.

Hopman, M. T., B. Oeseburg, R. A. Binkhorst. 1993. Cardiovascular responses in persons with paraplegia to prolonged arm exercise and thermal stress. *Med. Sci. Sports Exerc.* 25 (5):577–83.

Huckaba, C. E., D. B. Frewin, J. A. Downey, H. S. Tam, R. C. Darling, H. Y. Cheh. 1976. Sweating responses of normal, paraplegic and anhidrotic subjects. *Arch. Phys. Med. Rehabil.* 57 (6):268–74.

Huitinga, I., C. J. De Groot, P. Van der Valk, W. Kamphorst, F. J. Tilders, D. F. Swaab. 2001. Hypothalamic lesions in multiple sclerosis. *J. Neuropath. Exp. Neurol.* 60 (12):1208–18.

Ito, T., T. Itoh, T. Hayano, K. Yamauchi, A. Takamata. 2005. Plasma hyperosmolality augments peripheral vascular response to baroreceptor unloading during heat stress. *Am. J. Physiol. Regul. Integr. Comp. Physiol.* 289 (2):R432–40.

Iwamoto, G. A., J. H. Mitchell, M. Mizuno, N. H. Secher. 1987. Cardiovascular responses at the onset of exercise with partial neuromuscular blockade in cat and man. *J. Physiol. (Lond)* 384:39–47.

Johansson, J. E. 1895. Uber die einwirkung der muskelthatigkeit auf die athmung und die herzthatigkeit. *Skand Arch. Physiol.* 5:20–66.

Johnson, J. M. 1986. Nonthermoregulatory control of human skin blood flow. *J. Appl. Physiol.* 61 (5):1613–22.

Johnson, J. M. 2010. Exercise in a hot environment: The skin circulation. *Scand. J. Med. Sci. Sports* 20(Suppl. 3):29–39.

Johnson, J. M., D. L. Kellogg, Jr. 2010. Local thermal control of the human cutaneous circulation. *J. Appl. Physiol.* 109 (4):1229–38.

Johnson, J. M., M. K. Park. 1981. Effect of upright exercise on threshold for cutaneous vasodilation and sweating. *J. Appl. Physiol.* 50 (4):814–18.

Johnson, J. M., M. K. Park. 1982. Effect of heat stress on cutaneous vascular responses to the initiation of exercise. *J. Appl. Physiol.* 53 (3):744–9.

Johnson, J. M., D. W. Proppe. 1996. Cardiovascular adjustments to heat stress. In M. J. Fregly and C. M. Blatteis (Eds.), *Handbook of physiology: Sect. 4 Environmental Physiology*. New York: Oxford University Press, pp. 215–43.

Journeay, W. S., F. D. Reardon, C. R. Martin, G. P. Kenny. 2004. Control of cutaneous vascular conductance and sweating during recovery from dynamic exercise in humans. *J. Appl. Physiol.* 96 (6):2207–12.

Kahn, R. H. 1904. Uber die erwarmung des carotidenblutes. *Arch. Anat. Physiol.* Suppl 81:134.

Kamijo, Y., Y. Okada, S. Ikegawa, K. Okazaki, M. Goto, H. Nose. 2011. Skin sympathetic nerve activity component synchronizing with cardiac cycle is involved in hypovolaemic suppression of cutaneous vasodilatation in hyperthermia. *J. Physiol. (Lond)* 589 (Pt 24):6231–42.

Kamijo, Y., T. Okumoto, Y. Takeno, K. Okazaki, M. Inaki, S. Masuki, H. Nose. 2005. Transient cutaneous vasodilatation and hypotension after drinking in dehydrated and exercising men. *J. Physiol. (Lond)* 568 (Pt 2):689–98.

Katz, A., K. Ekberg, B. L. Johansson, J. Wahren. 2001. Diminished skin blood flow in Type I diabetes: Evidence for non-endothelium-dependent dysfunction. *Clin. Sci. (Lond)* 101 (1):59–64.

Keller, D. M., S. L. Davis, D. A. Low, M. Shibasaki, P. B. Raven, C. G. Crandall. 2006. Carotid baroreceptor stimulation alters cutaneous vascular conductance during whole-body heating in humans. *J. Physiol. (Lond)* 577 (Pt 3):925–33.

Kellogg, D. L., Jr. 2006. In vivo mechanisms of cutaneous vasodilation and vasoconstriction in humans during thermoregulatory challenges. *J. Appl. Physiol.* 100 (5):1709–18.

Kellogg, D. L., Jr., C. G. Crandall, Y. Liu, N. Charkoudian, J. M. Johnson. 1998. Nitric oxide and cutaneous active vasodilation during heat stress in humans. *J. Appl. Physiol.* 85 (3):824–9.

Kellogg, D. L., Jr., J. M. Johnson, W. L. Kenney, W. A. Kosiba, P. E. Pergola. 1993. Mechanisms of control of skin blood flow during prolonged exercise in humans. *Am. J. Physiol. Heart Circ. Physiol.* 265 (34):H562–8.

Kellogg, D. L., Jr., J. M. Johnson, W. A. Kosiba. 1989. Selective abolition of adrenergic vaso-constrictor responses in skin by local iontophoresis of bretylium. *Am. J. Physiol. Heart Circ. Physiol.* 257 (5 Pt 2):H1599–606.

Kellogg, D. L., Jr., J. M. Johnson, W. A. Kosiba. 1991. Competition between cutaneous active vasoconstriction and active vasodilation during exercise in humans. *Am. J. Physiol. Heart Circ. Physiol.* 261 (30):H1184–9.

Kennedy, W. R., X. Navarro. 1989. Sympathetic sudomotor function in diabetic neuropathy. *Arch. Neurol.* 46 (11):1182–6.

Kennedy, W. R., M. Sakuta, D. Sutherland, F. C. Goetz. 1984. The sweating deficiency in diabe-tes mellitus: methods of quantitation and clinical correlation. *Neurology* 34 (6):758–63.

Kenney, W. L., J. M. Johnson. 1992. Control of skin blood flow during exercise. *Med. Sci. Sports Exerc.* 24 (3):303–12.

Kenny, G. P., J. M. Stapleton, J. E. Yardley, P. Boulay, R. J. Sigal. 2013. Older adults with type 2 diabetes store more heat during exercise. *Med. Sci. Sports Exerc.* 45 (10):1906–14.

Khan, F., T. A. Elhadd, S. A. Greene, J. J. Belch. 2000. Impaired skin microvascular func-tion in children, adolescents, and young adults with type 1 diabetes. *Diabetes Care* 23 (2):215–20.

Kihara, M., T. L. Opfer-Gehrking, P. A. Low. 1993. Comparison of directly stimulated with axon-reflex-mediated sudomotor responses in human subjects and in patients with dia-betes. *Muscle Nerve* 16 (6):655–60.

Kondo, N., H. Tominaga, M. Shibasaki, K. Aoki, S. Koga, T. Nishiyasu. 1999. Modulation of the thermoregulatory sweating response to mild hyperthermia during activation of the muscle metaboreflex in humans. *J. Physiol. (Lond)* 515 (Pt 2):591–8.

Kondo, N., H. Tominaga, T. Shiojiri, M. Shibasaki, K. Aoki, S. Takano, S. Koga, T. Nishiyasu. 1997. Sweating responses to passive and active limb movements. *J. Therm. Biol.* 22 (4/5):351–6.

Krogh, A., J. Lindhard. 1913. The regulation of respiration and circulation during the initial stages of muscular work. *J. Physiol. (Lond)* 47:112–36.

Kuno, Y., ed. 1956. *Human Perspiration.* Springfield, IL: Charles Thomas.

Lafrenz, A. J., J. E. Wingo, M. S. Ganio, K. J. Cureton. 2008. Effect of ambient tempera-ture on cardiovascular drift and maximal oxygen uptake. *Med. Sci. Sports Exerc.* 40 (6):1065–71.

Leonard, B., J. H. Mitchell, M. Mizuno, N. Rube, B. Saltin, N. H. Secher. 1985. Partial neu-romuscular blockade and cardiovascular responses to static exercise in man. *J. Physiol. (Lond)* 359:365–79.

Low, P. A. 2004. Evaluation of sudomotor function. *Clin. Neurophysiol.* 115 (7):1506–13.

Lynn, A. G., D. Gagnon, K. Binder, R. C. Boushel, G. P. Kenny. 2012. Divergent roles of plasma osmolality and the baroreflex on sweating and skin blood flow. *Am. J. Physiol. Regul. Integr. Comp. Physiol.* 302 (5):R634–42.

Mack, G. W., D. Cordero, J. Peters. 2001. Baroreceptor modulation of active cutaneous vasodilation during dynamic exercise in humans. *J. Appl. Physiol.* 90 (4):1464–73.

Mack, G. W., H. Nose, E. R. Nadel. 1988. Role of cardiopulmonary baroreflexes during dynamic exercise. *J. Appl. Physiol.* 65 (4):1827–32.

McCord, G. R., C. T. Minson. 2005. Cutaneous vascular responses to isometric handgrip exercise during local heating and hyperthermia. *J. Appl. Physiol. (1985)* 98 (6):2011–18.

McGibbon, B., W. V. Beaumont, J. Strand, F. X. Paletta. 1973. Thermal regulation in patients after the healing of large deep burns. *Plast. Reconstr. Surg.* 52 (2):164–70.

McGregor, I. A. 1950. The regeneration of sympathetic activity in grafted skin as evidence by sweating. *Br. J. Plast. Surg.* 3:12–27.

Mitchell, J. H. 1985. Cardiovascular control during exercise: central and reflex neural mechanisms. *Am. J. Cardiol.* 55 (10):34D–41D.

Mitchell, J. H. 1990. Neural control of the circulation during exercise. *Med. Sci. Sports Exerc.* 22 (2):141–54.

Moorhouse, V. H. K. 1911. Effect of increased temperature of the carotid blood. *Am. J. Physiol.* 28:223–34.

Nadel, E. R., S. M. Fortney, C. B. Wenger. 1980. Effect of hydration state on circulatory and thermal regulations. *J. Appl. Physiol.* 49 (4):715–21.

Nadel, E. R., J. W. Mitchell, B. Saltin, J. A. J. Stolwojk. 1971. Peripheral modifications to the central drive for sweating. *J. Appl. Physiol.* 31 (6):828–33.

Nagashima, K., H. Nose, A. Takamata, T. Morimoto. 1998. Effect of continuous negative-pressure breathing on skin blood flow during exercise in a hot environment. *J. Appl. Physiol.* 84 (6):1845–51.

Nakamura, K., K. Matsumura, T. Hubschle, Y. Nakamura, H. Hioki, F. Fujiyama, Z. Boldogkoi, M. Konig, H. J. Thiel, R. Gerstberger, S. Kobayashi, T. Kaneko. 2004. Identification of sympathetic premotor neurons in medullary raphe regions mediating fever and other thermoregulatory functions. *J. Neurosci.* 24 (23):5370–80.

Nakamura, K., S. F. Morrison. 2008. A thermosensory pathway that controls body temperature. *Nat. Neurosci.* 11 (1):62–71.

Nakamura, K., S. F. Morrison. 2010. A thermosensory pathway mediating heat-defense responses. *Proc. Natl. Acad. Sci. U. S. A.* 107 (19):8848–53.

National Multiple Sclerosis Society. 2014. *Who Gets MS? (Epidemiology).* Accessed December 15, 2014. http://www.nationalmssociety.org/What-is-MS/Who-Gets-MS.

Nicotra, A., M. Asahina, C. J. Mathias. 2004. Skin vasodilator response to local heating in human chronic spinal cord injury. *Eur. J. Neurol.* 11 (12):835–7.

Nielsen, B., M. Nielsen. 1965. On the regulation of sweat secretion in exercise. *Acta Physiol. Scand.* 64:314–22.

Nielsen, M. 1938. Die regulation der körpertemperatur bei muskelarbeit. *Skand. Arch. Physiol.* 79:193–230.

Nishiyasu, T., X. G. Shi, G. W. Mack. 1991. Effect of hypovolemia on forearm vascular resistance control during exercise in the heat. *J. Appl. Physiol.* 71 (4):1382–6.

Noronha, M. J., C. J. Vas, H. Aziz. 1968. Autonomic dysfunction (sweating responses) in multiple sclerosis. *J. Neurol. Neurosurg. Psychiatry* 31 (1):19–22.

Nose, H., G. W. Mack, X. R. Shi, K. Morimoto, E. R. Nadel. 1990. Effect of saline infusion during exercise on thermal and circulatory regulations. *J. Appl. Physiol.* 69 (2):609–16.

Ott, I. 1877. Heat center in the brain. *J. Nerv. Ment. Dis.* 14:152.

Petrofsky, J., S. Lee. 2005. The effects of type 2 diabetes and aging on vascular endothelial and autonomic function. *Med. Sci. Monit.* 11 (6):CR247–54.

Petrofsky, J. S. 1992. Thermoregulatory stress during rest and exercise in heat in patients with a spinal cord injury. *Eur. J. Appl. Physiol. Occup. Physiol.* 64 (6):503–7.

Petrofsky, J. S. 2012. Resting blood flow in the skin: Does it exist, and what is the influence of temperature, aging, and diabetes? *J. Diabetes Sci. Technol.* 6 (3):674–85.

Ponten, B. 1960. Grafted skin. *Acta Chir. Scand. Suppl.* 257:1–78.

Price, M. J. 2006. Thermoregulation during exercise in individuals with spinal cord injuries. *Sports Med.* 36 (10):863–79.

Price, M. J., I. G. Campbell. 1997. Thermoregulatory responses of paraplegic and able-bodied athletes at rest and during prolonged upper body exercise and passive recovery. *Eur. J. Appl. Physiol. Occup. Physiol.* 76 (6):552–60.

Price, M. J., I. G. Campbell. 1999a. Thermoregulatory and physiological responses of wheel-chair athletes to prolonged arm crank and wheelchair exercise. *Int. J. Sports Med.* 20 (7):457–63.

Price, M. J., I. G. Campbell. 1999b. Thermoregulatory responses of spinal cord injured and able-bodied athletes to prolonged upper body exercise and recovery. *Spinal Cord.* 37 (11):772–9.

Robertshaw, D. 2006. Mechanisms for the control of respiratory evaporative heat loss in panting animals. *J. Appl. Physiol.* 101 (2):664–8.

Roskind, J. L., J. Petrofsky, A. R. Lind, F. X. Paletta. 1978. Quantitation of thermoregulatory impairment in patients with healed burns. *Ann. Plast. Surg.* 1 (2):172–6.

Saltin, B., A. P. Gagge, J. A. J. Stolwijk. 1970. Body temperatures and sweating during thermal transients caused by exercise. *J. Appl. Physiol.* 28:318–27.

Sato, K., W. H. Kang, K. Saga, K. T. Sato. 1989. Biology of sweat glands and their disorders. I. Normal sweat gland function. *J. Am. Acad. Dermatol.* 20 (4):537–63.

Schmiedel, O., M. L. Schroeter, J. N. Harvey. 2007. Microalbuminuria in type 2 diabetes indicates impaired microvascular vasomotion and perfusion. *Am. J. Physiol. Heart Circ. Physiol.* 293 (6):H3424–31.

Shapiro, Y., Y. Epstein, C. Ben-Simchon, H. Tsur. 1982. Thermoregulatory responses of patients with extensive healed burns. *J. Appl. Physiol.* 53 (4):1019–22.

Shastry, S., N. M. Dietz, J. R. Halliwill, A. S. Reed, M. J. Joyner. 1998. Effects of nitric oxide synthase inhibition on cutaneous vasodilation during body heating in humans. *J. Appl. Physiol.* 85 (3):830–4.

Shastry, S., C. T. Minson, S. A. Wilson, N. M. Dietz, M. J. Joyner. 2000. Effects of atropine and L-NAME on cutaneous blood flow during body heating in humans. *J. Appl. Physiol.* 88 (2):467–72.

Shibasaki, M., K. Aoki, K. Morimoto, J. M. Johnson, A. Takamata. 2009. Plasma hyperosmolality elevates the internal temperature threshold for active thermoregulatory vasodilation during heat stress in humans. *Am. J. Physiol. Regul. Integr. Comp. Physiol.* 297 (6):R1706–12.

Shibasaki, M., C. G. Crandall. 2001. Effect of local acetylcholinesterase inhibition on sweat rate in humans. *J. Appl. Physiol.* 90 (3):757–62.

Shibasaki, M., N. Kondo, C. G. Crandall. 2001. Evidence for metaboreceptor stimulation of sweating in normothermic and heat-stressed humans. *J. Physiol. (Lond)* 534 (2):605–11.

Shibasaki, M., P. Rasmussen, N. H. Secher, C. G. Crandall. 2009. Neural and non-neural control of skin blood flow during isometric handgrip exercise in the heat stressed human. *J. Physiol. (Lond)* 587 (Pt 9):2101–7.

Shibasaki, M., M. Sakai, M. Oda, C. G. Crandall. 2004. Muscle mechanoreceptor modulation of sweat rate during recovery from moderate exercise. *J. Appl. Physiol.* 96 (6):2115–19.

Shibasaki, M., N. H. Secher, J. M. Johnson, C. G. Crandall. 2005. Central command and the cutaneous vascular response to isometric exercise in heated humans. *J. Physiol. (Lond)* 565 (Pt 2):667–73.

Shibasaki, M., N. H. Secher, C. Selmer, N. Kondo, C. G. Crandall. 2003. Central command is capable of modulating sweating from non-glabrous human skin. *J. Physiol. (Lond)* 553 (Pt 3):999–1004.

Sokolnicki, L. A., S. K. Roberts, B. W. Wilkins, A. Basu, N. Charkoudian. 2007. Contribution of nitric oxide to cutaneous microvascular dilation in individuals with type 2 diabetes mellitus. *Am. J. Physiol. Endocrinol. Metab.* 292 (1):E314–18.

Sokolnicki, L. A., N. A. Strom, S. K. Roberts, S. A. Kingsley-Berg, A. Basu, N. Charkoudian. 2009. Skin blood flow and nitric oxide during body heating in type 2 diabetes mellitus. *J. Appl. Physiol.* 106 (2):566–70.

Stansberry, K. B., M. A. Hill, S. A. Shapiro, P. M. McNitt, B. A. Bhatt, A. I. Vinik. 1997. Impairment of peripheral blood flow responses in diabetes resembles an enhanced aging effect. *Diabetes Care* 20 (11):1711–16.

Stapleton, J. M., J. E. Yardley, P. Boulay, R. J. Sigal, G. P. Kenny. 2013. Whole-body heat loss during exercise in the heat is not impaired in type 1 diabetes. *Med. Sci. Sports Exerc.* 45 (9):1656–64.

Stephens, D. P., A. R. Saad, L. A. Bennett, W. A. Kosiba, J. M. Johnson. 2004. Neuropeptide Y antagonism reduces reflex cutaneous vasoconstriction in humans. *Am. J. Physiol. Heart Circ. Physiol.* 287 (3):H1404–9.

Strom, N. A., L. W. Meuchel, D. W. Mundy, J. R. Sawyer, S. K. Roberts, S. M. Kingsley-Berg, N. Charkoudian. 2011. Cutaneous sympathetic neural responses to body cooling in type 2 diabetes mellitus. *Auton. Neurosci.* 159 (1–2):15–19.

Strom, N. A., J. R. Sawyer, S. K. Roberts, S. M. Kingsley-Berg, N. Charkoudian. 2010. Local sensory nerve control of skin blood flow during local warming in type 2 diabetes mellitus. *J. Appl. Physiol.* 108 (2):293–7.

Takamata, A., G. W. Mack, C. M. Gillen, A. C. Jozsi, E. R. Nadel. 1995. Osmoregulatory modulation of thermal sweating in humans: Reflex effects of drinking. *Am. J. Physiol. Regul. Integr. Comp. Physiol.* 268 (37):R414–22.

Takamata, A., K. Nagashima, H. Nose, T. Morimoto. 1997. Osmoregulatory inhibition of thermally induced cutaneous vasodilation in passively heated humans. *Am. J. Physiol. Regul. Integr. Comp. Physiol.* 273 (1):R197–204.

Tam, H. S., R. C. Darling, H. Y. Cheh, J. A. Downey. 1978. Sweating response: A means of evaluating the set-point theory during exercise. *J. Appl. Physiol. Respir. Environ. Exerc. Physiol.* 45 (3):451–8.

Taylor, W.F., J. M. Johnson, D. S. O'Leary, M. K. Park. 1984. Modification of the cutaneous vascular response to exercise by local skin temperature. *J. Appl. Physiol.* 57 (6):1878–84.

Theisen, D., Y. Vanlandewijk, X. Sturbois, M. Francaux. 2000. Cutaneous vasomotor adjustments during arm-cranking in individuals with paraplegia. *Eur. J. Appl. Physiol.* 83 (6):539–44.

Theisen, D., Y. Vanlandewijck, X. Sturbois, M. Francaux. 2001. Cutaneous vascular response and thermoregulation in individuals with paraplegia during sustained arm-cranking exercise. *Int. J. Sports Med.* 22 (2):97–102.

Totel, G. L. 1974. Physiological responses to heat of resting man with impaired sweating capacity. *J. Appl. Physiol.* 37 (3):346–52.

Totel, G. L., R. E. Johnson, F. A. Fay, J. A. Goldstein, J. Schick. 1971. Experimental hyperthermia in traumatic quadriplegia. *Int. J. Biometeorol.* 15 (2):346–55.

Uhthoff, W. 1889. Untersuchungen uber die bei der multiplen Herdsklerose vorkommenden Augenstorungen. *Archiv. für Psychiatrie und Nervenkrankheiten.* 20:55.

Uno, H. 1977. Sympathetic innervation of the sweat glands and piloarrector muscles of macaques and human beings. *J. Invest. Dermatol.* 69 (1):112–20.

van Beaumont, W., R. W. Bullard. 1963. Sweating: its rapid responses to muscular work. *Science* 141:643–6.

Vas, C. J. 1969. Sexual impotence and some autonomic disturbances in men with multiple sclerosis. *Acta Neurol. Scand.* 45 (2):166–82.

Vissing, S. F., U. Scherrer, R. G. Victor. 1994. Increase of sympathetic discharge to skeletal muscle but not to skin during mild lower body negative pressure in humans. *J. Physiol. (Lond)* 481 (1):233–41.

Wallin, B. G., L. Stjernberg. 1984. Sympathetic activity in man after spinal cord injury. Outflow to skin below the lesion. *Brain* 107 (Pt 1):183–98.

White, M. D. 2006. Components and mechanisms of thermal hyperpnea. *J. Appl. Physiol.* 101 (2):655–63.

Wick, D. E., S. K. Roberts, A. Basu, P. Sandroni, R. D. Fealey, D. Sletten, N. Charkoudian. 2006. Delayed threshold for active cutaneous vasodilation in patients with type 2 diabetes mellitus. *J. Appl. Physiol.* 100 (2):637–41.

Wilson, S. B., P. E. Jennings, J. J. Belch. 1992. Detection of microvascular impairment in type I diabetics by laser Doppler flowmetry. *Clin. Physiol.* 12 (2):195–208.

Wilson, T. E., R. Carter, 3rd, M. J. Cutler, J. Cui, M. L. Smith, C. G. Crandall. 2004. Active recovery attenuates the fall in sweat rate but not cutaneous vascular conductance after supine exercise. *J. Appl. Physiol.* 96 (2):668–73.

Wilson, T. E., J. Cui, C. G. Crandall. 2001. Absence of arterial baroreflex modulation of skin sympathetic activity and sweat rate during whole-body heating in humans. *J. Physiol. (Lond)* 536 (2):615–23.

Wilson, T. E., J. Cui, C. G. Crandall. 2005. Mean body temperature does not modulate eccrine sweat rate during upright tilt. *J. Appl. Physiol.* 98 (4):1207–12.

World Health Organization. 2014. *Diabetes Fact Sheet.* Last Modified November 2014. http://www.who.int/mediacentre/factsheets/fs312/en/.

Yaggie, J. A., T. J. Niemi, M. J. Buono. 2002. Adaptive sweat gland response after spinal cord injury. *Arch. Phys. Med. Rehabil.* 83 (6):802–5.

Yamazaki, F. 2002. Vasomotor responses in glabrous and nonglabrous skin during sinusoidal exercise. *Med. Sci. Sports Exerc.* 34 (5):767–72; discussion 773.

Yamazaki, F., R. Sone, H. Ikegami. 1994. Responses of sweating and body temperature to sinusoidal exercise. *J. Appl. Physiol.* 76 (6):2541–5.

Section II

Effects of Fluid Imbalance on Body Functions and Performance

4 Cardiovascular Responses to Body Fluid Imbalance

Ricardo Mora-Rodriguez and Juan F. Ortega

CONTENTS

4.1 INTRODUCTION

During exercise the cardiovascular system supplies contracting skeletal muscles with nutrients and oxygen in accordance with its energy demands. It also removes waste products and importantly serves to conduct heat from the contracting muscles to the skin and from there to the surrounding environment. Heat dissipation involvement of the cardiovascular system is crucial to prevent core temperature from increasing to levels that compromise exercise capacity and health (i.e., heat exhaustion and heat stroke, respectively). Sweat evaporation is the main mechanism to dissipate heat during exercise (0.58 kcals g^{-1} of sweat evaporated). Sweat production rate is nicely matched to exercise intensity (Buono et al. 2008; Hamouti et al. 2011) and thus to heat production. However, as subjects dehydrate during continuous exercise, heat dissipation by sweat falls behind heat production and heat accumulates in tissues (i.e., hyperthermia).

On the other hand, the composition and amount of the cardiovascular system vehicle (i.e., the blood) is markedly affected by dehydrating exercise. Sweat filtrate in the sweat glands originates from blood plasma and interstitial fluid, and prolonged profuse sweating results in plasma fluid losses and hemoconcentration. This challenges the cardiovascular system to cope with the demand of perfusing both the contracting muscles and the skin while blood volume is reduced (i.e., dehydration-induced cardiovascular strain). Both dehydration and hyperthermia developed during prolonged exercise not only strain the cardiovascular system (Gonzalez-Alonso et al. 1995) but linked to this produce central neuromuscular fatigue (Nybo et al. 2014) and affect the sources of energy that muscles use during contraction (i.e., affect muscle metabolism) (Febbraio 2001). Fluid intake during exercise prevents some of the consequences of exercise dehydration (i.e., hyperthermia) unless exercise is too intense (Mora-Rodriguez et al. 2010) and heat production surpasses the capacity of the environment to accept heat (i.e., uncompensable environment) (Mora-Rodriguez 2012).

Readers interested in learning more about fluid replacement are directed to the 2007 American College of Sports Medicine position stand which provides guidance on fluid replacement to hydrate individuals during exercise (Sawka et al. 2007). This chapter deals with the consequences of exercise-induced dehydration with emphasis on the cardiovascular system strain that appears during prolonged submaximal intensity exercise. This strain is better understood when measuring the full cardiovascular response to exercise (i.e., arterial and central venous pressure, cardiac output, heart rate [HR], and stroke volume [SV]). *Cardiovascular drift* (i.e., CDV drift) is the term that details the response of these individual components during prolonged dehydrating exercise at a fixed workload. In this chapter we review the possible causes of CDV drift and the potential impact on the adaptations sought when exercising to improve athletic performance or health.

4.2   CARDIOVASCULAR DRIFT

CDV drift is the progressive change in the cardiovascular system that occurs during prolonged fixed-load aerobic submaximal exercise. Although some salient circulatory changes like increase in HR and reduction in SV are common to all definitions of CDV drift, cardiac output ($\dot{Q}$) and blood pressure (BP) could be either maintained or decreased depending on the exercise intensity, duration, percent dehydration, and hyperthermia incurred during exercise. Most studies designed to unveil the causes and mechanisms of CDV drift use a control exercise bout where they try to prevent the drift. The different means to prevent CDV drift include oral rehydration, body cooling (either full body or only skin), intravenous infusion of fluids (rehydrating either the intravascular or all fluid spaces), reducing exercise intensity, pacing the heart with cardioselective ß-adrenergic receptor blockers, or trying to improve venous return with the use of compressive bandage (Grimby et al. 1966). In this chapter, we will only present studies in which at least HR, cardiac output, SV, and BP are presented describing the response of the components of CDV drift. Most of the studies discussed in this chapter investigate the response to

TABLE 4.1

Cardiovascular Measurements and Calculated Components of CDV Drift in the Studies Presented in This Chapter

Variable Measured during Exercise	Measurement Technique or Calculation
Cardiac output ($\dot{Q}$)	Non-invasive re-breathing
	Thermodilution
	Indocyanide green dye
	Doppler echocardiography
HR	ECG
	HR monitors
SV	Calculated as $\dot{Q}/HR$
BP	Non-invasive sphygmomanometry
	Pressure gauged intravenous or arterial catheters
Systemic vascular resistance (SVR)	Calculated as $BP/\dot{Q}$
Skin blood flow (SkBF)	Venous occlusion plethysmography
	Laser Doppler flowmetry in the forearm
Cutaneous vascular conductance (CVC)	Calculated as SkBF/BP
Rate of O_2 consumption ($\dot{V}O_2$)	Indirect calorimetry
Oxygen pulse	Calculated as $\dot{V}O_2/HR$

exercise and different manipulations on the measurements and calculated variables presented in Table 4.1.

4.2.1 Cardiovascular Drift: Traditional View

One of the most researched areas of thermal physiology is the cardiovascular response to fluid deficit and hyperthermia. Studies around the Second World War showed that dehydration during prolonged walking in the heat caused exhaustion accompanied by a continuous drift up in HR despite maintenance of the workload (Pitts and Consolazio 1944; Ladell 1955). These authors noticed fairly good positive correlations between the increases in rectal temperature and the elevations in HR (Ladell 1955). However, the common mechanisms that linked hyperthermia and tachycardia were not clear at the time.

In the seventies, Rowel and collaborators conducted a series of elaborated studies with BP measurements at the superior vena cava and descending aorta while also measuring cardiac output using indocyanide green dye. In one of their experiments at rest, they found that when the skin was warmed using water perfusion suits, the increases in HR paralleled a rise in systemic cardiac output (i.e., $\dot{Q}$ in L min^{-1}) (Rowell et al. 1969). At rest, without the influence of muscle contraction, the skin and the splanchnic organs are the only candidates to receive that increased $\dot{Q}$ and cause the HR drift. Thus, the idea of skin vasodilation being responsible for initiating CDV drifts set out from this study at rest. At the time, there were studies documenting that during exercise, renal (Radigan and Robinson 1949) and splanchnic (Rowell et al. 1965) blood flow decrease when heating up the body compared to when exercising in a cold environment (Rowell et al. 1969). This suggested that the increase in cardiac output observed during mild exercise in the heat (Rowell et al. 1967) was actually the consequence of blood flow redistribution to the skin vasculature, a circulatory bed that was not thought at that time to be able to hold such a large increase in systemic $\dot{Q}$.

Cardiac output is not always increased during exercising in the heat as a consequence of the increased SkBF. Rowell and coworkers observed that $\dot{Q}$ actually decreased during exercise in a hot environment (43°C) compared to a thermoneutral environment (26°C) when exercise intensity ranged from moderate to high (Rowell et al. 1966). These researchers concluded that when the workload demands upon the circulation exceed a certain level, it seems that blood delivery to muscles takes preference over thermal requirements for heat transfer. The increased core temperature in the hot environment (i.e., 39.4°C vs. 38.2°C) could also reduce $\dot{Q}$ during exercise as it has been reported during exercise in a hot environment (Gonzalez-Alonso et al. 1997, 1998).

In a subsequent study, Rowell and coworkers monitored several cardiovascular responses to rapid changes in skin temperature during mild and intense exercise (Rowell et al. 1969). They noticed that the relocation of blood to the skin during prolonged exercise provoked a drop in central venous pressure, aortic pressure, and SV. They hypothesized that the reduction in the pace and pressure of blood returning to the heart triggered baroreceptors activity (Rowell 1986) which in turn generate chronotropic signals to the heart. All these responses were reversed when the skin was cooled and vasoconstrictor tone was regained during the same exercise session. These early experiments (Rowell et al. 1969; Rowell 1974; Shaffrath and Adams 1984) coined the definition of *classical CDV drift* and pointed to the fact that while SkBF serves an unavoidable heat dissipatory role, skin vasodilation has the potential to markedly perturb the systemic circulation during exercise in the heat. As we will relate below, the sequence of events described by Rowell and coworkers, increases in SkBF inducing reduction in central venous pressure, SV, and increased HR, do not seem to hold during dehydration-induced CDV drift (Coyle 1998; Coyle and Gonzalez-Alonso 2001). In addition, recent experiments pacing the heart (β-adrenergic receptor blockade) during exercise in thermoneutral and hot environments provide new perspectives on the causes of CDV drift.

4.2.2 CARDIOVASCULAR DRIFT: ALTERNATIVE EXPLANATIONS

A series of studies from Dr. Edward Coyle laboratory at the University of Texas at Austin in the nineties explored the effects of fluid replacement on the CDV drift that develops during prolonged exercise. In the first study of this series, subjects cycled for 120 min at 70% of their $\dot{V}O_{2\,max}$ without or with full oral fluid replacement (Hamilton et al. 1991). Fluid replacement prevented the decline in $\dot{Q}$ (7%), SV (15%), and the increase in HR (10%) observed when no fluid was allowed and subjects dehydrated (2.9%) while their rectal temperature escalated to 38.9°C. However, skin temperature was not measured in this initial study to establish if skin vasodilation had a major role in CDV drift as Rowell (1986) proposed. This experiment was conducted in a thermoneutral environment (22°C) and subjects were ventilated at airflow of 3 m s^{-1}. Thus, it is unlikely that augmented flow to the skin was behind the CDV drift in these experimental conditions. At the time, the authors hypothesized that reduced plasma volume (9%) and increased core temperature in the no-fluid trial could have triggered CDV drift.

In a subsequent experiment, Dr. Coyle's group explored the dose–response relationship between dehydration and CDV drift. They found that during prolonged exercise in a hot environment (i.e., 120 min at 33°C), HR increased and SV declined in proportion to the amount of dehydration (Montain and Coyle 1992b) when tested in a range from 1.1% to 4.2% dehydration. SkBF was 20%–22% higher in the trial where the largest amount of fluid was ingested and subjects dehydrated little (i.e., 1.1%) compared to when low or no fluid was provided (3.4% and 4.2% dehydration) (Montain and Coyle 1992b). Therefore, SkBF and CDV drift responded in dissociation, since in the trial with the higher SkBF (i.e., 1.1% dehydration) CDV drift was the lowest. Investigators searched for alternative explanations to CDV drift beyond blood relocation in the skin. Below we present studies focusing on alternative factors that could be causing CDV drift during prolonged exercise.

4.2.3 Role of Blood Volume

Accompanying progressive dehydration during exercise in the heat there is a progressive loss of plasma volume. Plasma volume actively participates in the sweat and respiratory losses of water during prolonged exercise. Since CDV drift pertains to the circulatory system and it has its origin in a decrease of central blood volume (Rowell 1974), it is logical to hypothesize that plasma volume reduction could be a major factor in the development of CDV drift. In one experiment, Dr. Coyle team had subjects dehydrated by losing 2.5% of body weight in the heat prior to a bout of exercise in a thermoneutral 21°C environment. Dehydration was followed by 2 h of oral rehydration or no fluid ingestion (i.e., no-fluid trial). During these 2 h, plasma volume recovered to pre-exercise levels even in the trial where no-fluid was allowed (Heaps et al. 1994).

We and others (Nose et al. 1988; Hamouti et al. 2013) have observed this phenomenon of plasma volume recovery despite withholding fluids. The recovery of plasma volume despite maintained dehydration is possible thanks to the transfer of fluid from the interstitial and intracellular space into the vasculature (Mora-Rodriguez et al. 2015). Apparently, plasma volume is high ranked in the hierarchy of fluid recovery and even active muscle water is used to recover most PV within 1 h after finishing exercise (Mora-Rodriguez et al. 2015). In the Heaps et al. (1994) study, they observed that despite full endogenous plasma volume recovery during the trial without rehydration, CDV drift ensued during subsequent exercise. These observations suggested that CDV drift during exercise in a 21°C environment could be influenced by intracellular and interstitial dehydration and not only by reductions in plasma volume (intravascular dehydration).

During dehydrating exercise, plasma volume accounts for up to 10%–20% of the fluid losses (Mora-Rodriguez et al. 2015). Therefore, extravascular dehydration from intracellular and interstitial spaces ought to importantly contribute to fluid losses during exercise. We have recently reported that skeletal muscle water content declines during the recovery of dehydrating exercise (Mora-Rodriguez et al. 2015). The increases in metabolite concentration in the muscle could alter mechanoreceptor activity and sympathetic nerve activity and indirectly contribute to the upward HR drift in subsequent bout of exercise (Fernández-Elías et al. 2015). However, when the hyperthermia and blood dehydration induced by prolonged exercise in the heat are prevented by exercising in a cold environment with plasma volume expansion, SV, cardiac output, and mean arterial pressure are well maintained. The prevention of CDV drift despite 3–4 L body fluid deficit (Gonzalez-Alonso et al. 1997) suggests that intravascular fluid loss is the only fluid space relevant for CDV drift (Figure 4.1).

Looking at the problem from a different angle, we recently studied CDV drift after a regular blood bank donation (i.e., 450 mL blood withdrawal in less than 15 min) (Mora-Rodriguez et al. 2012). Blood donation reduced blood volume by 9% in our 74 kg subjects. Interestingly, 3% of that loss was endogenously recovered in the 2 h after the donation and before the start of exercise. During the 60 min of exercise in the heat that followed the donation (35°C at 60% of $\dot{V}O_{2\,max}$), HR increased compared to a control trial performed several days before donation (150 vs. 138 beats min^{-1} at the end of exercise). Blood donation also caused hyperthermia from the beginning of exercise, and thus, it is unclear what percent of the observed CDV drift was attributable to the 6% lower blood volume and what portion was due to hyperthermia.

The precise contribution of the reduction in plasma volume during exercise on CDV drift was addressed by Montain and Coyle (1992a). In this study, the investigators prevented the normal plasma volume contraction that takes place during 120 min of dehydrating exercise in the heat (33°C) by intravenous infusion of a plasma expander (i.e., Dextran®). This expander only rehydrates the vasculature without recovering the extravascular fluid space unlike experiments infusing saline that rehydrates all

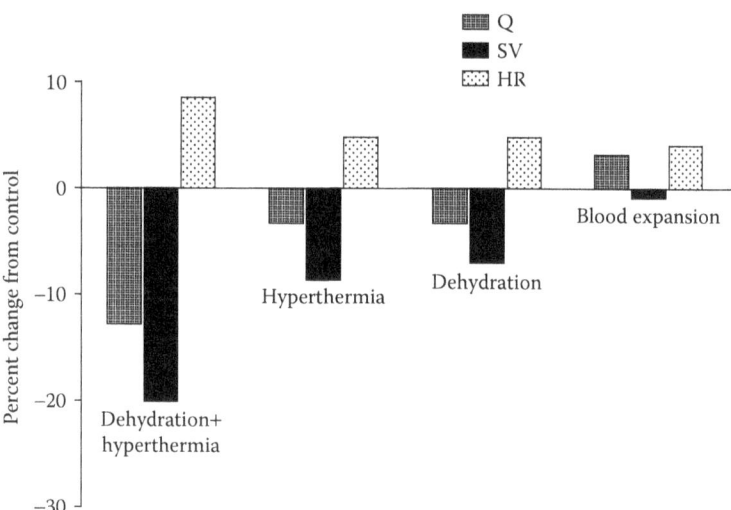

FIGURE 4.1 Separated contribution of hyperthermia, dehydration, and plasma volume losses to the CDV drift that develops during exercise in a hot environment. (Data from Gonzalez-Alonso, J. et al., *J. Appl. Physiol.*, 82, 1229–36, 1997.)

fluid spaces (Deschamps et al. 1989; Nose et al. 1990). This manipulation allowed to factor out the role of blood volume in CDV drift. Interestingly, plasma volume expansion did not prevent core temperature from increasing similarly to when no fluid was ingested during exercise. However, the plasma expander prevented two-thirds of the increases in HR and one-third of the declines in SV observed in the no-fluid trial. Importantly, this study revealed that when hyperthermic, plasma volume reductions account for some but not all the CDV drift (Deschamps et al. 1989; Nose et al. 1990). In contrast, when all body fluid spaces were rehydrated by ingesting 2.4 L of a sports drink, SkBF increased, hyperthermia was greatly attenuated and so was CDV drift (Montain and Coyle 1992a). This study suggests that a portion of CDV drift is related to blood volume losses but another important portion is due to hyperthermia.

Other investigators have tried to expand plasma volume before prolonged exercise by natural means avoiding banned substances or intravenous infusions. We and others have used ingestion of salt and water to achieve plasma volume expansion while measuring HR drift (Greenleaf et al. 1997; Coles and Luetkemeier 2005; Sims et al. 2007; Hamouti et al. 2014). To our knowledge, only one study reported the full cardiovascular response to exercise in the heat after this *ecological* means of expanding plasma volume (Hamouti et al. 2014). We observed that similar to the results obtained with the blood expanders, pre-exercise ingestion of saline (i.e., 82 mM of Na$^+$) attenuated the SV, and cardiac output declines were observed when only water was ingested. HR was also somewhat lower (149 vs. 153 beats min^{-1}; $p = .06$) with the saline ingestion, and thus, CDV drift was blunted despite no effects on lowering core temperature. Interestingly, the reduction in cardiovascular strain resulted in alleviation of fatigue and improved performance (Hamouti et al. 2014). Other investigators have proposed that the alleviation of CDV drift would improve performance independently of effects on core temperature (Cheuvront et al. 2010). Although it is unclear if rehydration of all fluid spaces is required to prevent CDV drift, it seems important to improve performance since replenishing of the vascular space only has no ergogenic effect (Watt et al. 1999).

4.2.4   ROLE OF SKIN BLOOD FLOW

4.2.4.1   Effects of Body Position

During exercise in a hot environment with progressive dehydration, SkBF is maintained or reduced while HR progressively increases (Montain and Coyle 1992a; Gonzalez-Alonso et al. 1995, 1997, 2000). This dissociation between the progression of CDV drift upward and the plateau response of SkBF is the main argument to negate the traditional mechanistic explanation for CDV drift. However, the fact that we can dissociate SkBF from CDV drift does not negate that increases in SkBF augment the demands in the circulatory system likely contributing to the first stages of CDV drift during prolonged exercise. We should keep in mind that increases in SkBF allow higher heat dissipation reducing hyperthermia and preventing the portion of CDV drift associated with the increases in core temperature. Under this perspective, the dissociation between the responses of CDV drift and SkBF is not that surprising.

Some experiments suggest that during prolonged exercise in a hot environment, manipulations to raise central venous pressure permit higher skin circulation, preventing excessive hyperthermia and CDV drift. Maybe the clearest study describing this mechanism was conducted by Gonzalez-Alonso and coworkers (1999). They compared supine and upright cycling at the same VO_2 when subjects were similarly dehydrated and hot. In this experiment, supine exercise in the heat (35°C) allowed the maintenance of higher cutaneous blood flow, $\dot{Q}$, SV, while plasma catecholamine concentration and HR were lower compared to cycling upright. Likely, the augmentation in central blood volume during supine exercise allowed higher skin blood while preventing SV reductions (Poliner et al. 1980). Supine position increases central blood volume inhibiting the baroreflex, sympathetic nerve, and hormonal signals (i.e., plasma norepinephrine) that prevent skin vasodilation during upright cycling (Kellogg et al. 1990). Thus, semi-recumbent or supine exercise has a larger effect on preventing CDV drift than plasma volume expansion in a dehydrated condition (Montain and Coyle 1992a).

4.2.4.2   Effects of Airflow Restriction

Shaffrath and Adams (1984) investigated the effects of preventing airflow in CDV drift during prolonged (70 min) moderately intense (60% $\dot{V}O_{2max}$) upright cycling in a thermoneutral environment (24°C). They observed that when airflow was reduced from 4.3 to 0.2 m s^{-1}, SkBF and HR increased while SV and BP decreased. This was consistent with the views of Rowell in which redistribution of blood volume into the skin reduces central blood volume and initiates CDV drift (Rowell 1986). During 60 min of exercise in a hot environment (36°C), we have found that oral rehydration (i.e., 2 L) does not prevent CDV drift when airflow is not allowed (Mora-Rodriguez et al. 2007). Like in the data of Shaffrath and Adams (1984), we found a tendency for SkBF to be higher when no airflow was allowed during exercise. Wingo and coworkers (2005) also reported that when endurance-trained subjects exercise in a 35°C environment without airflow, fluid ingestion does not attenuate CDV drift. Thus, when airflow is prevented, relocation of blood to the skin lowers central blood volume becoming a likely cause of the CDV drift (Rowell 1974; Shaffrath and Adams 1984; Mora-Rodriguez et al. 2007).

4.2.4.3   Effects of Exercise in Cold versus Hot Environments

Rowell and coworkers (1969) found that rapid cooling of skin from 39°C (a temperature that fully vasodilates the skin) to 27°C during 30 min bouts of exercise reversed the drop in SV and the increases in HR. Therefore, venous pooling induced by heating the skin can certainly cause CDV drift during short-term exercise. However, it is unclear if the CDV drift that ensues during prolonged exercise is due to progressive venous pooling into the cutaneous circulation. In agreement with Rowell (1974), we found that during the first stages of submaximal exercise (65% $\dot{V}O_{2max}$) in a euhydrated condition, $\dot{Q}$ is 1.2 L min^{-1} higher in the heat (33°C) than in a cold (4°C) environment (Gonzalez-Alonso et al. 2000). This difference in $\dot{Q}$ was accompanied by a 4.5-fold higher cutaneous blood flow in the hot environment which was likely causing these increases. The increased $\dot{Q}$ was driven by an increase in HR (150 to 161 beats min^{-1}) while SV was similar in both trials (135 mL blood per heart beat) (Gonzalez-Alonso et al. 2000).

This situation of increased HR and $\dot{Q}$ in parallel to the increases in SkBF is however altered when subjects are dehydrated. When subjects are 4% dehydrated, the reduction in blood volume and the increase in core temperature and HR drive SV and $\dot{Q}$ down (Gonzalez-Alonso et al. 2000) independently of SkBF. The increasing level of dehydration as prolonged exercise progresses eventually results in a reduction in SkBF despite progressive hyperthermia and thermal drive for skin vasodilation. In fact, stepwise regression analysis excludes SkBF as a factor related to the reduction in SV with progressive dehydration (Gonzalez-Alonso et al. 2000). Of note, SkBF reductions during progressive dehydration could be mediated by the increase in plasma catecholamine concentrations acting upon alpha adrenergic receptors in the skin vasculature (Mora-Rodriguez et al. 1996). In summary, when subjects are normothermic and euhydrated (at the start of prolonged exercise), skin warming elicits a simultaneous increase in HR and $\dot{Q}$ with maintenance in SV. However, the parallelism between HR and SkBF ceases during prolonged exercise when dehydration and hyperthermia develop.

4.2.5 ROLE OF METABOLIC DRIFT

Together with increase in HR during prolonged exercise, there is also an increase in oxygen consumption despite unaltered work rate. This drift in $\dot{V}O_2$ is named *metabolic drift* or *slow component of* $\dot{V}O_2$. Initially, this drift was thought to be caused by increased hepatic oxygen consumption with exercise duration to provide energy for the increased rates of liver glucose output and gluconeogenesis. However, Poole and coworkers (1991) circumscribed this phenomenon to increase metabolic demand of the exercising muscle as exercise progresses and muscle fibers become glycogen depleted. Others have confirmed that this drift originates in the contracting muscle (Gonzalez-Alonso et al. 1998), and it is related to a decreased economy of movement by recruiting extra muscle fibers during prolonged exercise (Dick and Cavanagh 1987; Westerlind et al. 1992).

It is tempting to hypothesize that the drift in O_2 consumption (slow $\dot{V}O_2$ component) is demanding an increase in $\dot{Q}$ to deliver more oxygen in this condition of prolonged exercise. Rowland and coworkers (2008) observed during prolonged submaximal exercise (65% $\dot{V}O_{2peak}$) in prepubertal boys a parallel rise in $\dot{V}O_2$ and $\dot{Q}$ (8% and 10%, respectively). They defend that the increase in muscle temperature during exercise augments $\dot{V}O_2$ by the Q10 effect resulting in increases in systemic blood flow ($\dot{Q}$). However, during exercise in the heat when subjects are hydrated, systemic $\dot{Q}$ increases to supply blood to the vasodilated skin, a bed that does not consume much oxygen, and thus, $\dot{V}O_2$ does not rise in parallel with $\dot{Q}$ (Gonzalez-Alonso et al. 2000).

In an experiment, Hamilton and coworkers (1991) prevented metabolic drift by intravenous infusion of glucose to maintain blood hyperglycemia (i.e., 10 mM) during 2 h of intense cycling (70% $\dot{V}O_{2max}$) in a thermoneutral environment. Despite total avoidance of $\dot{V}O_2$ drift with glucose infusion, $\dot{Q}$ was not different to when subjects rehydrated with plain water (Hamilton et al. 1991). Very prolonged exercise (i.e., ultra-endurance) is another situation where CDV drift and $\dot{V}O_2$ drift do not increase in parallel, which will be discussed in Section 4.2.8. Thus, this dissociation between metabolic and CDV drift in several experiments suggests an unlikely link between them.

4.2.6 ROLE OF HEART RATE

Experiments pacing the heart at rest and during exercise indicate that increases in HR could by itself be responsible for the reductions in SV (Ross et al. 1965; Bevegard et al. 1967). Thus, it is possible that increases in HR associated with elevations in core temperature during exercise contribute to reductions in SV characteristic of CDV drift. Plasma adrenaline level is an important extrinsic cardiac factor that affects HR. In one experiment, $\dot{Q}$, SV, and HR were measured during prolonged exercise at 65% $\dot{V}O_{2max}$ in a hot environment. In one of the trials, subject's blood adrenaline was raised by intravenous infusion to levels expected when exercising at 85% $\dot{V}O_{2max}$ (Mora-Rodriguez et al. 1996). Adrenaline infusion increased exercise HR by 6–8 beats compared to the control saline infusion trial. Although there was a tendency for SV to be reduced after 60 min of exercise (119 vs. 127 mL beat^{-1}), this difference did not reach statistical significance. The adrenaline infused in saline replaced 35% of fluid losses and likely prevented the reductions in central venous volume required to observe the declines in SV of CDV drift. Other experiments, reducing rather than increasing HR, have been more successful at altering CDV drift by pacing the heart (Fritzsche et al. 1999; Trinity et al. 2012).

Supporting a strong influence of HR on SV during prolonged exercise, Fritzsche et al. (1999) demonstrated that blunting the 11% increase in HR during 60 min of exercise in a thermoneutral environment (27°C) with the ingestion of β-adrenergic receptor blockade (i.e., atenolol) totally abolished the declines in SV from 15 to 60 min of moderate intensity exercise (~60% $\dot{V}O_{2max}$). However, the small dose of atenolol provided (0.1 mg kg^{-1}) did not affect BP core temperature, skin temperature, or cutaneous blood flow compared to the placebo trial. This experiment suggested that the reductions in SV during prolonged exercise were due to the increases in HR and independent of cutaneous blood flow. However, this study was conducted in a thermoneutral environment, and core temperature only increases 0.3°C from resting values and did not reach 38°C. Maybe larger hyperthermia could upset this direct effect of HR on SV.

In a recent study from Dr. Coyle's laboratory, they elevated subject's core temperature to examine the SV responses to slowing HR (i.e., β-adrenergic receptor blockade) during hyperthermic conditions. In this experiment, atenolol dose was doubled compared to the previous study (0.2 mg kg^{-1}). To attain the hyperthermic condition, subjects worn a vinyl rain jacket during exercise. Dehydration was prevented in all trials by ingestion of fluid before and after 15, 30, and 45 min of exercise (Trinity et al. 2010). The study confirmed that preventing HR elevations when subjects were hyperthermic by atenolol ingestion maintained SV at the 10 min levels during the 60 min of exercise. As planned, HR was the same during the placebo normothermic trial compared to the β-blockade hyperthermic trial.

However, $\dot{Q}$ and SV were higher in the β-blockade trial than in the placebo trial despite higher core temperature. Thus, when the effects of hyperthermia on increasing HR are prevented by β-blockade, hyperthermia alone does not seem to reduce SV. In the hyperthermic trial along with the higher SkBF, they found reduced BP. Seemingly, a reduced afterload due to skin vasodilation in the condition of hyperthermia plus β-blockade could have raised SV (Trinity et al. 2010).

In summary, studies with β-blockade during prolonged exercise revealed that increases in HR determine the reductions in SV. β-blockade in a hyperthermic condition prevents SV from decreasing despite a high SkBF. These two studies suggest that high SkBF *per se* does not reduce SV during prolonged submaximal exercise. In contrast, hyperthermia raises HR, which in turn seems to lower SV.

4.2.7 ROLE OF MUSCLE MASS AND CONTRACTION VELOCITY

Nassis and Geladas (2002) have suggested that the amount of contracting muscle mass could play a role in the magnitude of CDV drift. They originally observed a larger CDV drift (reduction in $\dot{Q}$ and SV after 85 min of exercise) during cycling versus running. However, differences in core temperature (higher when running), SkBF (higher when cycling), and similar HR during cycling and running make difficult the interpretation of this study. In a latter experiment (Kounalakis et al. 2008b), they found larger CDV drift when cycling with two legs compared to one-legged cycling. They set the workload during one-leg cycling to elicit half of the $\dot{V}O_2$ attained when cycling with two legs. Thus, during the exercise with one leg the initial values for $\dot{Q}$, SV, and HR were lower than when the larger muscle mass of two legs were involved. Although in absolute values HR and SV drifted more in the two-legged cycling trial, when expressed as percent of their initial values, CDV drift was similar in both trials.

In a subsequent study (Kounalakis et al. 2008a), these authors explored the effect of the *muscle pump* on CDV drift. They reasoned that a stronger muscle pump during exercise would improve blood return to the heart, increasing diastolic filling pressure, reducing the unloading of baroreceptors, and thus limiting the upward drift in HR. To elicit different level of muscle pump, subjects cycled during 90 min at either 40 or 80 pedal-revolutions per minute while matching $\dot{V}O_2$ during both trials. They found that the reductions in SV and the increases in HR were similar between trials. The authors concluded that muscle pump is not an important factor for the development of CDV drift during cycling at least in the conditions where $\dot{Q}$ and BP do not decrease during exercise (Kounalakis et al. 2008a). In summary, the role of the amount of muscle mass recruited during exercise in the development of CDV drift remains unclear.

4.2.8 ROLE OF EXERCISE DURATION

During very prolonged exercise at a steady work rate, HR drifts during the first 6 h of exercise but thereafter there is no further increase and even a reduction toward initial exercise values has been reported (Mattsson et al. 2010, 2011). Although the myocardium has been thought to be fatigue resistant, this reduction in HR could be due to limitations in left ventricular function during very prolonged exercise (>6 h). Studies examining left ventricular function pre- and postprolonged exercise have the confounding effect of changes in preload due to the associated dehydration of prolonged exercise (Goodman et al. 2001). In a recent study, Dawson and coworkers (2007) maintained heart preload and central venous pressure stable by infusing saline during 3 h of semi-recumbent cycling. Unlike previous reports (Saltin and

Stenberg 1964), they found no evidence of cardiac fatigue in left ventricular contractility indices measured by laser echocardiography. However, they found reduction in peak diastolic filling velocity that however did not affect left ventricular internal diameter at diastole. So, beyond this subtle decrease in diastolic function (Dawson et al. 2005) it does not seem that the reversal of the upward HR drift during ultra-endurance exercise could be due to cardiac fatigue.

The reversal of HR drift during ultra-endurance exercise after 6 h of exercise is not accompanied by a reduction in $\dot{V}O_2$ drift, and as a consequence, the ratio between $\dot{V}O_2/HR$ increases above initial levels. Some authors have interpreted this to mean an improved distribution of blood flow during exercise (Mattsson et al. 2010). However, it is difficult to interpret this ratio when the drift in each of the components may be caused by different factors.

The causes of the reversed HR drift during very prolonged exercise remain elusive. It could be associated with a desensitization of the heart's adrenergic receptors due to the prolonged presence of elevated levels of catecholamines. However, reduction in catecholamine stimulation of the heart would also cause a decrease in contractility and SV was not reduced but rather increased in this study (Mattsson et al. 2011). Another possibility is the redistribution of blood flow from the periphery (skin) or other visceral organs to the working muscle with prolonged exercise, although this possibility seems remote. Finally, these authors found an increase in blood concentration of the vasodilatory cardiac hormone brain natriuretic peptide which could act to improve cardiac filling (Mattsson et al. 2011). Further ultra-endurance experiments are needed to confirm this reversal of HR drift in conditions where hydration, body mineral balance, and circadian rhythms are tightly controlled.

4.2.9 ROLE OF FITNESS LEVEL

The important cardiovascular adaptations derived from aerobic training (e.g., increased blood volume, muscle capillarity, reduction in systemic peripheral resistance, increase in myocardial size, and contractility) could be thought to preserve the functionality of the CDV system when body fluid is reduced by dehydration and CDV ensues. In a recent series of experiments, our group studied the physiological responses to dehydrating exercise in the heat (34°C) comparing aerobically trained to untrained subjects (Mora-Rodriguez et al. 2010, 2013; Mora-Rodriguez 2012). We achieved a percent dehydration from 1.4% to 2.3% depending on the exercise intensity, and in another set of trials, dehydration was offset by fluid ingestion (Mora-Rodriguez et al. 2013).

In a thermoneutral environment, it has been previously shown that the HR response are similar when trained and untrained subjects exercise at the same percentage of their maximal aerobic capacity (i.e., 50%–90% $\dot{V}O_{2peak}$) (Fritzsche and Coyle 2000; Gant et al. 2004). In a hot environment, we observed that when work rates were equaled to a percent of the individual maximal aerobic capacity (% $\dot{V}O_{2peak}$), the increases in HR and BP were similar between groups (Mora-Rodriguez et al. 2013). This similar HR response took place despite larger reductions in body fluid and plasma volume in the trained than in the untrained individuals due to their higher absolute workload and sweat rate (see Table 4.2). In this view, the fact that HR was

TABLE 4.2

Cardiovascular Responses at the End of Exercise (40%, 60%, and 80% $\dot{V}O_{2\,peak}$) in Aerobically Trained (Tr) and Untrained (UTr) Subjects

Responses	Subjects	40% $\dot{V}O_{2\,peak}$	60% $\dot{V}O_{2\,peak}$	80% $\dot{V}O_{2\,peak}$
Heart rate (beats min[-1])	Tr	128 ± 12	156 ± 11^b	$182 \pm 9^{b,c}$
	UTr	132 ± 13	160 ± 12^b	$183 \pm 8^{b,c}$
Mean arterial pressure (mmHg)	Tr	97 ± 6	100 ± 12	116 ± 15^c
	UTr	93 ± 13	101 ± 12	115 ± 17^c
Plasma volume (% of resting)	Tr	-9.6 ± 3.7^a	$-13.7 \pm 2.6^{a,b}$	-13.4 ± 4.4^c
	UTr	-7.6 ± 3.0	-10.0 ± 2.5	$-12.8 \pm 2.4^{b,c}$
Sweat rate (L h[-1])	Tr	0.8 ± 0.2	1.2 ± 0.3^b	$1.6 \pm 0.5^{a,b,c}$
	UTr	0.7 ± 0.2	1.0 ± 0.3^b	$1.3 \pm 0.4^{b,c}$
Dehydration (%)	Tr	2.3 ± 0.3^a	2.1 ± 0.3^a	$2.1 \pm 0.3^{a,c}$
	UTr	1.9 ± 0.1	1.6 ± 0.1^b	1.4 ± 0.1^c

Source: Mora-Rodriguez, R. et al., *Eur. J. Appl. Physiol.,* 109(5), 973–81, 2010.

Data are average for 10 subjects in each group $\pm$ SD.

[a] Difference between Tr and UTr subjects.

[b] Difference from previous exercise intensity within the same group of subjects.

[c] Difference from 40% $\dot{V}O_{2\,peak}$ ($p < .05$).

similar in the trained individuals compared to the untrained subjects despite exercising with a higher body fluid deficit is remarkable. Furthermore, it suggests that the cardiovascular system of trained individuals adapts to prevent a HR overshoot response despite large body fluid losses.

Fluid replacement may have different impact on CDV drift in trained and untrained individuals. We and other investigators have reported that trained individuals derive larger thermoregulatory benefits from rehydration than untrained (lower core temperature) at least when exercising at a moderate intensity (i.e.,~60% $\dot{V}O_{2peak}$) (Merry et al. 2010; Mora-Rodriguez et al. 2013). Figure 4.2 shows the effects of fluid replacement on the cardiovascular responses in trained and untrained subjects during cycling exercise at 60% $\dot{V}O_{2peak}$ and when exercising at the same absolute intensity.

When exercising at the same absolute work rate (180 W; see panel a of Figure 4.2) and thus similar heat production, $\dot{Q}$ was initially (15 min of exercise) similar between the two groups. However, the untrained group had larger reduction in $\dot{Q}$ when they were not rehydrated (i.e., see the long arrow down in Figure 4.2). At 180 W, HR was higher and SV lower in the untrained than in the trained group. Likely, the lower HR in trained subjects allowed longer heart filling time improving their SV. Both groups benefited from rehydration but the drift (height of the bars) on SV and HR was smaller in the trained subjects.

When comparing the cardiovascular responses at the same percent of the maximal aerobic capacity (60% $\dot{V}O_{2peak}$, see panel b of Figure 4.2), $\dot{Q}$ was very different due to the higher absolute workload in the trained subjects. However, HR was

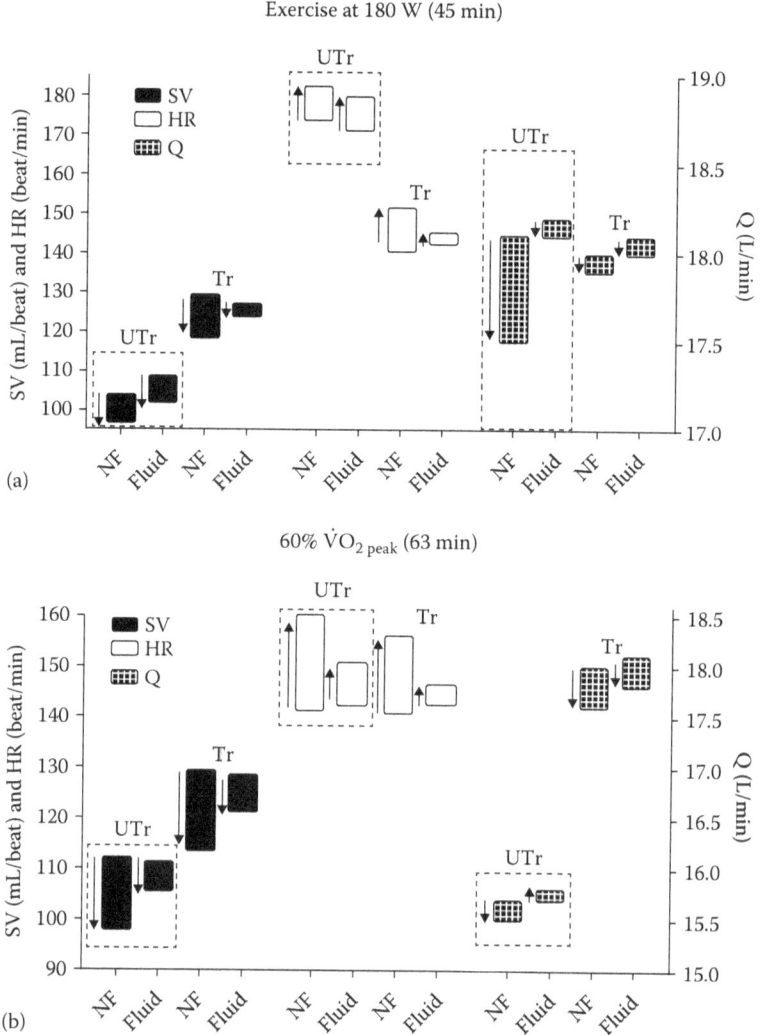

FIGURE 4.2 Cardiovascular responses during exercise (a) at the same absolute exercise intensity (180 W) and (b) at the same relative intensity (60% $\dot{V}O_{2\,peak}$) at the beginning and end of exercise (height of the bars) in aerobically trained (Tr) and untrained (UTr) subjects when dehydrated or when fluid was replaced in full. (Data from Mora-Rodriguez, R. et al., *Appl. Physiol. Nutr. Metab.*, 38, 73–80, 2013.)

similar between groups. When no fluid was allowed, the drift in all variables ($\dot{Q}$, SV, and HR) was not different between trained and untrained subjects. Both groups of subjects benefited from fluid replacement. However, the drift in HR was lower in the trained group after rehydration. In summary, it seems that along with the thermoregulatory responses, trained individuals benefit more from rehydration reducing their HR drift further than their untrained counterparts.

4.3 CARDIOVASCULAR DRIFT AND TRAINING FOR PERFORMANCE OR HEALTH

Athletes and coaches interpret a lowering in HR for a given performance velocity (i.e., running, cycling, rowing, or swimming) as an evidence of attainment of aerobic adaptations to training. However, the reduction in HR of an athlete that trained hard to get aerobic adaptation could be masked by CDV drift. The most salient characteristic of CDV drift is the increase in HR during exercise. Thus, the HR drift resulting from exercising in the heat could counterbalance the HR lowering effect of training. Coaches testing their athletes in a hot summer environment may conclude that training adaptations were not obtained if they do not have a pre-training control in similar environmental conditions. Training and heat acclimation reduce CDV drift by naturally expanding plasma volume and enhancing the functionality of the heat dissipatory mechanisms. Heat acclimation reduces but it does not prevent CDV drift and the stress response of the cardiovascular system.

Coaches may decide to abandon HR as an index to measure the cardiovascular adaptations to training when influenced by CDV drift (i.e., situations of prolonged dehydrating exercise). However, disregarding the HR increases during prolonged exercise could lead to premature fatigue. The HR drift that occurs during prolonged exercise is always accompanied by increases in the rate of perceived exertion. This tight relationship occurs if exercise takes place either in a hot (Gonzalez-Alonso et al. 1995, 1997) thermoneutral environment (Hamilton et al. 1991; Heaps et al. 1994) or with plasma volume expansion (Montain and Coyle 1992b) as it can be appreciated in Figure 4.3.

On the other hand, the prevention of CDV drift could reduce fatigue even when core temperature is not lowered. Two studies conducted in a thermoneutral environment reported improvements in performance when ingesting water and salt prior to exercise without affecting core temperature (Greenleaf et al. 1997; Coles and Luetkemeier 2005). In a hot environment (33°C), we found 7.4% improvement

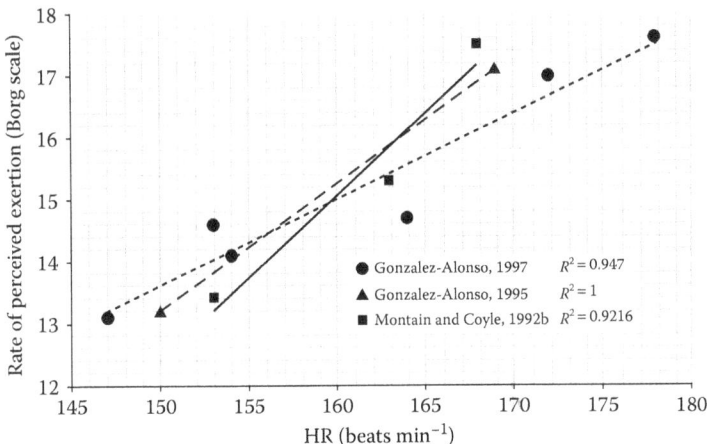

FIGURE 4.3 Correlations between HR and rate of perceived exertion (RPE, Borg scale).

in a cycling time trial after ingestion of water and salt despite no change on heat accumulation compared to a trial where only water was ingested pre-exercise (Hamouti et al. 2014). Other investigators have linked the reduced cycling performance during prolonged exercise in the heat to the cardiovascular strain and the increases in perceived exertion (Cheuvront et al. 2010).

Hyperthermia induced by warming the skin and core with a water-perfused suit (10°C increase in skin and 1°C increase in core) results in a 26% reduction in the time to exhaustion in a graded exercise test. The increases in HR in the hyperthermic trial were associated with a reduction in $\dot{V}O_{2max}$ (Gonzalez-Alonso and Calbet 2003). Maximal cardiac output and thus maximal SV are the limiting factors for oxygen delivery and the attainment of maximal oxygen consumption $\dot{V}O_{2max}$ (Saltin and Strange 1992). Since SV reduction is one of the constant features of CDV drift, it is logical to infer that CDV drift may compromise $\dot{V}O_{2max}$ (Saltin 1964). In fact the reduction in $\dot{V}O_{2max}$ has been observed not only during incremental exercise to exhaustion but also after 45 min of constant load exercise in a hot environment (Wingo et al. 2005) coinciding with CDV drift. Furthermore, CDV drift and the reductions in $\dot{V}O_{2max}$ are avoided when hyperthermia is prevented by exercising in a 22°C environment or by air cooling (Wingo and Cureton 2006a; Lafrenz et al. 2008).

CDV drift may not only affect performance but also alter training load when using target HR as an index during the workouts. Wingo et al. (2012) have nicely described the decisional crossroad faced by endurance athletes when training in a hot-dehydrating environment using HR as index for training intensity. As described above, dehydration and hyperthermia result in reductions in $\dot{V}O_{2max}$ and maximal workload during a graded exercise test to fatigue. The upward drift in HR during prolonged exercise is proportional to the increase in relative metabolic intensity due to the decrease in $\dot{V}O_{2max}$. Thus, if subjects are instructed to set the workload to their target HR during the first stages of exercise and maintain that workload, the increasing HR (i.e., %HR max) would still correspond to an increased %$\dot{V}O_{2max}$ since absolute $\dot{V}O_{2max}$ would be reduced. Alternatively, preserving HR at target level would require a marked reduction in workload as exercise progresses (e.g., 40% during 45 min of exercise at 35°C) (Wingo and Cureton 2006b). This reduction in workload results in a 24% lower $\dot{V}O_2$ and a 10% lower SV. This reduction in the metabolic and cardiovascular demands may compromise the attainment of the adaptations of aerobic training in those physiological systems (Figure 4.4).

While the consequences and alternatives to the occurrence of CDV drift during training for performance are analyzed by Wingo and coworkers (2005), the occurrence of CDV drift in people that exercise to improve their fitness is not well established. One study following 326 overweight women that trained for fitness 3–5 times per week, 30–90 min at 50% $\dot{V}O_{2max}$ reported small increases in HR during these workouts (i.e., 1–4 beats). Furthermore, in less than 1% of the sessions intensity was reduced to prevent CDV drift (Mikus et al. 2009). Thus, during continuous aerobic exercise training to improve fitness, CDV drift may not be an important workload modifier when using HR as an index to prescribe exercise intensity. Of note, in the study by Mikus et al. (2009), subjects rehydrated ad libitum and had access to fan

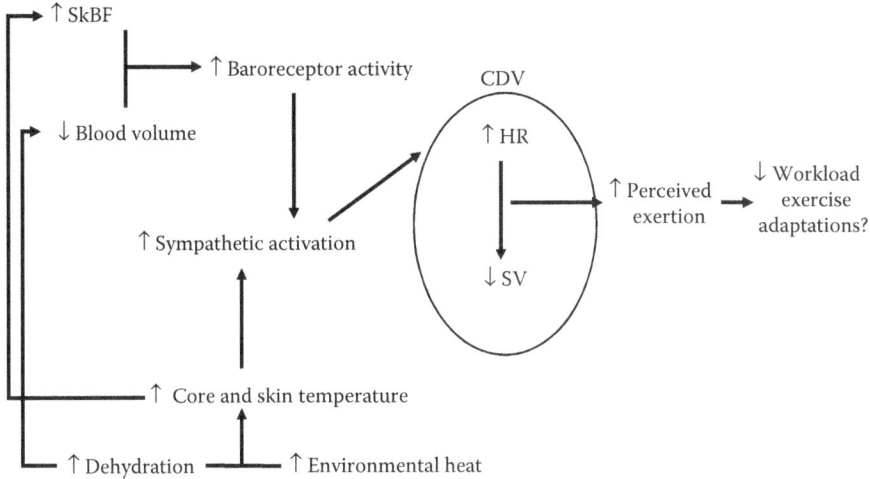

FIGURE 4.4 Schematic diagram of factors affecting CDV drift and its consequences.

cooling in all training bouts. It is currently unclear if CDV drift may be important in health-oriented training programs that use more intense bouts of exercise (Mora-Rodriguez et al. 2015) or when airflow and rehydrating fluid are not available.

4.4 SUMMARY

In summary, CDV drift (increase in HR and concomitant reduction in SV) is the most visible sequence of cardiovascular events that takes place during prolonged submaximal exercise. Its prevalence during exercise in a hot environment causing dehydration by sweating is high. Its causes and consequences have been investigated for more than 50 years. Although initially linked to the increases in SkBF, it also takes place despite unchanged cutaneous circulation. Sympathetic stimulation of HR linked to hyperthermia, dehydration, or hypovolemia is the more likely cause of CDV drift according to recent experiments. CDV drift has consequences for performance and could potentially limit the adaptations and health outcomes sought by training. Prevention of CDV drift involves rehydration and cooling strategies that although long time known are habitually neglected.

REFERENCES

Bevegard, S., B. Jonsson, I. Karlof, H. Lagergren, and E. Sowton. 1967. Effect of changes in ventricular rate on cardiac output and central pressures at rest and during exercise in patients with artificial pacemakers. *Cardiovasc. Res.* 1(1): 21–33.

Buono, M. J., R. Claros, T. Deboer, and J. Wong. 2008. Na+ secretion rate increases proportionally more than the Na+ reabsorption rate with increases in sweat rate. *J. Appl. Physiol.* 105(4): 1044–8.

Cheuvront, S. N., R. W. Kenefick, S. J. Montain, and M. N. Sawka. 2010. Mechanisms of aerobic performance impairment with heat stress and dehydration. *J. Appl. Physiol.* 109(6): 989–1995.

Coles, M. G., and M. J. Luetkemeier. 2005. Sodium-facilitated hypervolemia, endurance performance, and thermoregulation. *Int. J. Sports. Med.* 26(3): 182–7.

Coyle, E. F. 1998. Cardiovascular drift during prolonged exercise and the effects of dehydration. *Int. J. Sports Med.* 19(Suppl. 2): S121–4.

Coyle, E. F., and J. Gonzalez-Alonso. 2001. Cardiovascular drift during prolonged exercise: new perspectives. *Exerc. Sport Sci. Rev.* 29(2): 88–92.

Dawson, E. A., R. Shave, K. George et al. 2005. Cardiac drift during prolonged exercise with echocardiographic evidence of reduced diastolic function of the heart. *Eur. J. Appl. Physiol.* 94(3): 305–9.

Dawson, E. A., R. Shave, G. Whyte et al. 2007. Preload maintenance and the left ventricular response to prolonged exercise in men. *Exp. Physiol.* 92(2): 383–90.

Deschamps, A., R. D. Levy, M. G. Cosio, E. B. Marliss, and S. Magder. 1989. Effect of saline infusion on body temperature and endurance during heavy exercise. *J. Appl. Physiol.* 66(6): 2799–804.

Dick, R. W., and P. R. Cavanagh. 1987. An explanation of the upward drift in oxygen uptake during prolonged sub-maximal downhill running. *Med. Sci. Sports Exerc.* 19(3): 310–17.

Febbraio, M. A. 2001. Alterations in energy metabolism during exercise and heat stress. *Sports Med.* 31(1): 47–59.

Fernández-Elías, V. E., F. J. Ortega, N. Hamouti, and R. Mora-Rodríguez. 2015. Hyperthermia but not muscle water deficit increases glycogen use during intense exercise. *Scand. J. Med. Sci. Sports* 25(Suppl. 1): 126–34.

Fritzsche, R. G., and E. F. Coyle. 2000. Cutaneous blood flow during exercise is higher in endurance-trained humans. *J. Appl. Physiol.* 88(2): 738–44.

Fritzsche, R. G., T. W. Switzer, B. J. Hodgkinson, and E. F. Coyle. 1999. Stroke volume decline during prolonged exercise is influenced by the increase in heart rate. *J. Appl. Physiol.* 86(3): 799–805.

Gant, N., C. Williams, J. King, and B. J. Hodge. 2004. Thermoregulatory responses to exercise: relative versus absolute intensity. *J. Sports Sci.* 22(11–12): 1083–90.

Gonzalez-Alonso, J., and J. A. Calbet. 2003. Reductions in systemic and skeletal muscle blood flow and oxygen delivery limit maximal aerobic capacity in humans. *Circulation* 107(6): 824–30.

Gonzalez-Alonso, J., J. A. Calbet, and B. Nielsen. 1998. Muscle blood flow is reduced with dehydration during prolonged exercise in humans. *J. Physiol.* 513 (Pt 3): 895–905.

Gonzalez-Alonso, J., R. Mora-Rodriguez, P. R. Below, and E. F. Coyle. 1995. Dehydration reduces cardiac output and increases systemic and cutaneous vascular resistance during exercise. *J. Appl. Physiol.* 79(5): 1487–96.

Gonzalez-Alonso, J., R. Mora-Rodriguez, P. R. Below, and E. F. Coyle. 1997. Dehydration markedly impairs cardiovascular function in hyperthermic endurance athletes during exercise. *J. Appl. Physiol.* 82(4): 1229–36.

Gonzalez-Alonso, J., R. Mora-Rodriguez, and E. F. Coyle. 1999. Supine exercise restores arterial blood pressure and skin blood flow despite dehydration and hyperthermia. *Am. J. Physiol.* 277(2 Pt 2): H576–83.

Gonzalez-Alonso, J., R. Mora-Rodriguez, and E. F. Coyle. 2000. Stroke volume during exercise: Interaction of environment and hydration. *Am. J. Physiol. Heart Circ. Physiol.* 278(2): H321–30.

Goodman, J. M., P. R. McLaughlin, and P. P. Liu. 2001. Left ventricular performance during prolonged exercise: Absence of systolic dysfunction. *Clin. Sci. (Lond.)* 100(5): 529–37.

Greenleaf, J. E., R. Looft-Wilson, J. L. Wisherd, M. A. McKenzie, C. D. Jensen, and J. H. Whittam. 1997. Pre-exercise hypervolemia and cycle ergometer endurance in men. *Biol. Sport* 14(2): 103–14.

Grimby, G., N. J. Nilsson, and H. Sanne. 1966. Serial determinations of cardiac output at rest. *Br. Heart J.* 28(1): 118–21.

Hamilton, M. T., J. Gonzalez-Alonso, S. J. Montain, and E. F. Coyle. 1991. Fluid replacement and glucose infusion during exercise prevent cardiovascular drift. *J. Appl. Physiol.* 71(3): 871–7.

Hamouti, N., J. Del Coso, and R. Mora-Rodriguez. 2013. Comparison between blood and urinary fluid balance indices during dehydrating exercise and the subsequent hypohydration when fluid is not restored. *Eur. J. Appl. Physiol.* 113(3): 611–20.

Hamouti, N., J. Del Coso, J. F. Ortega, and R. Mora-Rodriguez. 2011. Sweat sodium concentration during exercise in the heat in aerobically trained and untrained humans. *Eur. J. Appl. Physiol.* 111(11): 2873–81.

Hamouti, N., V. E. Fernandez-Elias, J. F. Ortega, and R. Mora-Rodriguez. 2014. Ingestion of sodium plus water improves cardiovascular function and performance during dehydrating cycling in the heat. *Scand. J. Med. Sci. Sports* 24(3): 507–18.

Heaps, C. L., J. Gonzalez-Alonso, and E. F. Coyle. 1994. Hypohydration causes cardiovascular drift without reducing blood volume. *Int. J. Sports Med.* 15(2): 74–9.

Kellogg, D. L., Jr., J. M. Johnson, and W. A. Kosiba. 1990. Baroreflex control of the cutaneous active vasodilator system in humans. *Circ. Res.* 66(5): 1420–6.

Kounalakis, S. N., M. E. Keramidas, G. P. Nassis, and N. D. Geladas. 2008a. The role of muscle pump in the development of cardiovascular drift. *Eur. J. Appl. Physiol.* 103(1): 99–107.

Kounalakis, S. N., G. P. Nassis, M. D. Koskolou, and N. D. Geladas. 2008b. The role of active muscle mass on exercise-induced cardiovascular drift. *J. Sports Sci. Med.* 7(3): 395–401.

Ladell, W. S. 1955. The effects of water and salt intake upon the performance of men working in hot and humid environments. *J. Physiol.* 127(1): 11–46.

Lafrenz, A. J., J. E. Wingo, M. S. Ganio, and K. J. Cureton. 2008. Effect of ambient temperature on cardiovascular drift and maximal oxygen uptake. *Med. Sci. Sports Exerc.* 40(6): 1065–71.

Mattsson, C. M., J. K. Enqvist, T. Brink-Elfegoun, P. H. Johansson, L. Bakkman, and B. Ekblom. 2010. Reversed drift in heart rate but increased oxygen uptake at fixed work rate during 24 h ultra-endurance exercise. *Scand. J. Med. Sci. Sports* 20(2): 298–304.

Mattsson, C. M., M. Stahlberg, F. J. Larsen, F. Braunschweig, and B. Ekblom. 2011. Late cardiovascular drift observable during ultraendurance exercise. *Med. Sci. Sports Exerc.* 43(7): 1162–8.

Merry, T. L., P. N. Ainslie, and J. D. Cotter. 2010. Effects of aerobic fitness on hypohydration-induced physiological strain and exercise impairment. *Acta Physiol. (Oxf.)* 198(2): 179–90.

Mikus, C. R., C. P. Earnest, S. N. Blair, and T. S. Church. 2009. Heart rate and exercise intensity during training: Observations from the DREW Study. *Br. J. Sports Med.* 43(10): 750–5.

Montain, S. J., and E. F. Coyle. 1992a. Fluid ingestion during exercise increases skin blood flow independent of increases in blood volume. *J. Appl. Physiol.* 73(3): 903–10.

Montain, S. J., and E. F. Coyle. 1992b. Influence of graded dehydration on hyperthermia and cardiovascular drift during exercise. *J. Appl. Physiol.* 73(4): 1340–50.

Mora-Rodriguez, R. 2012. Influence of aerobic fitness on thermoregulation during exercise in the heat. *Exerc. Sport Sci. Rev.* 40(2): 79–87.

Mora-Rodriguez, R., R. Aguado-Jimenez, J. Del Coso, and E. Estevez. 2012. A standard blood bank donation alters the thermal and cardiovascular responses during subsequent exercise. *Transfusion* 52(11): 2339–47.

Mora-Rodriguez, R., J. Del Coso, R. Aguado-Jimenez, and E. Estevez. 2007. Separate and combined effects of airflow and rehydration during exercise in the heat. *Med. Sci. Sports Exerc.* 39(10): 1720–6.

Mora-Rodriguez, R., J. Del Coso, N. Hamouti, E. Estevez, and J. F. Ortega. 2010. Aerobically trained individuals have greater increases in rectal temperature than untrained ones during exercise in the heat at similar relative intensities. *Eur. J. Appl. Physiol.* 109(5): 973–81.

Mora-Rodriguez, R., V. E. Fernandez-Elias, N. Hamouti, and J. F. Ortega. 2015. Skeletal muscle water and electrolytes following prolonged dehydrating exercise. *Scand. J. Med Sci. Sports* 25(3): e274–82.

Mora-Rodriguez, R., J. Gonzalez-Alonso, P. R. Below, and E. F. Coyle. 1996. Plasma catecholamines and hyperglycaemia influence thermoregulation in man during prolonged exercise in the heat. *J. Physiol.* 491(Pt 2): 529–40.

Mora-Rodriguez, R., N. Hamouti, J. Del Coso, and J. F. Ortega. 2013. Fluid ingestion is more effective in preventing hyperthermia in aerobically trained than untrained individuals during exercise in the heat. *Appl. Physiol. Nutr. Metab.* 38(1): 73–80.

Mora-Rodriguez, R., J. F. Ortega, N. Hamouti et al. 2014. Time-course effects of aerobic interval training and detraining in patients with metabolic syndrome. *Nutr. Metab. Cardiovasc. Dis.* 24(7): 792–8.

Nassis, G. P., and N. D. Geladas. 2002. Cardiac output decline in prolonged dynamic exercise is affected by the exercise mode. *Pflugers Arch.* 445(3): 398–404.

Nose, H., G. W. Mack, X. R. Shi, K. Morimoto, and E. R. Nadel. 1990. Effect of saline infusion during exercise on thermal and circulatory regulations. *J. Appl. Physiol.* 69(2): 609–16.

Nose, H., G. W. Mack, X. R. Shi, and E. R. Nadel. 1988. Shift in body fluid compartments after dehydration in humans. *J. Appl. Physiol.* 65(1): 318–24.

Nybo, L., P. Rasmussen, and M. N. Sawka. 2014. Performance in the heat-physiological factors of importance for hyperthermia-induced fatigue. *Compr. Physiol.* 4(2): 657–89.

Pitts, G. J., and F. C. Consolazio.1944. Work in the heat as affected by intake of water, salt and glucose. *Am. J. Physiol.* 142: 253–9.

Poliner, L. R., G. J. Dehmer, S. E. Lewis, R. W. Parkey, C. G. Blomqvist, and J. T. Willerson. 1980. Left ventricular performance in normal subjects: A comparison of the responses to exercise in the upright and supine positions. *Circulation* 62(3): 528–34.

Poole, D. C., W. Schaffartzik, D. R. Knight et al. 1991. Contribution of exercising legs to the slow component of oxygen uptake kinetics in humans. *J. Appl. Physiol.* 71(4): 1245–60.

Radigan, L. R., and S. Robinson. 1949. Effects of environmental heat stress and exercise on renal blood flow and filtration rate. *J. Appl. Physiol.* 2(4): 185–91.

Ross, J. J., J. W. Linhart, and E. Brauwald. 1965. Effects of changing heart rate in man by electrical stimulation of the right atrium. studies at rest, during exercise, and with isoproterenol. *Circulation* 32(4): 549–58.

Rowell, L. B. 1974. Human cardiovascular adjustments to exercise and thermal stress. *Physiol. Rev.* 54(1): 75–159.

Rowell, L. B. 1986. Circulatory adjustments to dynamic exercise and heat stress: Competing controls. In L. Rowell (Ed.), *Human Circulation: Regulation During Physical Stress.* New York: Oxford University Press, pp. 363–406.

Rowell, L. B., J. R. Blackmon, R. H. Martin, J. A. Mazzarella, and R. A. Bruce. 1965. Hepatic clearance of indocyanine green in man under thermal and exercise stresses. *J. Appl. Physiol.* 20(3): 384–94.

Rowell, L. B., G. L. Brengelmann, and J. A. Murray. 1969. Cardiovascular responses to sustained high skin temperature in resting man. *J. Appl. Physiol.* 27(5): 673–80.

Rowell, L. B., K. K. Kraning, 2nd, J. W. Kennedy, and T. O. Evans. 1967. Central circulatory responses to work in dry heat before and after acclimatization. *J. Appl. Physiol.* 22(3): 509–18.

Rowell, L. B., H. J. Marx, R. A. Bruce, R. D. Conn, and F. Kusumi.1966. Reductions in cardiac output, central blood volume, and stroke volume with thermal stress in normal men during exercise. *J. Clin. Invest.* 45(11): 1801–16.

Rowell, L. B., J. A. Murray, G. L. Brengelmann, and K. K. Kraning, 2nd. 1969. Human cardiovascular adjustments to rapid changes in skin temperature during exercise. *Circ. Res.* 24(5): 711–24.

Rowland, T., D. Pober, and A. Garrison. 2008. Cardiovascular drift in euhydrated prepubertal boys. *Appl. Physiol. Nutr. Metab.* 33(4): 690–5.

Saltin, B. 1964. Circulatory response to submaximal and maximal exercise after thermal dehydration. *J. Appl. Physiol.* 19: 1125–32.

Saltin, B., and J. Stenberg. 1964. Circulatory response to prolonged severe exercise. *J. Appl. Physiol.* 19: 833–8.

Saltin, B., and S. Strange. 1992. Maximal oxygen uptake: "old" and "new" arguments for a cardiovascular limitation. *Med. Sci. Sports Exerc.* 24(1): 30–7.

Sawka, M. N., L. M. Burke, E. R. Eichner, R. J. Maughan, S. J. Montain, and N. S. Stachenfeld. 2007. American College of Sports Medicine position stand. Exercise and fluid replacement. *Med. Sci. Sports Exerc.* 39(2): 377–90.

Shaffrath, J. D., and W. C. Adams. 1984. Effects of airflow and work load on cardiovascular drift and skin blood flow. *J. Appl. Physiol. Respir. Environ. Exerc. Physiol.* 56(5): 1411–17.

Sims, S. T., L. van Vliet, J. D. Cotter, and N. J. Rehrer. 2007. Sodium loading aids fluid balance and reduces physiological strain of trained men exercising in the heat. *Med. Sci. Sports Exerc.* 39(1): 123–30.

Trinity, J. D., J. F. Lee, M. D. Pahnke, K. C. Beck, and E. F. Coyle. 2012. Attenuated relationship between cardiac output and oxygen uptake during high-intensity exercise. *Acta Physiol. (Oxf.)* 204(3): 362–70.

Trinity, J. D., M. D. Pahnke, J. F. Lee, and E. F. Coyle. 2010. Interaction of hyperthermia and heart rate on stroke volume during prolonged exercise. *J. Appl. Physiol.* 109(3): 745–51.

Watt, M. J., M. A. Febbraio, A. P. Garnham, and M. Hargreaves. 1999. Acute plasma volume expansion: Effect on metabolism during submaximal exercise. *J. Appl. Physiol.* 87(3): 1202–6.

Westerlind, K. C., W. C. Byrnes, and R. S. Mazzeo. 1992. A comparison of the oxygen drift in downhill vs. level running. *J. Appl. Physiol.* 72(2): 796–800.

Wingo, J. E., and K. J. Cureton. 2006a. Body cooling attenuates the decrease in maximal oxygen uptake associated with cardiovascular drift during heat stress. *Eur. J. Appl. Physiol.* 98(1): 97–104.

Wingo, J. E., and K. J. Cureton. 2006b. Maximal oxygen uptake after attenuation of cardiovascular drift during heat stress. *Aviat. Space Environ. Med.* 77(7): 687–94.

Wingo, J. E., M. S. Ganio, and K. J. Cureton. 2012. Cardiovascular drift during heat stress: Implications for exercise prescription. *Exerc. Sport Sci. Rev.* 40(2): 88–94.

Wingo, J. E., A. J. Lafrenz, M. S. Ganio, G. L. Edwards, and K. J. Cureton. 2005. Cardiovascular drift is related to reduced maximal oxygen uptake during heat stress. *Med. Sci. Sports Exerc.* 37(2): 248–55.

5 Thermal Strain and Exertional Heat Illness Risk

Total Body Water and Exchangeable Sodium Deficits

Michael F. Bergeron

CONTENTS

5.1 INTRODUCTION

Hydration is arguably among the foremost priorities athletes habitually consider during and after any workout, be it training or competition—that is, drinking regularly to offset often obvious extensive losses of sweat and intrinsic prompts of thirst. Informed athletes typically appreciate the influence of hydration on athletic performance and how inadequate hydration can ultimately be a determining factor in a loss or underachievement during training or competition. It is increasingly well recognized, as well, that inadequate hydration and resultant measurable sweat-induced body water and sodium deficits can also be a danger to one's health and safety—particularly, during strenuous physical activity in the heat and humidity. Notably, complete post-exercise rehydration involves more than just ample water intake, as often extensive sweat sodium losses need to be replaced as well to achieve optimum hydration status—that is, the most advantageous distribution and retention of water to all body fluid compartments. Ideally, the preferred emphasis is to ensure a well-hydrated and sodium-balanced state prior to practice, training, or play and regular fluid (sometimes including added sodium, depending on individual needs) intake during these activities to sufficiently offset the ongoing sweat loss. However, for myriad athletes and athletic scenarios, it is impractical to avoid accumulating and sometimes measurable total body water and exchangeable sodium deficits during extended or repeated same-day bouts of training or competition. In addition to the potential effect on performance, these deficits can significantly increase thermal strain and exertional heat illness risk.

5.2   THERMAL STRAIN

As environmental heat stress and the intensity and duration of physical activity increase, metabolic heat production and the associated level of thermal strain are progressively and proportionally amplified in parallel. Evaporation of sweat, along with additional heat exchange through radiation (from the skin to the air) and convective heat loss (from air moving across the skin), is usually enough to amply offset metabolic heat production. Accordingly, an athlete's body core temperature is typically kept from increasing excessively during training and competition across a broad range of diverse environments and thermal challenges. In certain scenarios, however, the workload and consequent metabolic heat production rate exceed the capacity of sweating and other mechanisms of environmental heat exchange. Consequently, body core temperature progressively increases. If the level of physical activity inappropriately (for the prevailing environmental conditions) continues or increases, the degree of undue thermal strain could readily reach and exceed a dangerous and potentially life-threatening threshold. This effect and resultant challenge to an athlete are readily hastened and magnified when the humidity is high, as the high water content of the air acts as a barrier to sweat evaporation and thus evaporative cooling. In such conditions, an athlete will continue to sweat, but the sweat does not evaporate as effectively. The water from the sweat glands is left to collect on the skin or fall to the ground with little-to-no release of body heat to the air. In weather where the air temperature approaches an athlete's skin temperature, the progressively less favorable thermal gradient reduces the capacity for heat loss through radiation from the skin. And at certain times of the day, the passive solar heat gain can increase the total net thermal load even further from the constant exposure of the body to the sun's electromagnetic energy. Moreover, when there is no wind, there is also a dramatic reduction in convective heat loss. Accordingly, from the perspective of potentially reaching dangerous levels of thermal strain, the scenario presenting the greatest risk during outdoor vigorous and/or sustained athletic exertion is when the air temperature and humidity are high, exposure to the sun is constant and strong, with additional reflective solar energy coming off the ground, and there is little to no breeze or wind. Notably, increasing levels of environmental heat stress generally have a more measurable negative effect on the less fit athletes (Ely et al. 2007).

The human body has an effective thermoregulatory capacity for minimizing the rise in body core temperature during physical activity. However, with greater levels of ambient temperature and/or humidity, metabolic heat production can exceed the capacity for sufficient environmental heat exchange and thus prompt an excessive and dangerous level of thermal strain.

5.3 TOTAL BODY WATER DEFICITS

A greater dependence in evaporative cooling via sweating parallels increasing workloads and climatic heat stress, though sweating rates vary considerably between individuals, even in identical environmental conditions and athletic settings. Myriad recreational and competitive athletes are capable of sweating well in excess of a liter per hour while training or competing, particularly in the heat. For some older adolescents or adult athletes, sweat loss can reach 2.5 L or more per hour with strenuous physical activity in a wide range of environmental conditions. Considering these sweat loss rates, it is easy to appreciate how some individuals can readily incur a sizable total body water deficit during training and competition. Even when fluid is consumed voluntarily and regularly as desired or just to thirst, a post-exercise body water deficit is often significant following prolonged and/or repeated sessions of strenuous physical activity, especially for those individuals who sweat considerably. In fact, in certain sports and extended-duration athletic scenarios, these post-exercise body water deficits can be very substantial—2–4 L or more—even with ample fluid availability, opportunities to rehydrate, and regular consumption throughout the activity session (Bergeron 2003, 2014). Notably, there is a particular challenge in minimizing total body water deficits with same-day repeated bouts of physical activity—such as with tournaments and multiple same-day training sessions—when sweat losses are extensive and repeat-bout recovery time is insufficient (Bergeron 2009).

Sweating rates vary considerably, and post-exercise body water deficits can be considerable, even when water and/or other fluids are consumed regularly during activity.

5.4 EXCHANGEABLE SODIUM DEFICITS

As with sweating rate, the sodium content of sweat is widely variable among individuals. Moreover, as an athlete's rate of sweating increases, there is a concomitant increase in the rate of sweat electrolyte loss (particularly sodium) in that same individual, owing to the resultant larger total volume of sweat and the parallel higher sweat sodium concentration. A sufficient period of progressive exercise-heat exposure and consequent acclimatization to exercising in the heat typically lowers one's sweat sodium concentration; however, sweat sodium losses can still be considerable, even for an athlete who is well acclimatized to the heat. With pre- or early pubescent athletes, a sweat-induced sodium deficit incurred during a single training/practice session or competition bout alone is not likely to be substantial, because of a comparatively (to adults) low sweating rate and total sweat sodium loss (Meyer et al. 2012). A normal diet surrounding these bouts of activity will typically be sufficient to maintain daily electrolyte balance, even if only water is consumed during and after each practice or game/match. This often is not the case with many older adolescents and adults who generally sweat considerably more and can lose via sweating 2000–5000 mg or more of sodium per hour. For extended or same-day repeated vigorous training/practice sessions or competition bouts in the heat, total body exchangeable sodium deficits can reach up to 20,000 mg or more in a single day. The same effect may also evolve less dramatically (i.e., more incrementally over time) after several successive days of training/practice or competition.

As with sweating rate, electrolyte (especially sodium) losses from sweating vary considerably too. Depending on the rate of sweat sodium loss, a measurable total body exchangeable sodium deficit can be evident from a single session of extensive sweating or after several successive same-day sessions or multiple days of physical exertion.

5.5 EFFECTS OF TOTAL BODY WATER AND EXCHANGEABLE SODIUM DEFICITS

Body core temperature, heart rate, and overall physiological and perceptual strain are proportionally greater with incremental levels of total body water deficits (e.g., 1%–4% from a normally hydrated state), during fixed-intensity exercise (Gonzalez-Alonso et al. 1997; Moran et al. 1998; Nadel et al. 1980). A reduced capacity for heat transfer and dissipation may be the result of blood volume decrements and accompanying changes in plasma osmolality resulting in alterations in cardiovascular and thermoregulatory control (Gonzalez-Alonso et al. 1997; Harrison et al. 1978; Nadel et al. 1980). While these examined levels of hypohydration are representative of body water deficits incurred during training and competition in many typical athletic scenarios, the invariable constant workload of laboratory experiments may not necessarily elicit the same responses to dehydration as in sport- and field-specific situations when there is an opportunity for self-regulatory variation in workload and rest cycles. However, in most athletic situations, athletes can safely tolerate a small total body water deficit (3% or less, or 2% of body mass) that develops by the end of a practice/training session or competitive event without incurring undue thermal or cardiovascular strain and associated medical complications, so long as pre-activity hydration status is good.

In compensatory response to the typical decrease in plasma volume and increase in plasma osmolality during vigorous exercise, which can be more readily exacerbated by extensive sweating, water from the interstitial fluid compartment shifts to the intravascular space (Costill et al. 1976; Nguyen and Kurtz 2006; Nose et al. 1988b; Sanders et al. 1999). With ongoing exercise and sweating and a concomitant escalating body water deficit, the interstitial fluid compartment becomes increasingly contracted (Costill et al. 1976). This can go on after exercise, as sweating lingers for some time while body core temperature returns to its pre-exercise level (Nose et al. 1988b). Even with continuous exercise maintained at a moderate level, persistent sweating will contribute to a slowly accumulating body water deficit and a parallel decrease in circulating plasma volume. Accordingly, a progressive upward drift in heart rate to maintain cardiac output and a concomitant increase in body core temperature and perceived exertion are often noted, even while maintaining the same level of sustainable physical activity. This, in due course, considerably challenges an athlete's physiology, perception of effort and well-being, and performance and typically translates into an unplanned progressive decrease in intensity and effort with an overall slowing of activity. These same effects and responses are also generally evident with intermittent vigorous physical activity, as is characteristic of many individual and team sports, in the heat. Moreover, a proposed cascade of additional contributing triggers associated with hyperthermic fatigue and exhaustion, affecting the central nervous system (CNS) and voluntary neuromuscular activation and control, can further lessen the body's capacity and even the desire to continue exercising (Cheung and Sleivert 2004; Gonzalez-Alonso 2007).

Consequent to a contracted interstitial compartment from exercise and extensive sweating, certain neuromuscular junctions in the primary activated musculature can become hyperexcitable by mechanical deformation and exposure of the unmyelinated nerve terminals and post-synaptic membranes to increased levels of excitatory extracellular constituents such as acetylcholine, electrolytes (Na^+ and K^+), and exercise-related metabolites in the surrounding extracellular spaces. This can trigger affected nerve fibers to spontaneously fire or independently prompt end-plate currents and excitatory post-synaptic potentials (Layzer 1994; Sjogaard et al. 1985). Fasciculations and exertional muscle cramps often begin in the legs (often notably first in the quadriceps and/or hamstring muscles) (Bergeron 1996, 2003), which is not surprising given that the interstitial fluid compartment in the more highly active muscle group regions is likely to be more strongly challenged by concomitant osmotic and metabolic forces that help to maintain circulatory (Costill et al. 1976; Nguyen and Kurtz 2006; Nose et al. 1988b; Sanders et al. 1999) and intracellular (Nguyen and Kurtz 2006; Nygren and Kaijser 2002; Sjogaard et al. 1985) volumes, respectively. As more water is shifted from the interstitial compartment to the intravascular space, adjacent and other nerve terminals and post-synaptic membranes may be similarly affected and the consequent muscle spasms (cramps) would then characteristically spread or jump around with various muscle fibers and bundles alternately contracting and relaxing (Hubbard and Armstrong 1988). The etiology and presentation of exertional muscle cramping related to a whole-body exchangeable sodium deficit and contracted interstitial compartment are notably different in all respects from overload and fatigue-related muscle cramps that come on suddenly and remain localized.

The total body sodium deficit threshold required to prompt muscle cramping is not well described; however, an estimated sweat-induced loss of 20%–30% of the exchangeable Na^+ pool has been noted with severe muscle cramping (Bergeron 1996; McCance 1990). How readily this occurs depends on sweating rate (Buono et al. 2007), sweat sodium concentration (typically 20–80 mmol L^{-1}) (Bergeron 2003; Costill 1977; Maughan et al. 2004), and dietary intake (Maughan et al. 1996). Sweating rate remains fairly consistent during such long-term activity, and serum sodium concentration is typically maintained or elevated, along with potential changes in sweat gland function or sympathetic nervous system activity that would tend to increase sweat sodium concentration (Morgan et al. 2004). Accordingly, a high sweat sodium concentration generally stays high, even as whole-body water and sodium deficits progressively increase, with continuous or intermittent physical activity over an extended period of time (e.g., 3–4 h or more). Moreover, plasma osmolality and circulating electrolyte concentrations will be maintained or somewhat elevated during and after exercise as water shifts from the extravascular space to *defend* central volume and free water loss (primarily from sweating) continues, even as considerable sodium is lost via sweating (Fortney et al. 1981; Nguyen and Kurtz 2006; Nose et al. 1988b; Sanders et al. 1999, 2001). Accordingly, a total body exchangeable sodium deficit is *not* apparent from a venous blood measure.

An accumulating total body water deficit from extensive sweating can increasingly and measurably exacerbate cardiovascular and thermal strain and thus negatively affect exercise-heat tolerance, perception of effort, performance, and safety. Moreover, an accumulating sodium deficit can prompt a contracted interstitial compartment and spontaneous and debilitating muscle cramping.

5.6 MITIGATING THERMAL STRAIN AND EXERTIONAL HEAT ILLNESS RISK

Adequate hydration (i.e., appropriately minimizing a total body water and exchangeable sodium deficit) is integral to athlete safety in the heat. However, the primary determinant of thermal strain is the combined effect of intensity and duration of physical activity and environmental heat stress, and the consequent capacity for effective heat transfer and dissipation and the resultant thermal load. While more optimal exercise economy can help to lessen metabolic heat production and the ensuing thermoregulatory requirement thus minimizing thermal strain (Smoljanic et al. 2014), self-determined and self-controlled pacing—that is, intentional slowing down and decreasing the intensity of effort in advance of getting overheated—is the more practical effective choice in minimizing physiological, perceptual, and thermal strain; averting premature fatigue; and safely extending the sustainable duration of one's athletic activity (Tucker 2009; Tucker and Noakes 2009). Eventually, however, the unrelenting environmental burden and progressively shrinking cardiovascular reserve, increasing thermal load, and energy depletion become overwhelming, prompting an athlete to voluntarily stop or be compelled to discontinue. Heat acclimatization by way of graduated exercise-heat exposure (including a progressive increase in exercise intensity and duration) over several days or more can improve exercise-heat tolerance and reduce the risk of exertional heat illness (Armstrong et al. 2007; Garrett et al. 2011). Unfortunately, this is not always practical or sufficiently applied. Many athletes are also frequently in situations where there is strong self-driven or external motivation to keep going or keep up (e.g., in team practices with strong *encouragement* from a coach), and deliberate slowing down is not a seemingly viable option to the affected athlete during these intense conditioning or competition challenges, even when the lack of capacity to safely continue is evident. Extensive or repeated overexertion, particularly in the heat and high humidity, also increases the risk for rhabdomyolysis in the absence or presence of dehydration, as a result of the consequent breakdown of skeletal muscle fibers (Armstrong et al. 2007; Clarkson 2007; Cleary et al. 2007; Sauret et al. 2002). Conversely, an increase in kidney stress and thermal strain can be particularly evident during subsequent physical activity in the heat following a bout of muscle-damaging exercise (Fortes et al. 2013; Junglee et al. 2013)—another cautionary consideration to athletes with respect to excessive workloads or training and competing in the heat without sufficient recovery from extended or repeated periods of challenging physical exertion.

Other contributing factors that raise body metabolism and heat production (e.g., certain medications or fever) or promote body water loss (e.g., recent illness involving diarrhea) also warrant caution, prior to participating in strenuous physical activity in the heat. All of these scenarios underscore the shared responsibility of the athlete, parent (for youth), and coach to be aware of these and other contributing risk factors to incurring undue thermal strain and avoid, limit, or modify practice, training, or competition in the heat accordingly. However, excessive thermal and cardiovascular strain and exertional heat illness that are primarily due to inadequate hydration—that is, incurring a measurable total body water and/or exchangeable sodium deficit—are readily preventable (Table 5.1).

With brief to relatively short-duration athletic activities (especially in cool environmental conditions where sweat loss is minimal), athletes may not necessitate any fluid intake *while* training or competing, so long as they are amply hydrated at the start. Any resultant deficit in total body water will be negligible and readily restored with subsequent meals and casual fluid consumption. For extended training sessions and competition bouts, 100–250 mL (about 3–8 oz.) every 20 min for young adolescents and *up to* 1.0 L (about 35 oz.) or a little more per hour for older adolescents and adults is generally enough to offset sweat losses or sufficiently minimize sweating-induced body water deficits incurred during these activities. But for many athletes—men and women *and* numerous adolescent boys and girls—even this rate of fluid intake is paralleled by a progressively increasing body water deficit that can be

---

TABLE 5.1

Singular and Collective Effects of Offsetting Measures to Address Inadequate Hydration That Yields a Measurable Total Body Water and/or Exchangeable Sodium Deficit

Condition	Effects	Preventive Measures
Inadequate hydration Total body water and/or Total body exchangeable sodium deficit	↑ Thermal and cardiovascular strain ↑ Heart rate ↑ Body core temperature ↑ Perception of effort Premature fatigue/exhaustion ↓ Performance Contracted interstitial fluid compartment ↑ Risk of muscle cramping	Ensure adequate daily fluid and sodium intake based on individual exercise- and activity-related sweat losses – Prior to exercise/activity, drink regularly throughout the day – During exercise/activity, consume fluid regularly to appropriately offset body water and sodium losses from sweating – After exercise/activity, drink ~120% of any remaining body water deficit (indicated by a change in body weight) over the rest of the day and before the next day's workout – Consume additional sodium daily via foods and fluids when sweating extensively

sizable, especially when sweating nears or exceeds 2 Lh⁻¹. Following these common instances of resulting body water deficits, post-exercise fluid intake should be 1–1.2 L for every kg (16–20 oz. per pound) of a remaining post-exercise body water deficit, as indicated by a pre- to post-activity decrease in body weight. This will be potentially modulated by how much time is available for rehydration before, for example, the start of the next practice/training session. In tournament scenarios with multiple same-day rounds of competition, there may only be enough time to safely and comfortably partially rehydrate, so as not to overconsume fluid in a short amount of time which can lead to gastric discomfort or even hyponatremia—a dangerous and potentially deadly complication from overconsumption of fluids (Hew-Butler et al. 2005).

With rehydration, plasma volume is preferentially restored (Mitchell et al. 2000; Sanders et al. 1999, 2001) prompting a reduced drive to drink and increase in renal free water clearance often before complete restoration of the interstitial spaces; thus, the interstitial fluid compartment remains somewhat contracted, even though the athlete is no longer thirsty and increased urine production (especially after activity) deceptively suggests sufficient whole-body water recovery. This particularly occurs when plain water or very low-sodium fluid is consumed alone (Nose et al. 1988a). Accordingly, athletes who lose a great deal of sodium from sweating need to be deliberate in offsetting the bulk of this abundant electrolyte loss to more fully rehydrate—that is, better retain and distribute the required large volume of ingested water to rehydrate all body fluid compartments (Mitchell et al. 2000; Sanders et al. 1999, 2001; Shirreffs and Maughan 1998)—by appreciably increasing sodium intake during certain activities and more closely matching daily dietary sodium with total sweat sodium loss. Certain dietary choices (e.g., canned soup, tomato juice and sauce, cheese, salted pretzels and peanut butter, and pizza) throughout the day can readily enhance daily sodium replenishment and aid in retention and distribution of ingested fluid. For those athletes who sweat extensively, it is often essential to regularly consume a low-sodium—80 mg or less per 0.23 L (8 oz.) serving—commercial sport drink with added salt (e.g., 1.5–3.0 g of salt to 1 L) or a commercial rehydration beverage with inherently higher levels of sodium that can exceed 800 mg per serving during and after vigorous training sessions and competition bouts. Notably, pre-exercise sodium loading and regular intake of sodium during activity coincident with ample fluid intake can to some extent modulate thermal strain during exercise in the heat, likely owing to greater water retention and distribution (Bergeron et al. 2006; Sims et al. 2007a,b).

Importantly, the physiological carryover effects and impact on the next day's athletic activity can be significant, if the previous day's total body water and exchangeable sodium deficits (estimated by comparing sweat sodium loss to dietary sodium intake) are not sufficiently restored. Accordingly, practical and individualized strategies to encourage adequate pre-activity hydration and sodium balance and also sufficient fluid and sodium intake to appropriately offset sweat loss (Bergeron 2007, 2008; Eichner 2014) during and following lengthier training sessions and competitions are essential in maintaining hydration and minimizing day-to-day total body water and exchangeable sodium deficits and consequent thermal strain and exertional heat illness risk. Unfortunately, determining individual sodium loss rates and needs requires expertise and laboratory facilities that are not widely available, except at select universities and comprehensive medical and sport research/training centers (Patterson et al. 2000).

While the primary determinant of thermal strain is the combined effect of intensity and duration of physical activity and environmental heat stress, sufficient rehydration and sodium intake are essential in minimizing total body water and exchangeable sodium deficits and thus mitigating thermal strain and exertional heat illness risk.

5.7 MANAGING EXERTIONAL HEAT ILLNESSES RELATED TO TOTAL BODY WATER AND EXCHANGEABLE SODIUM DEFICITS

Most episodes of exertional heat illness prompted by extensive sweating and a consequent total body water and/or exchangeable sodium deficit are preventable. And while the best approach is to focus on prevention, unanticipated circumstances sometimes happen or an athlete might underestimate the imminent or current hydration (including sodium loss and replenishment) and/or environmental challenges. Therefore, as a competitor or one who oversees on-site medical care and implements first aid or an emergency action plan in treating exertional heat illness, it is essential to know the primary contributing factors and most effective clinical response. Importantly, during training or competition, any significant deterioration in performance with notable signs of struggling and developing exertional heat illness should be sufficient reason to immediately stop participation and promptly seek appropriate medical attention. Moreover, any athlete experiencing exertional heat illness should not continue training or competing after treatment or participate again that same day, even with rehydration and complete signs and symptoms resolved.

5.7.1 EXERTIONAL HEAT (MUSCLE) CRAMPS

Exertional heat cramps occur during or after extended or repeated exertion and concomitant extensive sweat losses and a resultant total body exchangeable sodium deficit (Armstrong et al. 2007; Bergeron 2008; Eichner 2007; Valentine 2007). These involuntary muscle spasms can occur in any environment and affected athletes are generally not overheated. Subtle indications (slight muscle cramping or twitches) are often the first signs of exertional heat cramping, prior to progressively developing to more severe and widespread (often bilaterally) intermittent and eventually debilitating muscle spasms. In contrast, comparatively sudden-onset exercise-associated muscle cramping that is localized (e.g., affecting solely the calf or hamstring muscles), constant, asymmetric, and responsive to passive stretching is highly likely to have been prompted by muscle overload and fatigue (Bergeron 2008; Schwellnus et al. 1997). The athlete with exertional heat cramps may be dehydrated, but the overriding issue that needs to be urgently addressed is the sweat-induced exchangeable sodium deficit. Accordingly, prompt treatment with an oral high-salt solution (>200 mg per serving) or intravenously (Givan and Diehl 2012) is imperative, whereas immediate management of exercise-associated muscle cramping related to overload and fatigue should include rest and passive stretching to assist in relaxing the muscles and relieving some of the spasms (Miller et al. 2010; Schwellnus et al. 2008).

5.7.2　Heat Exhaustion

Heat exhaustion is a moderate-severity exertional heat illness that is characterized by hypotension and cardiovascular insufficiency prompted in large part by extensive sweating and a sizable total body water deficit. For an athlete, heat exhaustion during training or competition typically results from the combination of strenuous and/or long-duration physical exertion, environmental heat stress, acute dehydration, energy depletion, and central fatigue, and it often results in physical collapse. Notable evolving signs and symptoms that can alert an athlete include weakness, dizziness, nausea, syncope, and headache. Excessive thermal strain is typically not evident, as body core temperature is usually less than 40°C (104°F). The athlete should promptly stop activity and move to a shaded or air-conditioned area. Any excess clothing should be removed, and the athlete should assume or be placed in the supine position with legs elevated. To help address the body water deficit, oral fluids can be consumed, if the athlete is conscious and able to swallow (Armstrong et al. 2007; Roberts 2007). Other conservative cooling therapy measures can improve medical and perceptual status as well. If trained medical support is available, heart rate, respiratory rate, and CNS status should be monitored closely, and blood pressure and rectal temperature should also be checked. If a significantly elevated body core temperature is suspected and/or there are signs of CNS dysfunction, exertional heatstroke (EHS) cooling therapy should be immediately initiated and emergency medical services activated.

5.7.3　Exertional Heatstroke

Excessive thermal strain during training or competition can rapidly progress to EHS. EHS is the most severe exertional heat illness—a clear medical emergency affecting multiple body systems. EHS is characterized by CNS abnormalities such as delirium, convulsions or coma, endotoxemia, circulatory failure, thermoregulatory dysregulation, and, potentially, multiple organ and tissue dysfunction and damage (e.g., brain, heart, liver, kidneys, spleen, and muscle), which results from an excessively elevated body core temperature induced by strenuous exercise and typically high environmental heat stress. Excessive (for the current environmental conditions) intensity and/or duration of physical activity—single or repeated bout(s)—is the primary driver of resultant excessive thermal strain. However, this can be prompted earlier and exacerbated by insufficient hydration and a significant total body water deficit. Nevertheless, the highest priority is *cooling*—not rehydration. An athlete with EHS will typically have a rectal temperature greater than 40°C (>104°F), and this should prompt immediate on-site whole-body rapid cooling using proven techniques (cold- or ice-water immersion is the preferred most effective method, although applying ice packs to the neck, axillae, and groin and rotating ice-water-soaked towels to all other areas of the body can be effective as well) (Armstrong et al. 2007; Casa et al. 2007, 2012). This process should be continued until rectal temperature reaches just under 39°C (approximately 102°F) or the victim exhibits noticeable clinical improvement. If rectal temperature cannot be assessed in an athlete who has clinical signs or symptoms suggestive of EHS, rapid cooling for 10–15 min should be promptly initiated, while awaiting the arrival of additional medical assistance. Importantly, emergency medical services communication should be initiated immediately for any athlete who collapses or exhibits moderate or severe CNS dysfunction or encephalopathy.

5.8 WRAPPING IT UP

Significant total body water and exchangeable sodium deficits can have a measurable impact on increasing thermal strain and exertional heat illness risk. However, appropriate preparatory steps and prompt responsive actions can mitigate the related clinical risks and potential negative effects on well-being and performance.

- All athletes must plan in advance and adjust to on-site for the environmental conditions and current circumstances that contribute to greater thermal strain and exertional heat illness risk. Ensuring a well-hydrated and sodium-balanced state prior to practice, training or play begins with adequate recovery of water and electrolyte deficits incurred from the previous activity session(s).
- Total body water and exchangeable sodium deficits are often inevitable during vigorous training and competition. However, knowing one's personal rates of body water and electrolyte (sodium) losses (as discussed further in Chapter 2) from sweating is a key step in minimizing these often significant deficits and sufficiently recovering before the next bout. Managing these nutrient losses more optimally day-to-day and same-day session-to-session can reduce the risk of undue thermal strain and exertional heat illness.
- It is critical to recognize and initiate a prompt and appropriate response, including slowing down or immediately stopping and seeking appropriate medical attention and treatment, at the earliest signs of developing exertional heat illness, even if the prevailing indications suggest that oral restoration of water and sodium can mitigate the evolving danger and risk.

REFERENCES

Armstrong, L.E., D.J. Casa, M. Millard-Stafford, D.S. Moran, S.W. Pyne, and W.O. Roberts. 2007. American College of Sports Medicine position stand: Exertional heat illness during training and competition. *Med Sci Sports Exerc* 39(3): 556–72.

Bergeron, M.F. 1996. Heat cramps during tennis: A case report. *Int J Sport Nutr* 6(1): 62–8.

Bergeron, M.F. 2003. Heat cramps: Fluid and electrolyte challenges during tennis in the heat. *J Sci Med Sport* 6(1): 19–27.

Bergeron, M.F. 2007. Exertional heat cramps: Recovery and return to play. *J Sport Rehabil* 16(3): 190–6.

Bergeron, M.F. 2008. Muscle cramps during exercise: Is it fatigue or electrolyte deficit? *Curr Sports Med Rep* 7(4): S50–5.

Bergeron, M.F. 2009. Youth sports in the heat: Recovery and scheduling considerations for tournament play. *Sports Med* 39(7): 513–22.

Bergeron, M.F. 2014. Hydration and thermal strain during tennis in the heat. *Br J Sports Med* 48(Suppl 1): i12–7.

Bergeron, M.F., J.L. Waller, and E.L. Marinik. 2006. Voluntary fluid intake and core temperature responses in adolescent tennis players: Sports beverage versus water. *Br J Sports Med* 40(5): 406–10.

Buono, M.J., K.D. Ball, and F.W. Kolkhorst. 2007. Sodium ion concentration vs. sweat rate relationship in humans. *J Appl Physiol* 103(3): 990–4.

Casa, D.J., L.E. Armstrong, G.P. Kenny, F.G. O'Connor, and R.A. Huggins. 2012. Exertional heat stroke: New concepts regarding cause and care. *Curr Sports Med Rep* 11(3): 115–23.

Casa, D.J., B.P. McDermott, E.C. Lee, S.W. Yeargin, L.E. Armstrong, and C.M. Maresh. 2007. Cold water immersion: The gold standard for exertional heatstroke treatment. *Exerc Sport Sci Rev* 35(3): 141–9.

Cheung, S.S., and G.G. Sleivert. 2004. Multiple triggers for hyperthermic fatigue and exhaustion. *Exerc Sport Sci Rev* 32(3): 100–6.

Clarkson, P.M. 2007. Exertional rhabdomyolysis and acute renal failure in marathon runners. *Sports Med* 37(4–5): 361–3.

Cleary, M., D. Ruiz, L. Eberman, I. Mitchell, and H. Binkley. 2007. Dehydration, cramping, and exertional rhabdomyolysis: A case report with suggestions for recovery. *J Sport Rehabil* 16(3): 244–59.

Costill, D.L. 1977. Sweating: Its composition and effects on body fluids. *Ann N Y Acad Sci* 301: 160–74.

Costill, D.L., R. Cote, and W. Fink. 1976. Muscle water and electrolytes following varied levels of dehydration in man. *J Appl Physiol* 40(1): 6–11.

Eichner, E.R. 2007. The role of sodium in "heat cramping." *Sports Med* 37(4–5): 368–70.

Eichner, E.R. 2014. The salt paradox for athletes. *Curr Sports Med Rep* 13(4): 197–8.

Ely, M.R., S.N. Cheuvront, W.O. Roberts, and S.J. Montain. 2007. Impact of weather on marathon-running performance. *Med Sci Sports Exerc* 39(3): 487–93.

Fortes, M.B., U. Di Felice, A. Dolci, N.A. Junglee, M.J. Crockford, L. West, R. Hillier-Smith, J.H. Macdonald, and N.P. Walsh. 2013. Muscle-damaging exercise increases heat strain during subsequent exercise heat stress. *Med Sci Sports Exerc* 45(10): 1915–24.

Fortney, S.M., E.R. Nadel, C.B. Wenger, and J.R. Bove. 1981. Effect of blood volume on sweating rate and body fluids in exercising humans. *J Appl Physiol* 51(6): 1594–600.

Garrett, A.T., N.J. Rehrer, and M.J. Patterson. 2011. Induction and decay of short-term heat acclimation in moderately and highly trained athletes. *Sports Med* 41(9): 757–71.

Givan, G.V., and J.J. Diehl. 2012. Intravenous fluid use in athletes. *Sports Health* 4(4): 333–9.

Gonzalez-Alonso, J. 2007. Hyperthermia impairs brain, heart and muscle function in exercising humans. *Sports Med* 37(4–5): 371–3.

Gonzalez-Alonso, J., R. Mora-Rodriguez, P.R. Below, and E.F. Coyle. 1997. Dehydration markedly impairs cardiovascular function in hyperthermic endurance athletes during exercise. *J Appl Physiol* 82(4): 1229–36.

Harrison, M.H., R.J. Edwards, and P.A. Fennessy. 1978. Intravascular volume and tonicity as factors in the regulation of body temperature. *J Appl Physiol Respir Environ Exerc Physiol* 44(1): 69–75.

Hew-Butler, T., C. Almond, J.C. Ayus, J. Dugas, W. Meeuwisse, T. Noakes, S. Reid, A. Siegel, D. Speedy, K. Stuempfle, J. Verbalis, and L. Weschler. 2005. Consensus statement of the 1st international exercise-associated hyponatremia consensus development conference, Cape Town, South Africa 2005. *Clin J Sport Med* 15(4): 208–13.

Hubbard, R.W., and L.E. Armstrong. 1988. The heat illnesses: Biochemical, ultrastructural, and fluid-electrolyte considerations. In *Human Performance Physiology and Environmental Medicine at Terrestrial Extremes*, edited by K.B. Pandolf, M.N. Sawka, and R.R. Gonzalez, 305–59. Indianapolis, IN: Benchmark Press.

Junglee, N.A., U. Di Felice, A. Dolci, M.B. Fortes, M.M. Jibani, A.B. Lemmey, N.P. Walsh, and J.H. Macdonald. 2013. Exercising in a hot environment with muscle damage: Effects on acute kidney injury biomarkers and kidney function. *Am J Physiol Renal Physiol* 305(6): F813–20.

Layzer, R.B. 1994. The origin of muscle fasciculations and cramps. *Muscle Nerve* 17(11):1243–9.

Maughan, R.J., J.B. Leiper, and S.M. Shirreffs. 1996. Restoration of fluid balance after exercise-induced dehydration: Effects of food and fluid intake. *Eur J Appl Physiol Occup Physiol* 73(3–4): 317–25.

Maughan, R.J., S.J. Merson, N.P. Broad, and S.M. Shirreffs. 2004. Fluid and electrolyte intake and loss in elite soccer players during training. *Int J Sport Nutr Exerc Metab* 14(3): 333–46.

McCance, R.A. 1990. Proceedings of the Royal Society of London. Series B—Biological Sciences, Volume 119, 1935–1936: Experimental sodium chloride deficiency in man. *Nutr Rev* 48(3): 145–7.

Meyer, F., K.A. Volterman, B.W. Timmons, and B. Wilk. 2012. Fluid balance and dehydration in the young athlete: Assessment considerations and effects on health and performance. *Am J Lifestyle Med* 6(6): 489–501.

Miller, K.C., M.S. Stone, K.C. Huxel, and J.E. Edwards. 2010. Exercise-associated muscle cramps: Causes, treatment, and prevention. *Sports Health* 2(4): 279–83.

Mitchell, J.B., M.D. Phillips, S.P. Mercer, H.L. Baylies, and F.X. Pizza. 2000. Postexercise rehydration: Effect of Na^+ and volume on restoration of fluid spaces and cardiovascular function. *J Appl Physiol* 89: 1302–9.

Moran, D.S., S.J. Montain, and K.B. Pandolf. 1998. Evaluation of different levels of hydration using a new physiological strain index. *Am J Physiol* 275(3 Pt 2): R854–60.

Morgan, R.M., M.J. Patterson, and M.A. Nimmo. 2004. Acute effects of dehydration on sweat composition in men during prolonged exercise in the heat. *Acta Physiol Scand* 182(1): 37–43.

Nadel, E.R., S.M. Fortney, and C.B. Wenger. 1980. Effect of hydration state on circulatory and thermal regulations. *J Appl Physiol* 49(4): 715–21.

Nguyen, M.K., and I. Kurtz. 2006. Quantitative interrelationship between Gibbs-Donnan equilibrium, osmolality of body fluid compartments, and plasma water sodium concentration. *J Appl Physiol* 100(4): 1293–300.

Nose, H., G.W. Mack, X.R. Shi, and E.R. Nadel. 1988a. Role of osmolality and plasma volume during rehydration in humans. *J Appl Physiol* 65(1): 325–31.

Nose, H., G.W. Mack, X.R. Shi, and E.R. Nadel. 1988b. Shift in body fluid compartments after dehydration in humans. *J Appl Physiol* 65(1): 318–24.

Nygren, A.T., and L. Kaijser. 2002. Water exchange induced by unilateral exercise in active and inactive skeletal muscles. *J Appl Physiol* 93(5): 1716–22.

Patterson, M.J., S.D. Galloway, and M.A. Nimmo. 2000. Variations in regional sweat composition in normal human males. *Exp Physiol* 85(6): 869–75.

Roberts, W.O. 2007. Exercise-associated collapse care matrix in the marathon. *Sports Med* 37(4–5): 431–3.

Sanders, B., T.D. Noakes, and S.C. Dennis. 1999. Water and electrolyte shifts with partial fluid replacement during exercise. *Eur J Appl Physiol Occup Physiol* 80(4): 318–23.

Sanders, B., T.D. Noakes, and S.C. Dennis. 2001. Sodium replacement and fluid shifts during prolonged exercise in humans. *Eur J Appl Physiol* 84(5): 419–25.

Sauret, J.M., G. Marinides, and G.K. Wang. 2002. Rhabdomyolysis. *Am Fam Physician* 65(5): 907–12.

Schwellnus, M.P., E.W. Derman, and T.D. Noakes. 1997. Aetiology of skeletal muscle "cramps" during exercise: A novel hypothesis. *J Sport Sci* 15: 277–85.

Schwellnus, M.P., N. Drew, and M. Collins. 2008. Muscle cramping in athletes--risk factors, clinical assessment, and management. *Clin Sports Med* 27(1): 183–94, ix–x.

Shirreffs, S.M., and R.J. Maughan. 1998. Volume repletion after exercise-induced volume depletion in humans: Replacement of water and sodium losses. *Am J Physiol* 274(5Pt 2):F868–75.

Sims, S.T., N.J. Rehrer, M.L. Bell, and J.D. Cotter. 2007a. Preexercise sodium loading aids fluid balance and endurance for women exercising in the heat. *J Appl Physiol* 103(2): 534–41.

Sims, S.T., L. van Vliet, J.D. Cotter, and N.J. Rehrer. 2007a. Sodium loading aids fluid balance and reduces physiological strain of trained men exercising in the heat. *Med Sci Sports Exerc* 39(1): 123–30.

Sjogaard, G., R.P. Adams, and B. Saltin. 1985. Water and ion shifts in skeletal muscle of humans with intense dynamic knee extension. *Am J Physiol* 248(2 Pt 2): R190–6.

Smoljanic, J., N.B. Morris, S. Dervis, and O. Jay. 2014. Running economy, not aerobic fitness, independently alters thermoregulatory responses during treadmill running. *J Appl Physiol* 117(12): 1451–9.

Tucker, R. 2009. The anticipatory regulation of performance: The physiological basis for pacing strategies and the development of a perception-based model for exercise performance. *Br J Sports Med* 43(6): 392–400.

Tucker, R., and T.D. Noakes. 2009. The physiological regulation of pacing strategy during exercise: A critical review. *Br J Sports Med* 43(6): e1.

Valentine, V. 2007. The importance of salt in the athlete's diet. *Curr Sports Med Rep* 6(4): 237–40.

6 Gastrointestinal and Metabolic Responses to Body Fluid Imbalance during Exercise

G. Patrick Lambert

CONTENTS

6.1 INTRODUCTION

Proper gastrointestinal (GI) function is vitally important to the maintenance of health and performance during prolonged athletic events. This is because fluids (i.e., water) and nutrients (e.g., carbohydrate (CHO); sodium; and potassium) ingested during exercise must be absorbed by the GI tract. Unfortunately, the GI tract is susceptible to dysfunction during such exercise. Reductions in GI blood flow, gastric emptying rate (GER), intestinal absorption, and GI barrier function, accompanying increases in GI symptoms, can occur, and dehydration likely exacerbates these problems.

Impaired GI function during prolonged exercise accompanying dehydration would also likely affect metabolic function. Reduced capacity to absorb fluid and nutrients would ultimately promote even greater dehydration, increased hyperthermia, and greater cardiovascular strain, which may ultimately affect muscle metabolism. Furthermore, the reduced availability of exogenous energy for working muscles (due to delayed absorption of such nutrients) would also likely have a negative influence on muscle metabolic function. Such factors will be discussed in detail in the upcoming sections of this chapter. The effects of hyperhydration on GI function during exercise will not be discussed as no studies have been conducted in this area.

6.2 ROLE OF THE GI SYSTEM DURING EXERCISE

The most obvious role of the GI system during exercise is to absorb fluids and nutrients ingested prior to and during an event. Importantly, all of the major processes and functions of the GI tract must be functioning normally for proper absorption to occur. For example, if motility (e.g., gastric emptying) is inhibited, ingested fluids and nutrients will not be delivered at an optimal rate to the proximal small intestine for rapid absorption to take place; if motility is enhanced (e.g., colonic motility), increased urge to defecate or diarrhea may occur. Similarly, if intestinal secretions are inhibited, proper digestion of certain larger molecules (e.g., maltodextrins) may be reduced leading to delayed or inefficient absorption, and if intestinal absorption is inhibited, replacement of water and nutrients will be reduced.

Normally, the GI tract is highly efficient (~99%) at absorbing all of the fluid presented to it each day. However, GI dysfunction during exercise could lead to malabsorption, possible GI symptoms (e.g., fullness, nausea, cramps, and diarrhea), and further dehydration because of such symptoms. Finally, the role of the GI barrier must not be overlooked. The GI barrier consists of physical (i.e., tight junctions, enterocyte membranes, and mucus) and immune factors (i.e., tissue macrophages) that prevent harmful (e.g., pathogenic, immunogenic, antigenic, and enzymatic) substances from entering the internal environment of the body. It is known that during prolonged, strenuous exercise involving heat stress and/or dehydration, this barrier can dysfunction (i.e., *leaky gut*) leading to local and systemic immune reactions and GI symptoms.

6.3 EFFECT OF EXERCISE-INDUCED DEHYDRATION ON GI BLOOD FLOW

A number of studies have shown that GI blood flow declines during exercise in humans and other animals (Bell et al. 1983; Bradley 1949; Flaim et al. 1979; Iwao et al. 1995; Manohar 1986; Perko et al. 1998; Puvi-Rajasingham et al. 1997; Qamar and Read 1987; Rehrer et al. 2001, 2005; Rowell et al. 1964, 1965; Wade et al. 1956). Furthermore, this has been shown to be exacerbated by heat stress in humans (Rowell et al. 1965), and by dehydration both in rodents (Horowitz and Samueloff 1988; Horowitz et al. 1985; Massett et al. 1996) and dogs (Adar et al. 1975, 1976). No studies have directly examined the role of dehydration on GI blood flow in humans. It seems reasonable based on the animal data, and the fact that hypohydration during exercise heat stress in humans increases cardiovascular strain, that dehydration likely promotes greater splanchnic and mesenteric vasoconstriction during exercise and thereby further reduces GI blood flow. This effect could then lead to alterations in other GI functions.

6.4   EFFECT OF EXERCISE-INDUCED DEHYDRATION ON GASTRIC EMPTYING

GER is controlled by a number of neural, hormonal, and local factors (Lambert et al. 2012). Furthermore, there are large differences in GER among individuals as well as gender (women emptying more slowly than men) (Notivol et al. 1984). With regard to exercise and dehydration, the primary concern is how quickly ingested fluids can empty the stomach to allow for rapid absorption by the small intestine and thus rehydration. Previous studies have shown that hypohydrated individuals may have reduced gastric emptying. Neufer and colleagues (1989) found that GER was lower during exercise in the heat when hypohydrated. It is unclear from that study, however, whether the reduction in GER was caused by the heat stress or hypohydration. Rehrer and colleagues (1990a) also observed a reduced GER during exercise in individuals dehydrated to ~4% body weight by a dehydration-exercise regimen compared to dehydration-rest (i.e., sauna), euhydration-exercise, or euhydration-rest regimens. Again, though, the dehydration-exercise regimen resulted in higher core body temperatures making it difficult to determine whether the dehydration or the heat stress caused the slowing of the GER.

There are a number of other very important variables to consider with regard to GER of fluids during exercise. The most important variable in this regard may be the volume and timing of fluid ingestion. Accordingly, Ryan et al. (1998) showed that repeated drinking (~200 ml every 10 min) to maintain a constant yet comfortable stomach in hypohydrated (3% of body mass) individuals maintains a GER comparable to when euhydrated during prolonged exercise (Figure 6.1). Thus, while dehydration

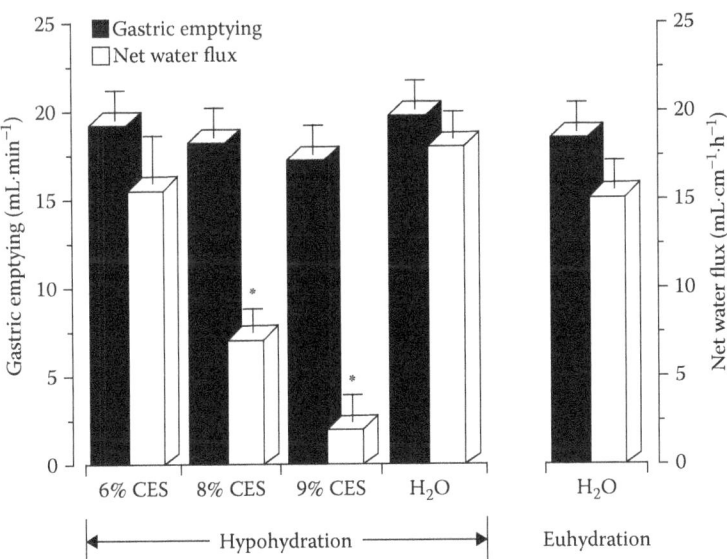

FIGURE 6.1   Effect of hypohydration on gastric emptying and intestinal water absorption. (Reprinted with permission from Ryan, A.J. et al., *J. Appl. Physiol.*, 84, 1581–8. Copyright 1998 American Physiological Society.)

TABLE 6.1
Primary Variables Influencing GER during Exercise

Variable	Influence
Gastric volume	Larger volumes (up to ~600 ml) empty faster than smaller volumes and maintenance of a relatively high gastric volume throughout exercise maintains higher GER.
Beverage composition	Increased CHO content (above 7%) reduces GER; increased osmolality (at higher CHO concentrations) reduces GER; carbonation does not affect GER among solutions of similar CHO content.
Exercise intensity	No effect below 75% VO_2max; slower GER above this level; slower GER during intermittent, high-intensity exercise.
Exercise mode	Reduced GER during running compared to cycling at 70% VO_2max after 40 min.
Heat stress	Reduced GER during exercise in the heat.
Hydration state	Reduced GER when exercising hypohydrated in the heat; no effect of hypohydration on GER when exercising in cooler environments. No effect of hypohydration with repeated drinking to maintain constant stomach volume during exercise.

has been shown to reduce GER in some studies, this effect can potentially be offset by repeated drinking during exercise.

Other key factors to consider are as follows: (1) exercise intensity (Costill and Saltin 1974; Fordtran and Saltin 1967; Leiper et al. 2001a,b, 2005), (2) exercise mode (Houmard et al. 1991; Rehrer et al. 1990b), (3) heat stress (Neufer et al. 1989; Owen et al. 1986), and (4) characteristics of the ingested beverage (Brouns et al. 1995; Costill and Saltin 1974; Coyle et al. 1978; Fordtran and Saltin 1967; Foster et al. 1980; Lambert et al. 1993; Mitchell et al. 1988, 1989; Murray et al. 1994, 1999; Neufer et al. 1986; Owen et al. 1986; Rogers et al. 2005; Ryan et al. 1989, 1998; Shi et al. 2000; Vist and Maughan 1994, 1995; Zachwieja et al. 1992). Table 6.1 summarizes the most important factors that influence GER.

6.5 EFFECT OF EXERCISE-INDUCED DEHYDRATION ON INTESTINAL ABSORPTION

Similar to gastric emptying, intestinal absorption is affected by a number of factors. With regard to exercise and dehydration, the intestinal absorption rate of ingested fluids is of primary importance. Intestinal perfusion studies in humans have found that exercise at intensities below ~70%–80% maximal oxygen consumption (VO_2max) does not appear to affect intestinal absorption of infused solutions (Fordtran and Saltin 1967; Gisolfi et al. 1991). Studies utilizing ingestion of non-metabolizable CHO's such as D-xylose (passively absorbed) and 3-O-methyl-D-glucose (3MG) (absorbed via sodium-linked secondary active transport) have shown that prolonged, low-intensity exercise in the heat (4.5 h walking at 3 mph; 38°C dry bulb, 27°C wet bulb) (Williams et al. 1964) and higher intensity exercise (70% VO_2max

TABLE 6.2
Primary Variables Influencing Intestinal Absorption during Exercise

Variable	Influence
GER	Slower GER will reduce absorption rate of ingested water and nutrients.
Beverage composition	Multiple transportable substrates (e.g., glucose and fructose) enhance solute and water absorption compared to single substrates.
	Increasing CHO concentration (>7%) reduces water absorption.
	Increasing osmolality inhibits water absorption.
Exercise intensity	Exercise intensity at >70% VO_2max can inhibit CHO absorption.
	Lower intensity, prolonged exercise in the heat can inhibit CHO absorption.
Hydration state	Hypohydration (3% body mass) does not affect water or solute absorption during prolonged exercise in a cool environment, unknown if higher levels of dehydration may influence intestinal absorption.

cycling for 1 h) (van Nieuwenhoven et al. 1999) reduced urinary excretion of 3MG indicating reduced absorption via sodium-linked secondary active transport. Lang and colleagues (2006) observed significant reductions in both d-xylose and 3MG absorption during running at 70% VO_2max for 1 h suggesting that both active and passive absorption may be inhibited with this type and intensity of exercise. It should be noted however that such studies do not directly measure intestinal absorption but are also influenced by GER of the ingested probes. Thus, lower excretion rates may be a reflection of slower GERs.

Very little research has studied the direct effect of dehydration during exercise on intestinal absorption. Of the studies that have been published, only one has directly measured intestinal absorption of fluid replacement beverages. Ryan and colleagues (1998), using the modified segmental perfusion technique developed by Lambert and colleagues (1996), observed that hypohydrated (3% of body mass) individuals did not have reduced intestinal water absorption rates during prolonged cycling exercise (65% VO_2max for 85 min) (Figure 6.1). However, it should be remembered that if GER is reduced by dehydration then intestinal absorption will be delayed.

Of greatest importance regarding the intestinal absorption rate of ingested beverages (that have similar GER) is the beverage composition. Factors such as CHO type and concentration and beverage osmolality are particularly important. These variables along with the previously discussed factors are summarized in Table 6.2.

6.6 EFFECT OF EXERCISE-INDUCED DEHYDRATION ON GI BARRIER FUNCTION

The GI barrier is basically composed of the enterocyte membranes, tight junctions, mucus, and tissue macrophages. Studies have found that during prolonged, strenuous exercise, when GI blood flow may be compromised and GI temperatures may be high, this barrier can become disrupted leading to increases in GI permeability. This condition is also known as *leaky gut* (Bosenberg et al. 1988; Brock-Utne et al. 1988; Jeukendrup et al. 2000; Lambert et al. 1999, 2008; Moses et al. 1991; Oktedalen et al. 1992; Pals et al. 1997; Smetanka et al. 1999). Essentially, this is a situation in which the GI barrier no longer is able to keep normally restricted substances in the GI lumen from entering the internal environment (i.e., GI tissue, interstitial fluid, and blood) of the body. Such substances include endotoxin (i.e., lipopolysaccharide [LPS]), dietary antigens, and hydrolytic enzymes. The result can be immune responses and tissue damage leading to local and systemic inflammation. Passage of LPS into the systemic circulation may also result in impaired thermoregulation during exercise via the production of *endogenous pyrogens* (e.g., interleukin-1, interleukin-6, tumor necrosis factor-α) (Lambert 2008).

The role dehydration plays in exercise-induced GI barrier dysfunction has not been well studied. Only two studies have been conducted to date examining this variable. In the first study, van Nieuwenhoven and colleagues (2000) did not observe any change in intestinal permeability compared to euhydrated conditions after dehydrating individuals to 3% of body mass (via sauna regimen) and then having them perform cycle exercise at 70% maximal exercise intensity. In contrast, Lambert and colleagues (2008) observed that running without fluid replacement (i.e., dehydration of 1.5% body mass) for 60 min at 70% VO_2max elicited increased gastroduodenal and small intestinal permeability (Figure 6.2). These results indicate that differences in exercise mode may influence GI permeability during exercise (i.e., running vs. cycling) and that even low levels of dehydration during exercise may significantly affect GI barrier function.

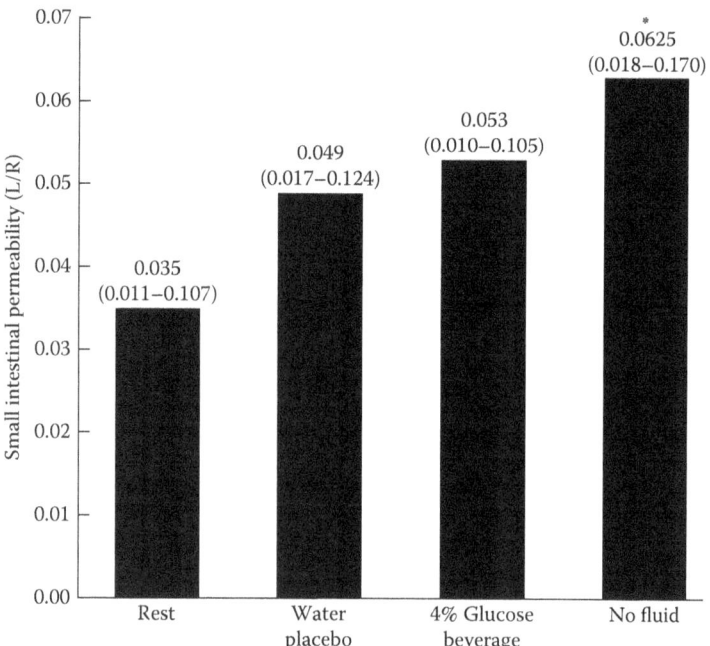

FIGURE 6.2 Fluid restriction during prolonged exercise increases small intestinal permeability. *Note:* * indicates $p < .05$. (Modified from Lambert, G.P. et al., Fluid restriction during running increases GI permeability, *Int. J. Sports Med.*, 29, 194–8. With permission. Copyright 2008 Georg Thieme.)

6.7   EFFECT OF EXERCISE-INDUCED DEHYDRATION ON GI SYMPTOMS

Both upper and lower GI symptoms are common in endurance athletes. Peters et al. (1999) found that during distance running, up to 95% of triathletes experience lower GI complaints during run training. Rehrer et al. (1990a) observed that 37.5% of subjects reported GI symptoms when exercising while dehydrated (4% body weight loss) compared to the euhydrated-exercise, euhydrated-rest, or dehydrated-rest conditions. GI symptoms can also be influenced by the composition of the beverage consumed during exercise. In this regard, Shi et al. (2004) found in a group of adolescent and adult athletes performing prolonged (4 × 12 min) circuit-training exercises that beverages containing 8% CHO produced more stomach upset and side ache compared to drinking an equivalent amount of 6% CHO. Likewise, Morton et al. (2004) observed that ingestion of reconstituted fruit juice (10.4% CHO) produced more exercise-related transient abdominal pain and bloating compared to flavored water, a commercially available sports drink (6% CHO) or no fluid. van Nieuwenhoven et al. (2005) also found that drinking water produced less GI complaints than when drinking a sports drink (6.9% CHO with added vitamins, electrolytes, amino acids, myoinositol, and choline) or the same sports drink with caffeine (150 mg/L) during an 18-km competitive run in cool environmental conditions. These studies suggest that solutions with higher CHO concentrations (>~6%) and higher osmolalities most likely increase GI distress when ingested during exercise.

6.8   METABOLIC CONSEQUENCES OF GI DYSFUNCTION DUE TO EXERCISE-INDUCED DEHYDRATION

Based on the previous discussion, dehydration during exercise can potentially lead to reductions in GERs, delayed intestinal absorption, and GI symptoms. The result would therefore be delayed rehydration and delayed availability of exogenous energy sources. The overall effect during exercise would likely be a greater degree of hyperthermia, increased cardiovascular strain, altered metabolic function, and central nervous system dysfunction (Sawka et al. 2007).

With regard to whole-body metabolism, it has been shown that dehydration decreases maximal oxygen uptake (VO_2max) (Sawka and Coyle 1999). Therefore, GI dysfunction during exercise could impair aerobic exercise performance via effects on rehydration rate. Furthermore, skeletal muscle blood flow may be further reduced by delayed rehydration that could inhibit muscular free fatty acid uptake, increased reliance on muscle glycogen, and higher blood lactate levels during intense exercise (Gonzalez-Alonso et al. 1999). Logan-Sprenger et al. (2013) have shown that even low levels of dehydration (<2% body mass) cause increased muscle glycogenolysis and blood lactate levels during prolonged cycling (120 min at 65% peak VO_2max). In addition, dysfunction of the GI barrier allowing passage of LPS into the blood may alter metabolism. Kamisoglu et al. (2013) have shown that low-dose LPS administration to healthy humans results in increased plasma fatty acids (mostly monounsaturated and polyunsaturated) and decreased plasma amino acids. It is believed these responses are the result of the inflammatory response (increased lipolysis and increased hepatic uptake of amino acids from the plasma). Increased lipolysis is likely related to increased catecholamine release as a result of infection and also increased production of cytokines that are known to affect metabolism. The increased hepatic uptake of amino acids is likely due to greater need for substrates used in the synthesis of antioxidant molecules and acute phase proteins.

6.9 SUMMARY

As summarized in Figure 6.3, body fluid imbalance (i.e., dehydration/hypohydration) can negatively affect GI function during exercise-heat stress. Increased GI temperature and reductions in GI blood flow will likely be exacerbated by dehydration further decreasing gastric emptying and, thus, intestinal absorption rates. This will delay rehydration. Potential impairments of the GI barrier will also likely be worsened by dehydration. This may lead to greater hyperthermia as increased permeability to LPS (i.e., endotoxemia) can cause the release of endogenous pyrogens. Endotoxemia may also lead to altered metabolism. GI symptoms can result from reduced gastric emptying, intestinal absorption, and GI barrier dysfunction that will likely inhibit fluid intake, also promoting higher core body temperatures and modified metabolism. Ultimately, this could negatively affect exercise performance.

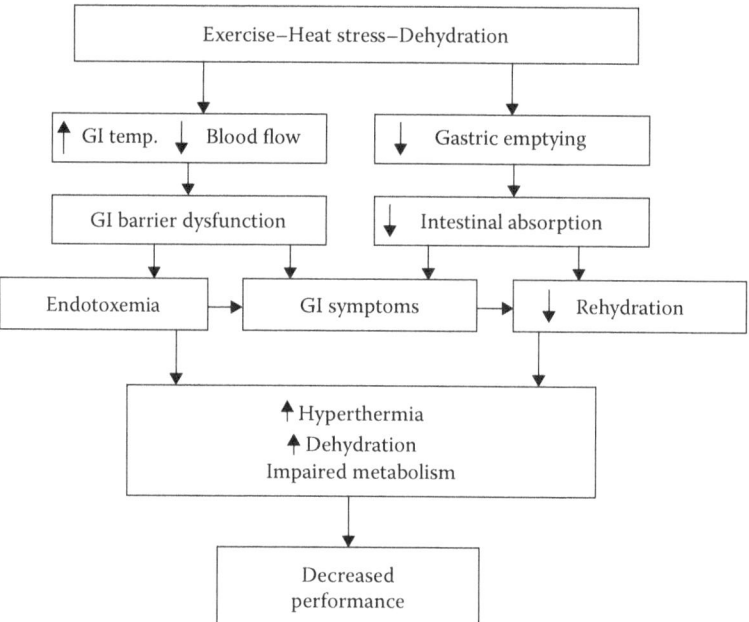

FIGURE 6.3 Summary of the effects of dehydration on GI function and metabolism during exercise-heat stress.

REFERENCES

Adar, R., A. Franklin, and E. W. Salzman. 1975. Disproportionate reduction in superior mesenteric artery flow during dehydration and cardiac tamponade. *Surg. Forum* 26:295–7.

Adar, R., A. Franklin, R. F. Spark, C. B. Rosoff, and E. W. Salzman. 1976. Effect of dehydration and cardiac tamponade on superior mesenteric artery flow: Role of vasoactive substances. *Surgery* 79:534–43.

Bell, A. W., J. R. S. Hales, R. B. King, and A. A. Fawcett. 1983. Influence of heat stress on exercise-induced changes in regional blood flow in sheep. *J. Appl. Physiol.: Respirat. Environ. Exercise Physiol.* 55:1916–23.

Bosenberg, A., J. Brock-Utne, S. Gaffin, M. Wells, and G. Blake. 1988. Strenuous exercise causes systemic endotoxemia. *J. Appl. Physiol.* 65:106–8.

Bradley, S. E. 1949. Variations in hepatic blood flow in man during health and disease. *N. Engl. J. Med.* 240:456–61.

Brock-Utne, J., S. Gaffin, M. Wells, P. Gathiram, E. Sohar, M. James, D. Morrell, R. Norman. 1988. Endotoxemia in exhausted runners after a long-distance race. *S. African Med. J.* 73:533–6.

Brouns, F., J. Senden, E. J. Beckers, and W. H. M. Saris. 1995. Osmolarity does not affect the gastric emptying rate of oral rehydration solutions. *J. Parent. Ent. Nutr.* 19:403–6.

Costill, D. L., and B. Saltin. 1974. Factors limiting gastric emptying during rest and exercise. *J. Appl. Physiol.* 37:679–83.

Coyle, E. F., D. L. Costill, W. J. Fink, and D. G. Hoopes. 1978. Gastric emptying rates for selected athletic drinks. *Res. Quart.* 49:119–24.

Flaim, S. F., W. J. Minteer, D. P. Clark, and R. Zelis. 1979. Cardiovascular response to acute aquatic and treadmill exercise in the untrained rat. *J. Appl. Physiol.: Respirat. Environ. Exercise Physiol.* 46:302–8.

Fordtran, J. S., and B. Saltin. 1967. Gastric emptying and intestinal absorption during prolonged severe exercise. *J. Appl. Physiol.* 23:331–5.

Foster, C., D. L. Costill, and W. J. Fink. 1980. Gastric emptying characteristics of glucose and glucose polymer solutions. *Res. Quart. Exer. Sport* 51:299–305.

Gisolfi, C. V., K. J. Spranger, R. W. Summers, H. P. Schedl, and T. L. Bleiler. 1991. Effects of cycle exercise on intestinal absorption in humans. *J. Appl. Physiol.* 71:2518–27.

Gonzalez-Alonso, J., J. A. L. Calbet, and B. Nielsen. 1999. Metabolic and thermodynamic responses to dehydration-induced reductions in muscle blood flow in exercising humans. *J. Physiol.* 520(2):577–89.

Horowitz, M., D. H. Bar-Ilan, and S. Samueloff. 1985. Redistribution of cardiac output in anesthetized thermally dehydrated rats. *Comp. Biochem. Physiol.* 81A:103–7.

Horowitz, M., and S. Samueloff. 1988. Cardiac output distribution in thermally dehydrated rodents. *Am. J. Physiol.* 254:R109–16.

Houmard, J. A., P. C. Egan, R. A. Johns, P. D. Neufer, T. C. Chenier, and R. G. Israel. 1991. Gastric emptying during 1 h of cycling and running at 75% VO_2max. *Med. Sci. Sports Exerc.* 23:320–5.

Iwao, T., A. Toyanaga, M. Ikegami, M. Sumino, K. Oho, M. Sakaki, H. Shigemori, K. Tanikawa, and J. Iwao. 1995. Effects of exercise-induced sympathoadrenergic activation on portal blood flow. *Dig. Dis. Sci.* 40:48–51.

Jeukendrup, A. E., K. Vet-Joop, A. Sturk, J. H. Stegen, J. Senden, W. H. Saris, and A.J. Wagenmakers. 2000. Relationship between gastro-intestinal complaints and endotoxemia, cytokine release and the acute-phase reaction during and after a long-distance triathlon in highly trained men. *Clin. Sci. (Lond.)* 98:47–55.

Kamisoglu, K., K. E. Sleight, S. E. Calvano, S. M. Coyle, S. A. Corbett, and I. P. Androulakis. 2013. Temporal metabolic profiling of plasma during endotoxemia in humans. *Shock* 40:519–26.

Lambert, G. P. 2008. Intestinal barrier dysfunction, endotoxemia and gastrointestinal symptoms: The "canary in the coal mine" during exercise-heat stress? In: F. E. Marino (ed.), *Thermoregulation and Human Performance: Physiological and Biological Aspects.* Med. Sport. Sci. No. 53. Basel, Switzerland: Karger, pp. 61–73.

Lambert, G. P., T. L. Bleiler, R.-T. Chang, A. K. Johnson, and C. V. Gisolfi. 1993. Effects of carbonated and noncarbonated beverages at specific intervals during treadmill running in the heat. *Int. J. Sport Nutr.* 3:177–93.

Lambert, G. P., R. T. Chang, D. Joensen, X. Shi, R. W. Summers, H. P. Schedl, and C. V. Gisolfi. 1996. Simultaneous determination of gastric emptying and intestinal absorption during cycle exercise in humans. *Int. J. Sports Med.* 17:48–55.

Lambert, G. P., J. Lang, A. Bull, P. C. Pfeifer, J. Eckerson, G. Moore, S. Lanspa, and J. O'Brien. 2008. Fluid restriction during running increases GI permeability. *Int. J. Sports Med.* 29:194–8.

Lambert, G. P., R. Murray, D. Eddy, W. Scott, R. Laird, and C. V. Gisolfi. 1999. Intestinal permeability following the 1998 Ironman triathlon. *Med. Sci. Sports Exerc.* 31:S318.

Lambert, G. P., X. Shi, and R. Murray. 2012. The gastrointestinal system. In: P. A. Farrell, M. J. Joyner, and V. J. Caiozzo (eds.), *ACSM's Advanced Exercise Physiology.* Philadelphia, PA: Wolter Kluwer/Lippincott Williams & Wilkins, pp. 348–62.

Lang, J. A., C. V. Gisolfi, and G. P. Lambert. 2006. Effect of exercise intensity on active and passive glucose absorption. *Int. J. Sport Nutr. Exerc. Metab.* 16:485–93.

Leiper, J. B., N. Broad, and R. Maughan. 2001a. Effects of intermittent high-intensity exercise on gastric emptying in man. *Med. Sci. Sports Exerc.* 33:1270–8.

Leiper, J. B., C. W. Nicholas, A. Ali, C. Williams, and R. J. Maughan. 2005. The effect of intermittent high-intensity running on gastric emptying of fluids in man. *Med. Sci. Sports Exerc.* 37:240–7.

Leiper, J. B., A. S. Prentice, C. Wrightson, and R. J. Maughan. 2001b. Gastric emptying of a carbohydrate-electrolyte drink during a soccer match. *Med. Sci. Sports Exerc.* 33:1932–8.

Logan-Sprenger, H. M., G. J. F. Heigenhauser, G. L. Jones, and L. L. Spriet. 2013. Increase in skeletal-muscle glycogenolysis and perceived exertion with progressive dehydration during cycling in hydrated men. *Int. J. Sport Nutr. Exerc. Metab.* 23:220–9.

Manohar, M. 1986. Blood flow to the respiratory and limb muscles and to abdominal organs during maximal exertion in ponies. *J. Physiol.* 377:25–35.

Massett, M. P., D. G. Johnson, and K. C. Kregel. 1996. Cardiovascular and sympathoadrenal responses to heat stress following water deprivation in rats. *Am. J. Physiol.* 270: R652–9.

Mitchell, J. B., D. L. Costill, J. A. Houmard, W. J. Fink, R. A. Robergs, and J. A. Davis. 1989. Gastric emptying: Influence of prolonged exercise and carbohydrate concentration. *Med. Sci. Sports Exerc.* 21:269–74.

Mitchell, J. B., D. L. Costill, J. A. Houmard, M. G. Flynn, W. J. Fink, and J. D. Beltz. 1988. Effects of carbohydrate ingestion on gastric emptying and exercise performance. *Med. Sci. Sports Exerc.* 20:110–15.

Morton, D. P., L. F. Aragon-Vargas, and R. Callister. 2004. Effect of ingested fluid composition on exercise-related transient abdominal pain. *Int. J. Sport Nutr. Exerc. Metab.* 14:197–208.

Moses, F., A. Singh, B. Smoak, D. Hollander, and P. Deuster. 1991. Alterations in intestinal permeability during prolonged high intensity running. *Gastroenterology* 100:A472.

Murray, R., W. Bartoli, J. Stofan, M. Horn, and D. Eddy. 1999. A comparison of the gastric emptying characteristics of selected sports drinks. *Int. J. Sport Nutr.* 9:263–74.

Murray, R., D. E. Eddy, W. P. Bartoli, and G. L. Paul. 1994. Gastric emptying of water and isocaloric carbohydrate solutions consumed at rest. *Med. Sci. Sports Exerc.* 26:725–32.

Neufer, P. D., D. L. Costill, W. J. Fink, J. P. Kirwan, R. A. Fielding, and M. G. Flynn. 1986. Effects of exercise and carbohydrate composition on gastric emptying. *Med. Sci. Sports Exerc.* 18:658–62.

Neufer, P. D., A. J. Young, and M. N. Sawka. 1989. Gastric emptying during exercise: Effects of heat stress and hypohydration. *Eur. J. Appl. Physiol.* 58:433–9.

Notivol, R., I. Carrio, L. Cano, M. Estorch, and F. Vilardell. 1984. Gastric emptying of solid and liquid meals in healthy young subjects. *Scand J. Gastroenterol.* 19:1107–13.

Oktedalen, O., O. C. Lunde, P. K. Opstad, L. Aabakken, and K. Kvernebo. 1992. Changes in the gastrointestinal mucosa after long distance running. *Scand. J. Gastroenterol.* 27:270–4.

Owen, M. D., K. C. Kregel, P. T. Wall, and C. V. Gisolfi. 1986. Effects of ingesting carbohydrate beverages during exercise in the heat. *Med. Sci. Sports Exerc.* 18:568–75.

Pals, K. L., R. T. Chang, A. J. Ryan, and C. V. Gisolfi. 1997. Effect of running intensity on intestinal permeability. *J. Appl. Physiol.* 82:571–6.

Perko, M. J., H. B. Nielsen, C. Skak, J. O. Clemmesen, T. V. Schroeder, and N. H. Secher. 1998. Mesenteric, coeliac, splanchnic blood flow in humans during exercise. *J. Physiol.* 513(3):907–13.

Peters, H. P. F., M. Bos, L. Seebregts, L. M. A. Akkermans, G. P. van Berge Henegouwen, E. Bol, W. L. Mosterd et al. 1999. Gastrointestinal symptoms in long-distance runners, cyclists, and triathletes: Prevalence, medication, and etiology. *Am. J. Gastroenterol.* 94:1570–81.

Puvi-Rajasingham, S., B. Wijeyekoon, P. Natarajan, and C. J. Mathias. 1997. Systemic and regional (including superior mesenteric) haemodynamic responses during supine exercise while fasted and fed in normal man. *Clin. Autonomic Res.* 7:149–54.

Qamar, M. I., and A. E. Read. 1987. Effects of exercise on mesenteric blood flow in man. *Gut* 28:583–7.

Rehrer, N. J., E. J. Beckers, F. Brouns, F. TenHoor, and W. H. M. Saris. 1990a. Effects of dehydration on gastric emptying and gastrointestinal distress while running. *Med. Sci. Sports Exerc.* 22:790–5.

Rehrer, N. J., F. Brouns, E. J. Beckers, F. TenHoor, and W. H. M. Saris. 1990b. Gastric emptying with repeated drinking during running and bicycling. *Int. J. Sports Med.* 11:238–43.

Rehrer, N. J., E. Goes, C. DuGardeyn, H. Reynaert, and K. DeMeirleir. 2005. Effect of carbohydrate on portal vein blood flow during exercise. *Int. J. Sports Med.* 26:171–6.

Rehrer, N. J., A. Smets, H. Reynaert, E. Goes, and K. DeMeirleir. 2001. Effect of exercise on portal vein blood flow in man. *Med. Sci. Sports Exerc.* 33:1533–7.

Rogers, J., R. W. Summers, and G. P. Lambert. 2005. Gastric emptying and intestinal absorption of a low-carbohydrate sport drink during exercise. *Int. J. Sport Nutr. Exerc. Metab.* 15:220–35.

Rowell, L. B., J. R. Blackmon, and R. A. Bruce. 1964. Indocyanine green clearance and estimated hepatic blood flow during mild to maximal exercise in upright man. *J. Clin. Invest.* 43:1677–90.

Rowell, L. B., J. R. Blackmon, R. H. Martin, J. A. Mazzarella, and R. A. Bruce. 1965. Hepatic clearance of indocyanine green in man under thermal and exercise stresses. *J. Appl. Physiol.* 20:384–94.

Ryan, A. J. et al. 1998. Effect of hypohydration on gastric emptying and intestinal absorption during exercise. *J. Appl. Physiol.* 84:1581–8.

Ryan, A. J., T. L. Bleiler, J. E. Carter, and C. V. Gisolfi. 1989. Gastric emptying during prolonged cycling in the heat. *Med. Sci. Sports Exerc.* 21:51–8.

Sawka, M. N., L. M. Burke, E. R. Eichner, R. J. Maughan, S. J. Montain, and N. S. Stachenfeld. 2007. American College of Sports Medicine Position Stand. Exercise and fluid replacement. *Med. Sci. Sports Exerc.* 39:377–90.

Sawka, M. N., and E. F. Coyle. 1999. Influence of body water and blood volume on thermoregulation and exercise performance in the heat. In: J. O. Holloszy (ed.), *Exercise and Sport Sciences Reviews* No. 27. Philadelphia, PA: Lippincott Williams & Wilkins, pp. 167–218.

Shi, X., W. Bartoli, M. Horn, and R. Murray. 2000. Gastric emptying of cold beverages in humans: Effect of transportable carbohydrates. *Int. J. Sport Nutr.* 10:394–403.

Shi, X., M. K. Horn, K. L. Osterberg, J. R. Stofan, J. J. Zachwieja, C. A. Horswill, D. H. Passe, and R. Murray. 2004. Gastrointestinal discomfort during intermittent high-intensity exercise: Effect of carbohydrate-electrolyte beverage. *Int. J. Sport Nutr. Exerc. Metab.* 14:673–83.

Smetanka, R. D., G. P. Lambert, R. Murray, D. Eddy, M. Horn, and C. V. Gisolfi. 1999. Intestinal permeability in runners in the 1996 Chicago Marathon. *Int. J. Sports Nutr.* 9:426–33.

van Nieuwenhoven, M. A., F. Brouns, and R. J. Brummer. 1999. The effect of physical exercise on parameters of gastrointestinal function. *Neurogastroenterol. Motil.* 11:431–9.

van Nieuwenhoven, M. A., F. Brouns, and E. M. Kovacs. 2005. The effect of two sports drinks and water on GI complaints and performance during an 18 km run. *Int. J. Sports Med.* 26:281–5.

van Nieuwenhoven, M. A., B. E. P. J. Vriens, R.-J. M. Brummer, and F. Brouns. 2000. Effect of dehydration on gastrointestinal function at rest and during exercise in humans. *Eur. J. Appl. Physiol.* 83:578–84.

Vist, G. E., and R. J. Maughan. 1994. Gastric emptying of ingested solutions in man: Effect of beverage glucose concentration. *Med. Sci. Sports Exerc.* 26:1269–73.

Vist, G. E., and R. J. Maughan. 1995. The effect of osmolality and carbohydrate content on the rate of gastric emptying of liquids in man. *J. Physiol.* 486:523–31.

Wade, O. L., B. Combes, A. W. Childs, H. O. Wheeler, A. Cournand, and S. E. Bradley. 1956. The effect of exercise on the splanchnic blood flow and splanchnic blood volume in normal man. *Clin. Sci.* 15:457–63.

Williams, J. H., M. Mager, and E. D. Jacobson. 1964. Relationship of mesenteric blood flow to intestinal absorption of carbohydrates. *J. Lab. Clin. Med.* 63:853–63.

Zachwieja, J. J., D. L. Costill, G. C. Beard, R. A. Robergs, D. D. Pascoe, and D. E. Anderson. 1992. The effects of a carbonated carbohydrate drink on gastric emptying, gastrointestinal distress, and exercise performance. *Int. J. Sport Nutr.* 2:239–50.

7 Role of Fluid Intake in Endurance Sports

Louise M. Burke

CONTENTS

7.1 INTRODUCTION

There is irony in the observation that endurance sports have received the most attention from sports science and nutrition experts in terms of the development of guidelines for fluid intake, yet they remain the focus of controversy and continued discussion. For the purposes of this chapter, endurance sports are defined as continuous activities of greater than 90 min duration that can potentially span into ultra-endurance events (>4 h) and multiday competitions. The mode of locomotion can include running, cycling, swimming, skiing, or padding/rowing, and sometimes combinations of several of these within the same event, and the field of play is usually outdoors with exposure to the prevailing environmental conditions. In line with the duration of these sports, the majority of the work-rate is undertaken at submaximal intensities. However, it should be remembered that in many events, critical pieces that shape the outcome of the event involve high-intensity exercise: for example, the breakaway, the surge up a hill, and the sprint to the finish line. Furthermore, in some events there is an element of skill and technique that requires good coordination of central and peripheral function.

As is the case with other sporting activities, the goal of drinking fluids is to prevent the fluid deficit that accrues due to sweat losses from reaching avoidably problematic level. However, other considerations for consuming fluids during endurance sporting event include the ingestion of common drink ingredients known to enhance performance such as carbohydrate, electrolytes and caffeine, as well as the contribution of cool or icy fluids to comfort and thermoregulation during exercise. In multiday endurance events, fluids consumed during the activity may contribute a substantial proportion of the energy, protein, and micronutrients needed for general health, as well as specific guidelines for optimal performance and recovery. A common element to endurance sports is that fluid intake during an event is primarily achieved *on the run*—or while the athlete is actually exercising. This feature means that the *cost* of achieving the activity must be balanced against the *benefits* of its effects. These costs include the effort to have drinks available to consume, the time lost in slowing down to obtain and consume them, and the risk of incurring gastrointestinal discomfort or upsets as a result.

Guidelines for hydration practices in endurance sports have evolved from prescriptive recommendations to consume a certain fluid volume during exercise to the adoption of a practiced and individualized plan that can partially replace sweat losses as well as provide other ingredients or characteristics previously mentioned. This advice is not universally embraced; however, alternative views are that thirst should dictate the need for, and volume of, fluid replacement during exercise and that *ad libitum* drinking is sufficient to address fluid needs during sport. This chapter will review the background to the debate over fluid needs for endurance events, covering the evolution of the guidelines and the points of differences in current views. In proposing recommendations to find some common and practical middle-ground, the literature on self-chosen fluid intakes of competitors in a variety of endurance sports will be presented. While such data do not necessarily always represent optimal practice, at least they show what is logistically possible and valued by athletes under real-life competition conditions (Garth and Burke 2013).

The nutritional strategies that can be of benefit to endurance sports involve the integration of a number of elements including the intake of nutrients and evidence-based ergogenic aids, and the manipulation of stomach comfort and temperature regulation. The science behind these strategies and practical ways to achieve them are covered in a number of sections and chapters in this book. The principle focus of the current chapter is to consider how they can be incorporated into a plan in which the intake of fluid is the dominant or binding factor. This plan must be underpinned by an understanding of both the specific issues around hydration and the practical issues of achieving this within endurance sports.

7.2　EVOLUTION OF GUIDELINES FOR FLUID INTAKE DURING ENDURANCE SPORTS

Since the 1970s, expert groups have issued guidelines for fluid intake during sport and exercise. These guidelines have evolved over the years due to our increasing knowledge as well as some justifiable criticisms of previous ideas and education messages. The first position stands on fluids for exercise were focused on distance running and were mostly targeted to race organizers and medical support teams (American College of Sports Medicine [ACSM] 1975, 1987). They were prepared in response both to the increased community participation in marathons and fun runs with the running boom of the 1970s and a desire to update the prevailing rules of the IAAF (the international governing body of athletics) that no aid stations should be provided prior to the 15 km mark of a distance race.

Although these guidelines recommended more access to aid stations throughout races, they proved to be problematic in several ways: First, they continued to prescribe water as the only fluid that should be available (ACSM 1975, 1987), failing to recognize the accumulating evidence of the benefits of consuming carbohydrate during exercise of >1 h duration (see Chapter 19). Importantly, the recommendations were prescriptive and impractical to translate into the varying characteristics of distance running, let alone other endurance sports. Although many athletes like the simplicity of prescriptive advice, even apparently simple guidelines often fail the test of scrutiny. For example, suggestion that "distance runners should drink 100–200 ml of water at aid stations provided every 2–3 km in a race" (ACSM 1987) sounds reasonable. However, if taken literally, it could span the intake of 330 ml/h by a slow runner to 2 L/h by a fast runner. This would range from inadequate to suitable for slower runners depending on their actual sweat rates, but the high rate would be impossible to achieve when running at high speeds and any such attempt would be likely to cause gastrointestinal discomfort and substantial time loss at aid stations.

The updated ACSM guidelines published in 1996 increased the general scope of interest to fluid intake during exercise and recognized the benefits of including carbohydrate and electrolytes in beverages for endurance sports. However, they continued to promote a formulaic intake of fluid with the goal of minimizing the mismatch between fluid intake and sweat losses: "During exercise, athletes should start drinking early and at regular intervals in an attempt to consume fluids at a rate sufficient to replace all the water lost through sweating (i.e., body weight loss), or consume the maximal amount that can be tolerated" (ACSM et al. 2007). Concern that these guidelines could be interpreted out of context as "consume the maximal amount of fluid possible" and contribute to observations of hyponatremia/water intoxication caused by the excessive intake of fluids by some (usually recreational) participants in endurance/ultra-endurance sports activities (Noakes et al. 2005; Noakes and Speedy 2006) is an important factor in later guidelines.

The most recent guidelines by groups such as the ACSM (ACSM et al. 2007), the National Athletic Trainers Association (Casa 2007), and the International Olympic Committee (Shirreffs and Sawka 2011) have tried to accommodate the specificity of needs and opportunities to hydrate during exercise across a range of sports, as well as the unique needs of each participant. They recommend that each athlete develop an individualized fluid plan based on an appreciation of his or her likely sweat rates and knowledge of opportunities to drink during the exercise session. They typically maintain the goal of defending, where possible, a *gold standard* of hydration, suggested as loss of <2% body mass (BM) over the event, but warn against overdrinking as shown by a gain in BM. However, even these recommendations have been subjected to the (often vitriolic) criticism of being unnecessary, complicated, and driven by commercial interests (Beltrami et al. 2008; Noakes 2012). Instead, it has been argued that athlete should simply drink according to their thirst (Noakes 2007).

7.3 ISSUES AND CONTROVERSIES

There are two different considerations involved in setting guidelines for fluid intake during endurance sport. The first consideration is the level of hydration that is needed to achieve optimal performance (or from the obverse point of view, what is the largest level of fluid deficit that can be tolerated before performance is impaired?).

Second, the guidelines need to consider how the athlete can be educated to achieve this hydration level/avoid a fluid deficit larger than this. Both of these considerations have been the focus of challenge and discussion over the recent decade.

Optimal performance in endurance sports requires the integration of the muscle's ability to produce the power needed to achieve optimal movement patterns and the central nervous system's role in choosing appropriate pacing strategies and execute optimal technique, decision making, and skill. Historically, it has been considered that impairments of *aerobic* performance undertaken in a hot environment can be detected when there is a fluid deficit of ~2% or greater, with the magnitude of effects increasing as the degree of dehydration increases. Furthermore, the effects of dehydration are less obvious when aerobic exercise is undertaken in a cool environment and it may take a fluid deficit of 3%–5% before they are detectable (Sawka 1992; Cheuvront et al. 2003, 2010; Sawka and Noakes 2007; Cheuvront and Kenefick 2014). However, in recent times, there have been debates about whether these concepts are correct, with arguments in both directions that they are too simplistic to capture the real essence of the effects of dehydration on sports performance.

The case that endurance athletes can tolerate larger fluid deficits without problem includes the following ideas:

- The evidence that dehydration impairs endurance performance is mostly based on studies conducted in laboratory sessions which don't adequately replicate real-life conditions of sport, including factors such as the cooling effect of the air movement from outdoors (Saunders et al. 2005) or the effect of motivation and reward for successful outcomes.
- Some reviews of laboratory-based studies have concluded that dehydration of <4% is unlikely to impair the performance of time-trial type protocols (Goulet 2013).
- Studies and anecdotal observations show that successful athletes can tolerate high levels of dehydration (Beis et al. 2012), with the winners in a race often being the most dehydrated.
- Competitive athletes learn to accommodate the physiological decrements associated with laboratory-based observations of dehydration.
- Successful performance does not always require optimal outputs; rather, the athlete only needs to be better than his or her competitors.

There are counter-arguments to these suggestions, including the following:

- Laboratory studies of graded levels of dehydration show that the effects on a range of physiological factors and perceived effort of exertion increase in a continuum (Montain and Coyle 1992); it is also likely that the effect on performance gradually increases until the point when, according to the sensitivity of the protocol, it becomes measurable. Failure to see a performance impairment in a study may represent a Type II error due to underpowering of sample size or poor reliability of monitoring performance.
- Some field-based studies have reported that mild fluid deficits of even 2%–3% impair performance in the heat (Casa et al. 2010; Bardis et al. 2013).

- In many outdoor sports, athletes deliberately reduce the potential for convective cooling by adopting aerodynamic positions (riding/running in slipstream of other competitors, choosing special equipment and clothing).
- The margins of winning and losing in sport are typically very small, making it unlikely that we can measure all the factors that make meaningful changes in the outcomes.
- The effort required to overcome the physiological and psychological decrements associated with dehydration may manifest in ways other than a measurable drop in work-rates: for example, loss of concentration during the event and greater difficulty with recovery between events.
- Not all studies report that the race winners are the most dehydrated. Some studies show no correlation between loss of mass over event and performance (Speedy et al. 1999) and others show an inverse relationship with most successful athletes being the best hydrated (Ross et al. 2014). In any case, observations cannot tell cause or effect, or how an athlete would have performed with a different fluid intake strategy.

In summing up these views, this author feels that endurance athletes should make a calculated risk about how much of their relative or absolute performance is needed on the specific day of any race after undertaking a cost-benefit analysis of their potential opportunities for fluid intake in the event. In other words, they should counterbalance the time loss or risk of gut discomfort associated with drinking at the available opportunities with the impairment of perception of effort and performance associated with a fluid deficit. The outcome of this analysis is likely to vary according to factors such as the individual athlete, the environmental conditions, the logistics of drinking, and the type of event. This will be discussed further in the conclusion of this chapter.

The second topic of recent debate has been the need for guidelines for fluid intake. Although the expert position stands have moved away from prescriptive guidelines to favor the development of an individualized fluid intake plan, some sports scientists have heavily criticized even this approach. Instead, they argue that humans can optimize their performance and health during a sporting event by drinking according to behavior that is innate and spontaneous rather than planned. They advocate that hydration can be managed by *drinking to thirst* or *ad libitum fluid intake* (Noakes 2010). The hypothesis that drinking to thirst is the best way to approach education around fluid intake is based on several assumptions (Noakes 2010) given as follows:

- That the human thirst response is highly developed in guiding athletes to the need to drink in the short-term and long-term and that the athlete can respond to thirst by drinking appropriate volumes of fluid
- That humans have evolved to tolerate a certain level of thirst and fluid deficit in the short-term without impairment of health or performance
- That this behavior will err on the side of allowing a fluid deficit rather than overload, thus preventing the development of hyponatremia in individuals and groups

Concern about over-hydration during exercise has been at the forefront of the revisions and concerns over guidelines for fluid intake during sport. This is rare, at least historically, in non-endurance sports where athletes exercise at high intensities with commensurately high sweat rates. However, observational studies of endurance and ultra-endurance sports over the past two decades have found that some individuals are overzealous with their fluid intake and drink at a rate that substantially exceeds their sweat losses and their ability to excrete fluid via urine. Excessive fluid intake is a major contributor to the potentially fatal condition of hyponatremia, although other factors include the retention of excess fluid because of inadequate suppression of antidiuretic hormone secretion (Noakes et al. 2005; Siegel et al. 2007). Risk factors for over-hydrating include undertaking endurance and ultra-endurance events at a slow pace (lower sweat rates and greater likelihood of stopping to drink at aid stations) and being female (smaller body size with lower sweat rates) (Almond et al. 2005). Mathematical modeling also shows that mild levels of hyponatremia may also occur as a result of large salt losses in individuals who excrete sweat that is salty or simply high in volume (Montain et al. 2006). However, overdrinking underpins the development of severe hyponatremia and the sequelae of encephalopathies that have caused several unfortunate and preventable deaths among athletes and military personnel. As a result, new fluid intake guidelines are clear in their warnings against overdrinking during exercise (ACSM et al. 2007; Casa 2007; Shirreffs and Sawka 2011).

Although this author believes that thirst provides a reasonable starting point or contribution to the development of fluid plan for endurance sports, there are some problems in making this a universal or single approach. First, it is unclear what *drinking to thirst* really means, since it can be interpreted in different ways which would require different behavior and lead to different outcomes in terms of fluid intake and fluid balance. For example, interpretations include (1) drink every time you are thirsty; (2) when presented with fluids, only drink if you are thirsty; or (3) drink during the opportunities in sport so that thirst doesn't develop. Although some people use *drinking to thirst* interchangeably with the term *ad libitum fluid intake*, this is not strictly true since the latter implies that fluid intake can be undertaken whenever and in whatever volumes the athlete desires.

In addition to the potential confusion around what the term actually means, there are several types of athletes or situations, however, where there may be benefits from building a more calculated approach than *drinking if you are thirsty*. It is noted that the following conditions apply to many endurance and ultra-endurance sports:

- In events where opportunities for fluid intake are limited, the athlete may need to drink at the available opportunities early in an event (i.e., *ahead of their thirst*) to better pace the total fluid intake over the session.
- In events where performance can benefit from the regular intake of other nutrients that can be delivered in fluids—in particular, carbohydrate. As long as it doesn't require an excessive fluid intake, athletes may develop a plan to consume beverages such as sports drinks to contribute to refueling targets.
- For athletes who are unable to respond to thirst or hunger due to personal characteristics or the confusion about appropriate volumes to consume due to exposure to a food environment with continually upsized food portions.

In these situations, there may be benefits from developing a personalized plan that can be manipulated according to circumstances or characteristics of each event.

7.4 WHAT DO ENDURANCE ATHLETES CURRENTLY DRINK?

A range of factors that influence the intake of fluids during competitive endurance events; these include thirst, genetic predisposition to be avid or reluctant drinker, beliefs regarding dehydration or over-hydration, access/availability of fluids, opportunity to drink, palatability of drink, gastrointestinal comfort, wish to avoid the need to urinate, desire to reduce BM over the race, and the desire for other ingredients (e.g., carbohydrate and caffeine) or characteristics (e.g., temperature) of fluid. The available literature on observed practices of athletes in endurance competitions is now summarized (Tables 7.1 through 7.3), to discuss the apparent importance of these factors.

7.4.1 SINGLE-DAY ENDURANCE EVENTS (EVENTS LASTING <~3 H FOR TOP COMPETITORS)

An unusual characteristic of many endurance events (marathons and Olympic distance triathlons) is that elite and recreational athletes often compete in the same race, meaning that the event will include participants with a large range of finishing times. As in all sports, the rates and total volume of sweat loss vary with the intensity and duration of the event, with potential for large differences between athletes, even those in the same race. The outdoor setting increases the potential for large differences in sweat losses between events of the same type, according to specific characteristics such as the event terrain and environmental conditions (heat, humidity, altitude, and wind).

We have previously identified a range of features that influence fluid intake during competitive endurance events (Garth and Burke 2013). A key characteristic of the hydration opportunities in endurance events is that athletes must drink while *on the move*. Access to fluids during the majority of events is typically governed by a network of drink stations/feed zones, although this can be supplemented or replaced in other endurance sports by the transport of fluids by the individual athlete. Elite athletes are often able to provide themselves their own specific race supplies at aid stations, while in mass participation events, the provisions at feed stations available to general competitors are governed by the race organizer. Opportunities to drink must consider the time lost in obtaining and consuming fluid, and the potential for gut discomfort due to drinking while exercising at relatively high intensities. Practicing drinking during event-simulating training sessions may facilitate the development of appropriate skills and gut tolerance in some athletes. Devices such as fluid containing backpacks and spill-proof bottles may also enhance access to fluid and opportunities to drink during some endurance sports. However, in other sports, technique requirements such as bike handling during downhill mountain bike riding or maintaining an aerodynamic position during road cycling time trials may interfere with opportunities to obtain or consume fluids. Similarly, pacing strategies and race tactics may interfere with the athlete's opportunities to drink, since they may choose to surge past the aid stations. Some endurance athletes may deliberately or subconsciously restrict fluid intake

TABLE 7.1
Fluid Balance Characteristics of Single-Day Endurance Events

Study	Subjects	Event	Duration (min)[a]	Environment (°C, %)	Sweat Rate (L/h)[a]	Δ Body Mass (%)[a]	Fluid Intake (L/h)[a]
Beis et al. (2012)	10 M Elite	13 Olympic and Big City Marathons	126 ± 1	Air: 0–30 Humidity: 39–89	N/A	N/A	0.55 ± 0.34[c]
Kipps et al. (2011)	53 M, 35 F Mixed caliber	London Marathon	252 ± 43 (Nor) 266 ± 48 (EAH)	Air: 9–12 Humidity: 73 Raining	NR	N/A	0.45 (Nor) 0.84 (EAH)
Tam et al. (2011)	12 M, 9 F Mixed caliber	Two Oceans Half Marathon, South Africa	129 ± 24	Air: 18–24 Humidity: 50–70	NR	−1.9 (−1.4 ± 0.6 kg)	0.33 ± 0.18
Zouhal et al. (2011)	560 M, 83 F Mixed caliber	Mont Saint-Michel Marathon France	NR	Air: 9–16 Humidity: 60–80	NR	−2.3 (All) [range −8 to +5] −3.1 (finish <3 h) −2.5 (finish 3–4 h) −1.8 (finish >4 h)	N/A
van Rooyen et al. (2010)	4 M, 5 F Elite	Athens Olympic Marathon	NR	Air: 30–33 Humidity: 31–39	N/A	N/A	Range: 0.43–1.30[c] (F) 0.30–0.35[c] (M)
Au-Yeung et al. (2010)	240 M, 32 F Mixed caliber	Hong Kong Marathon	255	Air: 12–19 Humidity: 59–88 Raining	NR	N/A	0.40

(Continued)

TABLE 7.1 (Continued)
Fluid Balance Characteristics of Single-Day Endurance Events

Study	Subjects	Event	Duration (min)[a]	Environment (°C, %)	Sweat Rate (L/h)[a]	Δ Body Mass (%)[a]	Fluid Intake (L/h)[a]
Mettler et al. (2008)	128 M, 39 F Mixed caliber	Zurich Marathon	220 ± 32 (M) 245 ± 23 (F)	Air: ~10 Humidity: NR Raining	NR	−0.8 ± 0.8 (M) −0.2 ± 0.8 (F)	0.47 (M) 0.36 (F)
Hew (2005)	63 M, 54 F Mixed caliber	Houston Marathon	269 ± 45 (M) 303 ± 54 (F)	NR	NR	−2.1 (M) (−1.7 ± 1.8 kg) −1.0 (F) (−0.6 ± 1.1 kg)	0.74 (M) 0.68 (F)
Myhre et al. (1982)	3 M Mixed caliber	Marathon Southern United States	216	Air: 15.5–24.5 Humidity: NR Raining	1.24 (1.06–1.17)	−4.7 (range: −3.4 to −6.7%)	1.33 (range: 0.65–1.90)

Source: Garth, A.K., and Burke, L.M., *Sports Med.*, 43(7), 539–64, 2013.

M, male; F, female; N/A, data excluded from the table due to use of inappropriate methodology; NR, not reported.

[a] Data are reported as mean ± SD (if provided) unless otherwise stated.

TABLE 7.2

Fluid Balance Characteristics of Single-Day Ultra-Endurance Events

Study	Subjects	Event	Duration (min)[a]	Environment (°C, %)	Sweat Rate (L/h)[a]	Δ Body Mass (%)[a]	Fluid Intake (L/h)[a]
			Multisport				
Schwellnus et al. (2011)	209 M + F Mixed caliber	South Africa IM	759 ± 96 (CR) 795 ± 93 (NC)	Air: 20 Humidity: 70	NR	−3.1 ± 1.9 (CR) −2.8 ± 1.8 (NC)	NR
Pahnke et al. (2010)	26 M, 20 F Mixed caliber	Hawaii IM	NR	Air: 27.6 Humidity: NR	Race day data NR	−2.1 ± 2.1	1.00 ± 0.30 (All) 0.85 ± 0.30 (M) 1.05 ± 0.30 (F)
Laursen et al. (2006)	10 M Mixed caliber	Busselton IM	611 ± 49	Air: 23.3 ± 1.9; Water: 19.5; Humidity: 60	NR	−3.0 ± 1.5	NR
Sulzer et al. (2005)	20 M + F Mixed caliber	South Africa IM	661 ± 78 (CR) 685 ± 49 (NC)	Air: 20.5; Water: 16 Humidity: 68	NR	−3.4 ± 1.3 (CR) −3.9 ± 2.0 (NC)	NR
Sharwood et al. (2002)	311 M, 45 F Mixed caliber	South Africa IM	757 ± 100	Air: 20.5; Water: 16 Humidity: 68	NR	−5.2 ± 2.2	NR
Speedy et al. (2002) (*data = median*)	11 M, 7 F Mixed caliber	New Zealand IM	738	Air: 21; Water: 20.7 Humidity: 91	0.81 (bike) 1.02 (run)	−3.5 [−6.1 to +2.5%] −1.0 kg (swim): +0.5 kg (bike); −2.0 kg (run)	0.72 0.89 (bike) 0.63 (run)
Speedy et al. (1999)	292 M, 38 F Mixed caliber	New Zealand IM	734	Air: 21; Water: 20.7 Humidity: 91	NR	−4.3 ± 2.3 (M) −2.7 ± 3.1 (F)	NR
O'Toole et al. (1995)	26 M, 4 F Mixed caliber	Hawaii IM	711 ± 105	Air: 22–31; Water: 26 Humidity: 40–85	NR	−2.6 (Nor) −0.6 (EAH)	NR

(Continued)

TABLE 7.2 (Continued)
Fluid Balance Characteristics of Single-Day Ultra-Endurance Events

Study	Subjects	Event	Duration (min)[a]	Environment (°C, %)	Sweat Rate (L/h)[a]	Δ Body Mass (%)[a]	Fluid Intake (L/h)[a]
Speedy et al. (1997)	46 M + 2 F Mixed caliber	Coast to Coast New Zealand	879 ± 83	Air: 7.5–19.6 Humidity: 56–94	NR	−3.1 ± 2.1	NR
Rogers et al. (1997)	13 M Mixed caliber	South Africa IM Ultra-triathlon	620 ± 64	Air: 28.0 ± 4.9 Humidity: 48	0.94 ± 0.16	−4.6 ± 1.8	0.74 ± 0.14
van Rensberg et al. (1986)	23 M Mixed caliber	Rand Daily Mail-Nutri-Sport Triathlon	687	Air: 24.7–33.8 Humidity: NR	NR	−4.5	NR
Stuempfle et al. (2003)	17 M, 3 F Mixed caliber	Susitna 100 mile alpine multisport (run, cycle, or ski) Alaska	2292	Air: −14 to −2 Humidity: NR Snow	NR	−1.6	0.30
		Ultra-Running					
Bracher et al. (2012)	50 M Mixed caliber	"100 km Lauf Biel" Switzerland Start time 2200	All finished within 735	Air: 15.6–21.7 Humidity: 52–69	NR	−2.5	0.58
Tam et al. (2011)	9 M, 3 F Mixed caliber	Two Oceans 56 km South Africa	340 ± 64	Air: 18–24; Humidity: 50–70	NR	−3.5 (−2.5 ± 1.1 kg)	0.54 ± 0.36
Knechtle et al. (2011b)	27 M Mixed caliber	"100 km Lauf Biel" Switzerland	689 ± 119	Air: 8–18; Humidity: NR; Start time 2200	NR	−2.6 (−1.9 ± 1.4 kg)	0.52 ± 0.18
Knechtle et al. (2011c)	145 M Mixed caliber	"100 km Lauf Biel" Switzerland	640 ± 74 (EAH) 710 ± 120 (Nor)	Air: 8–28; Humidity: NR; Start time 2200	NR	−2.4 ± 1.8[c] (All) −2.6 (EAH) −2.4 (Nor)	0.58 ± 0.23 (EAH) 0.65 ± 0.30 (Nor)
Knechtle et al. (2010)	11 F Mixed caliber	"100 km Lauf Biel" Switzerland	All finished within 762 ± 91	Air: 8–18; Humidity: NR; Start time 2200	NR	−2.4 (−1.5 ± 1.1 kg)	0.30 ± 0.10

(Continued)

TABLE 7.2 (Continued)
Fluid Balance Characteristics of Single-Day Ultra-Endurance Events

Study	Subjects	Event	Duration (min)[a]	Environment (°C, %)	Sweat Rate (L/h)[a]	Δ Body Mass (%)[a]	Fluid Intake (L/h)[a]
Lebus et al. (2010)	35 M, 10 F Mixed caliber	Rio Del Lago 100 mile California	1547 ± 190	Air: 12.2–37.6 Humidity: NR	NR	−2.9[c]	NR
Kao et al. (2008)	17 M, 1 F Mixed caliber	Soochow University International 12 h	720 (89.7 ± 11.7 km)	Air: 11.5–14.6 Humidity: 55–60	NR	−2.9 ± 1.6	NR
Kao et al. (2008)	19 M, 4 F Mixed caliber	Soochow University International 24 h	1440 (199.4 ± 37 km)	Air: 11.5–14.6 Humidity: 55–60	NR	−5.1 ± 2.3	NR
Kruseman et al. (2005)	39 M, 3 F Mixed caliber	44 km Mountain Marathon Switzerland	423 ± 77	Air: 18–30 Humidity: 34–92	NR	−4.0 (−2.9 ± 1.1 kg)	0.55 ± 0.16
Glace et al. (2002)	13 M + F Mixed caliber	160 km trail run Start time 0430	1572 ± 216	Air: 21–38 Humidity: NR	NR	−0.5 (−0.5 ± 1.5 kg)	0.74
Fallon et al. (1998)	7 M Mixed caliber	100 km road run	629 ± 113	Air: 2–17 Humidity: 45	0.86 ± 0.15	−3.3 ± 1.1	0.54 ± 0.21
Rehrer et al. (1992)	158 M, 12 F Mixed caliber	Swiss Alpine Marathon (67 km)	498 (M) 536 (F)	Air: 7–11 Humidity: 64–72		−3.3 (M) −4% (F)	0.40 (M) 0.31 (F)
Cycling							
Armstrong et al. (2012)	42 M, 6 F Mixed caliber	164 km cycle event USA	546 ± 72 (M) 540 ± 12 (F)	Air: 34.5 ± 5.0 Humidity: 53	1.13 (n = 20 M)	N/A	0.65 (M) 0.52 (F)
Hew-Butler et al. (2010)	26 M, 7 F Mixed caliber	109 km cycle race South Africa	296	Air: 24.9 Humidity: 50	NR	−1.5	0.44
Knechtle et al. (2009)	37 M Mixed caliber	Swiss MTB Bike Masters 120 km	540 ± 80	Air: 11 (at start) Humidity: NR	NR	−1.9 ± 1.6	0.7 ± 0.2

(Continued)

TABLE 7.2 (*Continued*)
Fluid Balance Characteristics of Single-Day Ultra-Endurance Events

Study	Subjects	Event	Duration (min)[a]	Environment (°C, %)	Sweat Rate (L/h)[a]	Δ Body Mass (%)[a]	Fluid Intake (L/h)[a]
				Swimming			
Wagner et al. (2012)	25 M, 11 F Mixed caliber	26.4 km swim Switzerland	528 (M) 599 (F)	Air: 18.28; Humidity: 42–93; Water: 23–24	NR	−0.5 ± 1.1 (M) −0.1 ± 1.6 (F)	0.56 ± 0.22 (M) 0.44 ± 0.17 (F)

Source: Garth, A.K., and Burke, L.M., *Sports Med.*, 43(7), 539–64, 2013.

M, male; F, female; IM, Ironman triathlon; N/A, data excluded from the table due to use of inappropriate methodology; NR, not reported; Nor, normotremic; EAH, exercise associated hyponatremia; CR, cramps reported; NC, no cramps.

[a] Data are reported as mean ± SD (if provided) unless otherwise stated.

TABLE 7.3

Fluid Balance Characteristics of Multiday Stage Events

Study	Subjects	Event	Environment (°C, %)	Sweat Rate (L/h)[a]	Δ Body Mass (%)[a]	Fluid Intake (L/h)[a]
Ross et al. (2012)	5 M: Elite Australian National Road Series team	Tour of Gippsland cycling stage race (9 stages in 5 days)	Air: 15.8 ± 1.4 Humidity: 54 ± 12	1.1 ± 0.3	−1.5 ± 0.3 (road) −1.1 ± 0.2 (crit)	0.41 ± 0.19 (road) 0.24 ± 0.19 (crit)
Ross et al. (2012)	5 M: Elite Australian National Road Series team	Tour of Geelong cycling (6 stages over 5 days)	Air: 13.2 ± 2.1 Humidity: 80 ± 8			0.56 ± 0.14 (road) 0.27 ± 0.21 (crit)
Rust et al. (2012)	65 M: Mixed caliber	Swiss Cycling Marathon (720 km in ~3 days)	Air: 9–25 Humidity: NR	NR	−1.5 ± 1.7[c]	0.67 ± 0.23
Ebert et al. (2007)	8 M: Elite Professional team	Tour Down Under cycling stage race (719 km in 6 days)	Air: 20.2–32.9 Humidity: 14–69	1.60 ± 0.10	−2.8	1.00 ± 0.10
Ebert et al. (2007)	6 F: Elite Australian national squad	Tour De L'Aude (788 km in 10 days)	Air: 7.7–27.8 Humidity: 29–76	0.90	−2.6	0.40 ± 0.06
Garcia-Roves et al. (1998)	10 M: Elite Professional team	3 × 24 h periods during the 3-week Tour of Spain	NR	NR	NR	1.26 ± 0.55L[b] =1.03 L/h W =0.23 L/h SD
Rose et al. (2010)	18 M: Mixed caliber	Sani2C MTB race (248 km over 3 stages)	Air: 9–22 Humidity: 43–100 Rain stage: 1	NR	−1.4 (Stage 1) −2.0 (Stage 2) −1.0 (Stage 3)	0.34 (Stage 1) 0.41 (Stage 2) 0.55 (Stage 3)
Schenk et al. (2010)	25 M: Mixed caliber	Transalp MTB (665 km in 8 stages)	Air: 4–32 Humidity: NR Rain stage: 2, 3, 8	NR	−0.17 to −1.44	0.49 to 0.75[b] (range)

(Continued)

TABLE 7.3 (Continued)
Fluid Balance Characteristics of Multiday Stage Events

Study	Subjects	Event	Environment (°C, %)	Sweat Rate (L/h)[a]	Δ Body Mass (%)[a]	Fluid Intake (L/h)[a]
Singh et al. (2012)	5 M, 7 F: Mixed caliber	Three Cranes Challenge (95 km trail run over 3 stages)	Air: 11.5–22.8 Humidity: 54–97	NR	−3.1[b] (−2.06 ± 0.57 kg)	NR
Knechtle et al. (2011a)	25 M: Mixed caliber	Swiss Jura Marathon (350 km in 7 stages)	Air: Not stated Humidity: NR	NR	−1.4 ± 2.0[c]	0.54–0.75[c] (range)

Source: Garth, A.K., and Burke, L.M., *Sports Med.*, 43(7), 539–64, 2013.

M, male; F, female; NR, not reported; road, road race stage; crit, criterium race stage; W, water; SD, carbohydrate-electrolyte sports drink.

[a] Data are reported as mean ± SD (if provided).
[b] Per stage.
[c] Total event.

during events in the belief that accrual of a fluid deficit may enhance performance, particularly in hilly terrain, due to the effect of a lower BM in increasing the economy of movement and improving power to weight ratio. Finally, fluid intake by some endurance athletes may be driven by their desire to consume other ingredients found in everyday drinks or specialized sports beverages such as carbohydrate, caffeine, and electrolytes, or by the desire to regulate body temperature via the intake of cool drinks.

The available literature on fluid intake during endurance sports has been limited to marathon and half-marathon running races conducted in mild to warm environments (Table 7.1).

Notwithstanding the limitations of methodologies used in such studies, which have been previously acknowledged (Garth and Burke 2013), studies in which data were collected immediately pre- and post-race showed that the typical change in BM across the event was a deficit of ~1%–2%; however, the range across the subjects in the same race spanned a deficit of >2% BM to a gain in mass. Indeed, in the study which involved the largest number (>600) of participants, individual BM changes over the marathon ranged from −8% to +5% (Zouhal et al. 2011). Although the authors noted a relationship between finishing time and BM losses, with the faster runners incurring a greater fluid deficit, BM change only accounted for 4.7% of the variance in race time, suggesting a complex relationship. Several studies in this category noted drinking behavior between the group who were characterized as having normal plasma sodium concentrations and those who developed mild/asymptomatic hyponatremia, with the hyponatremic group drinking more fluid and losing less BM (or gaining) over the course of the event (Hew 2005; Kipps et al. 2011).

The only observations on fluid intake by elite marathon runners were gathered by an innovative but largely unvalidated technique of retrospectively examining television footage of the behavior of the leading runner at the race drinking stations at the 2004 Athens Olympics (van Rooyen et al. 2010) and at 13 Olympic or Big City marathons (Beis et al. 2012). The footage revealed that they spent a total of 2–51 s, representing less than 1% of race time, engaged in drinking activities. The estimated (maximum) intake of fluid by male marathon winners, the majority of which are likely to be East African athletes, was claimed to be an average of 550 ± 340 ml/h with a range of 30–1090 ml/h (Beis et al. 2012). There were no correlations between fluid intake and either ambient conditions or running speed among these observations. Indeed, in similar environmental conditions, runners can behave differently in different races as illustrated by the athlete who ran the Berlin marathon in 2006 (12°C) and 2008 (16°C) with an estimated fluid intake of 1839 ml for the first year (2:03:59 finishing time) and 1098 ml for the second (2:06:08).

7.4.2 SINGLE-DAY ULTRA-ENDURANCE EVENTS (TOP COMPETITORS FINISH IN >3 H)

Events such as ultra-marathons, 50 km race walking, cycling road races, and half Ironman and Ironman triathlons also involve mass participation with a mixture of elite to recreational competitors. They also share the characteristics of endurance sports with regard to opportunities for fluid intake during the event, and the variable effect of the outdoor environment on sweat rates. Since the intensity of the event is reduced compared with endurance events, sweat rates are theoretically lower and there may be increased opportunity for fluid intake. However, the extended duration of the race may also increase the absolute fluid deficit or gain if there is a mismatch between sweating and fluid intake. Fluids may contribute a substantial amount of the carbohydrate needed to meet sports nutrition recommendations for extended events: for example, Speedy and colleagues reported that ~2/3 of the fluid consumed by Ironman triathletes contained carbohydrate (sports drink and cola drinks) and can contribute ~50% of the carbohydrate consumed during the race (Speedy et al. 2001; Kimber et al. 2002).

Table 7.2 summarizes the results of studies of ultra-endurance events involving running, cycling, and multisport combinations conducted over 5–24 h as a single-day race. Observations of fluid intake during ultra-endurance events noted mean intakes ranging from 300 to 1000 ml/h with large individual variations in these rates. Factors contributing to differences in fluid intake include the mode of activity: greater rates of intake were typically observed during cycling activities (400–900 ml/h) than running events (300–700 ml/h).

Overall, mean BM loss over the race ranged from 1.5% to 5.2%, with individual outcomes spanning a loss of >7% to a gain of 5% BM. Correlations between sweat loss and finishing time were unclear, with faster athletes recording a greater total loss of BM over the race in some studies (Lebus et al. 2010; Bracher et al. 2012) while the slowest athletes reported greatest losses in others (Speedy et al. 1999). Again, many investigations were focused on the incidence of hyponatremia, which occurred in 0%–51% of the study participants and occurred mostly in asymptomatic forms. As in the endurance events, weight gain was associated with hyponatremia (Speedy et al. 1999, 2001) particularly in the case of severe decreases in serum sodium concentrations. However, hyponatremia was also reported in individuals who maintained (Speedy et al. 1997) or even lost BM (Speedy et al. 1999, 2001) including substantial changes of a 9% BM loss (Speedy et al. 2001). Thus, the etiology of hyponatremia is complex.

7.4.3 FLUID BALANCE IN MULTIDAY ULTRA-ENDURANCE EVENTS

Sports such as cycling, mountain biking, running, and single or multisport adventure racing include multiday competition formats, with events lasting from two days to three weeks. Events can be further divided into those in which competitors are required to complete the course in a continuous manner of their own choosing, where the periods taken to sleep or eat are included in the finishing time, and those in which competitors complete a number of stages each day with these individual performances accumulating to produce the final results. Access to nutritional support may come from a variety and combination of sources including self-sufficiency, official feed zones, sporadic checkpoints for supplies, and assistance from team support crew. The determinants of sweat losses and fluid intake vary as discussed in the previous section on single-day endurance and ultra-endurance sports, with the additional challenges that intake during the event may need to contribute to substantial requirements for fluid, carbohydrate, and energy over the duration of the whole event and that deficits from one day may carry over to the next.

Table 7.3 summarizes observations from 10 separate multiday events, including four involving elite cyclists of international caliber, and formats ranging from a continuous road cycling format to events involving road cycling, mountain biking, or ultra-running with one or more stages each day. Studies reported mean fluid intakes across a stage of ~300–1000 ml/h and mean BM changes of 0.2%–3% BM, with the likelihood that elite athletes recorded a fluid deficit exceeding 3% BM in hot weather races. Fluid intake in some events was correlated to the temperature at the start of the stage (Schenk et al. 2010) and the duration of the stage (Ross et al. 2014). In addition, the format of a cycling race was seen to influence fluid intake, with road cyclists drinking less during criterium and individual time-trial formats than road races (Ross et al. 2014). Explanations for this observation include the briefer length of the race as well as reduced access to fluids (lack of feed zones) and opportunity to drink (the conflict between taking time to drink and the need to ride aggressively or in a streamlined position). However, it was also noted that the rules and culture of road cycling have evolved to promote greater opportunities for fluid and energy intake during the road race format. In addition to feed zones in which all cyclists can obtain food and fluid supplies from their support crews, designated riders within a cycling team (*domestiques*) assume a role of ferrying food and drinks supplies from the team car throughout the race to the cyclists who are in contention to win (Ross et al. 2014). Nevertheless, there are obstacles to drinking during the stage including the need to keep hands on the handlebars during steep ascents and descents in road cycling (Ebert et al. 2007; Ross et al. 2014), difficult terrain in mountain biking (Schenk et al. 2010) as well as aggressive riding tactics and the *breakaway* whereby the lead rider is distant to the support of the domestiques (Ebert et al. 2007; Ross et al. 2014).

There were reports of both a negative correlation between the finishing time within a stage and fluid intake or level of deficit (Ebert et al. 2007) and a lack of association or even positive correlation between success in a race and fluid intake/BM maintenance (Ross et al. 2014). Although it is intuitive that the fastest athletes in a race might incur the greatest fluid deficit as a result of a higher sweat rate and less opportunity or desire to obtain or drink fluids at high speed, in some sports there are unique factors that may change this relationship. For example, in one cycling study, the fastest competitors within each stage were shown to have incurred the smallest losses of BM (Ross et al. 2014). This was explained by the team tactic in road cycling whereby the *protected rider* (cyclists who are deemed to have the best chance of winning) spends much of the race riding within the slipstream of the peloton or their team mates, thus reducing their power outputs (and sweat rates) while allowing them to achieve greater intakes of fluid and energy. Further studies on such events, including those involving elite competitors, may provide further insights into cultural, behavioral, and logistical determinants of fluid intake.

7.5 SUMMARY AND FINDING MIDDLE GROUND ON FLUID GUIDELINES

Further studies of real-life hydration practices during competitive events including information on motives for drinking or not, along with intervention studies that simulate the actual nature of real-life sport, are needed before conclusions can be made about ideal drinking strategies for endurance and ultra-endurance sports (Garth and Burke 2013). In any case, it is likely that a range of drinking strategies may be appropriate and that athletes need to have an individualized and flexible approach to their hydration practices. This approach should also incorporate goals for other nutrition strategies known to enhance fluid palatability and voluntary consumption, thermoregulation, or performance and may dictate a desirable volume and pattern of intake that is independent of thirst. There may be benefits associated with a *paced* approach to drinking during sport, in which the athlete plans to spread their intake of these nutrients as well as a reasonable replacement of their sweat losses across the opportunities that their event provides to consume fluids.

An important step in developing messages about hydration practices in endurance sport is to recognize that a single approach is unlikely to be successful. Rather, it is likely that needs, challenges, and opportunities will vary between athletes within any single event, as well as changing from event to event. Indeed, the range of experiences between and within endurance/ultra-endurance sports is likely to include athletes whose fluid plan needs to increase their habitual/natural intake because sweat rates are likely to greatly exceed opportunities to drink, as well as athletes with the opposite characteristics. We have recently tried to conceptualize a model (Figure 7.1) that explains why a spectrum of approaches needs to coexist. Hopefully, this will help to find middle ground in the current debate about fluid guidelines for competitive sports.

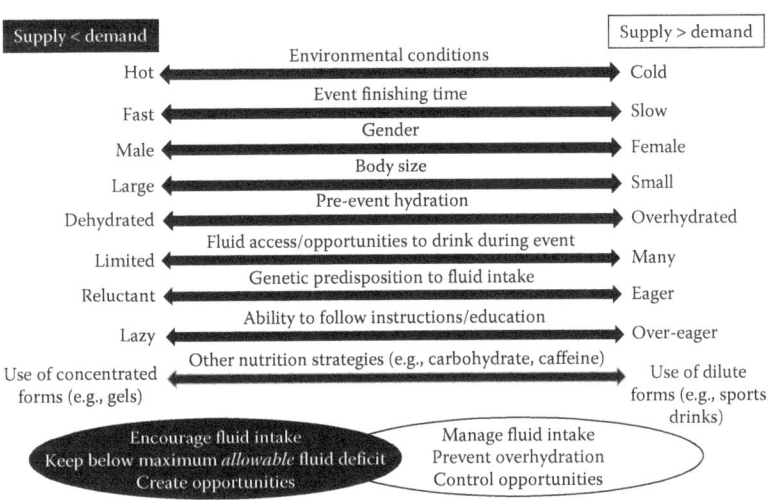

The spectrum of issues that need to be considered in education and event logistics regarding fluid intake during competitive events includes consideration of the balance between supply (fluids) and demand (sweat rates)

FIGURE 7.1 Fluid guidelines for endurance and ultra-endurance sports should recognize that a range of approaches to encouraging or managing drinking behavior is needed. (Adapted from Burke, L.M., *Aspetar Sports Med. J.*, 2, 86–93, 2012.)

REFERENCES

Almond, C.S.D., A.Y. Shin, E.B. Fortescue et al. 2005. Hyponatremia among runners in the Boston marathon. *N. Engl. J. Med.* 352: 1550–6.

American College of Sports Medicine. 1975. Position statement of the American College of Sports Medicine: Prevention of heat injuries during distance running. *Med. Sci. Sports Exerc.* 7: vii–ix.

American College of Sports Medicine. 1987. Position stand on the prevention of thermal injuries during distance running *Med. Sci. Sports Exerc.* 19: 529–33.

American College of Sports Medicine, M.N. Sawka, L.M. Burke et al. 2007. American college of sports medicine position stand. Exercise and fluid replacement. *Med. Sci. Sports Exerc.* 39: 377–90.

Armstrong, L.E., D.J. Casa, H. Emmanuel et al. 2012. Nutritional, physiological, and perceptual responses during a summer ultraendurance cycling event. *J. Strength Cond. Res.* 26: 307–18.

Au-Yeung, K.L., W.C. Wu, W.H. Yau et al. 2010. A study of serum sodium level among hong kong runners. *Clin. J. Sport Med.* 20: 482–7.

Bardis, C.N., S.A. Kavouras, G. Arnaoutis, D.B. Panagiotakos, L. S. Sidossis. 2013. Mild dehydration and cycling performance during 5-kilometer hill climbing. *J. Athl. Train.* 48(6): 741–7.

Beis, L.Y., M. Wright-Whyte, B. Fudge et al. 2012. Drinking behaviors of elite male runners during marathon competition. *Clin. J. Sport Med.* 22: 254–61.

Beltrami, F.G., T. Hew-Butler, T.D. Noakes. 2008. Drinking policies and exercise-associated hyponatraemia: Is anyone still promoting overdrinking? *Br. J. Sports Med.* 42(10): 796–501.

Bracher, A., B. Knechtle, M. Gnädinger et al. 2012. Fluid intake and changes in limb volumes in male ultra-marathoners: Does fluid overload lead to peripheral oedema? *Eur. J. Appl. Physiol.* 112: 991–1003.

Burke, L.M. 2012. Fluid facts and fads. *Aspetar Sports Med J.* 2: 86–93.

Casa, D.J. 2007. *Proper Hydration for Distance Running—Identifying Individual Fluid Needs. A USA Track and Field Advisory 2007*. [online]. Available from: http://www.usatf.org/groups/Coaches/library/2007/hydration/ProperHydrationForDistanceRunning.pdf (Accessed April 1, 2015).

Casa, D.J., R.L. Stearns, R.M. Lopez et al. 2010. Influence of hydration on physiological function and performance during trail running in the heat. *J. Ath. Train.* 45(2): 147–56.

Cheuvront, S.N., R. Carter, M.N. Sawka. 2003. Fluid balance and endurance exercise performance. *Curr. Sports Med. Rep.* 2: 202–8.

Cheuvront, S.N., R.W. Kenefick. 2014. Dehydration: Physiology, assessment, and performance effects. *Compr. Physiol.* 4(1): 257–85.

Cheuvront, S.N., R.W. Kenefick, S.J. Montain, M.N. Sawka. 2010. Mechanisms of aerobic performance impairment with heat stress and dehydration. *J. Appl. Physiol.* 109: 1989–95.

Ebert, T.R., D.T. Martin, B. Stephens et al. 2007. Fluid and food intake during professional men's and women's road-cycling tours. *Int. J. Sports Physiol. Perform.* 2: 58–71.

Fallon, K.E., E. Broad, M.W. Thompson et al. 1998. Nutritional and fluid intake in a 100-km ultramarathon. *Int. J. Sport Nutr.* 8: 24–35.

Garcia-Roves, P.M., N. Terrados, S.F. Fernandez et al. 1998. Macronutrients intake of top level cyclists during continuous competition—change in the feeding pattern. *Int. J. Sports Med.* 19: 61–7.

Garth, A.K., L.M. Burke. 2013. What do athletes drink during competitive sporting activities? *Sports Med.* 43(7): 539–64.

Glace, B.W., C.A. Murphy, M.P. McHugh. 2002. Food intake and electrolyte status of ultramarathoners competing in extreme heat. *J. Am. Coll. Nutr.* 21: 553–9.

Goulet, E.D. 2013. Effect of exercise-induced dehydration on endurance performance: Evaluating the impact of exercise protocols on outcomes using a meta-analytic procedure. *Br. J. Sports Med.* 47: 679–86.

Hew, T.D. 2005. Women hydrate more than men during a marathon race. Hyponatremia in the Houston marathon: A report on 60 cases. *Clin. J. Sport Med.* 15: 148–53.

Hew-Butler., T., J.P. Dugas, T.D. Noakes et al. 2010. Changes in plasma arginine vasopressin concentrations in cyclists participating in a 109-km cycle race. *Br. J. Sports Med.* 44: 594–7.

Kao, W.-F., C.-L. Shyu, X.-W. Yang et al. 2008. Athletic performance and serial weight changes during 12- and 24-hour ultra-marathons. *Clin. J. Sport Med.* 18: 155–8.

Kimber, N.E., J.J. Ross, S.L. Mason et al. 2002. Energy balance during an ironman triathlon in male and female triathletes. *Int. J. Sport Nutr. Exerc. Metab.* 12: 47–62.

Kipps, C., S. Sharma, D.T. Pedoe. 2011. The incidence of exercise-associated hyponatraemia in the London marathon. *Br. J. Sports Med.* 45: 14–19.

Knechtle, B., M. Gnadinger, P. Knechtle et al. 2011a. Prevalence of exercise-associated hyponatremia in male ultraendurance athletes. *Clin. J. Sport Med.* 21: 226–32.

Knechtle, B., P. Knechtle, T. Rosemann et al. 2009. No dehydration in mountain bike ultra-marathoners. *Clin. J. Sport Med.* 19: 415–20.

Knechtle, B., P. Knechtle, T. Rosemann. 2011b. Low prevalence of exercise-associated hyponatremia in male 100 km ultra-marathon runners in switzerland. *Eur. J. Appl. Physiol.* 111: 1007–16.

Knechtle, B., O. Senn, R. Imoberdorf et al. 2010. Maintained total body water content and serum sodium concentrations despite body mass loss in female ultra-runners drinking ad libitum during a 100 km race. *Asia Pac. J. Clin. Nutr.* 19: 83–90.

Knechtle, B., O. Senn, R. Imoberdorf et al. 2011c. No fluid overload in male ultra-runners during a 100 km ultra-run. *Res. Sports Med.* 19: 14–27.

Kruseman, M., S. Bucher, M. Bovard et al. 2005. Nutrient intake and performance during a mountain marathon: An observational study. *Eur. J. Appl. Physiol.* 94: 151–7.

Laursen, P.B., R. Suriano, M.J. Quod, et al. 2006. Core temperature and hydration status during an ironman triathlon. *Br. J. Sports Med.* 40: 320–5.

Lebus, D.K., G.A. Casazza, M.D. Hoffman et al. 2010. Can changes in body mass and total body water accurately predict hyponatremia after a 161-km running race? *Clin. J. Sport Med.* 20: 193–9.

Mettler, S., C. Rusch, W.O. Frey et al. 2008. Hyponatremia among runners in the Zurich marathon. *Clin. J. Sport Med.* 18: 344–9.

Montain, S.J., E.F. Coyle. 1992. Influence of graded dehydration on hyperthermia and cardiovascular drift during exercise. *J. Appl. Physiol.* 73(4): 1340–50.

Myhre, L.G., G.H. Hartung, D.M. Tucker. 1982. Plasma volume and blood metabolites in middle-aged runners during a warm-weather marathon. *Eur. J. Appl. Physiol.* 48: 227–40.

Noakes, T.D. 2007. Hydration in the marathon: Using thirst to gauge safe fluid replacement. *Sports Med.* 37(4–5): 463–6.

Noakes, T.D. 2010. Is drinking to thirst optimum? *Ann. Nutr. Metab.* 57(Suppl 2): 9–17.

Noakes, T.D. 2012. *Waterlogged: The Serious Problem of Overhydration in Endurance Sports.* Human Kinetics Publishers, Champaign, IL.

Noakes, T.D., K. Sharwood, D. Speedy et al. 2005. Three independent biological mechanisms cause exercise-associated hyponatremia: Evidence from 2,135 weighed competitive athletic performances. *Proc. Natl. Acad. Sci. U. S. A.* 102(51): 18550–5.

Noakes, T.D., D.B. Speedy. 2006. Case proven: Exercise associated hyponatraemia is due to overdrinking. So why did it take 20 years before the original evidence was accepted? *Br. J. Sports Med.* 40(7): 567–72.

O'Toole, M.L., P.S. Douglas, R.H. Laird et al. 1995. Fluid and electrolyte status in athletes receiving medical care at an ultradistance triathlon. *Clin. J. Sport Med.* 5: 116–22.

Pahnke, M.D., J.D.Trinity, J.J. Zachwieja et al. 2010. Serum sodium concentration changes are related to fluid balance and sweat sodium loss. *Med. Sci. Sports Exerc.* 42: 1669–74.

Rehrer, N.J., Brouns, F., E.J. Beckers et al. 1992. Physiological changes and gastro-intestinal symptoms as a result of ultra-endurance running. *Eur. J. Appl. Physiol.* 64: 1–8.

Rogers, G., C. Goodman, C. Rosen. 1997. Water budget during ultra-endurance exercise. *Med. Sci. Sports Exerc.* 29: 1477–81.

Rose, S., E.M. Peters-Futre. 2010. Ad libitum adjustments to fluid intake during cool environmental conditions maintain hydration status during a 3-day mountain bike race. *Br. J. Sports Med.* 44: 430–6.

Ross, M.L., B. Stephens, C.R. Abbiss et al. 2014. Observations of fluid balance, carbohydrate ingestion and body temperature regulation during men's stage-race cycling in temperate environmental conditions. *Int. J. Sports Physiol. Perform.* 9(3): 575–82.

Rust, C.A., B. Knechtle, P. Knechtle et al. 2012. No case of exercise-associated hyponatraemia in top male ultra-endurance cyclists: The "Swiss Cycling Marathon." *Eur. J. Appl. Physiol.* 112: 689–97.

Saunders, A.G., J.P Dugas, R. Tucker, M.I. Lambert, T.D. Noakes. 2005. The effects of different air velocities on heat storage and body temperature in humans cycling in a hot, humid environment. *Acta Physiol. Scand.* 183(3): 241–55.

Sawka, M.N. 1992. Physiological consequences of hypohydration: Exercise performance and thermoregulation. *Med. Sci. Sports Exerc.* 24: 657–70.

Sawka, M.N., T.D. Noakes. 2007. Does dehydration impair exercise performance? *Med. Sci. Sports Exerc.* 39: 1209–17.

Schenk, K., H. Gatterer, M. Ferrari et al. 2010. Bike transalp: Liquid intake and its effect on the body's fluid homeostasis in the course of a multistage, cross-country, MTB marathon race in the central Alps. *Clin. J. Sport Med.* 20: 47–52.

Schwellnus, M.P., N. Drew, M. Collins. 2011. Increased running speed and previous cramps rather than dehydration or serum sodium changes predict exercise-associated muscle cramping: A prospective cohort study in 210 ironman triathletes. *Br. J. Sports Med.* 45: 650–6.

Sharwood, K., M. Collins, J. Goedecke et al. 2002. Weight changes, sodium levels, and performance in the South African Ironman Triathlon. *Clin. J. Sport Med.* 12: 391–9.

Shirreffs, S.M., M.N. Sawka. 2011. Fluid and electrolyte needs for training, competition, and recovery. *J. Sports Sci.* 29(Suppl 1): S39–46.

Siegel, A.J., J.G.Verbalis, S. Clement et al. 2007. Hyponatremia in marathon runners due to inappropriate arginine vasopressin secretion. *Am. J. Med.* 120(5): 461.e11–17.

Singh, N.R., E.C. Denissen, A.J. McKune et al. 2012. Intestinal temperature, heart rate, and hydration status in multiday trail runners. *Clin. J. Sport Med.* 22(4): 311–18.

Speedy, D.B., R. Campbell, G. Mulligan et al. 1997. Weight changes and serum sodium concentrations after an ultradistance multisport triathlon. *Clin. J. Sport Med.* 7: 100–3.

Speedy, D.B., T.D. Noakes, N.E. Kimber et al. 2001. Fluid balance during and after an ironman triathlon. *Clin. J. Sport Med.* 11: 44–50.

Speedy, D.B., T.D. Noakes, I.R. Rogers et al. 1999. Hyponatremia in ultradistance triathletes. *Med. Sci. Sports Exerc.* 31: 809–15.

Stuempfle, K.J., D.R. Lehmann, H.S. Case et al. 2003. Change in serum sodium concentration during a cold weather ultradistance race. *Clin. J. Sport Med.* 13: 171–5.

Sulzer, N.U., M.P. Schwellnus, T.D. Noakes. 2005. Serum electrolytes in ironman triathletes with exercise-associated muscle cramping. *Med. Sci. Sports Exerc.* 37: 1081–5.

Tam, N., H.W. Nolte, T.D. Noakes. 2011. Changes in total body water content during running races of 21.1 km and 56 km in athletes drinking ad libitum. *Clin. J. Sport Med.* 21: 218–55.

van Rooyen, M., T. Hew-Butler, T. Noakes. 2010. Drinking during marathon running in extreme heat: A video analysis study of the top finishers in the 2004 Athens Olympic marathons. *S. Afr. Med. J.* 22: 55–61.

Wagner, S., B. Knechtle, P. Knechtle et al. 2012. Higher prevalence of exercise-associated hyponatremia in female than in male open-water ultra-endurance swimmers: The "marathon-swim" in lake Zurich. *Eur. J. Appl. Physiol.* 112: 1095–106.

Zouhal, H., C. Groussard, G. Minter et al. 2011. Inverse relationship between percentage body weight change and finishing time in 643 forty-two-kilometer marathon runners. *Br. J. Sports Med.* 45: 1101–5.

8 Effect of Dehydration on Muscle Strength, Power, and Performance in Intermittent High-Intensity Sports

Brent C. Creighton, J. Luke Pryor,
Daniel A. Judelson, and Douglas J. Casa

CONTENTS

8.1 INTRODUCTION

Intermittent high-intensity sports encompass those sports requiring athletes to perform exercise, predominantly of anaerobic nature, interspersed with less demanding, active recovery and/or variable rest periods (basketball, soccer, american football, tennis, etc.). For athletes competing in high-intensity intermittent sports, strength and power critically influence performance and success. Strength, the absolute force that can be produced, differs from power, which characterizes the amount of force generated per unit time. Performance is more difficult to define, including quantifiable (e.g., points scored) and qualifiable (e.g., winning the match) components. All three measures are routinely used to assess an athlete's capability or potential success.

In order to facilitate an understanding of the relationships between hydration status and performance-related parameters, let us begin by defining some key terms. Exercise will be operationally defined within the context of athletes engaging in intermittent high-intensity sports. *Hydrate* by definition is the addition of water or moisture. In the context of exercise, *hydration* refers to the process of adding water to the body by consuming fluid. Hydration is also a global term used to reference total body water content. Fluid balance is dynamic and fluctuates constantly, especially during exercise as physical activity stimulates many factors influencing fluid turnover (addition through drinking and subtraction through sweat, exhalation, and urine) (Burke and Hawley 1997; Godek et al. 2005). To best characterize the various hydration states, Figure 8.1 visually depicts commonly used terms.

Euhydration indicates a normal or balanced body water content; *hypohydration* refers to a longer term, chronic total body water deficit (Armstrong 2007; Sawka et al. 2007); and *dehydration* (*being dehydrated*) refers to the acute loss of body water and

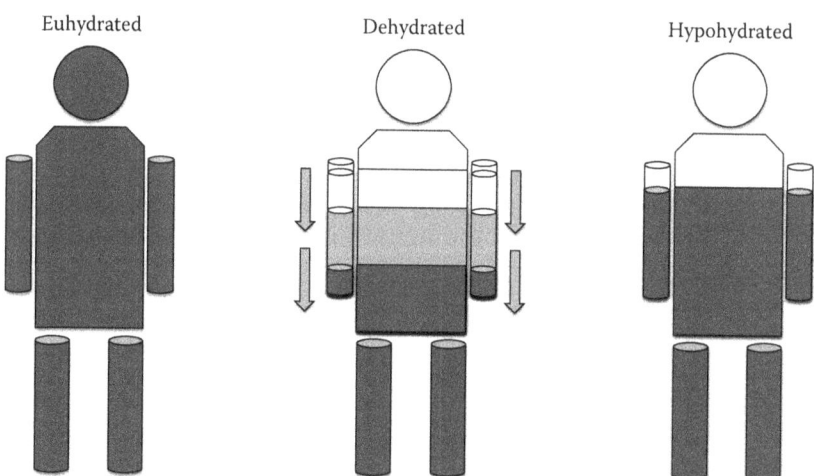

FIGURE 8.1 Variations in hydration state (water represented in gray). Euhydrated: normal or balanced hydration state; dehydrated: acute loss or the act of water loss from the body; hypohydrated: long term or chronic total body water deficit.

acute hypohydration (Sawka et al. 2007). Less commonly a concern, *hyperhydration* (*being hyperhydrated*) refers to a state of total body water excess. Finally, *rehydration* refers to the restoration of fluid following a dehydrating stimulus.

Exercise commonly occurs under suboptimal hydration states, either hypohydrated or dehydrated (Maughan and Shirreffs 2010a; Osterberg et al. 2009). It is unfortunately common for athletes of many sports (including basketball, tennis, and American football) to begin practice or competition in a hypohydrated state (Osterberg et al. 2009; Yeargin et al. 2010). Athletes are further challenged during exercise, when the average athlete drinking *ad libitum* will consume fluids equal to only ~66% of sweat losses (Greenleaf and Sargent 1965; Hubbard et al. 1984). Couple pre-exercise hypohydration with inadequate fluid intake during exercise and the stage is set for potentially significant body water deficits.

Adequate fluid balance is paramount for optimal physiological function in athletes (Cheuvront et al. 2003; Sawka et al. 2001; Shirreffs 2005). Much of our current understanding surrounding the effects of dehydration stems from research examining endurance exercise; dehydration decreases endurance performance (Armstrong et al. 1985; Cheuvront et al. 2003; Sawka et al. 2001; Shirreffs 2005). Far less is understood about how hypohydration and/or dehydration impact strength, power, and subsequent performance in intermittent high-intensity sports. Even so, a growing amount of research clearly demonstrates the negative consequences of poor hydration on these anaerobic variables of athletic measurement.

8.2 PHYSIOLOGY OF DEHYDRATION

Human metabolism is not 100% efficient. During exercise, humans use only ~20% of the energy liberated from the deconstruction of chemical bonds, releasing the remaining ~80% as heat. Muscular work (e.g., high-intensity exercise) dramatically increases metabolic demand, enhancing heat production. In an attempt to offset this heat production and homeostatically control core temperature, compensatory heat loss mechanisms activate. Sweating, the primary mechanism by which humans dissipate heat, contributes to thermoregulation by releasing fluid to the skin, causing an increase in evaporative cooling. Thus, body fluid is sacrificed to offset a rise in body temperature. Sweat rate is affected by various factors, including exercise intensity, training state, heat-acclimatization, environmental conditions, and clothing (King et al. 2002). Overall fluid loss (dehydration) is strongly influenced by individual sweat rate, but is also impacted by pre-exercise hydration state, duration of exercise, and degree of rehydration (Casa et al. 2010; Ladell 1955; Maughan and Shirreffs 2010a).

Far less research evaluates fluid loss during intermittent high-intensity sports than endurance exercise, suggesting that most fluid intake recommendations result from research on steady-state endurance exercise. Given the differences between continuous submaximal exercise and intermittent anaerobic exercise in thermal stresses and resulting fluid dynamics, a more thorough analysis of fluid needs for intermittent exercise athletes is necessary. The unpredictable and varied nature of many high-intensity intermittent sports also makes providing fluid intake recommendations difficult. Differing work-to-rest ratios, exercise intensities, mental stressors, clothing, protective gear, and environmental characteristics mean each intermittent sport requires an individual recommendation for fluid intake, unlike endurance-based sports, which share similar traits.

8.3 LIMITATIONS AND CONSIDERATIONS FOR EXERCISE/DEHYDRATION TESTING

Because reductions in total body water can impair endurance performance, we might logically extrapolate that dehydration and/or hypohydration must also negatively impact athletes competing in intermittent high-intensity sports. These relationships between fluid loss, strength, power and anaerobic performance, however, have yet to be fully mapped. Thus far, Judelson et al. (2007a) and Kraft et al. (2012) provided the two most extensive reviews describing how hydration influences strength, power, and performance. Both reviews describe the need for additional and better-controlled studies, as discrepancies in methodology and a disregard for confounding factors make interpretation of the current literature difficult. In many instances, research not specifically designed to assess dehydration or hypohydration has been inappropriately used to extrapolate findings on force parameters and performance. The following sections help outline the methodological factors, as reviewed by Judelson et al. (2007a), which contribute to the large variability of results seen in the literature.

8.3.1 EXACERBATING FACTORS

To properly assess the effects of dehydration or hypohydration on strength, power, and/or performance, hydration must be assessed independently of other factors that adversely affect the intended measure. For example, athletes competing in *weight class* sports such as boxing, weight lifting, or karate typically decrease body mass prior to competition by reducing total body water and limiting caloric intake. Using such athletes as research participants complicates the data because the ergolytic effects of dehydration cannot be easily separated from those due to caloric restriction (Maughan et al. 1997; McMurray et al. 1991; Rankin et al. 1996).

Similarly, studies employing an exercise-heat stress model to reduce total body water pose a challenge. Although many intermittent high-intensity athletes commonly dehydrate when they exercise in the heat, both increased muscle and core temperature can impair performance independent of hydration state (Cheung and Sleivert 2004; Thomas et al. 2006). Even implementing isolated exercise to reduce total body water causes fatigue, an additional factor independent of fluid loss that can alter strength, power, and anaerobic performance. Future studies assessing hydration should incorporate sufficient rest after the dehydrating stimulus to ensure (1) muscle and/or core temperature has fallen below hyperthermic levels (Morrison et al. 2004) and (2) fatigue from previous exercise no longer influences performance. Judelson et al. (2007a) acknowledge that diuretics are likely a superior tool for reducing total body mass, but suggest that exercise and heat exposure can be useful so long as sufficient time is given between the dehydration protocol and the performance test (Armstrong et al. 1985).

8.3.2 MASKING FACTORS

Fitness level and training status can considerably influence hydration state and rate of dehydration, subsequently impacting muscular performance. Chronic endurance training expands plasma volume, theoretically helping to minimize the impact of an absolute fluid loss. The findings of Caterisano et al. (1988) and Schoffstall et al. (2001) support this hypothesis. Thus, any manner of increasing total body water may minimize the effects of dehydration on strength, power, and/or performance.

Most studies examining the effects of dehydration and hypohydration on strength, power, or performance studied males. Of those studies evaluating women (Evetovich et al. 2002; Greenleaf et al. 1967; Gutiérrez et al. 2003), none accounted for menstrual status despite the endocrine impacts of cycle phase; fluid hormones vary across the reproductive cycle, promoting water retention during the luteal phase (White et al. 2011). This enhanced fluid pool might act synonymously to training-induced increases in plasma volume, although no scientific research confirms this hypothesis. Prudence suggests, however, menstrual cycle should be taken into account when planning a study that examines female athletes.

Finally, various studies (Bosco et al. 1968; Watson et al. 2005) have utilized measurements in which only body mass resisted the testing movement (e.g., vertical jump or short-distance sprinting). Dehydration decreases body mass, so implementing dependent variables that rely only on body mass can confuse the results. This point was elegantly demonstrated by Cheuvront et al. (2010) who compared force parameters during vertical jumping among three conditions: euhydrated, hypohydrated by ~4%, and hypohydated by ~4% while wearing a weighted vest containing a weight equal to that lost due to hypohydration. Outward performance (i.e., vertical jump height) did not change between euhydrated and hypohydrated conditions because reductions in body mass (promoting improved jump height) were offset by decreased ground reaction impulse (impairing jump height). The weight added during the weighted vest trial eliminated the ergogenic effects of decreased body mass and resulted in worse vertical jump performance. These findings highlight the importance of proper test selection when measuring dehydration and force output.

8.4 DEHYDRATION AND HYPOHYDRATION: EFFECTS ON STRENGTH AND POWER

Dehydration due to exercise, heat exposure, insufficient fluid consumption, diuretics, or some combination thereof is common in virtually all sports. Muscular strength and power are important components to many sports and likely indicate potential success (Pryor et al. 2014). High-quality literature reviews of older (Sawka and Pandolf 1990) and newer (Judelson et al. 2007a; Kraft et al. 2012) studies report dehydration or hypohydration up to 7% body mass loss (BML) inconsistently affects strength and power, both in frequency and magnitude of effect. Despite this variability, none of the reviewed studies documented a beneficial effect of hypohydration or dehydration on performance outcomes alone. Notwithstanding the methodological limitations outlined above, suboptimal hydration status ($\geq$3%–4% BML) appears to negatively influence both strength and power by approximately 2% and 3%, respectively (Judelson et al. 2007a).

Even with a majority of the current literature pointing toward an ergolytic effect of dehydration on overall strength and/or power (see Judelson et al. 2007a; Pandolf and Sawka 1990; Kraft et al. 2012 for reviews), many studies report mixed outcomes with dehydration up to 4% BML (Bigard et al. 2001; Evetovich et al. 2002; Montain et al. 1998). This variability might exist because a *critical degree* of dehydration or total body water loss that impairs strength and/or power is likely to exist, but appears individualized given the variability of results. Figures 8.2 and 8.3 show the large variability of findings reviewed by Judelson et al. (2007a) as well as more recent findings by Kraft et al. (2012) for studies measuring strength and power respectfully. Note that despite varied results, as fluid is lost (% dehydration increases), power ($R^2 = 0.02$) and strength ($R^2 = 0.25$) trend negatively.

Unfortunately, differing research methods, athletes' various hydration practices, and individual dehydration tolerances make predicting the effects of dehydration on any given individual difficult (Maughan et al. 2007). Hayes and Morse (2010) assessed isometric force production and isokinetic peak torque (30°/s) after progressive levels of heat and physical activity-mediated dehydration. Maximal isometric force production was diminished after 1.0%, 1.9%, 2.6%, 3.3%, and 3.9% BML, while isokinetic peak torque was reduced when dehydration exceeded 2.6% BML. No dose relationship was observed but the threshold for dehydration-mediated strength and power-related performance impairment was in line with the threshold of 3%–4% reported in a seminal review by Judelson et al. (2007a). With regard to power, slower compared to faster speed muscular contractions (30°/s vs. ≥120°/s) appear more sensitive to dehydration-induced power impairments after dehydration >2.5% (Ftaiti et al. 2001; Hayes and Morse 2010).

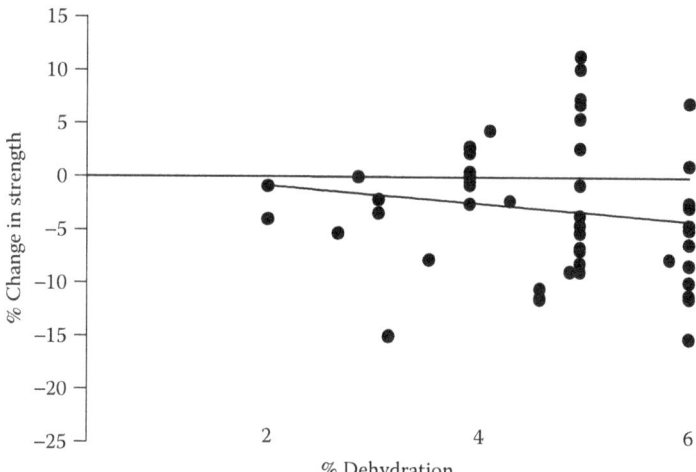

FIGURE 8.2 Relationship between fluid loss (% dehydration) and strength (% change in strength). Individual data points represent separate findings from studies reviewed. (Data from Judelson, D.A. et al., *Sports Med.* 37, 907–21, 2007a. Data from Kraft, J.A. et al., *Res. Q. Exerc. Sport.*, 83, 282–92, 2012.)

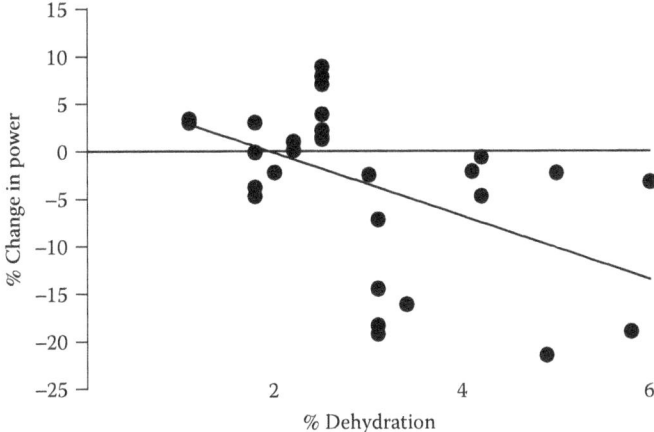

FIGURE 8.3 Relationship between fluid loss (% dehydration) and power (% change in power). Individual data points represent separate findings from studies reviewed. (Data from Judelson, D.A. et al., *Sports Med.* 37, 907–21, 2007a. Data from Kraft, J.A. et al., *Res. Q. Exerc. Sport.*, 83, 282–92, 2012.)

While single-effort measures of strength and power provide some insight into the dehydration–performance interaction, intermittent high-intensity sports frequently demand repeated strength and power generation. Only a few studies have examined how dehydration impacts serial measures of strength and power. On separate occasions, performance during a six set squat protocol ($\leq$10 repetitions per set, 80% 1RM, 2 min rest between sets) was reduced 2.4% during sets 2–3 when dehydrated 2.4% and 4.8% during sets 2–5 when dehydrated 4.8% (Judelson et al. 2007b). Similarly, Kraft et al. (2010) observed that 3% dehydration reduced (1) total work during three sets to failure for six resistance exercises (2 min rest between sets), and (2) mean and peak power during sets 3–6 of six 15-s sprints separated by 30 s of recovery (Kraft et al. 2011). The effects of dehydration might be particularly pronounced for these repeated work bouts whose rest periods introduce a greater aerobic contribution to energy production. Dehydration clearly alters aerobic metabolism (Cheuvront et al. 2003), potentially creating a butterfly effect that diminishes strength and power performance during subsequent bouts (Judelson et al. 2007b; Kraft et al. 2011). This delayed performance decrement might be particularly applicable to intermittent high-intensity sports because end of competition scenarios typically require critical, decisive play.

8.4.1 Strength

In the context of exercise and intermittent sports, *strength* refers to the maximal force a muscle or muscle group can generate at a specified velocity (Knuttgen and Kramer 1987). In sports, strength is required in countless situations: linemen in American football exert force against each other during blocking, basketball players must squeeze rebounds to avoid defenders slapping loose the ball, and gymnasts hold static positions that place enormous stresses on muscle and joints. Laboratory protocols commonly used to measure strength include but are not limited to single maximal effort isometric (no change in muscle length), isotonic (no change in force), and iso-kinetic (no change in speed of muscle action) actions of the upper (e.g., back exten-sion, bear hug, bench press, row, elbow extension, elbow flexion, forearm flexion, grip strength, shoulder abduction, shoulder adduction, and shoulder extension) or lower body (e.g., hip flexion, knee extension, and knee flexion) (Judelson et al. 2007a).

Arguably the most comprehensive review examining how hydration state influ-ences strength was written by Judelson et al. (2007a). Of the 70 publications reviewed, 21% demonstrated statistically significant hypohydration-induced reductions in strength (Judelson et al. 2007a). Additionally, more than two-thirds of well-controlled studies (i.e., those lacking exacerbating or masking factors; see Sections 8.3.1 and 8.3.2) showed negative effects on strength. Overall, Judelson et al. (2007a) suggest that a 3%–4% reduction in total body water can impair strength by ~2%.

Although the National Athletic Trainers' Association (NATA) and the American College of Sports Medicine (ACSM) highly recommend athletes to adequately drink throughout competition (Casa et al. 2000; Sawka et al. 2007), many athletes fail to appropriately hydrate. Drinking to reduce water loss during competition appears to help mitigate dehydration associated strength losses. Rodrigues et al. (2014) showed that following 90 min of cycling in the heat, those who lost 2% of their body mass displayed strength deficits of nearly 16% post-exercise in the exercised muscles com-pared to a euhydrated control group.

Conversely, others found that maintaining hydration might not adequately prevent strength deficits. Ali and Williams (2013) compared soccer athletes when they hydrated during 90 min of simulated soccer versus when they abstained from fluid intake during the same protocol. Whether fluid was consumed or not during the soccer game, both groups reduced body mass by 2.3% and 3.7%, respectively. Isokinetic and isometric knee strength deficits (measured as pre- to post-reductions) equaled 8.3% and 16.5%, respectively, and did not differ between the fluid ingestion and no fluid ingestion trials. A sport's specific metabolic demands and competitive requirements can sig-nificantly impact resulting fatigue and access to fluids (and hence, an athlete's rate of fluid consumption). For example, the intermittent nature of badminton allows players to consume fluids more regularly compared to soccer athletes.

8.4.2 POWER

Power is distinct from strength, as it considers both the force generated and the time over which that force was generated; increasing force and decreasing duration linearly increase power. Similar to strength, power critically contributes to athletic ability and is necessary in many intermittent high-intensity sports: swinging a base-ball bat, kicking a soccer ball, or jumping to spike a volleyball. Common laboratory tests used to quantify muscular power include vertical jump, Wingate/cycling tests, maximal knee extension, or the Margaria Kalamen power test (Judelson et al. 2007a).

Judelson et al. (2007a) also reviewed the literature describing how hydration state influenced power. Of the 47 studies measuring power, 19% showed hypohydra-tion significantly reduced power, suggesting 3%–4% BML reduces muscular power by ~3% (Judelson et al. 2007a). However, when considering only those studies which accounted for confounding variables (only four investigations), ~81% of the findings showed negative effects on power. Similarly, Jones et al. (2008) showed ~3% BML impaired mean and peak power, but to a much greater degree than suggested by previous work. Subjects in Jones et al. (2008) performed upper and lower body Wingate tests prior to and following a heat stress trial, once while rehydrating and once while abstaining from rehydration. Both the mean and peak power in dehydrated subjects significantly decreased by ~10% in the upper body and ~20% in the lower body. Such results suggest that hydration plays an important role in reducing anaerobic power.

Although research suggests the loss of body water impairs strength and power, if and how this reduction affects athletic performance remains in question. Cheuvront and Kenefick (2014) address the issue thoroughly, revisiting the findings published by Judelson et al. (2007a) and concluding that any effect of water loss on strength and power is either small or nonexistent. They support their claim by identifying the lack of a clear physiological or cellular mechanism by which dehydration may impair strength or power. Despite this point, we recommend athletes continue to appropriately hydrate during intermittent exercise and sports; at worst, appropriate hydration will do no harm, but the mixed results in the scientific literature suggest inappropriately dehydrating might impair strength and power.

8.5 DEHYDRATION: EFFECTS ON PERFORMANCE

Performance is an umbrella term most often describing important sport-specific outcomes. As such, performance can be variously defined and assessed: by achievement (improved finishing time, increased shots on goal, number of wins, etc.), cognitive function (tasks requiring the use of short- or long-term memory, decision making), or overall improved ability (increased skill or proficiency). Thus, the context in which we discuss the effects of dehydration on performance is important.

8.5.1 Effects of Dehydration on Sport-Specific Tasks and Skill

The complexities inherent to sports and sport-specific motor tasks make it difficult to conclusively state that dehydration always inhibits sport skill performance. Neuromuscular, cognitive, and metabolic influences (acting alone or in combination) affect any given sport skill or task, and in any given circumstance, dehydration might mitigate one, some, or all components. For example, athletic skill and motor function decrease proportionally with greater dehydration (Baker et al. 2007b; Edwards et al. 2007). Figure 8.4 illustrates how key basketball skills become impaired with elevated dehydration.

Subjects from that study (Baker et al. 2007) reported feelings of lightheadedness and greater fatigue when dehydrated 3%–4%, likely reducing skill performance. Similarly, McGregor et al. (1999) observed that soccer players dehydrated to 2.4% BML demonstrated a 5% reduction in performance of the Loughborough Intermittent Shuttle Test and impaired soccer dribbling ability.

Not all studies, however, report consistent sport-specific skill deficits with dehydration. Two percent and 4% dehydration in young males failed to alter basketball shooting accuracy and boxing-related performance, respectively (Hoffman et al. 1995; Smith et al. 2000). This variability, how dehydration impacts sport-specific task

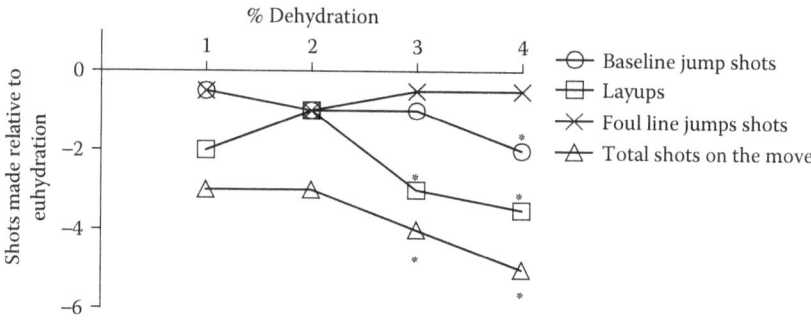

FIGURE 8.4 Basketball skill shots on the move made during 1%–4% dehydration relative to euhydration for individual drills and total (sum of baseline jump shots [8], layups [22], and foul-line jump shots [6]; single performance scores shots made [36]). Asterisks represent significantly different from euhydrated trial. (Modified from Baker, L.B. et al., *Med. Sci. Sports Exerc.* 39, 1114–23, 2007b.)

performance, likely results from a number influential confounding factors related to the dehydration (e.g., degree and mode of dehydration) and the task itself (e.g., task complexity and specificity). Further, the threshold for dehydration-induced performance impairment might vary across individuals due to the nature of the skill, dehydration tolerance, environment, and genetics. Thus, efforts should be guided toward (1) identifying the athletes and specific tasks most sensitive to dehydration, and (2) implementing appropriate hydration strategies to maintain performance.

8.5.2 MECHANISM OF DEHYDRATION INDUCES SPORT-SPECIFIC SKILL REDUCTION

Several potential mechanisms explain how dehydration impairs sport-specific skills. Although isolated dehydration up to 4% BML appears unable to impair dynamic balance (Seay et al. 2013), several balance measures suffer when dehydration is combined with exercise-induced fatigue and/or hyperthermia (Derave et al. 1998; Distefano et al. 2013). Alternately, studies report dehydration negatively alters mood, concentration, and cognition (Baker et al. 2007a,b, Cian et al. 2000, 2001; Gopinathan et al. 1988). These changes might directly result from an effect of hydration state on cognitive abilities (e.g., vigilance and decision making) or indirectly through discomfort and distraction associated with dehydration.

8.5.3 EFFECTS OF DEHYDRATION ON COGNITIVE PERFORMANCE

Numerous studies document dehydration negatively impacts cognitive function (Edwards et al. 2007; Ganio et al. 2011; Kempton 2011; Lieberman et al. 2005). This effect clearly applies to intermittent sport athletes, as many sport tasks require vigilance, mental acuity, short-term memory, working memory, decision-making, and focus.

In general, dehydration transiently and negatively impacts cognitive performance (Benton et al. 2008; Edwards et al. 2007; Ganio et al. 2011; Gopinathan et al. 1988; Grandjean and Grandjean 2007; Lieberman 2007; Sharma et al. 1986). Similar to

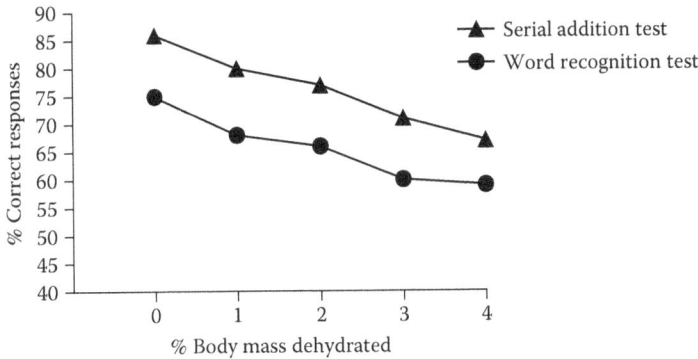

FIGURE 8.5 Cognitive skill is inversely related to the level of dehydration. Serial addition test evaluates capacity and rate of information processing and sustained and divided attention. Addition of the word recognition test. (Adapted from Lieberman, H.R., *J. Am. Coll. Nutr.* 26(Suppl 5), 555S–61S, 2007.)

strength, power, and other performance measures, cognitive deficits become significant when water loss reaches 1%–2% dehydration (Baker et al. 2007a; Edwards et al. 2007; Ganio et al. 2011; Gopinathan et al. 1988; Sharma et al. 1986) and progressively worsen with increasing BML (Baker et al. 2007a; Cian et al. 2001; Gopinathan et al. 1988; Shirreffs et al. 2004). For example, Figure 8.5 shows how the degree of cognitive performance reduction is proportional to body water loss, with diminished performance beginning around 1% BML.

Dehydration impairs cognition after passive and exercise-induced dehydration, in myriad hot and cold environments, across age groups, and among numerous occupations and sports (Benton et al. 2008; Lieberman 2007). Decision making, short-term memory, reasoning, skill achievement, speed of information processing, concentration, and hand-eye coordination all degrade following dehydration or while hypohydrated, with attention and vigilance being the first attributes affected (see Table 8.1) (Benton 2011; Cian et al. 2001; Ganio et al. 2011; Gopinathan et al. 1988; Lieberman et al. 2005).

Not all studies report dehydration induces cognitive impairment (Grandjean and Grandjean 2007). Different study parameters, including cognitive test selection, participant familiarization, degree of dehydration, exercise type, and other distractions, contribute to the often conflicting and equivocal findings (Lieberman 2007). Dehydration procedures might be especially important; independently and combined, exercise and heat exposure negatively affect cognitive performance (Cian et al. 2000). Resulting physical fatigue also impairs psychological performance. As dehydration enhances feelings of fatigue (Cian et al. 2000), distinguishing whether the ergolytic effect is primary to dehydration or secondary to dehydration-induced fatigue is difficult. Finally, the exact degree to which athletes suffer cognitive impairment appears highly individualistic (Baker et al. 2007a,b; Cian et al. 2000, 2001; Gopinathan et al. 1988; Grandjean and Grandjean 2007).

The effects of dehydration on cognitive performance appear transient, lasting only hours or minutes following dehydration in the heat and with exercise so long

TABLE 8.1

Psychological and Cognitive Performance Changes Due to Hypohydration (>2%–3% Body Mass)

Measures	Outcomes
Psychological	
Short-term memory	Decreased
Reasoning	Decreased
Attention	Decreased
Vigilance	Decreased
Processing speed	Decreased
Perceptual	
Effort	Increased
Motivation	Decreased
Vigor	Decreased
Mood	Altered

as fluid is adequately replenished (Cian et al. 2000). Therefore, proper and punctual rehydration during or after physical activity should reduce dehydration-related cognitive performance.

Research has yet to identify how dehydration induces cognitive impairment (Cheuvront and Kenefick 2014). Brain imaging studies have shown that brain fluid volume may shrink (Duning et al. 2005), enlarge (Kempton et al. 2009, 2011), or remain unchanged (Watson et al. 2010) by fluid restriction and/or exercise. Brain volume changes as a mechanism of dehydration-induced cognitive impairment seems unlikely as idiogenic osmole production has been shown to occur in the brain during progressive dehydration, conserving brain fluid (Cheuvront and Kenefick 2014; Gullans and Verbalis 1993); most cognition studies only dehydrate participants to 1%–5% BML. More likely, dehydration sequelae such as thirst, headache, mood shifts (irritability, confusion, anger, and depression), and/or discomfort distract athletes and subsequently impair cognitive prowess. Recent work demonstrates that mild hypohydration and/or dehydration of only 1%–2% affects both males and females, significantly degrading self-reported mood by increasing tiredness, perception of effort, task difficulty, headache symptoms, anxiety/tension, fatigue, and decreasing alertness (Armstrong et al. 2011; Ganio et al. 2011; Judelson et al. 2007b; Lieberman et al. 2005). Additionally, heart rate increases proportionally with dehydration and strongly correlates with the perceptual effort (Borg et al. 1987; Coyle 2001). Dehydration greater than 2%–3% decreases motivation and time to exhaustion even without compromising strength (Maresh et al. 2001). Taken together, deficits in subjective perceptions of exercise difficulty, motivation, fatigue, and mood due to hypohydration or dehydration can mitigate both cognitive and sport-specific skill performance.

8.6 MECHANISMS OF HYPOHYDRATION THAT MAY AFFECT STRENGTH, POWER, AND PERFORMANCE

Research clearly presents the probable physiological mechanisms explaining how dehydration and hypohydration impair endurance exercise, largely attributing deficits to cardiovascular impairment (Cheuvront et al. 2003). Conversely, how hydration state impacts strength and/or power in intermittent high-intensity sports is less understood. Cardiovascular, metabolic, and buffering mechanisms are unlikely culprits (Bigard et al. 2001; Gisolfi and Lamb 1991; Judelson et al. 2007a; Montain et al. 1998). A reduction in body water might negatively affect some aspects of the neuromuscular system (Ftaiti et al. 2001; Hoffman et al. 1995), possibly muscle membrane excitability (Costill et al. 1976) or the thermal influences on muscle function/elastic energy (Bennett 1984); however, evidence to support this hypothesis remains either inconclusive (Judelson et al. 2007a) or unknown (Cheuvront and Kenefick 2014). Additionally, increased permeability of the blood–brain barrier (Rapoport 2000) or a reduction in brain volume (Gullans and Verbalis 1993) might cause performance and strength/power declines, but these mechanisms lack thorough study. Changes in stress hormones might provide yet another possible explanation.

Yet another possible mechanism may lie with stress hormones; however, thus far, such a theory has not been provided a clear pathway by which dehydration affects performance. Stress hormones (epinephrine, norepinephrine, adrenocorticotropic hormone, and cortisol) increase in response to various stressors including exercise, heat, and dehydration/hypohydration. Unfortunately, most studies induce dehydration with heat and exercise, making it difficult to isolate dehydration and/or hypohydration as the sole stressor affecting these hormones. Despite this challenge, evidence shows that hypohydrated individuals undergoing alternating sessions of cycling and treadmill walking in 33°C (Castellani et al. 1997) and during exhaustive exercise (Casa et al. 2000) experience elevated concentrations of circulating stress hormones. Elevated norepinephrine and epinephrine might increase sympathetic nervous activity, offsetting plasma volume deficits through vasoconstriction and increased heart rate (Rowell 1974; Saltin 1964; Sawka et al. 1985). Furthermore, predicted changes in plasma volume and osmolality resulting from heat stress and dehydration elevate adrenocorticotropic hormone and cortisol (Brandenberger et al. 1986). Such hormones help regulate numerous physiological functions, but signal inflammation and stress when concentrations rise too high.

8.7 ASSESSING HYDRATION IN INTERMITTENT HIGH-INTENSITY SPORTS

Based on the previous review, total body water might importantly influence cognition, strength, power, aerobic, and anaerobic performance. Many athletes commonly experience fluid loss, heat stress, and fatigue, yet most fail to adequately rehydrate during or immediately following exercise (Boudou et al. 1987; Cheuvront and Haymes 2001; Hubbard et al. 1984; Maughan et al. 2004). This inadequate rehydration predisposes athletes to begin their next exercise session in a hypohydrated state (Osterberg et al. 2009; Stover et al. 2006; Volpe et al. 2009), creating a vicious cycle of dehydration, insufficient rehydration, and subsequent further dehydration, rarely allowing the exerciser to become euhydrated. The presence of this cycle is demonstrated by research that shows many athletes begin exercise in a hypohydrated state, further dehydrating while exercising (Osterberg et al. 2009; Stover et al. 2006; Volpe et al. 2009). Wrestlers, boxers, and others participating in weight class sports are especially of concern, as many of these athletes purposely dehydrate to compete in lower weight classes (Clark et al. 2004). Thus, medical and sport professionals should emphasize the development of proper hydration/rehydration strategies for athletes.

Several organizations (e.g., ACSM and NATA) offer general recommendations for (re)hydration strategies. However, universal hydration practices are giving way to more individualized guidelines tailored to fit specific athlete needs (Maughan and Shirreffs 2008, 2010b; Sawka et al. 2007). Numerous factors support the value of a personalized approach as nearly every situation represents a unique combination of physiological variability, exercise characteristics, clothing, and environmental conditions. Not surprisingly, few athletes (largely professional) have individualized hydration plans. The first step to any such plan is understanding methods to track hydration status. Although precise methods for assessing hydration status exist (see Chapter 1), these techniques are typically costly and impractical for field measurements. To maximize utility, the following Sections 8.7.1 through 8.7.5 highlight simple, cost-effective methods to track hydration levels (Maughan and Shirreffs 2010a).

8.7.1 Thirst

Thirst is likely an inadequate indicator of hydration, as most people experience thirst only after achieving 1%–2% dehydration (Armstrong 2007). Thus, thirst sensation occurs roughly coincident with the detrimental effects in strength, power, and performance, making thirst a poor indicator of real-time euhydration. Interestingly, ultra-endurance cyclists who drank to the dictates of thirst during a prolonged endurance event were no more dehydrated than their counterparts who drank *ad libitum* (Armstrong et al. 2014), suggesting that drinking *ad libitum* may not be superior than drinking to thirst. Regardless, both groups became dehydrated reinforcing the idea that thirst may be an inadequate queue if the goal is to minimalize BML (< 2%).

8.7.2 Body Mass Changes

During any exercise bout of a few hours or less, measurable body mass change is primarily due to sweat losses; recording body mass before and after exercise is therefore a simple method to closely estimate fluid loss (Pre-exercise mass − post-exercise mass/fluid consumption/urinary losses = mass lost as fluid). To completely estimate sweat rate, these calculations should be completed for a variety of exercise durations, exercise intensities, environmental conditions, and heat acclimation states. Players should consume fluids during exercise to prevent fluid losses from exceeding 1%–2% BML (Sawka et al. 2007). Following exercise, athletes should look to estimate fluid losses and attempt to replace their fluid deficit (fully accounting for obligatory urine losses) in order to return back to a euhydrated state (Shirreffs et al. 1996). Consuming excess fluid beyond what has been lost can help account for urine losses and normal fluid needs; however, athletes should not over consume fluids for risk of hyponatremia.

8.7.3 URINE VOLUME AND COLOR

Monitoring urine volume and color can help estimate hydration state (Armstrong et al. 2010). Decreased urine volume, reduced frequency of urination, attenuated urge to urinate, and darker than normal urine color might all indicate dehydration (acute) or hypohydration (chronic) (Armstrong et al. 1994, 1998, 2010). Urine color should be straw to lemonade in color and can easily be assessed using a field-expedient urine color chart (Armstrong et al. 1998). Urine color akin to apple juice or darker is suggestive of dehydration.

8.7.4 MONITORING SWEAT

Observing and estimating sweat composition can also assist in determining hydration status. Some athletes are *salty sweaters*, producing sweat of a higher salt concentration than other athletes. It has been thought that salt loss, fluid loss, and the onset of dehydration can lead to muscle cramps (Bergeron 2008); thus, assessing the relative saltiness sweat can be valuable for athletes to avoid potential performance decrements or clinical conditions (Casa et al. 2000). To qualitatively assess sweat composition, wear a dark colored t-shirt during exercise and look for *salt stains* afterward.

8.7.5 METHODS REQUIRING EQUIPMENT

Urine-specific gravity, the density of urine compared to the density of water, is a simple and quick technique that can easily be used in a laboratory of field settings to assess hydration status. Measurement requires the use of a refractometer, which is small, portable, and reasonably priced. Urine-specific gravity is more accurate than urine color, although not as reliable as more costly, less field-expedient lab measurements (Armstrong 2007).

8.8 SUMMARY

Success in sports is multifactorial. In intermittent-high intensity sports, strength, power, and performance typically contribute to an athlete's success or failure, and research suggests hydration state may impact these attributes, although considerable variation exists among the current findings. A definitive mechanism explaining how hydration impacts these key measures is currently unclear. To properly understand how hypohydration and dehydration impact strength, power, and performance, future research must account for possible confounding factors commonly seen with exercise in high-intensity intermittent sports. To maximize success, athletes should consume adequate fluids to eliminate the possibility that they might suffer dehydration-induced decrements in strength, power, and performance. Table 8.2 presents some of the key

TABLE 8.2

Degree of Water Loss (% Dehydration) at Which Performance and Cognitive Deficits Begin to Manifest

Factors Affected by Dehydration	Percent Dehydration					
	1	2	3	4	5	6
Strength						
Power						
High-intensity endurance						
Sport-specific tasks and skill						
Cognitive performance						
Perception of fatigue						

factors affected by dehydration and the percent dehydration at which they begin to manifest themselves.

Athletes in intermittent high-intensity sports can achieve euhydration with proper fluid consumption before, during, and after exercise. Fluid loss is highly individualistic; variations in genetic factors, training state, sporting demands, equipment requirements, and environmental conditions suggest each athlete should develop his or her own hydration plan.

REFERENCES

Ali, A., and C. Williams. 2013. Isokinetic and isometric muscle function of the knee extensors and flexors during simulated soccer activity: Effect of exercise and dehydration. *J. Sports Sci.* 31(8):907–16.

Armstrong, L. E. 2007. Assessing hydration status: The elusive gold standard. *J. Am. Coll Nutr.* 26(Suppl 5):575S–84S.

Armstrong, L. E., D. L. Costill, and W. J. Fink. 1985. Influence of diuretic-induced dehydration on competitive running performance. *Med. Sci. Sports Exerc.* 17(4):456–61.

Armstrong, L. E., E. C. Johnson, L. J. Kunces et al. 2014. Drinking to thirst versus drinking ad libitum during road cycling. *J. Athl. Train.* 49(5):624–31.

Armstrong, L. E., C. M. Maresh, J. W. Castellani et al. 1994. Urinary indices of hydration status. *Int. J. Sport Nutr. (United States)* 4(3):265–79.

Armstrong, L. E., A. C. Pumerantz, K. A. Fiala et al. 2010. Human hydration indices: Acute and longitudinal reference values. *Int. J. Sport Nutr. Exer Metab.* 20(2):145–53.

Armstrong, L. E., J. A. Soto, F. T. Hacker, D. J. Casa, S. A. Kavouras, and C. M. Maresh. 1998. Urinary Indices during dehydration, exercise, and rehydration. *Int. J. Sport Nutr. (United States)* 8(4):345–55.

Baker, L. B., D. E. Conroy, and W. L. Kenney. 2007a. Dehydration impairs vigilance-related attention in male basketball players. *Med. Sci. Sports Exerc.* 39(6):976–83.

Baker, L. B., K. A. Dougherty, M. Chow, and W. L. Kenney. 2007b. Progressive dehydration causes a progressive decline in basketball skill performance. *Med. Sci. Sports Exerc.* 39(7):1114–23.

Bennett, A. F. 1984. Thermal dependence of muscle function. *Am. J. Physiol.* 247(2 Pt 2):R217–29.

Bergeron, M. F. 2008. Muscle Cramps during exercise: Is it fatigue or electrolyte deficit? *Curr. Sports Med. Rep.* 7(4):S50–5.

Bigard, A. X., H. Sanchez, G. Claveyrolas, S. Martin, B. Thimonier, M. J. Arnaud. 2001. Effects of dehydration and rehydration on EMG changes during fatiguing contractions. *Med. Sci. Sports Exerc.* 33(10):1694–700.

Bosco, J. S., R. L. Terjung, and J. E. Greenleaf. 1968. Effects of Progressive hypohydration on maximal isometric muscular strength. *J. Sports. Med. Phys. Fitness.* 8(2):81–6.

Boudou, P., J. Fiet, C. Laureaux et al. 1987. Changes in several plasma and urinary components in marathon runners. *Ann. Bio Clin. (Paris)* 45(1):37–45.

Brandenberger, G., V. Candas, M. Follenius, J. P. Libert, and J. M. Kahn. 1986. Vascular fluid shifts and endocrine responses to exercise in the heat. Effect of rehydration. *Eur. J. Appl. Physiol. Occup. Physiol.* 55(2):123–9.

Burke, L. M., and J. A. Hawley. 1997. Fluid Balance in team sports: Guidelines for optimal practices. *Sports Med.* 24(1):38–54.

Casa, D. J., L. E. Armstrong, S. K. Hillman et al. 2000. National Athletic Trainers' Association position statement: Fluid replacement for athletes. *J. Athl. Train.* 35(2):212–4.

Casa, D. J., C. M. Maresh, L. E. Armstrong et al. 2000. Intravenous versus oral rehydration during a brief period: Stress hormone responses to subsequent exhaustive exercise in the heat. *Inter. J. Sport Nutr. Exer Metab.* 10(4):361–74.

Casa, D. J., R. L. Stearns, M. Lopez et al. 2010. Influence of hydration on physiological function and performance during trail running in the heat. *J. Athl. Train.* 45(2):147–56.

Castellani, J. W., C. M. Maresh, L. E. Armstrong et al. 1997. Intravenous vs. oral rehydration: Effects on subsequent exercise-heat stress. *J. App. Physiol.* 82(3):799–806.

Caterisano, A., D. N. Camaione, R. T. Murphy, and V. J. Gonino. 1988. The effect of differential training on isokinetic muscular endurance during acute thermally induced hypohydration. *Am. J. Sports Med.* 16(3):269–73.

Cheung, S. S., and G. G. Sleivert. 2004. Multiple triggers for hyperthermic fatigue and exhaustion. *Exerc. Sport Sci. Rev.* 32(3):100–6.

Cheuvront, S. N., R. Carter III, and M. N. Sawka. 2003. Fluid balance and endurance exercise performance. *Curr. Sports Med. Rep.* 2(4):202–8.

Cheuvront, S. N., and E. M. Haymes. 2001. Ad libitum fluid intakes and thermoregulatory responses of female distance runners in three environments. *J. Sports Sci.* 19(11):845–54.

Cheuvront, S. N., and R. W. Kenefick. 2014. Dehydration: Physiology, assessment, and performance effects. *Compr. Physiol.* 4(1):257–85.

Cheuvront, S. N., R. W. Kenefick, B. R. Ely et al. 2010. Hypohydration reduces vertical ground reaction impulse but not jump height. *Eur. J. Appl. Physiol.* 109(6):1163–70.

Cian, C., P. A. Barraud, B. Melin, and C. Raphel. 2001. Effects of fluid ingestion on cognitive function after heat stress or exercise-induced dehydration. *Int. J. Psychophysiol.* 42(3):243–51.

Cian, C., P. A. Barraud, and C. Raphel. 2000. Influence of variations in body hydration on cognitive function: Effect of hyperhydration, heat stress, and exercise-induced dehydration. *J. Psychophysiol.* 14(1):29–36.

Clark, R. R., C. Bartok, J. C. Sullivan, and D. A. Schoeller. 2004. Minimum weight prediction methods cross-validated by the four-component model. *Med. Sci. Sports Exerc.* 36(4):639–47.

Costill, D. L., R. Cote, and W. Fink. 1976. Muscle water and electrolytes following varied levels of dehydration in man. *J. Appl. Physiol.* 40(1):6–11.

Derave, W., D. D. Clercq, J. Bouckaert, and J. L. Pannier. 1998. The influence of exercise and dehydration on postural stability. *Ergonomics.* 41(6):782–9.

Distefano, L. J., D. J. Casa, M. M. Vansumeren et al. 2013. Hypohydration and hyperthermia impair neuromuscular control after exercise. *Med. Sci. Sports Exerc.* 45(6):1166–73.

Duning, T., S. Kloska, O. Steinstrater, H. Kugel, W. Heindel, and S. Knecht. 2005. Dehydration confounds the assessment of brain atrophy. *Neurology.* 64(3):548–50.

Edwards, A. M., M. E. Mann, M. J. Marfell-Jones, D. M. Rankin, T. D. Noakes, and D. P. Shillington. 2007. Influence of moderate dehydration on soccer performance: Physiological responses to 45 min of outdoor match-play and the immediate subsequent performance of sport-specific and mental concentration tests. *Br. J. Sports Med.* 41(6):385–91.

Evetovich, T. K., J. C. Boyd, S. M. Drake et al. 2002. Effect of moderate dehydration on torque, electromyography, and mechanomyography. *Muscle & Nerve* 26(2):225–31.

Ftaiti, F., L. Grélot, J. M. Coudreuse, and C. Nicol. 2001. Combined effect of heat stress, dehydration and exercise on neuromuscular function in humans. *Eur. J. Appl. Physiol.* 84(1–2):87–94.

Ganio, M. S., L. E. Armstrong, D. J. Casa et al. 2011. Mild dehydration impairs cognitive performance and mood of men. *Br. J. Nutr.* 106(10):1535–43.

Gisolfi, C. V., and D. R. Lamb. 1991. Perspectives in exercise science and sports medicine, volume 3: Fluid homeostasis during exercise. *Med. Sci. Sports Exerc.* 23(9):1116–308.

Godek, S. Fowkes., A. R. Bartolozzi, and J. J. Godek. 2005. Sweat rate and fluid turnover in American football players compared with runners in a hot and humid environment. *Br. J. Sports Med.* 39(4):205–11.

Gopinathan, P. M., G. Pichan, and V. M. Sharma. 1988. Role of dehydration in heat stress-induced variations in mental performance. *Arch. Environ. Health.* 43(1):15–7.

Grandjean, A. C., and N. R. Grandjean. 2007. Dehydration and cognitive performance. *J. Am. Coll. Nutr.* 26(Suppl 5):549S–54S.

Greenleaf, J. E., E. M. Prange, and E. G. Averkin. 1967. Physical performance of women following heat-exercise hypohydration. *J. Appl. Physiol.* 22(1):55–60.

Greenleaf, J. E., and F. Sargent. 1965. Voluntary dehydration in man. *J. Appl. Physiol.* 20(4):719–24.

Gullans, S. R., and J. G. Verbalis. 1993. Control of brain volume during hyperosmolar and hypoosmolar conditions. *Annu. Rev. Med.* 44:289–301.

Gutiérrez, A., J. L. M. Mesa, J. R. Ruiz, L. J. Chirosa, and M. J. Castillo. 2003. Sauna-induced rapid weight loss decreases explosive power in women but not in men. *Int. J. Sports Med.* 24(7):518–22.

Hayes, L. D., and C. I. Morse. 2010. The effects of progressive dehydration on strength and power: Is there a dose response? *Eur. J. Appl. Physiol.* 108(4):701–7.

Hoffman, J. R., H. Stavsky, and B. Falk. 1995. The effect of water restriction on anaerobic power and vertical jumping height in basketball players. *Int. J. Sports Med.* 16:214–18.

Hubbard, R. W., B. L. Sandick, W. T. Matthew et al. 1984. Voluntary dehydration and alliesthesia for water. *J. Appl. Physiol. Respir. Environ. Exerc Physiol.* 57(3):868–73.

Jones, L. C., M. A. Cleary, R. M. Lopez, R. E. Zuri, and R. Lopez. 2008. Active Dehydration impairs upper and lower body anaerobic muscular power. *J. Strength Cond. Res.* 22(2):455–63.

Judelson, D. A., C. M. Maresh, J. M. Anderson et al. 2007a. Hydration and muscular performance: Does fluid balance affect strength, power and high-intensity endurance? *Sports Med.* 37(10):907–21.

Judelson, D. A., C. M. Maresh, M. J. Farell et al. 2007b. Effect of hydration state on strength, power, and resistance exercise performance. *Med. Sci. Sports Exerc.* 39(10):1817–24.

Kempton, M. J., U. Ettinger, R. Foster et al. 2011. Dehydration affects brain structure and function in healthy adolescents. *Hum. Brain Mapp.* 32(1):71–9.

Kempton, M. J., U. Ettinger, A. Schmechtig et al. 2009. Effects of acute dehydration on brain morphology in healthy humans. *Hum. Brain Mapp.* 30(1):291–8.

King, G. A., C. J. Christie, and A. I. Todd. 2002. Effect of protective gear on skin temperature responses and sweat loss during cricket batting activity. *Sports Med.* 1–6.

Knuttgen, H. G., and W. J. Kraemer. 1987. Terminology and measurement in exercise performance. *J. Appl. Sport Sci. Res.* 1(1):1–10.

Kraft, J. A., J. M. Green, P. A. Bishop, M. T. Richardson, Y. H. Neggers, and J. D. Leeper. 2010. Impact of dehydration on a full body resistance exercise protocol. *Eur. J. Appl. Physiol.* 109(2):259–67.

Kraft, J. A., J. M. Green, P. A. Bishop, M. T. Richardson, Y. H. Neggers, and J. D. Leeper. 2011. Effects of heat exposure and 3% dehydration achieved via hot water immersion on repeated cycle sprint performance. *J. Strength Cond. Res.* 25(3):778–86.

Kraft, J. A., J. M. Green, P. A. Bishop, M. T. Richardson, Y. H. Neggers, and J. D. Leeper. 2012. The influence of hydration on anaerobic performance: A review. *Res. Q. Exerc. Sport.* 83(2):282–92.

Ladell, W. S. 1955. The effects of water and salt intake upon the performance of men working in hot and humid environments. *J. Physiol.* 127(1):11–46.

Lieberman, H. R. 2007. Hydration and cognition: A critical review and recommendations for future research. *J. Am. Coll. Nutr.* 26(Suppl 5):555S–61S.

Maughan, R. J., P. L. Greenhaff, J. B. Leiper, D. Ball, C. P. Lambert, and M. Gleeson. 1997. Diet composition and the performance of high-intensity exercise. *J. Sports Sci.* 15(3):265–75.

Maughan, R. J., S. J. Merson, N. P. Broad, and S. M. Shirreffs. 2004. Fluid and electrolyte intake and loss in elite soccer players during training. *Int. J. Sport Nutr. Exerc. Metab.* 14(3):333–46.

Maughan, R. J., and S. M. Shirreffs. 2008. Development of individual hydration strategies for athletes. *Int. J. Sport Nutr. Exerc. Metab.* 18(5):457–72.

Maughan, R. J., and S. M. Shirreffs. 2010a. Dehydration and rehydration in competitive sport. *Scand. J. Med. Sci. Sports.* 20(Suppl 3):40–7.

Maughan, R. J., and S. M. Shirreffs. 2010b. Development of hydration strategies to optimize performance for athletes in high-intensity sports and in sports with repeated intense efforts. *Scand. J. Med. Sci. Sports.* 20(Suppl 2):59–69.

Maughan, R. J., P. Watson, G. H. Evans, N. Broad, and S. M. Shirreffs. 2007. Water balance and salt losses in competitive football. *Int. J. Sport Nutr. Exerc Metab.* 17(6):583–94.

McGregor, S. J., C. W. Nicholas, H. K. Lakomy, and C. Williams. 1999. The influence of intermittent high-intensity shuttle running and fluid ingestion on the performance of a soccer skill. *J. Sports Sci.* 17(11):895–903.

McMurray, R. G., C. R. Proctor, and W. L. Wilson. 1991. Effect of caloric deficit and dietary manipulation on aerobic and anaerobic exercise. *Int. J. Sports Med.* 12(2):167–72.

Montain, S. J., R. P. Mattot, G. P. Zientara, F. A. Jolesz, and M. N. Sawka. 1998. Hypohydration effects on skeletal muscle performance and metabolism: A 31p-mrs study. *J. Appl. Physiol.* 84(6):1889–94.

Morrison, S., G. G. Sleivert, and S. S. Cheung. 2004. Passive hyperthermia reduces voluntary activation and isometric force production. *Eur. J. App. Physiol.* 91(5–6):729–36.

Osterberg, K. L., C. A. Horswill, and L. B. Baker. 2009. Pregame urine specific gravity and fluid intake by national basketball association players during competition. *J. Athl. Train.* 44(1):53–7.

Pryor, J. L., R. A. Huggins, D. J. Casa, G. A. Palmieri, W. J. Kraemer, and C. M. Maresh. 2014. A profile of a national football league team. *J. Strength Cond. Res.* 28(1):7–13.

Rankin, J. W., J. V. Ocel, and L. L. Craft. 1996. Effect of weight loss and refeeding diet composition on anaerobic performance in wrestlers. *Med. Sci. Sports Exerc.* 28(10):1292–99.

Rapoport, S. I. 2000. Osmotic opening of the blood–brain barrier: Principles, mechanism, and therapeutic applications. *Cell. Mol. Neurobiol.* 20(2):217–30.

Rodrigues, R., B. M. Baroni, M. G. Pompermayer et al. 2014. Effects of Acute dehydration on neuromuscular responses of exercised and non-exercised muscles after exercise in the heat. *J. Strength Cond. Res.* 28(12):3531–6.

Rowell, L. B. 1974. Human cardiovascular adjustments to exercise and thermal stress. *Physiol. Rev.* 54(1):75–159.

Saltin, B. 1964. Circulatory response to submaximal and maximal exercise after thermal dehydration. *J. Appl. Physiol.* 19:1125–32.

Sawka, M. N., L. M. Burke, R. E. Eichner, R. J. Maughan, S. J. Montain, and N. S. Stachenfeld. 2007. American College of Sports Medicine position stand. Exercise and fluid replacement. *Med. Sci. Sports Exerc.* 39(2):377–90.

Sawka, M. N., S. J. Montain, and W. A. Latzka. 2001. Hydration effects on thermoregulation and performance in the heat. *Comp. Biochem. Physiol. A. Mol. Integr. Physiol.* 128(4):679–90.

Sawka, M. N., and K. B. Pandolf. 1990. Effects of body water loss on physiological function and exercise performance. In: *Perspectives in Exercise Science and Sports Medicine.* Vol. 3: *Fluid Homeostasis During Exercise,* edited by C. V. Gisolfi, and D. R. Lamb, pp. 1–38. Carmel, IN: Benchmark Press.

Sawka, M. N., A. J. Young, R. P. Francesconi, S. R. Muza, and K. B. Pandolf. 1985. Thermoregulatory and blood responses during exercise at graded hypohydration levels. *J. Appl. Physiol.* 59(5):1394–401.

Schoffstall, J. E., J. D. Branch, B. C. Leutholtz, and D. E. Swain. 2001. Effects of dehydration and rehydration on the one-repetition maximum bench press of weight-trained males. *J. Strength Cond. Res.* 15(1):102–8.

Seay, J. F., B. R. Ely, R. W. Kenefick, S. G. Sauer, and S. N. Cheuvront. 2013. Hypohydration does not alter standing balance. *Motor Control.* 17(2):190–202.

Sharma, V. M., K. Sridharan, G. Pichan, and M. R. Panwar. 1986. Influence of heat-stress induced dehydration on mental functions. *Ergonomics* 29(6):791–9.

Shirreffs, S. M. 2005. The Importance of good hydration for work and exercise performance. *Nutr. Rev.* 63(6 Pt 2):S14–21.

Shirreffs, S. M., S. J. Merson, S. M. Fraser, and D. T. Archer. 2004. The effects of fluid restriction on hydration status and subjective feelings in man. *Br. J. Nutr.* 91(6):951–8.

Shirreffs, S. M., A. J. Taylor, J. B. Leiper, and R. J. Maughan. 1996. Post-exercise rehydration in man: Effects of volume consumed and drink sodium content. *Med. Sci. Sports Exerc.* 28(10):1260–71.

Smith, M. S., R. Dyson, T. Hale, J. H. Harrison, and P. McManus. 2000. The effects in humans of rapid loss of body mass on a boxing-related task. *Eur. J. Appl. Physiol.* 83(1):34–9.

Stover, E. A., H. J. Petrie, D. Passe, C. A. Horswill, B. Murray, and R. Wildman. 2006. Urine specific gravity in exercisers prior to physical training. *Appl. Physiol. Nutr. Metab.* 31(3):320–7.

Thomas, M. M., S. S. Cheung, G. C. Elder, and G. G. Sleivert. 2006. Voluntary muscle activation is impaired by core temperature rather than local muscle temperature. *J. Appl. Physiol.* 100(4):1361–9.

Volpe, S. L., K. A. Poule, and E. G. Bland. 2009. Estimation of prepractice hydration status of national collegiate athletic association division I athletes. *J. Athl. Train.* 44(6):624–9.

Watson, G., D. A. Judelson, L. E. Armstrong, S. W. Yeargin, D. J. Casa, and C. M. Maresh. 2005. Influence of diuretic-induced dehydration on competitive sprint and power performance. *Med. Sci. Sports Exerc.* 37(7):1168–74.

Watson, P., K. Head, A. Pitiot, P. Morris, and R. J. Maughan. 2010. Effect of exercise and heat-induced hypohydration on brain volume. *Med. Sci. Sports Exerc.* 42(12):2197–204.

White, C. P., C. L. Hitchcock, Y. M. Vigna, and J. C. Prior. 2011. Fluid retention over the menstrual cycle: 1-year data from the prospective ovulation cohort. *Obstet. Gynecol. Int.* 2011:138451–7.

Yeargin, S. W., D. J. Casa, D. A. Judelson et al. 2010. Thermoregulatory responses and hydration practices in heat-acclimatized adolescents during preseason high school football. *J. Athl. Train.* 45(2):136–46.

9 Effect of Dehydration on Cognitive Function, Perceptual Responses, and Mood

Dennis H. Passe

CONTENTS

9.1 INTRODUCTION

Hydration status may have an impact on a broad range of physiological functions including diseases (Popkin et al. 2010), athletic performance (Murray 2007), and cognitive function, perception, and mood (for reviews see Adan 2012; Armstrong 2012; Benton 2011; D'Anci et al. 2006; Edmonds 2012b; Grandjean and Grandjean 2007; Lieberman 2007, 2010, 2012; Maughan et al. 2007; Masento et al. 2014; McMorris et al. 2009; Popkin et al. 2010; Ritz and Berrut 2005; Secher and Ritz 2012; Shirreffs 2009; Wilson and Morley 2003). See also Tomporowski (2002) for a review of the effects of exercise on cognitive function. While water is essential for life and humans can survive only a matter of days without it, the experience of severe dehydration in daily living, sport, and exercise is rare. The range of dehydration commonly experienced (mild to moderate) and its effects on cognition, perception, and mood have been elusive to investigate. The magnitudes of the effects observed have often been moderate to small. Reliabilities in the research have been variable, likely due to inadequate statistical power related to experimental variability and small sample size. The relationship between dehydration and mental function is likely to be the result of interacting factors. There has been a paucity of well-controlled studies that have been able to investigate this complexity. The purpose of this chapter is to highlight the current status of research in this area (with an emphasis on the magnitudes of the effects [Cohen 1988] as opposed to p-values), to acknowledge design and method-ological challenges, and to bring additional focus to specific research design options now available to explore the key interactions that have been identified. The literature in the area of dehydration and mental function includes many different psychologi-cal measures with any given study, often including cognitive function (e.g., memory, learning, executive function, reaction time, accuracy, vigilance, and motor perfor-mance) and mood (e.g., Profile of Mood States, POMS; visual analogue scales, VAS).

For purposes of comparing study results, I will focus on effect size (ES) (Cohen 1988) which characterizes the magnitude of an effect (mean difference) as a simple proportion of the average standard deviation (pooled SD's associated with each of the means). To facilitate interpretation, Cohen's nomenclature of *small*, *medium*, and *large* ES will be adopted ($d = 0.2$, $d = 0.5$, $d = 0.8$ corresponding to *small*, *medium*, and *large*, respectively). In cases where correlations are reported, I have adopted Cohen's nomenclature for interpreting correlations of $r = .10$, $r = .30$, and $r = .5$ as *small*, *medium*, and *large* effects, respectively (Cohen 1988). These correlations attempt to bring further perspective to conclusions based on statistical reliability and p-values. Small ESs that are statistically significant may be less interesting than large ESs from a small study that failed to achieve traditional levels of reliability. Very few individual studies in the area of dehydration and cognitive function have discussed ESs of any kind with the exception of D'Anci et al. (2009) who reported correlations and eta-squares as magnitudes of effect.

9.2 IMPACT OF DEHYDRATION ON COGNITIVE FUNCTION AND PERCEPTUAL RESPONSES

A number of studies have looked at the relationship between dehydration and cogni-tive function using a variety of approaches including case study, field, and obser-vational (*adults*: Ackland et al. 2008; Doppelmayr et al. 2005; Lieberman et al. 2005; Suhr et al. 2004, 2010, *children*: Bar-David et al. 2005) and experimental (*adults*: Armstrong et al. 2012; Adam et al. 2008; Ainslie et al. 2002; Bandelow et al. 2010; Baker et al. 2007; Cian et al. 2000, 2001; D'Anci et al. 2009; Ely et al. 2013; Ganio et al. 2011; Gopinathan et al. 1988; Grego et al. 2005; Irwin et al. 2013; Leibowitz et al. 1972; Lindseth et al. 2013; Neave et al. 2001; Ogino et al. 2013; Petri et al. 2006; Serwah and Marino 2006; Sharma et al. 1986; Slaven and Windle 1999; Smith et al. 2012; Szinnai et al. 2005; Shirreffs et al. 2004, *children*: none). Several studies have looked at the effect of fluid intake on cognitive function from a naturally occurring state of hydration (euhydration, or at least no intentional induction of dehydration) (*adults*: Edmonds et al. 2013; Rogers et al. 2001, *chil-dren*: Benton and Burgess 2009; Booth et al. 2012; Edmonds 2012a; Edmonds and Burford 2009; Edmonds and Jeffes 2009; Fadda et al. 2012). A related area of active research investigates the impact of a glucose drink on mental function (e.g., Scholey et al. 2009). This and other nutritional interventions relative to cognitive function are outside the scope of this review and will not be covered here except to note that the perception of thirst (even in the absence of measured dehydration) may impact the effect of drinking on cognitive function (Scholey et al. 2009; Rogers et al. 2001).

9.2.1 CORRELATIONAL APPROACH

Suhr et al. (2004) explored available correlational data (from a larger study), using hierarchical regression to explore relationships between dehydration (measured by bioimpedance which was used to estimate percent total body water by weight, %TBW/WT) and neuropsychological measures of psychomotor processing speed and attention/memory (assessed by the summation of z-scores of batteries of psychological subtests measuring these constructs). Half of the subjects had been instructed to refrain from eating and drinking the night before. While group analysis failed to find a significant difference between the two groups in %TBW/WT (at the usual alpha level of 0.05), there was a tendency for the two groups to differ ($p = .07$), as might be expected. On this basis, the groups were combined for the regression modeling. However, the ES was 0.71, which may be viewed as at least of moderate magnitude. Preliminary correlational analysis revealed that lower %TBW/WT was related to slower psychomotor processing speed ($r = .49$, $p < .01$) and tended to be related to worse memory performance ($r = .34$, $p = .08$). Greater age was related to slower psychomotor processing speed ($r = .47$, $p < .01$). Systolic blood pressure (SBP) was related to poorer memory performance ($r = -.37$, $p = .06$). Two hierarchical models were then fit to the data to assess whether hydration would still be related to cognitive function after controlling for demographics (age and education) and a physiological variable (SBP). Results indicated that hydration status (%TBW/WT) remained reliably related to psychomotor processing ($p = .01$) and memory ($p = .03$). Interestingly, the absolute values of the standardized betas in the full model for psychomotor processing speed for age (0.39) and %TBW/WT (0.46) were of comparable magnitude suggesting that in this model the contributions of age and %TBW/WT were comparable. SBP was not a predictor of cognitive function in the full model. In a subsequent correlational analysis (Suhr et al. 2010) in 21 older women (60.3 ± 8.03 years) of naturally occurring hydration status to other measures of memory (auditory verbal learning test, AVLT, and auditory consonant trigrams, ACT), better hydration status (%TBW/WT) was also reliably related to these measures of memory ($r = .54$, $p = .01$; and $r = .47$, $p = .04$ for AVLT and ACT, respectively). In the current study (Suhr et al. 2010), unlike in the previous one (Suhr et al. 2004) age was not reliably related to %TBW/WT ($r = -.14$, $p = .50$). However, poorer hydration status was related to higher blood pressure (both diastolic blood pressure, DBP, and SBP), and higher blood pressure appeared to be a mediating factor for poorer immediate recall (AVLT: DBP $r = -.57$, $p = .009$, SBP $r = -.46$, $p < .05$) and poorer working memory (ACT: DBP $r = -.67$, $p = .002$, SBP $r = -.64$, $p < .003$). These studies, even though lacking an experimental frame of reference, demonstrate the importance of modeling available data to identify relationships and the potential interplay of interacting factors.

In a controlled prospective observational study, Ackland et al. (2008) examined the impact of dehydration induced by bowel prep (Citramag) in 38 men (age 63 ± 12) scheduled for elective colonoscopy with a control group of 14 men (age 60 ± 11)

matched for age, comorbidities, medications, education, and gender scheduled for sigmoidoscopy who did not undergo bowel prep. The confounding effects of surgery and anesthesia as well as exercise and exposure to heat were avoided in this study. The inclusion criteria for subjects were age 50 years or more, absence of cerebrovascular disease, visual or hearing impairments, acute or chronic psychosis, depression, learning disabilities or dementia, no current illness requiring hospitalization, and able to respond to written and spoken English. A power calculation was conducted to estimate the number of subjects required for alpha = 0.05 and power = 0.8 based on an estimated incidence of measureable cognitive decrement of 30% in this age group undergoing bowel prep (relative to a background cognitive dysfunction of 1% in the untreated population) and a relative frequency of patients presenting for colonoscopy versus sigmoidoscopy of 3 to 1. Weight loss (percent total body weight) for the experimental group was 2.0% (95% CI: 1.3%–2.6%) and a calculated median decrease of TBW of 2.6% (95% CI: 1.1%–4.8%), while the control group did not change in body weight (0.17 kg [−0.2 to 0.6kg]). Bioimpedance measures changed significantly for the bowel prep group (36 [Omega] [95% CI: 25–46]) but not for the control group (−1 [Omega] [95% CI: −13 to 10]). Neuropsychological tests measuring attention and executive functions (trail making test) and memory/learning (Rey AVLT) were conducted by an experienced psychologist blinded to the weight loss condition. Alternate forms of the tests were used to minimize practice effects. These tests were chosen because of demonstrated sensitivity to detect decrement in cognitive function following surgery. In addition to group comparisons of absolute scores, the results for each individual were compared to normative population data for these tests (criterion for impairment: >1 SD below the mean of the norm). In addition, since test anxiety may impact performance on neuropsychological tests, the Spielberger state-trait anxiety and subjective cognitive scales (self-report of perceived cognitive status) were also conducted. Tests were conducted twice, the second 3 days ± 1 day after the first (on the day of the procedure). There were no differences in neuropsychological test results between treatments and over time. The performance of all subjects was within 1 SD of the normative mean. There were no significant correlations between any of the hydration markers (% weight change, bioimpedance change, or calculated %TBW change) and changes in cognitive performance. However, there was a significant negative association between increasing age and neuropsychological scores at baseline ($r = −.47, p = .005$), and a significant positive association between education and better baseline neuropsychological performance ($r = .49, p = .003$).

The results of these correlational studies (Ackland et al. 2008, Suhr et al. 2004, 2010) are not consistent regarding the impact of hydration status, age, education, or blood pressure on memory. This may be related to the low sample sizes, potential restrictions of the range in the populations tested, and procedural differences within the measurement of memory. In spite of these short comings, the methods they employ may be able to make a substantial contribution to our understanding of the potential contribution of hydration status, in conjunction with many other factors, to cognitive function. They are ideally suited to the modeling of databases in biomedical research where routine and consistent measurement of cognitive function may already be occurring as part of ongoing research and where there is a substantial number of subjects to assess. However, requisite to this end is the implementation of at least simple measures of hydration status. This approach would also help to remedy the constrained nature of *stand-alone* studies (whether correlational or experimental) that must focus on a limited number of subjects.

9.2.2 EXPERIMENTAL APPROACH

Ely et al. (2013) in a well-controlled study investigated the impact of euhydration versus hypohydration (−4% body mass) under four different ambient temperature conditions (10°C, 20°C, 30°C, and 40°C). Thirty-two healthy, nonheat acclimated men (age 22 + 4.2 years) were assigned to four cohorts (of eight subjects each) matched for aerobic fitness. Five familiarization days in which a baseline was established for cognitive function, mood status, balance performance, and body weight were followed by a crossover design in which the order of EUH and HYP trials was randomly assigned (each trial of one day's duration separated by a week) controlling for time of the day and prior day's physical activity level. Both EUH and HYP trials involved exposure to alternating treadmill walking and seated rest at 50°C for 3 h either with (EUH) or without (HYP to achieve −4% body mass) fluid replacement followed by 90 min of controlled rest and then exposure to their respective environmental condition and cognitive, psychomotor, reaction-time, matching-to-sample, grammatical reasoning, and mood (POMS) tests. The dynamic balance test assessed subjects' ability to control their center of gravity and balance using visual feedback from an liquid crystal display screen. The cognitive tests chosen were previously shown to be sensitive to hydration status (Ganio et al. 2011). Subjects in EUH trials were within 1% of their 5-day body mass average. HYP groups were $-4.1 \pm 0.5\%$, $-4.2 \pm 0.2\%$, $-4.0 \pm 0.3\%$, and $-4.1 \pm 0.5\%$ body mass relative to baseline for 10°C, 20°C, 30°C, and 40°C ambient conditions, respectively. Thirst ratings were significantly higher ($p < .05$) during HYP than EUH trials for all ambient conditions. There were no effects of hydration or temperature on any cognitive measure and no interactions between hydration status and temperature. However, there was a significant effect of hydration status on the POMS total mood disturbance score, and for the anger/hostility, confusion/bewilderment, depression/dejection, and fatigue subscales (HYP > EUH, $p < .05$ for all comparisons), but not for the vigor/activity and tension/anxiety subscales. There was no interaction between hydration status and temperature for the POMS. Hydration status did not alter dynamic balance. This study is notable in its experimental control to mitigate the confounding effects of hydration status and heat exposure (by introducing a 90-min cool down period) yet investigating the impact of a range of environmental temperature conditions on cognition, mood, and a measure of physical performance. Care was also taken to control for cognitive learning effects by the use of an extensive baseline familiarization period (5 days). The percent reduction in body mass was substantial and well within the range of what would be expected to have an impact on cognitive function. While the ESs in the current study were generally S to M for the cognitive measures, the sample size of the individual cohorts ($n = 8$) was small, perhaps reducing power in spite of advanced experimental control. An additional factor may have been extra effort on the part of subjects to compensate for the additional stress from HYP and temperature conditions (Ely et al. 2013). As the quality of experimental control (and effective power) increases, *failure to find* becomes increasingly supportive of the conclusion that the effects being investigated are small and not reliable in the context of current test power.

Failure to find a reliable impact of dehydration on cognitive measures was also reported by Armstrong et al. (2012). In a controlled double-blind study, the authors investigated the effect of mild dehydration ($-1.36 \pm 0.16\%$ body mass) in 25 females. Treatment condition was disguised to both subjects and investigators via the use of a positive control condition in which dehydration was induced by a combination of both a diuretic and exercise. A pill (either a diuretic or placebo) was given at the beginning of every session, and fluid (volume disguised) was given during every session. The design was a crossover in which subjects experienced exercise-induced dehydration with a placebo capsule, exercise-induced dehydration plus a diuretic capsule, and euhydration plus a placebo capsule. Subjects as well as the investigators collecting the data in the environmental chamber were unaware of experimental conditions. Afterward, subjects were not able to identify hydrated versus dehydrated experimental conditions. Essentially, cognitive functions measured in this study were unaffected by mild dehydration; however, measures of mood and self-perception (of task difficulty, ability to concentrate, and head ache) were adversely affected.

9.2.3 FIELD STUDY APPROACH

In a study of football players in match play, the impact of varying rehydration and cooling strategies along with covariates of core temperature, body mass, plasma osmolality, and glucose levels on cognitive function were investigated (Bandelow et al. 2010). Data were collected from three matches immediately before, at half-time break, and immediately after play. Heat index apparent temperature for the three matches were $45°C \pm 2$, $46°C \pm 2$, and $43°C \pm 2$, respectively. Players were encouraged to follow their usual hydration practices before and during the first match. Starting with the second match, players were provided sports beverages and water and encouraged to increase fluid intake each day. Measured dehydration was mild to moderate ranging up to 2.5%. Cognitive tests (via laptop) included the visual sensitivity test (visuomotor reaction time, touch targets on screen as quickly as possible), finger tapping test (fine motor speed), Sternberg test (visual serial working memory, identification of target items in a list), Corsi block-tapping test (working memory capacity for spatial locations, repeat target location sequence on touch screen). Cognitive testing took approximately 15 min except for half-time at which shorter versions of the tests were used (8 min, Corsi block-tapping omitted at half-time testing). Mixed effects statistical modeling was conducted to assess the impact of hydration and cooling strategies along with body mass loss, plasma osmolality, plasma glucose levels, and core temperature on reaction time and accuracy in the cognitive tests. This approach allowed enhanced use of per-player detail in the modeling unlike factorial ANOVA-based methods relying on group-level information. The effects of dehydration (defined by body mass loss and plasma osmolality) were inconsistent across all four cognitive measures. Changes in plasma osmolarity had no significant effects on any cognitive test. Body mass loss only had significant effects on the finger tapping test (increase in speed as a function of body mass loss) and the Sternberg test of working memory (increase in reaction time and decrease in reaction time as a function of body mass loss depending on working memory load). Interestingly, higher core temperature had a significant negative impact on reaction time in all cognitive tests, and plasma glucose level had effects essentially opposite to those of core temperature. Higher glucose levels were associated with faster performance on all cognitive tests (where speed was measured) but this increase in speed did come at some cost to accuracy (in the Sternberg and Corsi test for working memory). In all instances where sports drink was related to cognitive function, the effect disappeared when plasma glucose levels were entered into the model suggesting that the sports drink effects in this study were driven by carbohydrate content. The impact of the use of the cooling tent was inconsistent on cognitive function and was not related to changes in core temperature. This investigation (Bandelow et al. 2010), which relied on relatively uncontrolled experimental manipulation in the field to increase the range of variability in a variety of independent measures, demonstrated via the use of extensive modeling of individual data the relative importance of dehydration to other factors, such as blood glucose. Bandelow et al. (2010) found that blood glucose may be more important than dehydration in impact on cognitive function. As Bandelow et al. (2010) point out, research on dehydration and cognitive function must include a range of potential moderating variables measured simultaneously.

In a small field study with 23 young children (6–7 years), Edmonds et al. (2009) compared cognitive performance (visual memory, visual search, and visuomotor performance) and mood (self-rated happiness) in two groups balanced for age, sex, and special education needs. The water group received 500 ml of water and was told they could drink as much as they wanted; the no-water group was not given any extra water and was unaware of the conditions relating to the water group. The water group drank on average 409.1 ml (SD = 130.2 ml). The two groups were tested at baseline and after their respective treatment or no treatment conditions and compared on the basis of change from baseline. The water group showed decreased ratings of thirst, increased ratings of happiness, and improved performance on visual attention, and visual search, but not visual memory or visuomotor performance. Interestingly, when thirst and mood were entered into subsequent exploratory analyses as covariates, they were not significant, indicating that neither thirst nor mood was the driver of the observed effects in the water group. The effects observed in Edmonds et al. (2009) suggest that in children, having a drink of water in the absence of intentional dehydration through exercise, heat or withholding fluids can impact cognitive function and mood. Additional testing and replication is warranted to verify these findings.

Even though the relationship between dehydration and cognitive function and mood is not an extensively investigated area in thirst physiology, there already is a diverse mix of approaches, dependent measures, and findings. See Table 9.1 for an ES summary of published dependent measures in studies investigating dehydration relative to cognitive function, mood, and self-perception. Although this is an attempt to include all published studies, it is recognized that this analysis may not be complete. However, it is unlikely that the observed trends will change substantially. Of 429 ES in this literature, 283 are cognitive measures, and 149 are mood or

TABLE 9.1

Summary of Findings of Studies Assessing the Effect of Dehydration on Cognitive Function, Mood, and Self Perception

Reference	Subjects	Dehydration Level (Method of Dehydration)	Cognitive Task/Mood Measurement	ES Cohen's *d*	ES Pearson *r*	Interpretation
Ackland et al. (2008)	23 adult M, 15 adult F	2% (bowel prep for colonoscopy)	Trail making test A	0.14		S
			Trail making test B	**0.16**		S
			Rey recall test	0.24		S
			Rey learning test	0.06		S
			Anxiety inventory	1.00		L
			Subjective cognition	0.40		M
Ainslie et al. (2002)	8 adult M	3% (exercise)	Choice reaction time		0.79	L
Armstrong et al. (2012)	25 adult F	>1% (exercise)	Scanning visual vigilance (at rest)			
			Correct responses	0.10		
			Reaction time	0.21		
			False alarms	0.38		
			Psychomotor vigilance test (at rest)			
			Correct hits	0.17		
			Premature errors	**0.21**		
			Reaction time	0.00		
			Four-choice reaction time (at rest)			
			Reaction time	0.00		
			Incorrect responses	**0.10**		
			Time-out errors	0.37		
			Matching to sample (at rest)			
			Correct responses	0.13		

(Continued)

TABLE 9.1 (Continued)
Summary of Findings of Studies Assessing the Effect of Dehydration on Cognitive Function, Mood, and Self Perception

Reference	Subjects	Dehydration Level (Method of Dehydration)	Cognitive Task/Mood Measurement	ES Cohen's d	ES Pearson r	Interpretation
			Time-out errors	0.37		
			Reaction time	**0.21**		
			Repeated acquisition			
			Incorrect responses	**0.02**		
			Time to complete	**0.18**		
			Grammatical reasoning (at rest)			
			Correct responses	0.16		
			Incorrect responses	0.16		
			No response	0.05		
			Reaction time	0.12		
			POMS (at rest)			
			Tension-anxiety	0.35		
			Depression-dejection	0.24		
			Anger-hostility	0.34		
			Vigor-activity	0.54		
			Fatigue-inertia	0.89		
			Confusion-bewilderment	0.47		
			Total mood disturbance	0.58		
			VAS (at rest)			
			Task difficulty	0.76		
			Concentration	0.70		
			Headache	0.60		

(Continued)

TABLE 9.1 (Continued)

Summary of Findings of Studies Assessing the Effect of Dehydration on Cognitive Function, Mood, and Self Perception

Reference	Subjects	Dehydration Level (Method of Dehydration)	Cognitive Task/Mood Measurement	ES Cohen's d	ES Pearson r	Interpretation
			Psychomotor vigilance test (exercise)			
			Correct hits, n	0.06		
			Premature errors, n	0.24		
			Reaction time, s	0.00		
			POMS (exercise)			
			Tension-anxiety	0.39		
			Depression-dejection	0.28		
			Anger-hostility	0.30		
			Vigor-activity	0.41		
			Fatigue-inertia	0.66		
			Confusion-bewilderment	0.52		
			Total mood disturbance	0.49		
			VAS (exercise)			
			Task difficulty	0.51		
			Concentration	0.62		
			Headache	0.58		
			RPE	0.15		
			Pain rating	0.10		
Baker et al. (2007)	11 adult M	1%–4% (exercise, heat)	Test of variables of attention (TOVA)			
			Target infrequent task: Sensitivity	1.04		
			Target infrequent task: Response time	1.00		

(Continued)

TABLE 9.1 (*Continued*)
Summary of Findings of Studies Assessing the Effect of Dehydration on Cognitive Function, Mood, and Self Perception

Reference	Subjects	Dehydration Level (Method of Dehydration)	Cognitive Task/Mood Measurement	ES Cohen's *d*	ES Pearson *r*	Interpretation
			Target infrequent task: Omission errors	1.10		
			Target frequent task: Omission errors	0.82		
			Target frequent task: Commission errors	1.09		
			Target frequent task: Sensitivity	0.76		
			Target frequent task: Response time	0.88		
			VAS: Subjective ratings			
			Lightheadedness exercise/heat	*1.90*		
			Lightheadedness, halftime	*1.90*		
			Lightheadedness, end	*1.08*		
			Hot/overheated exercise/heat	*1.32*		
			Hot/overheated, halftime	*1.24*		
			Hot/overheated, end	*1.04*		
			Total body fatigue exercise/heat	*1.54*		
			Total Body Fatigue, halftime	*1.04*		
			Total Body Fatigue, end	*0.70*		
Bar-David et al. (2005)	58 children	>800 mOsm/kg (observational)	AM testing			
			Hidden figures	0.03		
			Number span	0.54		
			Making groups	**0.11**		
			Verbal analogies	0.18		

(*Continued*)

TABLE 9.1 (*Continued*)

Summary of Findings of Studies Assessing the Effect of Dehydration on Cognitive Function, Mood, and Self Perception

Reference	Subjects	Dehydration Level (Method of Dehydration)	Cognitive Task/Mood Measurement	ES Cohen's d	ES Pearson r	Interpretation
			Number addition	0.23		
			Noon testing			
			Hidden figures	0.37		
			Number span	0.86		
			Making groups	0.37		
			Verbal analogies	0.50		
			Number addition	0.17		
Benton and Burgess (2009)	40 children	No dehydration reported, studied impact of added water	Recall of objects test (immediate)	0.29		
			Recall of objects test (delayed)	0.32		
Cian et al. (2000)	8 adult M	2.8% (heat)	*VAS: Subjective ratings*			
			Fatigue	*0.39*		
			Mood	*0.20*		
			Serial reaction time	0.05		
			Unstable tracking task	0.42		
			Perceptual comparison % correct	**0.11**		
			Perceptual comparison reaction time	0.37		
			Short-term memory	0.64		
			Free recall figure	**0.03**		
			Recognition	**0.12**		

(Continued)

TABLE 9.1 (*Continued*)
Summary of Findings of Studies Assessing the Effect of Dehydration on Cognitive Function, Mood, and Self Perception

Reference	Subjects	Dehydration Level (Method of Dehydration)	Cognitive Task/Mood Measurement	ES Cohen's *d*	ES Pearson *r*	Interpretation
		2.8% (exercise)	*VAS: Subjective ratings*			
			Fatigue	*0.89*		
			Mood	*0.17*		
			Serial reaction time	**0.22**		
			Unstable tracking task	0.49		
			Perceptual comparison % correct	**0.05**		
			Perceptual comparison reaction time	0.40		
			Short-term memory	0.74		
			Free recall Figure	0.08		
			Recognition	**0.18**		
Cian et al. (2001)	7 adult M	2.8% (heat)	*VAS: Subjective ratings*			
			Fatigue (Test 1)	0.67		
			Fatigue (Test 2)	0.71		
			Mood (Test 1)	0.19		
			Mood (Test 2)	0.21		
			Perceptual comparison % correct (Test 1)	0.25		
			Perceptual comparison % correct (Test 2)	0.31		
			Reaction time (Test 1)	0.31		
			Reaction time (Test 2)	0.20		
			Short-term memory (Test 1)	0.90		
			Short-term memory (Test 2)	0.52		
			Free recall (Test 1)	0.23		

(*Continued*)

TABLE 9.1 (Continued)
Summary of Findings of Studies Assessing the Effect of Dehydration on Cognitive Function, Mood, and Self Perception

Reference	Subjects	Dehydration Level (Method of Dehydration)	Cognitive Task/Mood Measurement	ES Cohen's d	ES Pearson r	Interpretation
		2.8% (exercise)	Free recall (Test 2)	0.23		
			VAS: Subjective ratings			
			Fatigue (Test 1)	0.79		
			Fatigue (Test 2)	0.87		
			Mood (Test 1)	0.28		
			Mood (Test 2)	0.27		
			Perceptual comparison % correct (Test 1)	0.35		
			Perceptual comparison % correct (Test 2)	0.26		
			Reaction time (Test 1)	0.27		
			Reaction time (Test 2)	0.28		
			Short-term memory (Test 1)	0.71		
			Short-term memory (Test 2)	0.50		
			Free recall (Test 1)	**0.10**		
			Free recall (Test 2)	0.15		
D'Anci et al. (2009)	16 adult M, 15 adult F	1.82% (exercise)	POMS			
			Anger		0.61	
			Fatigue		0.52	
			Depression		0.46	
			Tension		0.53	
			Confusion		0.53	

(Continued)

TABLE 9.1 (Continued)
Summary of Findings of Studies Assessing the Effect of Dehydration on Cognitive Function, Mood, and Self Perception

Reference	Subjects	Dehydration Level (Method of Dehydration)	Cognitive Task/Mood Measurement	ES Cohen's d	ES Pearson r	Interpretation
			Vigor		0.43	
	19 adult M	2% (exercise)	Choice reaction time	**0.21**		
			Choice reaction time errors	**0.74**		
	13 adult F	1.65% (exercise)	Choice reaction time	**0.27**		
			Choice reaction time errors	0.68		
	16 adult M, 13 adult F	1.82% (exercise)	Choice reaction time	**0.24**		
			Choice reaction time errors	**0.10**		
	8 adult M	2% (exercise)	Digit span	**1.37**		
	7 adult F	1.65% (exercise)	Digit span	**0.51**		
	8 adult M, 7 adult F	1.82% (exercise)	Digit span	**0.90**		
	16 adult M	2% (exercise)	Reaction time continuous performance: First 5 min	1.60		
			Reaction time continuous performance: Second 5 min	1.20		
			Reaction time continuous performance: Third 5 min	1.20		
	15 adult F	1.65% (exercise)	Reaction time continuous performance: First 5 min	0.00		
			Reaction time continuous performance: Second 5 min	0.80		

(Continued)

TABLE 9.1 (*Continued*)

Summary of Findings of Studies Assessing the Effect of Dehydration on Cognitive Function, Mood, and Self Perception

Reference	Subjects	Dehydration Level (Method of Dehydration)	Cognitive Task/Mood Measurement	ES Cohen's *d*	ES Pearson *r*	Interpretation
	16 adult M, 15 adult F	1.82% (exercise)	Reaction time continuous performance: Third 5 min	2.00		
			Reaction time continuous performance: First 5 min	1.50		
			Reaction time continuous performance: Second 5 min	1.50		
			Reaction time continuous performance: Third 5 min	2.00		
	6 adult M	1.44% (exercise)	Number of blanks left in map recall	1.03		
	6 adult M	1.44% (exercise), glucose	Number of blanks left in map recall	**1.23**		
	6 adult F	0.99% (exercise)	Number of blanks left in map recall	**1.84**		
	5 adult F	1.23% (exercise), glucose	Number of blanks left in map recall	1.21		
Edmonds and Burford (2009)	26 children M, 32 children F	No dehydration reported, studied impact of added water	Story memory	0.32		
			Letter cancellation	0.61		
			Spot the difference—Easy	0.66		
			Spot the difference—Hard	0.71		
			Visuomotor tracking task	0.21		

(*Continued*)

TABLE 9.1 (Continued)
Summary of Findings of Studies Assessing the Effect of Dehydration on Cognitive Function, Mood, and Self Perception

Reference	Subjects	Dehydration Level (Method of Dehydration)	Cognitive Task/Mood Measurement	ES Cohen's d	ES Pearson r	Interpretation
Edmonds and Jeffes (2009)	9 children M, 14 children F	No dehydration reported, studied impact of added water	*Happiness ratings*	**0.01**		
			Visual attention	1.11		
			Visual memory	0.88		
			Visual search	1.15		
			Visuomotor precision	0.15		
Edmonds et al. (2013)	15 adult M, 32 adult F	No dehydration reported, studied impact of added water	Letter cancellation Test 1: Expectancy	**0.46**		
			Forward digit span Test 1: Expectancy	**0.67**		
			Backward digit span Test 1: Expectancy	**0.39**		
			Simple reaction time mean Test 1: Expectancy	**1.05**		
			Letter cancellation Test 1: No Expectancy	**0.24**		
			Forward digit span Test 1: No expectancy	0.34		
			Backward digit span Test 1: No expectancy	0.33		

(Continued)

TABLE 9.1 (Continued)
Summary of Findings of Studies Assessing the Effect of Dehydration on Cognitive Function, Mood, and Self Perception

Reference	Subjects	Dehydration Level (Method of Dehydration)	Cognitive Task/Mood Measurement	ES Cohen's d	ES Pearson r	Interpretation
			Simple reaction time mean Test 1: No expectancy	**0.50**		
			VAS: Mood			
			Afraid: Expectancy	**0.18**		
			Confused: Expectancy	**0.02**		
			Sad: Expectancy	**0.12**		
			Angry: Expectancy	0.27		
			Energetic: Expectancy	0.09		
			Tired: Expectancy	**0.54**		
			Happy: Expectancy	**0.13**		
			Tense: Expectancy	0.14		
			Afraid: No expectancy	0.51		
			Confused: No expectancy	0.27		
			Sad: No expectancy	0.32		
			Angry: No expectancy	0.14		
			Energetic: No expectancy	0.26		
			Tired: No expectancy	0.16		
			Happy: No expectancy	**0.13**		
			Tense: No expectancy	**0.68**		
Ely et al. (2013)	32 adult M	4% (exercise, heat)	10°C test condition			
			PVT (reaction time)	0.05		
			4-Choice (correct responses)	0.09		

(Continued)

TABLE 9.1 (Continued)

Summary of Findings of Studies Assessing the Effect of Dehydration on Cognitive Function, Mood, and Self Perception

Reference	Subjects	Dehydration Level (Method of Dehydration)	Cognitive Task/Mood Measurement	ES Cohen's d	ES Pearson r	Interpretation
			Match (correct responses)	0.00		
			Gram (correct responses)	0.00		
			20°C test condition			
			PVT (reaction time)	**0.15**		
			4-Choice (correct responses)	0.69		
			Match (correct responses)	0.33		
			Gram (correct responses)	0.00		
			30°C test condition			
			PVT (reaction time)	**0.27**		
			4-Choice (correct responses)	**0.33**		
			Match (correct responses)	0.50		
			Gram (correct responses)	0.00		
			40°C test condition			
			PVT (reaction time)	**0.22**		
			4-Choice (correct responses)	0.44		
			Match (correct responses)	0.33		
			Gram (correct responses)	0.13		
			10°C test condition: POMS			
			TMD	*0.11*		
			Tension	**0.15**		
			Depression	*0.22*		
			Anger	*0.00*		

(Continued)

TABLE 9.1 (Continued)
Summary of Findings of Studies Assessing the Effect of Dehydration on Cognitive Function, Mood, and Self Perception

Reference	Subjects	Dehydration Level (Method of Dehydration)	Cognitive Task/Mood Measurement	ES Cohen's d	ES Pearson r	Interpretation
			Vigor	**0.15**		
			Fatigue	*0.50*		
			Confusion	**0.20**		
			20°C test condition: POMS			
			TMD	*0.56*		
			Tension	*0.22*		
			Depression	*0.00*		
			Anger	*0.00*		
			Vigor	*0.25*		
			Fatigue	*1.33*		
			Confusion	*0.40*		
			30°C test condition: POMS			
			TMD	*1.38*		
			Tension	*1.00*		
			Depression	*1.11*		
			Anger	*1.11*		
			Vigor	*0.33*		
			Fatigue	*0.89*		
			Confusion	*1.00*		
			40°C test condition: POMS			
			TMD	*0.32*		
			Tension	*0.40*		

(Continued)

TABLE 9.1 (*Continued*)

Summary of Findings of Studies Assessing the Effect of Dehydration on Cognitive Function, Mood, and Self Perception

Reference	Subjects	Dehydration Level (Method of Dehydration)	Cognitive Task/Mood Measurement	ES Cohen's *d*	ES Pearson *r*	Interpretation
Fadda et al. (2012)	82 children M, 86 children F	>800 mOsm/kg (observational)	*Depression*	*0.12*		
			Anger	*0.60*		
			Vigor	**0.15**		
			Fatigue	*0.20*		
			Confusion	*0.80*		
			Auditory number span		−0.56	
			Selective attention		−0.08	
			Number addition		−0.20	
			Verbal analogies		0.58	
			Visual spatial ability		−0.33	
			POMS vigor		−0.56	
			POMS fatigue		−0.12	
			POMS confusion		0.54	
Ganio et al. (2011)	26 adult M	>1% (exercise, diuretic)	Scanning visual vigilance: At rest			
			Correct responses	0.05		
			Reaction time (s)	0.26		
			False alarms	0.30		
			Psychomotor vigilance test: At rest			
			Correct hits	0.10		
			Premature errors	**0.12**		

(Continued)

TABLE 9.1 (*Continued*)

Summary of Findings of Studies Assessing the Effect of Dehydration on Cognitive Function, Mood, and Self Perception

Reference	Subjects	Dehydration Level (Method of Dehydration)	Cognitive Task/Mood Measurement	ES Cohen's *d*	ES Pearson *r*	Interpretation
			Reaction time (s)	0.00		
			Four-choice reaction time: At rest			
			Reaction time (ms)	0.18		
			Incorrect responses	0.16		
			Time-out errors	**0.08**		
			Matching to sample: At rest			
			Correct responses	**0.05**		
			Time-out errors	0.00		
			Response time (s)	0.26		
			Repeated acquisition: At rest			
			Incorrect responses	0.18		
			Time to complete (s)	**0.11**		
			Grammatical reasoning: At rest			
			Correct responses	0.12		
			Incorrect responses	0.12		
			Response time (s)	0.01		
			POMS: At rest			
			Tension/anxiety	*0.26*		
			Depression/dejection	*0.02*		
			Anger/hostility	**0.06**		
			Vigor/activity	*0.15*		
			Fatigue/inertia	*0.27*		

(*Continued*)

TABLE 9.1 (Continued)
Summary of Findings of Studies Assessing the Effect of Dehydration on Cognitive Function, Mood, and Self Perception

Reference	Subjects	Dehydration Level (Method of Dehydration)	Cognitive Task/Mood Measurement	ES Cohen's d	ES Pearson r	Interpretation
			Confusion/bewilderment	0.12		
			Total mood disturbance	0.16		
			VAS: At rest			
			Task difficulty	0.05		
			Concentration	0.00		
			Headache	0.11		
			Psychomotor vigilance test: During exercise			
			Correct hits	**0.22**		
			Premature errors	**0.36**		
			Reaction time (s)	**0.25**		
			POMS: During exercise			
			Tension/anxiety	0.05		
			Depression/dejection	0.02		
			Anger/hostility	**0.02**		
			Vigor/activity	0.12		
			Fatigue/inertia	0.19		
			Confusion/bewilderment	0.03		
			Total mood disturbance	0.09		
			VAS: During exercise			
			Task difficulty	0.10		
			Concentration	0.13		

(Continued)

TABLE 9.1 (Continued)
Summary of Findings of Studies Assessing the Effect of Dehydration on Cognitive Function, Mood, and Self Perception

Reference	Subjects	Dehydration Level (Method of Dehydration)	Cognitive Task/Mood Measurement	ES Cohen's d	ES Pearson r	Interpretation
Gopinathan et al. (1988)	11 adult M	1% (exercise, heat)	*Headache*	*0.00*		
			Serial addition test (% correct responses)	0.39		
		2% (exercise, heat)		0.63		
		3% (exercise, heat)		1.02		
		4% (exercise, heat)		1.27		
		1% (exercise, heat)	Trail marking test (mean speed in seconds)	0.45		
		2% (exercise, heat)		0.88		
		3% (exercise, heat)		1.30		
		4% (exercise, heat)		1.71		
		1% (exercise, heat)	Word recognition test (% correct words recalled)	0.44		
		2% (exercise, heat)		0.58		
		3% (exercise, heat)		0.99		
		4% (exercise, heat)		1.05		
Grego et al. (2005)	8 adult M	2%–4% (exercise)	Critical flicker fusion test (mdi)	0.45		
			Critical flicker fusion test (mtot)	0.30		
			Map recognition	0.21		
			Map recognition, errors	1.25		
Irwin et al. (2013)	16 adult M	2.5% (exercise)	Choice reaction, latency	**0.20**		
			Choice reaction time, % correct	**0.12**		
			Match to sample, reaction time	0.30		

(Continued)

TABLE 9.1 (*Continued*)

Summary of Findings of Studies Assessing the Effect of Dehydration on Cognitive Function, Mood, and Self Perception

Reference	Subjects	Dehydration Level (Method of Dehydration)	Cognitive Task/Mood Measurement	ES Cohen's *d*	ES Pearson *r*	Interpretation
			Match to sample, % correct	0.27		
			Stop signal task, reaction time	0.17		
			Stop signal task, errors	**0.18**		
			Spatial planning, number of problems solved	**0.19**		
			Spatial planning, mean number of moves	**0.23**		
Leibowitz et al. (1972)	2 lean adult M, 2 lean adult F	5% (exercise, heat)	Trial 1			
			Visual detection, central (lean subjects), *reaction time*	0.20		
			Visual detection, peripheral (lean subjects), *reaction time*	**0.02**		
			Trial 2			
			Visual detection, central (lean subjects), *reaction time*	0.42		
			Visual detection, peripheral (lean subjects), *reaction time*	0.00		
			Trial 3			
			Visual detection, central (lean subjects), *reaction time*	0.02		
			Visual detection, peripheral (lean subjects), *reaction time*	0.54		

(Continued)

TABLE 9.1 (*Continued*)

Summary of Findings of Studies Assessing the Effect of Dehydration on Cognitive Function, Mood, and Self Perception

Reference	Subjects	Dehydration Level (Method of Dehydration)	Cognitive Task/Mood Measurement	ES Cohen's *d*	ES Pearson *r*	Interpretation
	2 obese adult M, 2 obese adult F	5% (exercise, heat)	Trial 1			
			Visual detection, central (obese subjects), *reaction time*	**0.18**		
			Visual detection, peripheral (obese subjects), *reaction time*	1.32		
			Trial 2			
			Visual detection, central (obese subjects), *reaction time*	0.04		
			Visual detection, peripheral (obese subjects), *reaction time*	**0.29**		
			Trial 3			
			Visual detection, central (obese subjects), *reaction time*	0.03		
			Visual detection, peripheral (obese subjects), *reaction time*	0.28		
Lieberman et al. (2005)	31 adult M	3.6% (exercise, plus other stressors in military field study)	Four-choice reaction time			
			Reaction time (ms)	1.15		
			Incorrect responses	1.24		

(*Continued*)

TABLE 9.1 (Continued)

Summary of Findings of Studies Assessing the Effect of Dehydration on Cognitive Function, Mood, and Self Perception

Reference	Subjects	Dehydration Level (Method of Dehydration)	Cognitive Task/Mood Measurement	ES Cohen's d	ES Pearson r	Interpretation
			Time-out errors	0.75		
			Visual vigilance			
			Correct responses	1.22		
			Reaction time (s)	0.36		
			False alarms	0.65		
			Matching-to-sample			
			Correct responses	1.58		
			Time-out errors	1.03		
			Reaction time (s)	0.54		
			Repeated acquisition			
			Incorrect responses	1.01		
			Time to complete (s)	0.95		
			Grammatical reasoning			
			Correct responses	0.72		
			Incorrect responses	0.65		
			No response	0.45		
			Response time (s)	**0.29**		
			POMS			
			Tension	*0.55*		
			Depression	*0.72*		
			Confusion	*2.16*		
			Vigor	*2.73*		

(Continued)

TABLE 9.1 (*Continued*)

Summary of Findings of Studies Assessing the Effect of Dehydration on Cognitive Function, Mood, and Self Perception

Reference	Subjects	Dehydration Level (Method of Dehydration)	Cognitive Task/Mood Measurement	ES Cohen's *d*	ES Pearson *r*	Interpretation
Lindseth et al. (2013)	40 adults	1%–3% (dehydration induced by diet control)	*Fatigue*	*3.53*		
			Anger	*0.53*		
			General aviation flight simulator, errors	4.41		
			Sternberg short-term memory, response time	0.84		
			Vandenberg mental rotation situational awareness	0.77		
Neave et al. (2001)	12 adult M, 12 adult F	No dehydration reported, studied impact of added water	*VAS: Alertness*	*0.63*		
Ogino et al. (2013)	5 adult M	% dehydration not reported, dehydration urine osmolality = 784.6 mOsm/kg (Exercise, 12 h fasting of food and drink). Euhydration urine osmolality = 480.6 mOsm/kg	Arithmetic test speed and accuracy	1.16		

(Continued)

TABLE 9.1 (*Continued*)

Summary of Findings of Studies Assessing the Effect of Dehydration on Cognitive Function, Mood, and Self Perception

Reference	Subjects	Dehydration Level (Method of Dehydration)	Cognitive Task/Mood Measurement	ES Cohen's *d*	ES Pearson *r*	Interpretation
Petri et al. (2006)	10 adult M	No dehydration reported, 24 h water deprivation	Arithmetic test error rate	0.17		
			Total test solving time (aggregated across signal position, short-term memory, visual orientation, simple arithmetic, and complex motor coordination)			
			Comparison at 1400 h	0.34		
			Comparison at 1700 h	0.68		
			Comparison at 2000 h	0.76		
			Comparison at 2300 h	0.87		
			Minimum test solving time (aggregated across signal position, short-term memory, visual orientation, simple arithmetic, and complex motor coordination)			
			Comparison at 1100 h	0.17		
			Comparison at 1400 h	0.62		
			Comparison at 1700 h	0.87		
			Comparison at 2000 h	0.99		
			Comparison at 2300 h	1.30		
			Comparison at 0800 h (next day)	0.64		

(*Continued*)

TABLE 9.1 (Continued)

Summary of Findings of Studies Assessing the Effect of Dehydration on Cognitive Function, Mood, and Self Perception

Reference	Subjects	Dehydration Level (Method of Dehydration)	Cognitive Task/Mood Measurement	ES Cohen's d	ES Pearson r	Interpretation
Rogers et al. (2001)	10 adult M, 10 adult F	No dehydration reported, studied impact of added water	Rapid visual information processing test			
			120 ml water: Low thirst	**0.66**		
			330 ml water: Low thirst	**1.05**		
			120 ml water: High thirst	0.25		
			330 ml water: High thirst	0.69		
			VAS: Alertness rating			
			120 ml water: Low thirst	*0.13*		
			330 ml water: Low thirst	*0.57*		
			120 ml water: High thirst	*0.21*		
			330 ml water: High thirst	*0.63*		
Serwah and Marino (2006)	8 adult M	1.0% (exercise, heat)	Choice reaction time	0.14		
		1.7% (exercise, heat)	Choice reaction time	**0.09**		
Shirreffs et al. (2004)	9 adult M, 6 adult F	0.98%	*VAS: Subjective feelings*			
			How thirsty are you	*0.71*		

(Continued)

TABLE 9.1 (*Continued*)
Summary of Findings of Studies Assessing the Effect of Dehydration on Cognitive Function, Mood, and Self Perception

Reference	Subjects	Dehydration Level (Method of Dehydration)	Cognitive Task/Mood Measurement	ES Cohen's d	ES Pearson r	Interpretation
		1.78%		4.10		
		2.66%		1.43		
		0.98%	How hungry	0.37		
		1.78%		0.13		
		2.66%		0.09		
		0.98%	Mouth dryness	2.68		
		1.78%		3.89		
		2.66%		4.12		
		0.98%	Pleasant taste in mouth	0.00		
		1.78%		0.47		
		2.66%		1.01		
		0.98%	Sore head	0.76		
		1.78%		5.85		
		2.66%		2.66		
		0.98%	Ability to concentrate	0.44		
		1.78%		3.06		
		2.66%		2.27		

(*Continued*)

TABLE 9.1 (*Continued*)

Summary of Findings of Studies Assessing the Effect of Dehydration on Cognitive Function, Mood, and Self Perception

Reference	Subjects	Dehydration Level (Method of Dehydration)	Cognitive Task/Mood Measurement	ES Cohen's *d*	ES Pearson *r*	Interpretation
			How tired are you			
		0.98%		*2.09*		
		1.78%		*0.83*		
		2.66%		*1.44*		
			How alert are you			
		0.98%		*0.57*		
		1.78%		*0.95*		
		2.66%		*1.02*		
Smith et al. (2012)	7 adult M	1.5% (fluid restriction)	Golf shot distance			
			9 iron	1.08		
			7 iron	1.31		
			5 iron	1.27		
			Golf shot off-target accuracy			
			9 iron	2.00		
			7 iron	2.28		
			5 iron	1.87		
Suhr et al. (2004)	6 adult M, 22 adult F	No dehydration reported, correlational study of %TBW/WT and cog function	Psychomotor processing speed		0.49	

(Continued)

TABLE 9.1 (*Continued*)

Summary of Findings of Studies Assessing the Effect of Dehydration on Cognitive Function, Mood, and Self Perception

Reference	Subjects	Dehydration Level (Method of Dehydration)	Cognitive Task/Mood Measurement	ES Cohen's *d*	ES Pearson *r*	Interpretation
Suhr et al. (2010)	21 adult F	No dehydration reported, correlational study of %TBW/WT and cog function	Memory		0.34	
			Declarative memory (AVLT)		0.54	
			Working memory (ACT)		0.47	
Szinnai et al. (2005)	8 adult M	2.6% (fluid restriction)	Manual tracking test distance, pixels	0.05		
			PASAT accuracy %	**0.04**		
			PASAT verbal response time, ms	**0.11**		
			CRTT inter-stim interval, ms	**0.09**		
			CRTT reaction time, ms	**0.08**		
			Stroop script color naming verbal response time congruent, ms	**0.23**		
			Stroop script color naming verbal response time conflict, ms	**0.14**		
			Stroop effect, ms	0.23		
			Stroop word naming verbal response time congruent, ms	0.28		
			Stroop word naming verbal response time conflict, ms	**0.30**		

(Continued)

TABLE 9.1 (Continued)

Summary of Findings of Studies Assessing the Effect of Dehydration on Cognitive Function, Mood, and Self Perception

Reference	Subjects	Dehydration Level (Method of Dehydration)	Cognitive Task/Mood Measurement	ES Cohen's d	ES Pearson r	Interpretation
			P300 reaction time, ms	**0.20**		
			P300 amplitude, μv, Fz	0.20		
			P300 amplitude, μv, Cz	0.19		
			P300 amplitude, μv, PCz	0.19		
			P300 latency, μv, Fz	0.07		
			P300 latency, μv, Cz	0.04		
			P300 latency, μv, PCz	0.38		
	8 adult F		Manual tracking test distance, pixels	0.56		
			PASAT accuracy %	0.14		
			PASAT verbal response time, ms	0.18		
			CRTT inter stim interval, ms	0.15		
			CRTT reaction time, ms	0.14		
			Stroop script color naming verbal response time congruent, ms	**0.18**		
			Stroop script color naming verbal response time conflict, ms	**0.21**		
			Stroop effect, ms	0.04		
			Stroop word naming verbal response time congruent, ms	0.15		
			Stroop word naming verbal response time conflict, ms	0.27		
			P300 reaction time, ms	0.33		

(Continued)

TABLE 9.1 (*Continued*)
Summary of Findings of Studies Assessing the Effect of Dehydration on Cognitive Function, Mood, and Self Perception

Reference	Subjects	Dehydration Level (Method of Dehydration)	Cognitive Task/Mood Measurement	ES Cohen's *d*	ES Pearson *r*	Interpretation
	8 adult M, 8 adult F		P300 amplitude, μv, Fz	0.07		
			P300 amplitude, μv, Cz	0.10		
			P300 amplitude, μv, PCz	0.07		
			P300 latency, μv, Fz	0.27		
			P300 latency, μv, Cz	0.12		
			P300 latency, μv, PCz	0.15		
			VAS			
			Effort	*1.83*		
			Concentration	*2.55*		
			Tiredness	*1.58*		
			Alertness	*1.23*		

Mood and self-perception variables are in *italics*; ES in the direction of improved performance or mood with dehydration is indicated in bold. ES, effect size; M, males; F, females; POMS, Profile of Mood States; TMD, total mood disturbance; VAS, visual analogue scale; RPE, rating of perceived exertion; PVT, psychomotor vigilance test; PASAT, paced auditory serial addition task; CRTT, choice reaction time task; P300, auditory event-related potentials; Stroop, Stroop word color conflict test.

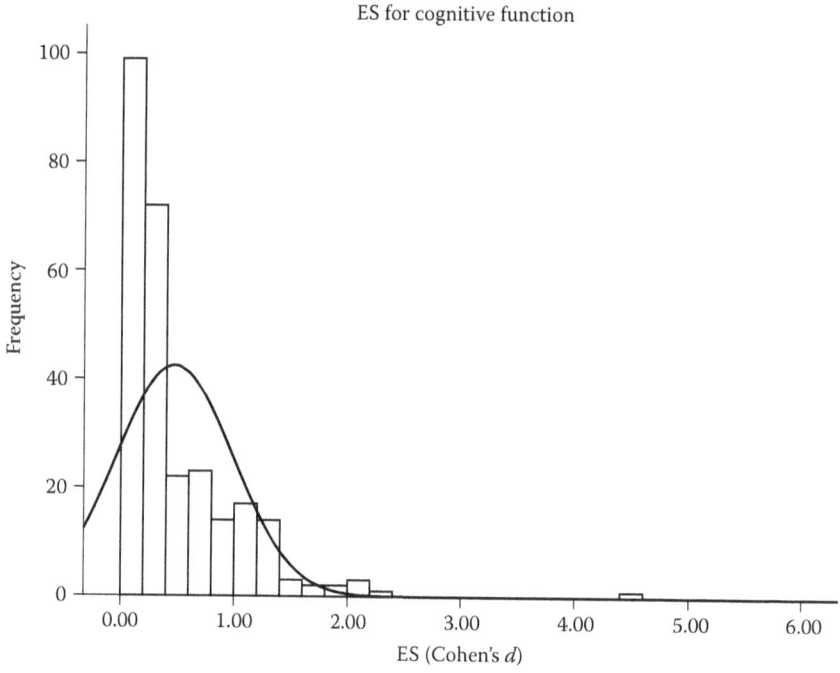

FIGURE 9.1   Frequency distribution of ES for measures of cognitive function.

subjective self-perception measures. The distribution of S, M, and L ES for cognitive and self-perception measures are 161, 43, 79, and 57, 29, and 60, respectively (chi square = 12.45, $df = 2$, $p = .002$). The preponderance of ES are Cohen's d (410) with the remaining being Pearson's r (19). The distributions of ES (Cohen's d) for cognitive and self-perception measures can be visualized in Figure 9.1 ($n = 273$) and Figure 9.2 ($n = 137$), respectively. The median ES (Cohen's d) for cognitive and self-perception measures are 0.27 (a small effect, 95% CI: 0.02, 1.34) and 0.51 (a medium effect, 95% CI: 0.02, 2.73), respectively. Most of the ES represent a decrement in performance related to dehydration; however, 79 are in the direction of improved functioning with a median ES of 0.18 (a small effect, 95% CI: 0.02, 1.05). The median ES (Cohen's d) for cognitive and self-perception measures in the direction of improved performance with dehydration are 0.20 (a small effect, 95% CI: 0.03, 1.18) and 0.14 (a small effect, 95% CI: 0.01, not available), respectively. All ES summaries are with absolute values of ES (no averaging of *negative* and *positive* values was done).

9.3   IMPACT OF DEHYDRATION ON MOOD

Very few studies have investigated the relationship between hydration and mood alone (Pross et al. 2013). Typically, investigations of mood have been embedded in studies of cognitive function. The area of mood is difficult to isolate precisely. Measurement of subjective feelings can focus on mood but also self-perception of alertness, ability to concentrate, and other subjective states that may relate to cognitive function

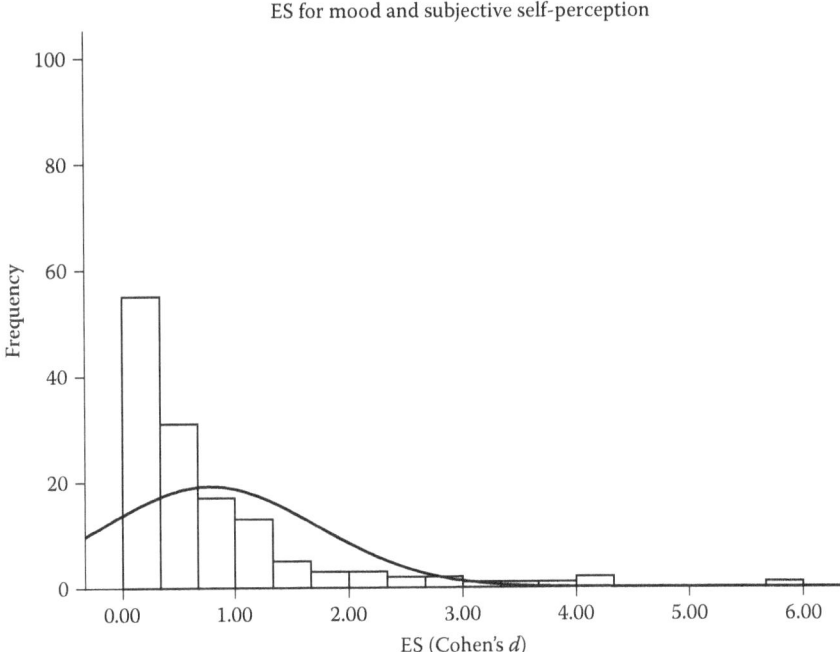

FIGURE 9.2 Frequency distribution of ES for measures of mood and self-perception.

even though no direct cognitive performance measures are taken (see Shirreffs et al. 2004 for an example of an investigation of dehydration on subjective feelings that covers both mood and self-perception of cognitive function). Armstrong et al. (2012) in a controlled double-blind study (see Section 9.2.2 for details) found an increase in anger-hostility ($p = .04$), fatigue-inertia ($p = .0003$), and total mood disturbance ($p = .01$) when women were dehydrated. In addition, self-reported perceived task difficulty, concentration, and headache were all adversely affected ($p = .004$, $.01$, and $.05$, respectively).

In general, it appears that the effect of dehydration on mood and self-perception is about twice that of the impact of dehydration on cognitive function (Table 9.1).

9.4 METHODOLOGICAL ISSUES

Methodological issues remain a fundamental challenge not only to the reliable measurement of cognitive function (and to a lesser extent mood and self-perception) but also to elucidating the relationship between hydration status and mental function (Adan 2012; Lieberman 2007, 2012). A wide variety of dependent measures has been used over the years (see Table 9.1) likely contributing to inconsistent results and diminishing our ability to draw firm conclusions. Many of the tests are drawn from the neuropsychological arena with its complexity of techniques for assessing attention, memory, motor control, learning, and executive function. These instruments vary in sensitivity and the conditions under which their use is most appropriate. Many are designed for detecting pathological conditions in clinical assessment, so healthy populations may fall outside the range of test specificity (Adan 2012). An alternative approach may be to focus the selection of measurement tools on those used in psychopharmacological research, which assesses cognitive change over a broad range of normal functioning (Adan 2012). As a general guide, Adan (2012) has suggested that those tasks requiring immediate memory, attention, psychomotor skills, and self-perception measures of subjective state are most negatively affected by dehydration and that long-term memory, working memory, and executive-function may be more resilient to the effects of dehydration.

Subject selection may impact the robustness of our tests, as well as our ability to generalize to other populations. Subjects are often athletes who are accustomed to exercise, heat, and varying levels of dehydration, which could possibly make their cognitive function less susceptible to these factors than in the general population (Adan 2012). Other characteristics such as diurnal variation in both physical and cognitive function related to circadian rhythm may impact neuropsychological test results (Adan 1993; Monk et al. 1997; Schmidt et al. 2007; Valdez 2012). Within-subject variation in cognitive function related to circadian rhythm has been estimated to be as high as 20% (Adan 2012). Additionally, personality characteristics such as level of anxiety, response to stress, motivation, tolerance of discomfort, and social characteristics such as education level should be considered.

More problematic may be the interaction of methods of achieving dehydration. Typically, heat, exercise, and fluid restriction or some combination thereof has been used to achieve dehydration. A potential issue, so far unresolved, is the possibility of these factors interacting in nonlinear fashion so that the impact of dehydration may be exacerbated by heat or exercise and possibly by other confounding factors such as caffeine or even motivation due to inadequate experimental control and blinding (Lieberman 2007, 2012).

Even the methods of defining dehydration are varied and not uniformly practicable across test settings. For instance, Armstrong (2007) reviews 13 different methods of assessing hydration status, pointing out the differences between laboratory and field method requirements. While TBW and plasma osmolality may be *gold standards* in the laboratory, simpler techniques such as body mass change, urine-specific gravity, and urine volume and color are more practical in the field. Popkin et al. (2012) remind us that it is methodological variability likely causing the inconsistent findings as opposed to hydration status per se having inconsistent effects on cognitive function.

A strong call has been made to adopt dose–response methods to help address some of these methodological shortcomings (Lieberman 2007, 2012). A dose–response approach has the benefit of providing internal replication if cognitive function is systematically related to hydration level. Including a range of hydration levels is also useful in providing a broad context in which to test since some cognitive measures may not be sensitive to lower levels of dehydration. Some research does suggest that hydration status is parametrically related to cognition and mental function (Gopinathan et al. 1988; Sharma et al. 2005). Gopinathan et al. (1988) in a study of 11 heat acclimatized soldiers exercising in the heat to 1%, 2%, 3%, and 4%

dehydration demonstrated clear dose–response effects for a serial addition test, trail marking test, and word recall test. A Spearman's rho nonparametric correlation between percent dehydration and Cohen's d ES ($n = 273$ pairs, see Table 9.1) was relatively low ($rho = 0.137$) although reliable ($p = .01$). This perhaps reflects both the mixed nature of the controlling variables involved and a substantial sample size.

9.5 RESEARCH DESIGN RECOMMENDATIONS

It is critical to address the issue of interactions among conditions used to induce dehydration (e.g., heat, exercise, and fluid restriction), dependent measure techniques (type of cognitive/mood/self-perception test) as well as extraneous factors often present (subject characteristics and motivation). The traditional approach of manipulating a single variable while holding all other conditions constant is insufficient for understanding interactions. A class of experimental designs aimed at understanding interactions are the response surface designs (Anderson and Whitcomb 2007; Box et al. 2005; Cornell 2002; Montgomery 2012), sometimes referred to as design of experiments. One example of such a design, applied to the investigation of dehydration and cognitive function and mood, is an I-optimal design with quadratic modeling (main effects, squared terms, and two-way interactions), sufficient replication points to establish satisfactory internal reliability, and design points to achieve adequate assessment of lack of fit in the model (Design Expert 9 Software, Minneapolis 2014) (see Table 9.2). In this design, dehydration (%), cognitive function, and mood are dependent measures. The independent measures (design factors) are time (min), exercise intensity (Borg scale rating of physical exertion), and temperature (°C) combined in a response surface design to produce a range of dehydration levels (0 to perhaps 4%). The values in the design in Table 9.2 are simply meant to convey the approach and are used essentially as place holders. (The decimal values for the Borg scale occur as a function of the algorithms selecting the design points. The important requirement is to select a wide enough Borg-scale range to produce exercise intensity effects.) Actual levels of the factors selected for research may be different. However, the rationale for factor level selection is to produce a wide enough range of dehydration (defined by the most *severe* combination) to allow assessment of effects (perhaps 4% dehydration). That is, the design point defined by time = 60 min, exercise intensity = a Borg scale score of 15, and temperature = 30°C is intended to produce the most dehydration (to achieve a reasonable upper range). Pilot work may be necessary to determine the exact mix of factor levels to achieve this. All other design points (less *severe* mix of conditions) can reasonably be expected to produce lower levels of dehydration. There are 20 points in this design space. The sequence of the experimental test conditions is given in the order column. A cohort of subjects (perhaps 10, but a number sufficient to produce a stable mean) are all exposed to the same sequence of 20 conditions (design points) in the order shown. Only aggregate data (means or medians) are analyzed. There are 20 values (one from each design point) in the entire analysis and modeling. No individual subject data are included in the analysis. Upon completion of the study, the data can then be modeled in such a way that all of the response space between the design points can be estimated. All factor

TABLE 9.2

Example of Response Surface Design Investigating the Impact of Time (min), Exercise Intensity (Borg-Scale Score), and Temperature (°C) on Dehydration, Cognitive Function, and Mood

Run Order	Time (min)	Exercise Intensity (Borg Scale)	Temp (°C)
1	38.4	12.7	23.5
2	60.0	11.4	22.8
3	20.0	6.0	30.0
4	20.0	9.2	18.0
5	60.0	6.0	30.0
6	36.2	11.4	30.0
7	36.2	11.4	30.0
8	43.4	9.7	18.0
9	20.0	9.7	25.1
10	45.6	15.0	18.0
11	36.0	6.0	22.8
12	60.0	10.5	30.0
13	60.0	6.0	18.0
14	38.4	12.7	23.5
15	20.0	15.0	25.8
16	46.8	8.5	26.5
17	20.0	9.2	18.0
18	60.0	15.0	30.0
19	36.0	6.0	22.8
20	60.0	11.4	22.8

main effects and two-way interactions are available in this design. Any combination of factor levels (even those not directly tested) can be modeled to estimate dehydration, cognitive function, and mood. Interestingly, this approach allows the possible discovery of enhanced cognitive function and mood. It also eliminates the issue of the confounding of time, exercise, and heat with dehydration seen in current research approaches that seek to induce dehydration and then in turn use it as an independent measure. Dehydration cannot be separated from the conditions used to induce it and in that sense is invariably a dependent measure. There in lies the impossibility of a clear interpretation of the impact of *dehydration* on cognitive function and mood apart from the factor conditions that operationally define the level of dehydration used as the independent variable. This problem is solved if dehydration and cognitive function, and mood are all investigated as a function of the conditions that produce dehydration and changes in cognitive function and mood. While time, exercise intensity, and heat are used in the current example, other factors could have been chosen.

There are a number of advantages of a designed approach. With a properly chosen design space, the need for subsequent experimentation is reduced due to the efficiency of capturing information about the interactions. In a single manipulation approach, or even a dose–response approach, one never knows how the results of a given experiment might change relative to the impact of other variables. Clamping levels of potentially impactful factors as the target factor is manipulated could be time consuming without any assurance of how long it will take to adequately explore the key relationships. The single manipulation approach, unlike a designed approach, offers little opportunity to optimize responses. Response optimization is available in a design of experiments because the modeling extends beyond the design points to the entire response surface allowing estimation of responses not actually measured by combinations of factors not actually tested (e.g., predicting the mix of factors producing the greatest decrement in cognitive function). Once the designed experiment is completed, the model can be quizzed many times for any combination of factors at any level desired (even levels not directly tested). This does not rule out the necessity for replication of specific predictions, but does provide a more efficient path to identify likely scenarios.

9.6 CONCLUSIONS AND RECOMMENDATIONS

The available research on the impact of dehydration on cognitive function and mood is variable but suggests some cognitive functions (immediate memory, attention, and psychomotor skills) are related to dehydration (but perhaps less so for long-term memory, working memory, and executive function). The effects of dehydration on mood and self-reported perceptions of well-being seem to be at least directionally greater than for cognitive function. Many researchers have pointed to the challenging nature of interacting factors involved in the study of dehydration and behavioral outcome measures. There may be opportunity to improve research design in this area by moving to a design of experiments approach, which directly and simultaneously explores main effects and interactions. In such an approach, dehydration as well as cognitive and mood outcomes are dependent measures. Modeling of interactions allows response optimization. Recognizing the incomplete and variable nature of the research in the area of dehydration and cognitive function, self-perception, and mood, prudent recommendations to athletes include paying attention to hydration status not only for physical performance optimization but to help prevent possible dehydration-related decrement in cognitive function, self-perception, and mood.

REFERENCES

Ackland, G.L., J. Harrington, P. Downie et al. 2008. Dehydration induced by bowel preparation in older adults does not result in cognitive dysfunction. *Anesth. Analg.* 106 (3):924–9.

Adam, G.E., R. Carter 3rd, S.N. Cheuvront et al. 2008. Hydration effects on cognitive performance during military tasks in temperate and cold environments. *Physiol. Behav.* 18:748–56.

Adan, A. 1993. Circadian variations in psychological measures: A new classification. *Chronobiologia* 20 (3–4):145–61.

Adan, A. 2012. Cognitive performance and dehydration. *J. Am. Coll. Nutr.* 31:71–8.

Ainslie, P.N., I.T. Campbell, K.N. Frayn et al. 2002. Energy balance, metabolism, hydration, and performance during strenuous hill walking: The effect of age. *J. Appl. Physiol.* 93 (2):714–23.

Anderson, M.J., P.J. Whitcomb. 2000. *DOE Simplified Practical Tools for Effective Experimentation*. Portland, OR: Productivity.

Armstrong, L.E. 2007. Assessing hydration status: The elusive gold standard. *J. Am. Coll. Nutr.* 26 (Suppl 5):575S–84S.

Armstrong, L.E., S.G. Matthew, D.J. Casa et al. 2012. Mild dehydration affects mood in healthy young women. *J. Nutr.* 142 (2):382–8.

Baker, L.B., D.E. Conroy. 2007. Dehydration impairs vigilance-related attention in male basketball players. *Med. Sci Sports Exerc.* 39 (6):976–3.

Bandelow, S., R. Maughan, S. Shirreffs et al. 2010. The effects of exercise, heat, cooling and rehydration strategies on cognitive function in football players. *Scand. J. Med. Sci. Sports* 20 (3):148–60.

Bar-David, Y., J. Urkin, E. Kozminsky. 2005. The effect of voluntary dehydration on cognitive functions of elementary school children. *Acta Paediatr.* 94 (11):1667–73.

Bar-David, Y., Urkin, J., D. Landau et al. 2009. Voluntary dehydration among elementary school children residing in a hot arid environment. *J. Hum. Nutr. Diet.* 22 (5):455–60.

Benton, D. 2011. Dehydration influences mood and cognition: A plausible hypothesis? *Nutrients* 3 (5):555–73.

Benton, D., N. Burgess. 2009. The effect of the consumption of water on the memory and attention of children. *Appetite* 53 (1):143–6.

Booth, P., C.J. Edmonds, B.G. Taylor. 2012. The effect of water consumption on cognitive and motor performance. *Appetite* 59 (2):621 (abstract).

Box, G.E.P., J.S. Hunter, W.G. Hunter. 2005. *Statistics for Experimenters: Design, Innovation, and Discovery*. 2nd edition. Hoboken, NJ: John Wiley & Sons.

Cian, C., P.A. Barraud, B. Melin et al. 2001. Effects of fluid ingestion on cognitive function after heat stress or exercise-induced dehydration. *Int. J. Psychophysiol.* 42 (3):243–51.

Cian, N., N. Koulmann, P.A. Barraud et al. 2000. Influences of variations in body hydration on cognitive function: Effect of hyperhydration, heat stress, and exercise-induced dehydration. *J. Psychophysiol.* 14 (1):29–36.

Cohen, J. 1988. *Statistical Power Analysis for the Behavioral Sciences*, 2nd edition. Hillsdale, NJ: Lawrence Erlbaum Associates.

Cornell, J.A. 2002. *Experiments with Mixtures. Designs, Models, and the Analysis of Mixture Data*. New York: John Wiley & Sons.

D'Anci, K.E., F. Constant, I.H. Rosenberg. 2006. Hydration and cognitive function in children. *Nutr. Rev.* 64 (10 Pt 1):457–64.

D'Anci, K.E., A. Vibhakar, J.H. Kanter. 2009. Voluntary dehydration and cognitive performance in trained college athletes. *Percept. Mot. Skills* 109 (1):251–69.

Doppelmayr, M.M., H. Finkernagel, H.I. Doppelmayr. 2005. Changes in cognitive performance during a 216 kilometer, extreme endurance footrace: A descriptive and prospective study. *Percept. Mot. Skills* 100 (2):473–87.

Edmonds, C.J. 2012a. The effect of water consumption on cognitive performance. *Appetite* 59 (2):624.

Edmonds, C.J. 2012b. Water, hydration status and cognitive performance. In *Nutrition and Mental Performance: A Lifespan Perspective*. L. Riby, M. Smith, J. Foster, Eds., pp. 193–211. New York: Palgrave Macmillan.

Edmonds, C.J., D. Burford. 2009. Should children drink more water? The effects of drinking water on cognition in children. *Appetite* 52 (3):776–9.

Edmonds, C.J., R. Crombie, H. Ballieux et al. 2013. Water consumption, not expectancies about water consumption, affects cognitive performance in adults. *Appetite* 60 (1):148–53.

Edmonds, C.J., B. Jeffes. 2009. Does having a drink help you think? 6–7-Year-old children show improvements in cognitive performance from baseline to test after having a drink of water. *Appetite* 53 (3):469–72.

Ely, B.R., K.J. Sollanek, S.N. Cheuvront et al. 2013. Hypohydration and acute thermal stress affect mood state but not cognition or dynamic postural balance. *Eur. J. Appl. Physiol.* 113 (4):1027–34.

Fadda, R., G. Rapinett, D. Grathwohl et al. 2012. Effects of drinking supplementary water at school on cognitive performance in children. *Appetite* 59 (3):730–7.

Ganio, M.S., L.E. Armstrong, D.J. Casa et al. 2011. Mild dehydration impairs cognitive performance and mood of men. *Br. J. Nutr.* 106 (10):1535–43.

Gopinathan, P.M., G. Pichan, V.M. Sharma. 1988. Role of dehydration in heat stress-induced variations in mental performance. *Arch. Environ. Health* 43 (1):15–17.

Grandjean, A.C., N.R. Grandjean. 2007. Dehydration and cognitive performance. *J. Am. Coll. Nutr.* 26 (Suppl 5):549S–54S.

Grego, F., J.-M. Vallier, M. Collardeau et al. 2005. Influence of exercise duration and hydration status on cognitive function during prolonged cycling exercise. *Int. J. Sports Med.* 26:27–33.

Irwin, C., M. Leveritt, D. Shum et al. 2013. The effects of dehydration, moderate alcohol consumption, and rehydration on cognitive functions. *Alcohol* 47 (3):203–13.

Kempton, M.J., U. Ettinger, R. Foster et al. 2011. Dehydration affects brain structure and function in healthy adolescents. *Hum. Brain Mapp.* 32:71–9.

Leibowitz, H.W., C.N. Abernethy, E.R. Buskirk et al. 1972. The effect of heat stress on reaction time to centrally and peripherally presented stimuli. *Human Factors: The Journal of the Human Factors and Ergonomics Society* 14 (2):155–60.

Lieberman, H.R. 2007. Hydration and cognition: A critical review and recommendations for future research. *J. Am. Coll. Nutr.* 26 (Suppl 5):555S–61S.

Lieberman, H.R. 2010. Hydration and human cognition. *Nutr. Today.* 45 (6):S33–6.

Lieberman, H.R. 2012. Methods for assessing the effects of dehydration on cognitive function. *Nutr. Rev.* 70 (Suppl 2):S143–6.

Lieberman, H.R., G.P. Bathalon, C.M. Falco et al. 2005. Severe decrements in cognition function and mood induced by sleep loss, heat, dehydration, and undernutrition during simulated combat. *Biol. Psychiatry.* 57 (4):422–9.

Lindseth, P.D., G.N. Lindseth, T.V. Petros et al. 2013. Effects of hydration on cognitive function of pilots. *Mil. Med.* 178 (7):792–8.

Masento, N.A., M. Golightly, D.T. Field et al. 2014. Effects of hydration status on cognitive performance and mood. *Br. J. Nutr.* 30:1–12.

Maughan, R.J. 2003. Impact of mild dehydration on wellness and on exercise performance. *Eur. J. Clin Nutr.* 57 (Suppl 2):S19–23.

Maughan, R.J., S.M. Shirreffs. 2012. Hydration and performance during Ramadan. *J. Sports Sci.* 30 (Suppl 1):S33–41.

Maughan, R.J., S.M. Shirreffs, K.T. Ozgünen et al. 2010. Living, training and playing in the heat: challenges to the football player and strategies for coping with environmental extremes. *Scand. J. Med. Sci. Sports.* 20:117–24.

Maughan, R.J., S.M. Shirreffs, P. Watson. 2007. Exercise, heat, hydration and the brain. *J. Am. Coll. Nutr.* 26 (Suppl 5):604S–12S.

McMorris, T. 2009. Exercise, dehydration and cognitive function. [Chapter] In *Exercise and Cognitive Function.* T. McMorris, P. Tomporowski, M. Audiffren, Eds., pp. 117–34. West Sussex: Wiley-Blackwell.

Monk, T., D. Buysse, C. Reynolds III et al. 1997. Circadian rhythms in human performance and mood under constant conditions. *J. Sleep Res.* 6 (1):9–18.

Montgomery, D.C. 2012. *Design and Analysis of Experiments*, 8th edition. New York: John Wiley & Sons.

Murray, B. 2007. Hydration and physical performance. *J. Am. Coll. Nutr.* 26 (5):542S–8S.

Neave, N., A.B. Scholey, J.R. Emmett et al. 2001. Water ingestion improves subjective alertness, but has no effect on cognitive performance in dehydrated healthy young volunteers. *Appetite* 37 (3):255–6.

Ogino, Y., T. Kakeda, K. Nakamura et al. 2014. Dehydration enhances pain-evoked activation in the human brain compared with rehydration. *Anesth. Analg.* 118 (6):1317–25.

Petri, N.M., N. Dropulic, G. Kardum. 2006. Effects of voluntary fluid intake deprivation on mental and psychomotor performance. *Croat. Med. J.* 47 (6):855–61.

Popkin, B.M., K.E. D'Anci, I.H. Rosenberg. 2010. Water, hydration, and health. *Nutr. Rev.* 68 (8):439–58.

Pross, N., A. Demazieres, N. Girard et al. 2013. Influence of progressive fluid restriction on mood and physiological markers of dehydration in women. *Br. J. Nutr.* 109 (2):313–21.

Ramirez, C., A. Garcia, P. Valdez. 2012. Identification of circadian rhythms in cognitive inhibition and flexibility using a Stroop task. *Sleep Biol. Rhythms* 10 (2):136–44.

Ritz, P., G. Berrut. 2005. The importance of good hydration for day-to-day health. *Nutr. Rev.* 63 (6 Pt 2):S6–13.

Rogers, P.J., A. Kainth, H.J. Smit. 2001. A drink of water can improve or impair mental performance depending on small differences in thirst. *Appetite* 36 (1):57–8.

Schmidt, C., F. Collette, C. Cajochen et al. 2007. A time to think: Circadian rhythms in human cognition. *Cogn. Neuropsychol.* 24 (7):755–89.

Scholey, A.B., S.I. Sünram-Lea, J. Greer et al. 2009. Glucose enhancement of memory depends on initial thirst. *Appetite* 53 (3):426–9.

Secher, M., P. Ritz. 2012. Hydration and cognitive performance. *J. Nutr. Health Aging* 16 (4):325–9.

Serwah, N., F.E. Marino. 2006. The combined effects of hydration and exercise heat stress on choice reaction time. *J. Sci. Med Sport* 9 (1):157–64.

Sharma, V.M., K. Sridharan, G. Pichan et al. 1986. Influence of heat-stress induced dehydration on mental functions. *Ergonomics* 29 (6):791–9.

Shirreffs, S.M. 2009. Symposium on "performance, exercise and health" hydration, fluids and performance. *Proc. Nutr. Soc.* 68 (1):17–22.

Shirreffs, S.M., S.J. Merson, S.M. Fraser et al. 2004. The effects of fluid restriction on hydration status and subjective feelings in man. *Br. J. Nutr.* 91 (6):951–8.

Slaven, G.M., C.M. Windle. 1999. Cognitive performance over 7 days in a distressed submarine. *Aviat. Space Environ. Med.* 70 (6):604–8.

Smith, M.F., A.J. Newell, M.R. Baker. 2012. Effect of acute mild dehydration on cognitive-motor performance in golf. *J. Strength Cond. Res.* 26 (11):3075–80.

Suhr, J.A., J. Hall, S.M. Patterson et al. 2004. The relation of hydration status to cognitive performance in healthy older adults. *Int. J. Psychophysiol.* 53 (2):121–5.

Suhr, J.A., S.M. Patterson, A.W. Austin et al. 2010. The relation of hydration status to declarative memory and working memory in older adults. *J. Nutr. Health Aging* 14 (10):840–3.

Szinnai, G., H. Schachinger, M.J. Arnaud et al. 2005. Effect of water deprivation on cognitive-motor performance in healthy men and women. *Am. J. Physiol. Regul. Integr. Comp. Physiol.* 289 (1):R275–80.

Tomporowski, P.D. 2003. Effects of acute bouts of exercise on cognition. *Acta Psychol. (Amst).* 112 (3):297–324.

Valdez, P., C. Ramírez, A. García. 2012. Circadian rhythms in cognitive performance: Implications for neuropsychological assessment. *ChronoPhysiol. Ther.* 2:81–92.

Wilson, M.M., J.E. Morley. 2003. Impaired cognitive function and mental performance in mild dehydration. *Eur. J. Clin. Nutr.* 57 (Suppl 2):S24–9.

Section III

Special Populations

10 Dehydration and the Young Athlete

Effects on Health and Performance

Anita M. Rivera-Brown

CONTENTS

10.1 INTRODUCTION

An optimal state of body hydration is necessary for adequate cardiovascular, thermoregulatory, and mental function during exercise. Young athletes may lose copious amounts of body fluid in sweat during training and competition in sports, especially in hot and humid environments. In addition to starting exercise in a state of fluid deficit, young athletes do not replace enough of the fluid lost through sweat when drinking water *ad libitum*. The mismatch between the sweat produced and the fluid ingested leads to dehydration which may result in decrements in sports performance and heat illness. The risk is exacerbated during exercise in the heat and by sports uniforms and protective gear that increases the metabolic load and impedes adequate heat dissipation. Of great concern is that dehydration has been implicated in deaths due to heat stroke in adolescent athletes.

Until recently, most of the studies about fluid balance in children and adolescents were performed in the laboratory with untrained subjects. During the last few years, more data have been collected in the field of play in team sports like basketball, soccer, American football, and ice hockey and in long-duration sports like tennis, triathlon, and swimming. This chapter will review data related to fluid and electrolyte deficits during exercise in young female and male athletes and its effects on thermal strain and decrements in endurance performance and sports skills. The benefits of carbohydrate–electrolyte drinks to promote a greater fluid intake and minimize dehydration will be discussed as well as impediments that young athletes face when trying to rehydrate, and ways to facilitate fluid intake before, during, and after exercise. General guidelines provided by leading sports and medical organizations will be presented and a description of the sweat test for the development of individual fluid intake recommendations according to the sweat rate and sweat sodium loss. For the purpose of the chapter, *young athletes* refers to children and adolescents between the ages of 9 and 18 who train and compete in sports at recreational and elite level.

10.2 SWEAT LOSS AND INVOLUNTARY DEHYDRATION DURING TRAINING AND COMPETITION

The need to cool the body by sweating can lead to a significant loss of body fluid during exercise. *Dehydration*, the reduction in body water due to the delay in restoring the fluid lost in sweat by drinking, is a common occurrence in young athletes of all ages who train and compete at recreational and elite levels. The risk is higher for those participating in prolonged duration sports in hot climate. In warm and humid environment, there is a greater need to dissipate internal heat and therefore higher sweat production and fluid loss (Yeargin et al. 2010). Sweat rates also vary depending on exercise intensity, level of heat acclimatization, maturity level of the athletes, and the type of clothing and protective equipment worn.

There is typically a mismatch between the sweat produced and the amount of fluid intake that leads to varying levels of body fluid deficit in young athletes in different sports and environmental conditions. Table 10.1 illustrates sweating rates and dehydration levels in young athletes studied in laboratory settings both indoors and outdoors, and in field conditions during training and competition in cool and hot weather. As seen in the table, the majority of these studies have examined non-acclimatized adolescent males ranging in age from 9 to 18 years and scant information is available for young female athletes. Soccer, the most played sport in the world, is where most data have been collected in the playing field in the past 10 years.

Reported sweat rate values are in the range of 0.5–2.0 L/h in young male athletes and 0.4–1.0 L/h in young female athletes. As evidenced by data in Table 10.1, even the youngest athletes may sweat as much as adults and may need as much fluid during training and competition to avoid dehydration. Furthermore, similar to adults dehydration levels may reach 2.5% during training and 3% during competition in hot and humid conditions.

10.2.1 MATURITY AND SEX-RELATED DIFFERENCES IN SWEAT LOSS

The lowest sweating rates have been reported in the youngest athletes (9–14 years old) both in laboratory settings (Bergeron 2009; Iuliano et al. 1998; Rivera-Brown et al. 1999) and during training in the playing field (McDermott et al. 2009; Perrone et al. 2011). Typical values in this age group range from 0.4 to 1.3 L/h, and no negative outcomes related to diminished heat dissipation because of low sweat production have been reported. There is variability in drinking patterns, and dehydration levels among the youngest athletes average about 1%. Differences between genders in this age group are minimal and may be explained, in part, by differences in fitness levels (Wilk et al. 2007). The highest sweating rates are observed in older male adolescents with reported values ranging from 0.7 to 1.7 L/h. The values for female adolescents range from 0.4 to 1.0 L/h. The available data indicate slightly higher sweat rates for male athletes before the age of 15 years and a larger difference between the genders as they mature.

10.2.2 EFFECTS OF ENVIRONMENTAL CONDITIONS AND HEAT ACCLIMATIZATION

In tropical countries, exercise in hot weather cannot be avoided. It is common to see young athletes training at any time of the day. Continued exposure to a hot environment induces physiological adaptations that cause an increased capacity and sensitivity of sweat glands to thermal stimuli that lead to a greater sweat production (Inoue et al. 1999) and a need to drink more fluid to avoid dehydration. Two investigations in the tropical island of Puerto Rico have documented sweat rate values during prolonged intermittent cycling in hot and humid environment in 11- to 14-year-old male

TABLE 10.1

Studies Reporting Sweat Loss and % Dehydration in Young Athletes during Testing in the Laboratory, and While Training and Competing in Different Sports in the Field

Laboratory Setting: Indoors and Outdoors

Study	Subjects/Age	Sport/Modality	Duration	Estimated Sweat Rate	% Dehydration	Climate (°C); RH; WBGT
Iuliano et al. 1998	7 M, 8 F 12.5–14.8 years Triathletes 9 M, 8 F 15–17 years	Simulated duathlon indoors	1 km run 8 km bike 2 km run 2 km run 12 km bike 4 km run	<15 years M = 0.6 L/h F = 0.5 L/h ≥ 15 years M = 1.3 L/h F = 0.7 L/h	<15 years M = 0.7 F = 0.7 >15 years M = 1.5 F = 1.2	Not reported Not reported
Rivera-Brown et al. 1999	12 M 13.4 ± 0.4 years Trained, heat-acclimatized	Stationary cycling outdoors	20 min exercise and 25 min rest × 4	0.5–0.6 L/h 398 mL/m²/h	W = 1 CED = 0	32.6°C–33.3°C; 56%–61% WBGT = 30.2°C–30.6°C
Horswill et al. 2005	9 M, 6 F 17–18 years High school athletes	Cycling, running and elliptical trainer indoors	3 × 20 min	M = 0.9 L/h F = 0.6 L/h	0	26.5°C; 27.3%
Rivera-Brown et al. 2006b	9 F 11.3 ± 0.3 years Trained, heat acclimatized	Stationary cycling outdoors	56.9 ± 6.3 (60% VO₂max until fatigue)	540 mL/m²/h	0 (forced drinking)	33.4°C; 55.1% WBGT = 29.9°C

(Continued)

TABLE 10.1 (Continued)

Studies Reporting Sweat Loss and % Dehydration in Young Athletes during Testing in the Laboratory, and While Training and Competing in Different Sports in the Field

Study	Subjects/Age	Sport/Modality	Duration	Estimated Sweat Rate	% Dehydration	Climate (°C); RH; WBGT
		Laboratory Setting: Indoors and Outdoors				
Rivera-Brown et al. 2008	12 F 10.6 ± 0.2 years Trained, heat acclimatized	Stationary cycling outdoors	20 min exercise and 25 min rest × 4	298 mL/m²/h	W = 1.1 FW = 1.0 CED = 0.7	33°C; 65%–68% WBGT = 30.9°C
Bergeron et al. 2009	6 M, 6 F 12–13 years 6 M, 6 F 16–17 years Soccer players	Treadmill and stationary cycling Climatic chamber	2 × 80 min (alternating 8 min treadmill and 8 min cycle)	0.7 L/h 1.0 L/h	0.9–1.0 0.8	33°C; 49%
Phillips et al. 2010	15 M 12.7 ± 0.8 years Soccer Rugby Field hockey	4 × 15 min shuttle run Intermittent run to exhaustion	60 min	0.7 L/h	1.5–1.6	Thermoneutral

(Continued)

TABLE 10.1 (Continued)

Studies Reporting Sweat Loss and % Dehydration in Young Athletes during Testing in the Laboratory, and While Training and Competing in Different Sports in the Field

Study	Subjects/Age	Sport/Modality	Field Setting-Training			Climate (°C; RH; WBGT)
			Duration	Estimated Sweat Rate	% Dehydration	
Broad et al.1996	32 M, 16–18 years	Soccer	Training	0.9 L/h	0.7	25°C; 9%–10%
Shirreffs and Maughan 2008	92 M 18 ± 1 years	Soccer	60–70 min	1.3–1.5 L/h	No fluids = 2 Water = 0	25°C–28°C; 50%–53%
Silva et al. 2011	20 M, 17 ± 0.5 years	Soccer	150 min	~1–1.1 L/h	1.8	33.1°C; 43.4% WBGT = 31.5°C
Gibson et al. 2012	34 F 15.7 ± 0.7 years	Soccer	90 min	~0.5 L/h	0.8	9.8°C; 63%
Perrone et al. 2011	21 M soccer 9.5 ± 1.1 years 13.1 ± 2.1 years 26 M futsal 9.3 ± 1.2 years 11.0 ± 1.4 years	Soccer and futsal	PP = 70 min P = 85 min PP = 70 min P = 75 min	PP = ~0.4 L/h P = ~0.6 L/h PP = ~0.4 L/h P = ~0.5 L/h	0.7 0.7 0.4 0.5	WBGT = 21.7°C–24.1°C WBGT = 24.6°C–24.8°C
Williams and Blackwell 2012	21 M 17.1 ± 0.7 years	Soccer	100 min	0.7 L/h	0.5	11.0°C; 50%
Phillips et al. 2014	14 M 16.9 ± 0.8 years	Soccer	75 min × 3	0.3–0.55 L/h	0.5–1.0	8.9°C–17.2°C; 50.3–76.8%
Bergeron et al. 2006	9 M 5 F 15.1 ± 1.4 years	Tennis	120 min × 2	~1.1 L/h	W = 0.9 CED = 0.5	WBGT = 26.3°C

(Continued)

TABLE 10.1 (Continued)

Studies Reporting Sweat Loss and % Dehydration in Young Athletes during Testing in the Laboratory, and While Training and Competing in Different Sports in the Field

			Field Setting-Training			
Study	Subjects/Age	Sport/Modality	Duration	Estimated Sweat Rate	% Dehydration	Climate (°C); RH; WBGT
McDermott et al. 2009	33 M, 12.0 ± 2.0 years	American football	120 min	0.6 L/h	Not Reported	25.8°C; 70%, WBGT = 25.6°C
Palmer et al. 2010	18 M, 17.6 ± 0.3 years	Ice hockey	95 min	W = 0.9 L/h, CED = 1.5 L/h	W = 0.9, CED = 1.1	11.4°C; 52%
Carvalho et al. 2011	12 M, 14.8 ± 0.4 years	Basketball	90 min, then 30 min drills	0.8–1.0 L/h	No fluid = 2.5, Water = 1.1, CED = 0.7	Not reported
Soler et al. 2003	9 M, 18 ± 1.7 years	Swimming	180 min	0.5 ± 0.1 L/h	2.5	Water = 26.6°C ± 0.3°C, WBGT = 29.8°C
Higham et al. 2009	18 M, 15.4 ± 1.5 years; 17 F, 15.4 ± 1.2 years	Swimming	4.5 km	Not reported	M = 1–1.5, F = 0.7	Water = ~28.5°C, WBGT = 24.5°C
Rivera-Brown and De Félix-Dávila 2012	14 M 10 F, 12–17 years	Judo	90 min	Mid pub = 0.4 L/h, Late pub = 0.8 L/h	Mid pub = 1.3, Late pub = 1.9	29.5°C; 78%
Rivera-Brown et al. 2006	55 F, 12.9 ± 1.8 years	Tennis, triathlon, soccer, running	101 ± 15.3 min	0.7 L/h	0–1.9	WBGT = 27.2°C
Arnaoutis et al. 2014	59 M, 15.2 ± 1.3 years	Basketball, gymnastics, swimming, running, canoeing	90 min	0.5–0.8 L/h	1.1–1.7	WBGT = 31.9°C

(Continued)

TABLE 10.1 (Continued)

Studies Reporting Sweat Loss and % Dehydration in Young Athletes during Testing in the Laboratory, and While Training and Competing in Different Sports in the Field

Study	Subjects/Age	Sport/Modality	Field Setting-Competition			
			Duration	Estimated Sweat Rate	% Dehydration	Climate (°C); RH; WBGT
Rico-Sanz et al. 1996	8 M 17.0 ± 0.6 years	Soccer	Game	1.3 L/h	2.8–3	26.8°C; 81% WBGT = 25.3°C
Broad et al. 1996	32 M 16–18 years	Soccer	Game	1.2 L/h	0.9	25°C; 9%–10%
Mao et al. 2001	13 M 16–18 years	Soccer	Game 60 min	1.5 ± 0.6 L/h		
Guerra et al. 2004	20 M 16.1 ± 1.1 years	Soccer	Game 75 min	0.9–1.4 L/h	CED = ~1.7 No Fluid = ~2.5	28°C
Gutierres et al. 2011	20 M 17.9 ± 1.3 years	Soccer	Game	0.9 L/h	1.3	29°C; 64%
Da Silva et al. 2012	10 M 17.0 ± 0.6 years	Soccer	Game 80 min	~1.7 L/h	1.6	31°C; 48% WBGT = 28.6°C–32°C
Coelho et al. 2012	24 M 15–17 years	Soccer	Game 51–80 min	1.5 L/h	0.8	WBGT 21.7°C–23.6°C
Bergeron et al.2007	8 M 13.9 ± 0.9 years	Tennis	Game	Singles = 1.5 ± 0.2 L Doubles = 1.9 ± 0.2 L	S = 0.9 D = 0.5	WBGT = 29.6°C–31.3°C

(Continued)

TABLE 10.1 (Continued)
Studies Reporting Sweat Loss and % Dehydration in Young Athletes during Testing in the Laboratory, and While Training and Competing in Different Sports in the Field

			Field Setting-Competition			
Study	Subjects/Age	Sport/Modality	Duration	Estimated Sweat Rate	% Dehydration	Climate (°C); RH; WBGT
McDermott et al. 2009	33 M 12.0 ± 2.0 years	American football	120 min	1.3 ± 0.6 L/h	Not Reported	25.8°C; 70% WBGT = 25.6°C
Logan-Sprenger et al. 2011	24 M 18.3 ± 0.3 years	Ice hockey	Game	3.2 ± 0.2 L ~1.6 L/h	1.3 (1/3 = 1.8–4.3)	10.8°C; 30%
Aragón-Vargas et al. 2013	54 M; 38 F 9–13 years 14–17 years	Triathlon	Competition M = 56–102 min F = 59–103 min	Not reported	JM = 1.2 SM = 1.3 JF = 1.3 SF = 1.7	34°C; 50% WBGT = >31°C

M, males; F, females; RH, relative humidity; WBGT, wet bulb globe temperature; PP, pre-pubertal; P, pubertal; W, water; FW, flavored water; CED, carbohydrate–electrolyte drink; JM, junior males; SM, senior males; JF, junior females; SF, senior females.

(Rivera-Brown et al. 1999) and 9- to 12-year-old female, heat-acclimatized athletes (Rivera-Brown et al. 2008) that are almost twice as high compared to those reported for non-acclimatized, untrained children studied with similar protocols (Bar-Or et al. 1980; Falk et al. 1992; Wilk and Bar-Or 1996; Wilk et al. 2007). A higher sweat production is beneficial since it helps cool the body during exercise, but there is also a greater need for fluid replenishment to avoid the adverse effects of dehydration.

10.2.3 Sports-Related Differences in Sweat Loss and Drinking Patterns

Intermittent intensity sports such as tennis and team sports like basketball, soccer, American football, and ice hockey are *stop and go* in nature with high intensity exercise bouts and periods of low intensity or no activity. Young athletes participating in these sports show higher sweat rates and dehydration levels during competition than during training, due to the higher intensities of exercise and less focus on hydration (Broad et al. 1996; McDermott et al. 2009). In soccer, for example, adolescent players may spend most of the game running at high intensity and sweat rates of 1.7 L/h have been reported during games in hot weather (Da Silva et al. 2012).

Sweat rates may be reduced in cool weather but several factors may modify that response such as the amount and type of clothing worn, protective equipment, and the intensity of exercise. Thick sports uniforms and heavy protective equipment that cover a large portion of the skin cause an elevation in metabolic rate and sweat production, and the potential for dehydration. In fact, sweat rate values as high as 1.6 L/h and dehydration levels of 1%–2% have been reported in young ice hockey players during competition (Logan-Sprenger et al. 2011).

In American football and combat sports like judo and taekwondo, athletes sweat profusely during intense competition in the heat. In these sports, uniforms and protective gear impede adequate heat dissipation. Mc Dermott et al. (2009) documented sweat rates of 1.3 L/h in American football players with a mean age of 12 ± 2 years during a game in full uniform in a warm and humid environment (WBGT = 25.6°C), a value that was twice the amount during training. Such high sweat rates often lead to significant dehydration levels especially in those who arrive to training in a state of body fluid deficit.

Deaths due to heat stroke have been reported in adolescent American football players. Young football players report to pre-season practice sessions unfit and unacclimatized to the heat which increases their risk of dehydration-related health problems. It is recommended that during the first week of practice, protective equipment be introduced in stages, starting with the helmet, progressing to the shoulder pads and helmet, and then to the full uniform (Bergeron et al. 2005). To avoid a pre-exercise fluid deficit, young athletes should drink enough fluid during the day. In judo, dehydration incurred during training in conditions of high heat stress, coupled with a pre-exercise fluid deficit, may result in a final level of dehydration higher than 3% (Rivera-Brown and De Felix-Dávila 2012).

Swimmers, water polo players, and other athletes in aquatic sports lose heat primarily by convection, and sweat rates during exercise in the water are not as high as in other sports. Since sweat is not noticeable in water, many young swimmers are not aware they are sweating during training and drink very little fluid.

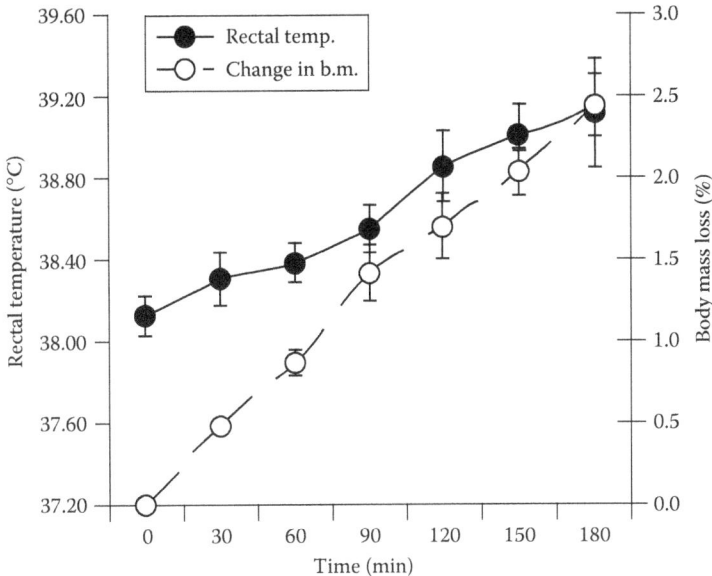

FIGURE 10.1 Progressive dehydration in adolescent swimmers during a 3-h training session, water temperature = 26.8°C ± 0.3°C. (Data from Soler, R. et al., *J. Strength Cond. Res.*, 17, 362–67, 2003. With permission.)

Soler et al. (2003) reported that adolescent heat-acclimatized swimmers (mean age = 18.0 ± 1.7 years) with a sweat rate of 0.5 L/h drank very little from their water bottles during an intense 3-h training session in a swimming pool in the tropics (water temperature = 26.8°C) and progressively dehydrated up to 2.5% (Figure 10.1). Young swimmers should be encouraged to observe symptoms of dehydration and drink water between laps and at the end of training.

10.2.4 SWEAT LOSS IN THE YOUNG FEMALE ATHLETE IN THE PLAYING FIELD

There is scant information about sweat loss in the playing field in young female athletes, especially those younger than 12 years. Most of the available information has been obtained during training sessions and has been reported in combination with data of young male athletes. It is surprising that compared to young males, very little data have been collected in young female soccer players, considering the fact that approximately 26 million women in 132 countries play soccer and of those, 2.9 million are younger than 18 years (FIFA 2007). Sweat rate values ranging from 0.6 to 0.7 L/h have been reported in female triathletes older than 15 years and in high school athletes exercising in laboratory conditions (Horswill et al. 2005; Iuliano et al. 1998) with dehydration levels of 0%–1.2%.

One investigation has provided data about fluid and electrolyte loss in young female soccer players during training in the field. Gibson et al. (2012) studied 34 junior elite Canadian players with a mean age of 15.7 years during two 90-min training sessions in a cool outdoors environment (9.8°C; 63% relative humidity).

The majority of players arrived to training in a state of fluid deficit. Similar to adult female players (Kilding et al. 2009), sweat rate was 0.5 L/h and they did not drink enough which led to a mean dehydration level of 0.8% with 8 players dehydrating to more than 2%.

Adaptations induced by chronic exposure to hot and humid environments are evident in the high sweat production reported for young female athletes indigenous to a tropical climate. Rivera-Brown et al. (2006a) reported an average sweat rate of 0.7 L/h in 55 heat acclimatized female athletes with a mean age of 12.9 ± 1.8 years during prolonged training sessions in tropical climate in tennis, triathlon, soccer, and track and field. This value is as high as that reported for female adult athletes.

Due to the few data available in females, gender-related comparisons should be made with caution. The few studies with young females report values that are about half that of males of similar age (Table 10.1). Sweating less may be an advantage for young female athletes during exercise in high heat stress environments since the amount of fluid they need to drink is lower and rehydration easier especially during long-duration training sessions and in *stop and go* sports.

10.2.5 ELECTROLYTE LOSS IN SWEAT

Sodium and chloride are the main electrolytes lost in sweat with other electrolytes like potassium, calcium, and magnesium present in low concentrations. Sodium helps regulate body water distribution and blood volume and is important for normal transmission of nerve impulses and skeletal muscle contraction. Some athletes called *salty sweaters* may lose considerable amounts of sodium in sweat. The use of absorbent patches has simplified the collection of sweat from one or various sites of the body for the measurement of electrolyte loss in young athletes in the field of play. A few recent studies have examined the sodium concentration and total sodium lost in sweat in athletes between the ages of 12–18 years during training and competition in various sports, using the sweat patch technique (Table 10.2). Most of the available data are in males, and scant information is available for young female athletes. Values range from 17 to 74 mmol/L, and the average value from the 10 studies in Table 10.2 is approximately 50 mmol/L, which is slightly higher than the average value of 40 mmol/L reported for adults (Baker et al. 2009).

In the only study that reports a maturation-related comparison in young athletes, Perrone et al. (2011) found no significant difference between prepubertal (58.7 mmol/L) and pubertal (65.0 mmol/L) heat-acclimatized male soccer and futsal players training in a warm climate. This is contrary to the findings of Meyer et al. (1992) who detected lower sodium and chloride values for non-trained prepubertal boys and girls compared to young adults and attributed the differences to the lower sweat rate in the children.

Total sweat sodium loss per hour in adolescent athletes may be quite variable, ranging from 0.3 to 3 g/h. The highest values are observed in adolescents with high sweat rates and are similar to adults (Logan-Sprenger et al. 2011; Palmer and Spriet 2008; Palmer et al. 2010). The Institute of Medicine (2005) advises that a sodium intake of 1.5–2.3 g/day is enough to cover sodium sweat losses in non-acclimatized individuals aged 9–50 years exposed to high temperature and who are physically active. However,

TABLE 10.2
Studies Reporting Sweat Sodium Concentration and Loss Using the Sweat Patch Technique in Young Athletes

Study	Subjects/Age	Sport	Sweat Site	Sweat [Na+] (mmol/L)	Sweat Na+ Loss	Environment Temperature and RH or WBGT
Shirreffs and Maughan 2008	92 M 18.0 ± 1.0 years	Soccer training	Forearm, chest, back, thigh	17.0 ± 7.0–20.0 ± 8.0	0.6–0.7 g/h 1.7 g NaCl	25°C–28°C; 50%–53%
Perrone et al. 2011	21 M soccer PP = 9.5 ± 1.1 years P = 13.1 ± 2.1 years 26 M futsal PP = 9.3 ± 1.2 years P = 11.0 ± 1.4 years	Soccer and futsal training	Scapula	PP = 58.7 ± 16.8 P = 65.0 ± 24.5	0.5–0.8 g/h 1.2–2.0 g NaCl	WBGT = 21.7°C–24.8°C
Gibson et al. 2012	34 F 15.7 ± 0.7 years	Soccer training	Forearm, chest, back, scapula, thigh	48.0 ± 12.0	0.7 g/h 1.7 g NaCl	9.8°C; 12%
Williams and Blackwell 2012	21 M 17.1 ± 0.7 years	Soccer training	Chest, back thigh, arm	67.8 ± 11.7	1.1 g/h 2.8 g NaCl	11°C; 50%
Palmer and Spriet 2008	44 M 18.4 ± 0.1 years	Ice hockey training	Forehead	54.2 ± 2.4	2.3 ± 0.2 g/h 5.7 g NaCl	13.9°C; 66%
Palmer et al. 2010	18 M 17.6 ± 0.3 years	Ice hockey training	Forehead	72.4 ± 5.6 73.0 ± 4.4	2.5 g/h 6.3 g NaCl	11.4°C; 52%
Logan-Sprenger et al. 2011	24 M 18.3 ± 0.3 years	Ice hockey game	Forehead	74 ± 4.0	3.1 ± 0.4 g/h 7.8 g NaCl	10.8°C; 30%
McDermott et al. 2009	33 M 12 ± 2 years	American football training	Forearm	32.1 ± 14.7	0.5–1.0 g/h 1.2–2.5 g NaCl	WBGT = 25.6°C

(Continued)

TABLE 10.2 (Continued)
Studies Reporting Sweat Sodium Concentration and Loss Using the Sweat Patch Technique in Young Athletes

Study	Subjects/Age	Sport	Sweat Site	Sweat [Na+] (mmol/L)	Sweat Na+ Loss	Environment Temperature and RH or WBGT
Yeargin et al. 2010	25 M 15.1 ± 1.0 years	American football training	Forearm	14–15 years 27.3 ± 17.2 16–17 years 40.4 ± 19.0	14–15 years 0.5 g/h; 1.2 g NaCl 16–17 years 0.7 g/h; .7 g NaCl	WBGT = 23.0°C
Rivera-Brown and De Felix-Dávila 2012	14 M, 10 F 12–17 years	Judo training	Forehead	44.5 ± 23.3	1.1 ± 0.8 g/h 2.8 g NaCl	WBGT = 27.2°C

following those recommendations may lead to body sodium deficit in young athletes who may lose the recommended intake amount in 1 h of exercise. For example, a young athlete who sweats profusely and loses 3 g of sodium in sweat per hour and exercises for 3 h may have a total sodium loss in sweat of more than 9 g and may need to ingest more than the recommended daily amount to replenish sodium lost in sweat.

Large sweat and sodium losses that are not replenished have been implicated in the development of muscle cramps during training and competition in adult athletes (Stofan et al. 2005). Similarly, young athletes with above average sweat rates and sodium loss could be more susceptible to cramping and impaired sports performance. Bergeron (2003) identified 6 adolescent tennis players between the ages of 15 and 18 years with sweat rates ≥2 L/h and sweat sodium losses between 1.4 and 4.8 g/h that a had a history of muscle cramps during tournament play. The extent of sodium loses in young athletes who sweat profusely (1–2 L/h) while participating in exercise sessions for several hours may warrant the presence of sodium in fluid replacement beverages to maintain sodium balance and reduce the risk of muscle cramping during prolonged exercise.

10.2.6 INVOLUNTARY DEHYDRATION DURING EXERCISE

The amount of fluid replaced is dependent on factors that may not be under the control of the athlete such as fluid availability, opportunities to drink during rest periods, fluid flavor and temperature, and limitations imposed by the rules of the game. For example, during cycling, athletes have many opportunities to drink and fluid availability, whereas in a game of soccer drinking opportunities are limited by the rules. In many team sports, young athletes depend on drink opportunities allowed by the coach and on the availability of fluids to be consumed in the training area.

Studies have shown that similar to adult athletes in all sports and climates, in spite of having many opportunities to drink and fluid readily available and palatable, the majority of young athletes will not drink enough during exercise and incur a body fluid deficit, or *involuntary dehydration*. As seen in Table 10.1, athletes of all ages dehydrate when participating in sports, especially during strenuous, prolonged exercise in the heat. Dehydration levels typically observed during exercise in young athletes range from 1% to 2% and may reach 3% during competition in the heat. Furthermore, dehydration incurred during training coupled with the fluid deficit before exercise may result in a level of dehydration higher than 3% (Rivera-Brown and De Felix-Dávila 2012). Dehydration levels as low as 1% exacerbate the rise in core temperature and impact negatively on sports performance especially in endurance events (Bardis et al. 2013). Dehydration has also been linked to augmented risk of heat illness among high school athletes (Yard et al. 2010).

10.3 PREVALENCE AND EFFECTS OF PRE-TRAINING FLUID DEFICIT

Adequate daily fluid intake is of utmost importance to arrive to practice and competition in optimal hydration state especially in hot and humid environments where copious sweating is expected. For many young athletes, the effects of dehydration

during exercise on physical and mental performance may result from the aggrava-
tion of a preexisting body fluid deficit before exercise. The urine-specific gravity
(USG), osmolality, and color have been used to examine body hydration status
before the start of sports training and/or competition in the field. The majority of
the available information is based on studies with males and only a few evaluated
females (Gibson et al. 2012; Higham et al. 2009; Kavouras et al. 2012; Kutlu and
Guler 2006; Lew et al. 2010; Rivera-Brown and De Felix-Dávila 2012). Results
indicate that young athletes in individual sports such as gymnastics, swimming,
running, canoeing, and tennis; in team sports such as American football, volleyball, basketball, ice hockey, and soccer; and in weight class sports such as taekwondo and judo arrive to practice and competition in a state of body fluid deficit,
with a USG ≥1.020 g/mL and/or urine osmolarity >700 mOsm/kg (Table 10.3).
Adequate fluid intake during games may not compensate for poor hydration status
before competition.

In sports like American football, soccer, and tennis, athletes may have repeated
training sessions or competitions in one day and need to improve hydration status both before the first exercise session and after, to be ready for the next session.
In soccer, young players have considerable fluid losses, few opportunities for fluid
intake during games, and poor rehydration habits (Da Silva et al. 2012). Da Silva
et al. (2012) reported that young heat acclimatized soccer players presented to training hypohydrated, and sensations of thirst and beverage palatability were not enough
to ensure sufficient intake to replace the fluid lost during exercise.

For young athletes in team sports of prolonged duration like soccer, a good
strategy may be to increase the daily fluid intake the week before games to
increase body water reserves. This strategy has been shown to improve temperature regulation in young soccer players during a match in the tropics (Rico-Sanz
et al. 1996). Coaches should instruct their players to drink more prior to competitions when large sweat losses and significant dehydration are expected, for
example, during exercise in the heat. Also, a post-training rehydration plan based
on individual sweat rates is needed to restore fluid balance in preparation for the
next training session.

10.3.1 Pre-Training Fluid Deficit and Thermal Strain

Many sports are played outdoors in the summer months or in tropical locations under
hot and humid conditions. Starting a game or training dehydrated or with a low level
of physical fitness magnifies the risk of uncontrolled rise of body temperature and
hyperthermia during exercise in the heat. The recent use of ingestible temperature
sensors has facilitated the examination of the influence of dehydration on the rate of
increase in core temperature in the playing field and the residual effect of dehydration. In a field study during tennis competition in the heat, Bergeron et al. (2007)
found that in junior male players the increase in core temperature during the match
was rapid (Figure 10.2) and strongly associated with the degree of fluid deficit, as
measured by the USG before singles competition. They also found that the core
temperature remained elevated even after 10 min of the end of the game. In those
players who went on to compete in a doubles match in the afternoon of the same day,

TABLE 10.3

Studies Reporting Pre-Exercise Hydration Status According to Urinary Indices in Young Athletes during Training and Competition

Study	Subjects/Age	Sport/Event	Time	Urine Osmolality (mOsmol/L)	Urine-Specific Gravity (mg/dL)	Estimated Dehydration Level[a]
Kutlu and Guler 2006	16 M, 19 ± 3 years 16 F, 17 ± 2 years	Taekwondo camp	Pre-training First morning urine Before competition	998 ± 171 1046 ± 205	1.017 ± 0.010 1.019 ± 0.010	Minimal dehydration
Stover et al. 2006	13 M 16.6 ± 0.4 years	American football training	Pre-training for 5 consecutive days	Not evaluated	1.022–1.024	Significant
Bergeron et al. 2007	8 M 13.9 ± 0.9 years	Tennis	Before singles Before doubles	Not evaluated	1.017 ± 0.002 1.025 ± 0.003	Minimal Significant
Decher et al. 2008	57 M, 12 ± 2 years 10 F, 13 ± 2 years	Soccer and American football camp	Pre-morning training	Not evaluated	1.022.5 ± 0.007 (all combined)	Significant
Palmer and Spriet 2008	44 M 18.4 ± 0.1 years	Ice hockey training	Pre-training	Not evaluated	11 players = 1.021–1.025 12 players = 1.026–1.030 1 player >1.030	Significant to serious
Higham et al. 2009	18 M, 15.4 ± 1.5 years 17 F, 15.4 ± 1.2 years	Swimming training	First morning urine pre-training	Not evaluated	1.024–1.026	Significant
McDermott et al. 2009	33 M 12 ± 2 years	American football camp	Days 2–5 of camp Each day before breakfast	796 ± 293	Not reported	Significant
Lew et al. 2010	36 M, 15 ± 0.9 years 30 F, 15 ± 1.0	Sports school athletes	Upon waking 5 days	Not evaluated	F = 1.021 ± 0.007 M = 1.021 ± 0.008	Significant

(Continued)

TABLE 10.3 (Continued)

Studies Reporting Pre-Exercise Hydration Status According to Urinary Indices in Young Athletes during Training and Competition

Study	Subjects/Age	Sport/Event	Time	Urine Osmolality (mOsmol/L)	Urine-Specific Gravity (mg/dL)	Estimated Dehydration Level[a]
Yeargin et al. 2010	25 M, 15 ± 1.0 years Young = 14 ± 1.0 Older = 17 ± 1.0	American football training	Pre-training First morning urine	881 ± 285 Average for 10 days	1.021–1.025	Significant
Kavouras et al. 2012	43 M, 49 F 13.8 ± 4.2	Volleyball Basketball	First morning urine	>700 in 85%	>1.020 in 93%	Significant
Gutierres et al. 2011	20 M 17.9 ± 1.3 years	Soccer	Pre-game	Not evaluated	1.023 ± 0.006	Significant
Logan-Sprenger et al. 2011	24 M 18.3 ± 0.3	Hockey	Pre-game	Not evaluated	1.016 ± 0.004 >1.020 in 41%	Minimal to significant
Perrone et al. 2011	21 M soccer 26 M futsal 9.5–13.1 years	Soccer and futsal training	Pre-training	Not evaluated	1.021 ± 0.006	Significant
Rivera-Brown and De Felix-Dávila 2012	14 M, 10 F 12–17 years	Judo training	Pre-training session and pre-next training session	Not evaluated	Mid-pubertal = 1.029 ± 0.004 Late pubertal = 1.024 ± 0.005	Significant
Da Silva et al. 2012	15 M 17.0 ± 0.6 years	Soccer	Pre-game	Not evaluated	1.021 ± 0.004	Significant
Gibson et al. 2012	34 F, 15.7 ± 0.7 years	Soccer	Pre-training	Not evaluated	45% of players = >1.020	Significant
Williams and Blackwell 2012	21 M 17.1 ± 0.7 years	Soccer training	Pre-training	1,319 ± 525	Not measured	Significant

(Continued)

TABLE 10.3 (Continued)

Studies Reporting Pre-Exercise Hydration Status According to Urinary Indices in Young Athletes during Training and Competition

Study	Subjects/Age	Sport/Event	Time	Urine Osmolality (mOsmol/L)	Urine-Specific Gravity (mg/dL)	Estimated Dehydration Level[a]
Coelho et al. 2012	24 M 15–17 years	Soccer	Pre-game	817 ± 169	1.020 ± 0.8 (1.012–1.032)	Minimal to significant
Arnaoutis et al. 2013	107 M 13.2 ± 2.6 years	Soccer camp	First morning urine	Not evaluated	89% >1.020	Significant
Arnaoutis et al. 2014	59 M 15.2 ± 1.3 years	Basketball, gymnastics, swimming, running, canoeing	First morning urine Pre-training	Not evaluated	1.026 ± 0.005 1.023 ± 0.007	Significant
Phillips et al. 2014	14 M 16.9 ± 0.8 years	Soccer training	First morning urine	Not evaluated	77% >1.020 on day 1 and day 3 62% on day 2	Significant

[a] Significant dehydration = >1.020; >700 mOsm/kg (Casa et al. 2000); minimal dehydration = 1.010–1.020 g/mL (Armstrong et al. 2007)

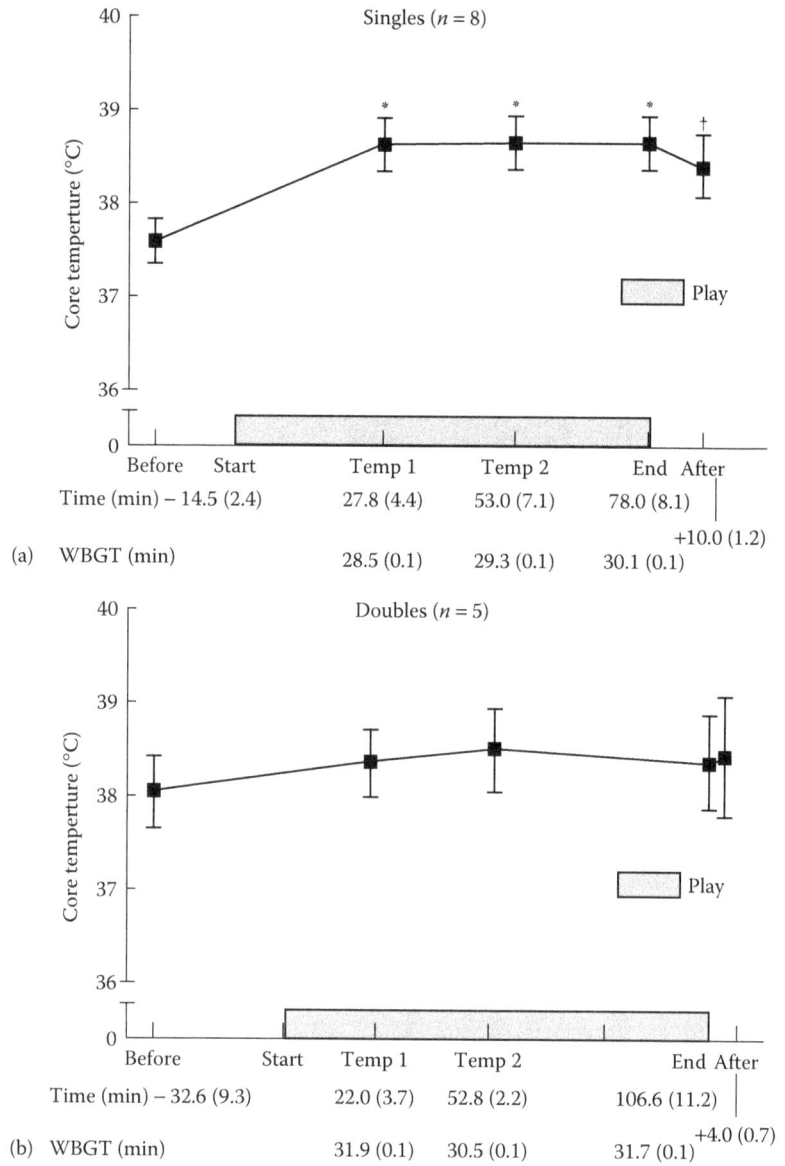

FIGURE 10.2   (a) Rise in core temperature during singles competition and (b) higher pre-play core temperature in doubles match later in the same day. (Data from Bergeron, M.F. et al., *Br. J. Sports Med.*, 41, 779–83, 2007. With permission.)

the pre-play USG was particularly high (1.025 g/mL) and they had a higher pre-play core temperature than before the previous singles match. These results indicate the need to drink more before and in between matches to avoid the negative effects of a preexisting fluid deficit on a first match and a *carryover* effect of dehydration on a second match.

10.3.2 Pre-Training Fluid Deficit and Deterioration in Sports Skills

Studies have shown deterioration in skills related to successful performance in team sports in young athletes who start exercise with a significant fluid deficit. The sports in which most of the recent data have been collected are basketball and soccer. Dougherty et al. (2006) studied 12- to 15-year-old male basketball players who started training after dehydrating by 2% of body mass to mimic the fluid deficits common in basketball and the typical pre-play hypohydration level. They found a deterioration in basketball shooting, sprinting, and lateral movement skills when players started in a hypohydrated state compared to when they were euhydrated. In another study with team sport players, Kavouras (2012) showed that when young male and female volleyball and basketball players improved their pre-exercise hydration status their time in a 600-m endurance run improved. However, no effects were observed in anaerobic skills such as vertical jump and sprint performance.

10.3.3 Persistent Fluid Deficit in Athletes Competing in Weight Category Sports

A state of persistent fluid deficit is a concern in athletes that train daily in hot and humid environments in weight category sports such as judo, taekwondo, boxing, and wrestling and restrict fluids intentionally with the purpose of *making weight*. For example, in a field study with heat acclimatized 12- to 18-year-old athletes during judo training in the tropics (Rivera-Brown and De Felix-Dávila 2012), 90% of the athletes arrived to practice with a USG >1.020 g/mL, drank very little during training, and finished the session with a USG >1.024 g/mL (Figure 10.3). When they

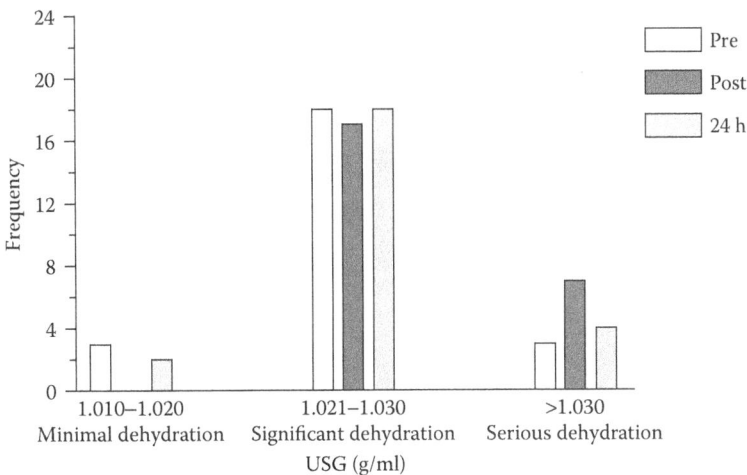

FIGURE 10.3 USG pre-, post-, and 24 h post a training session in a hot environment, in adolescent judo athletes indicating a persistent state of dehydration. (Data from Rivera-Brown, A.M. and De-Felix-Dávila, R.A. *Int. J. Sports Physiol. Perform.*, 7, 39–46, 2012. With permission.)

arrived to practice the next day, the USG was >1.027 g/mL indicating that the fluid deficit persisted from day to day.

In weight category sports, athletes frequently resort to routines of rapid weight loss in the days prior to competition to compete in a weight category that is usually lower than recommended for their body structure and composition. Dehydration results in increased cardiovascular strain, elevated heart rate, decreased venous return, and decreased stroke volume (Armstrong et al. 1997). Roshan et al. (2012) found a reduction in ventricular cavity size and increased amplitude of the P wave and QRS complexes after 3%–4% dehydration via sauna exposure in male high school wrestlers. It is common for young wrestlers to lose weight via dehydration in sauna prior to weigh-in, a practice that led to death in three young wrestlers in 1997 (CDC 1998) and should be avoided.

10.4 EFFECTS OF DEHYDRATION DURING EXERCISE ON HEALTH AND ATHLETIC PERFORMANCE

It is evident from studies with young athletes training and competing in athletic events that in addition to starting exercise in a state of fluid deficit, they generally do not replace enough of the fluid lost through sweat when drinking water *ad libitum*. There is variability in percent of the sweat loss replaced among athletes in different sports when the only choice is water: 28%–79% in soccer (Guerra et al. 2004; Gutierres et al. 2011); 55%–68% in futsal (Perrone et al. 2011); 68% in basketball (Carvalho et al. 2011); 58%–75% in tennis (Bergeron et al. 2006, 2007); 55%–58% in ice hockey (Logan-Sprenger et al. 2011; Palmer and Spriet 2008; Palmer et al. 2010); 32%–42% in judo (Rivera-Brown and De Felix-Dávila 2012), and only 5% in swimming (Soler et al. 2003). When athletes become dehydrated during exercise, they experience a greater physiologic strain. Studies with adult athletes have shown that even at low levels of dehydration, core temperature rises more rapidly compared to when well hydrated (Casa et al. 2010).

10.4.1 DEHYDRATION AND THE RISK OF HEAT ILLNESS DURING EXERCISE IN HOT WEATHER

Many competitions involving young athletes are played outdoors in the summer in hot and humid environments. A major concern is that when young athletes exercise in the heat at high intensity, while dehydrated body temperature may rise to dangerous levels, which may lead to heat injury. The National High School Sports-Related Injury Surveillance Study (Yard et al. 2010) indicated that during 2005–2009, a total of 118 time-loss heat illnesses (an average of 29.5 per school year) were reported by 100 participating schools in nine sports. These data correspond to an estimated annual average of 9,237 heat illnesses (dehydration, heat exhaustion, or heat stroke) that result in one or more days of time lost from athletic activity in U.S. high school athletes, most of them during pre-season American football practice. Factors that have been related to heat injuries in football players are a pre-exercise fluid deficit that worsens during training because of inadequate fluid intake; hot and humid environmental conditions, and insulation properties of the uniform. These factors predispose football players and other young athletes to hyperthermia which may lead to death from heat stroke. In fact, since 1995, 41 high school football players have died from heat stroke (Kucera et al. 2014) and many of these deaths have been linked to dehydration.

Studies with adult athletes have shown that even at low levels of dehydration, core temperature rises more rapidly compared to when well hydrated (Casa et al. 2010). For each 1% of body mass lost, the increase in core temperature is of 0.12°C–0.25°C (Buono and Wall 2000; Casa et al. 2010). The only information available specifically targeting the core temperature–dehydration relationship in children is from a study of Bar-Or et al. (1980), which suggests that the increase in core temperature is higher in dehydrated boys compared to men. In that study, core temperature increased by 0.28°C for each 1% of body mass lost in eleven 12-year-old boys cycling at moderate exercise intensity in a climatic chamber. Due to ethical constraints, no studies have specifically examined the core temperature–dehydration relationship in the laboratory or playing field in young athletes in the past 35 years, and there is a gap in our understanding of thermoregulatory strain due to dehydration in young athletes.

In any sport where young athletes compete several times during the same day, steps need to be taken to lower the core temperature (long recovery periods and cooling methods) and replace the fluid lost in the first exercise bout. Bergeron et al. (2009) showed that in young athletes, at least 1 h of complete rest, cool down, and rehydration are needed after prolonged strenuous exercise in the heat to eliminate carryover effects that may result in thermal and cardiovascular strain and poor performance in subsequent exercise bouts. An important consideration for tournament directors is to implement long recovery periods between matches played in hot and humid conditions to decrease the risk of heat illness.

10.4.2 IMPAIRMENT OF ENDURANCE PERFORMANCE

To what extent does the effect of dehydration influence sports performance in young athletes? At what dehydration level performance starts to deteriorate? In adult athletes, the consensus is that dehydration levels >2% lead to premature fatigue and impairment in exercise performance (Convertino et al. 1996). However, recent data indicate that dehydration levels as low as 1% can impair endurance performance in high level athletes (Bardis et al. 2013). If, as suggested in the pediatric literature, thermoregulatory strain in children is higher at a given level of dehydration, it is plausible that performance may be affected at lower levels of dehydration compared to adults. In the past few years, evidence has emerged that support that in young athletes, a preexisting fluid deficit and/or a low level of dehydration incurred during exercise impairs endurance performance as well as sports skills and that replacing lost fluid in timely fashion helps young athletes perform better.

A study with 10- to 12-year-old untrained boys exercising in a hot and humid environment in a climatic chamber found that 1% dehydration impaired cycling performance (Figure 10.4) possibly because of increased cardiovascular and thermoregulatory strain and reduced motivation to continue exercising due to intense thirst perception (Wilk et al. 2014).

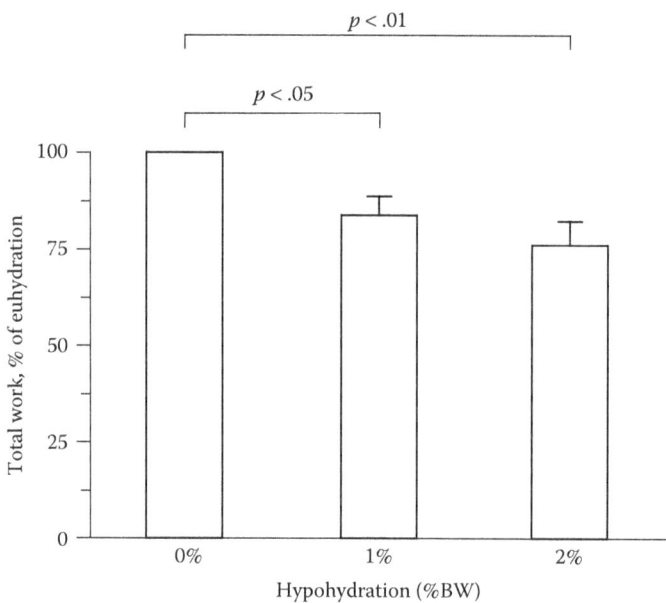

FIGURE 10.4 Reduction in total mechanical work performed during cycling with 1% and 2% dehydration expressed as a percentage of that performed with euhydration. (Data from Wilk, B. et al., *Eur. J. Appl. Physiol.*, 114, 707–13, 2014. With permission.)

The extent to which these novel findings in untrained children exercising in a climatic chamber can be generalized to field situations in youth sports is unclear at this time. Only one study to date has examined the influence of low levels of dehydration on endurance performance of child and adolescent athletes during a real-life competition. Aragón-Vargas et al. (2013) documented the relationship between final hypohydration level and race performance and the prevalence of self-reported symptoms of heat illness in 9- to 17-year-old heat-acclimatized male and female tri-athletes during a competition in a hot and humid climate. They found that 15%–20% of the younger athletes and 20%–35% of the older adolescents dehydrated between 2% and 3% with no ill effects. Interestingly, faster race times were achieved by athletes experiencing the highest degree of dehydration (Figure 10.5), a finding that has been reported before in adults (Laursen et al. 2006) and should not be interpreted to suggest an advantage of dehydrating while running. As noted by Aragón-Vargas et al. (2013), higher dehydration may be the result, not the cause, of faster racing. The fastest runners have difficulty drinking while running and those with high sweat rates are at a higher risk for dehydration. Data from Casa et al. (2010) prove that it is unlikely that low levels of dehydration help improve performance in runners. They showed that when adult runners raced at maximal effort while 2% dehydrated, they had higher core temperature and slower times compared to when euhydrated. More field studies in young athletes that include a control euhydration condition are needed before conclusions can be made on the effects of dehydration on endurance performance in prolonged duration sports.

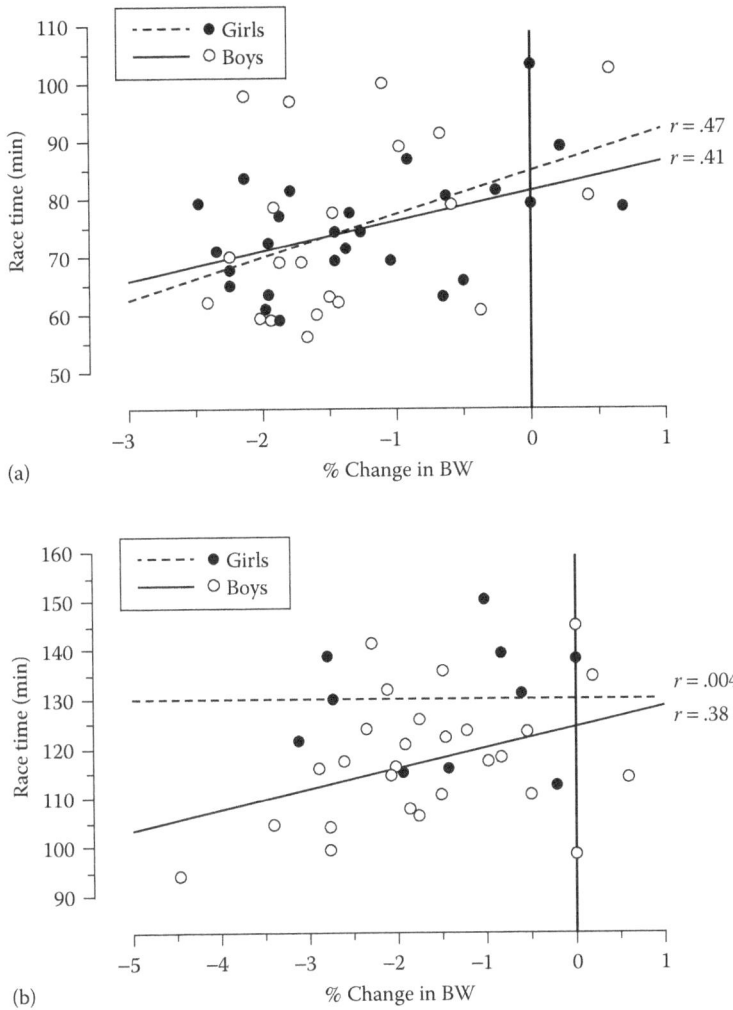

FIGURE 10.5 Relationship between race time and % change in body weight in (a) junior and (b) senior triathletes during a triathlon in Costa Rica. (Data from Aragón-Vargas, L.F. et al., *Eur. J. Appl. Physiol.*, 113, 233–9, 2013. With permission.)

10.4.3 IMPAIRMENT OF SPORTS SKILLS

Dehydration impairs sports skills and cognitive function required for good performance in sports with high technical and tactical components like soccer, volleyball, and basketball. Hoffman et al. (1995) found a 19% lower anaerobic power at the end of a basketball game and an 8.1% decrease in field goal and free throw percentage between the first and second half of the game when 17-year-old players were not allowed to drink (2% dehydration) compared to a game in which they drank *ad libitum*. In contrast, Carvalho et al. (2011) found a higher perceived exertion but no effect on basketball performance skills in adolescent basketball players when no fluid was ingested in a 90-min training session and dehydration reached 2.5%. However, they did not include a euhydration group as control, and therefore, definitive conclusions cannot be drawn about the effects of dehydration on performance skills.

10.5 FLUID INTAKE PATTERNS DURING TRAINING AND COMPETITION

An important question commonly posed by coaches of high-performance elite young athletes is if the athletes maintain a better hydration status will they perform better? Proper hydration minimizes the risk of excessive increase in core temperature, cardiovascular strain, and decrements in exercise performance and cognitive function associated with as little as 1% dehydration (Armstrong et al. 2007). Investigations show no maturational-related differences exist in cardiovascular and thermoregulatory responses, and exercise tolerance during exercise in the heat provided exercise intensity and body hydration are similar (Figure 10.6) (Rivera-Brown et al. 2006b).

An understanding of the amount, type, flavor, and temperature of fluids that promote an adequate intake in young athletes is needed to optimize their hydration regimes. In addition to plain water, young athletes consume a variety of beverages during exercise including sodas, fruit juices, enhanced water, milk, energy drinks, and sports drinks. Sodas and fruit juices are not recommended during exercise because of their high sugar content and carbonation which take longer to empty from the stomach and can cause stomach upset and interfere with proper hydration. Enhanced waters usually contain vitamins that are not necessary during exercise. Low-fat milk is effective for post-exercise fluid replacement because of its protein and electrolyte content that promotes fluid retention (Volterman et al. 2014). Energy drinks have become popular with high school students. They contain a large dose of caffeine which can have a greater stimulant effect in youngsters compared to adults. Often young athletes are not aware of the differences between sports and energy drinks and they consume energy drinks when their goal is to rehydrate. The American Academy of Pediatrics (2011b) and the Institute of Medicine (2014) have

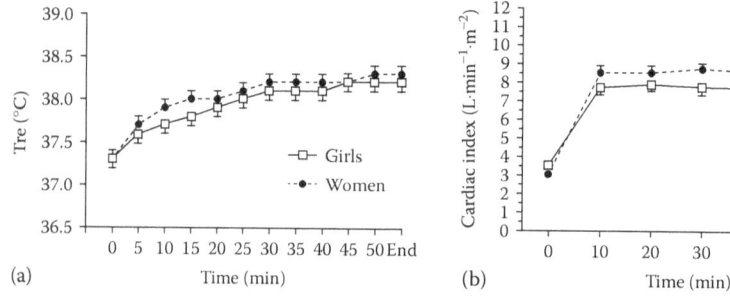

FIGURE 10.6   (a) Thermoregulatory and (b) Cardiovascular responses of 9- to 12-year-old heat-acclimatized trained girls and women exercising in hot and humid climate who maintained euhydration by drinking periodically to replace fluid lost in sweat. (Data from Rivera-Brown, A.M. et al., *Int. J. Sports Med.*, 27, 943–50, 2006b. With permission.)

advised against their use in children and adolescents due to potential health risks due to their stimulant content.

The carbohydrates in sports drinks help maintain blood glucose concentration and delay glycogen depletion in the muscles during prolonged exercise (>1 h). Young athletes are able to oxidize relatively more exogenous carbohydrates than adults and ingesting them in quantities found in commercially available sports drinks is advisable to help maintain blood glucose concentration (Timmons et al. 2003).

10.5.1 AD LIBITUM INTAKE OF WATER AND CARBOHYDRATE–ELECTROLYTE DRINKS

There is evidence that when young athletes are offered drinks that are flavored and contain carbohydrates and electrolytes (sports drinks), hydration status is better maintained and thermal stress and deterioration in sports skills are minimized. The effects have been attributed to increased palatability due to flavor, sodium, and carbohydrates and to the water-retaining qualities of sodium. Sex-related differences in drinking behavior have been identified in trained, heat-acclimatized young athletes.

Studies examining fluid intake patterns of youngsters were initially conducted in the laboratory, with males exercising in stationary bicycles in hot and humid conditions, first in a climatic chamber (Bar-Or et al. 1980) and replicated using an outdoor set up in tropical weather (Rodríguez-Santana et al. 1995). Results showed that young males do not drink enough when the only fluid available is water and they drink *ad libitum* during exercise in the heat. Later studies showed that both untrained, non-acclimatized (Wilk and Bar-Or 1996) as well as trained, heat-acclimatized (Rivera-Brown et al. 1999) 9- to 12-year-old boys increased their fluid intake and maintained euhydration when offered flavored water or flavored water with 6% carbohydrate and 18 mmol/L NaCl (sports drink) in a sports bottle within easy reach in different experimental conditions. With the addition of flavor alone, the increase in fluid intake in the untrained boys was of 45% compared to unflavored water. The trained, heat-acclimatized boys drank 32% more of the carbohydrate–electrolyte drink and no dehydration was observed (Figure 10.7).

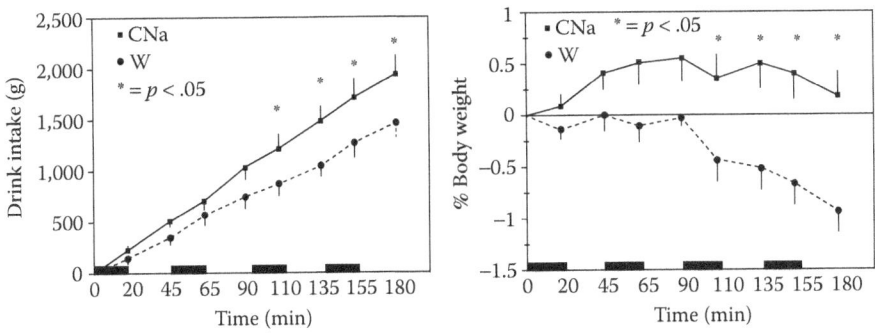

FIGURE 10.7 Fluid intake and dehydration during a prolonged exercise session in a hot and humid climate in trained, heat-acclimatized, 9- to 12-year-old boys when offered unflavored water or flavored water with 6% carbohydrate and 18 mmol/L NaCl (sports drink). (Data from Rivera-Brown, A.M. et al., *J. Appl. Physiol.*, 86, 78–84, 1999. With permission.)

A more recent study confirmed these findings in 17- to 18-year-old male and female high school athletes exercising for 1 h in laboratory conditions (Horswill et al. 2005). They observed an optimal drinking response and no dehydration when the only fluid available in a bottle within easy reach was a sports drink that had been previously cooled.

Similar studies with both young untrained non-acclimatized and trained heat-acclimatized females using identical protocols in controlled laboratory conditions revealed that drinking behaviors are different in female athletes (Rivera-Brown et al. 2008; Wilk et al. 2007). These studies showed that a 6% carbohydrate and 18 mmol/L NaCl sports drink was not more effective than water in increasing voluntary fluid intake. In the trained, heat-acclimatized girls (Rivera-Brown et al. 2008), there was a tendency toward less diuresis when drinking the carbohydrate–electrolyte drink possibly due to the fluid retaining effects of sodium. It has been suggested that young female athletes may be more motivated to drink what is available because they are more health and safety conscious (O'Neal et al. 2014).

The experimental model used in the laboratory has been transferred to the playing field in the sport of tennis to examine the effects of carbohydrate–electrolyte intake on performance. In tennis, many young athletes sweat profusely and for prolonged periods while training in hot and humid environments and face a high risk of dehydration and hyperthermia. Offering them sports drinks in addition to water may assure a high intake and minimize the thermal strain that accompanies dehydration, especially if the players arrive to training in a state of fluid deficit as is typical with young athletes. Young heat-acclimatized tennis players showed less dehydration and lower increase in core temperature when they drank a sports drink compared to when they ingested plain water only during a 2-h practice session in the heat (Bergeron et al. 2006) (Figure 10.8). These results should be interpreted with caution because training intensity may have differed during the sessions.

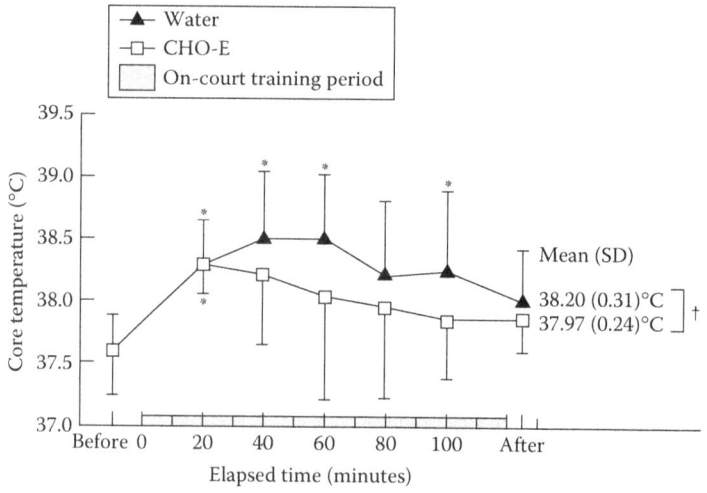

FIGURE 10.8 Lower final level of dehydration and lower increase in core temperature throughout a prolonged tennis session in the heat when drinking a sports drink. (Data from Bergeron, M.F. et al., *Br. J. Sports Med.*, 40, 406–10, 2006. With permission.)

The lower core temperature response while ingesting sports drink has not been confirmed in young female athletes in the lab or in field conditions. The only data available in females indicate that the increase in core temperature, heart rate, and perceptual variables did not differ when girls ingested a sports drink compared to water while stationary cycling in the heat (Rivera-Brown et al. 2008). Studies are needed to confirm these data in the training field.

The increased awareness of coaches about the importance of good hydration practices to optimize performance and protect the health of young athletes has resulted in a greater availability of sports drinks together with water in the playing field. However, some young athletes are unwilling to drink anything but water during training. When 18-year-old ice hockey players were given the choice of a sports drink and/or water to drink, all players chose to drink sports drinks before practice and only water during two 1-h intense practice sessions in a cool environment (Palmer and Spriet 2008). More than half of the players drank enough to minimize dehydration but one-third of the group dehydrated to more than 1%. In a separate study, a team of hockey players who were reluctant to drink anything but water were given a sports drink as the only option during a 90-min practice. Interestingly, the players drank a similar amount compared to a session in which the only choice was water and maintained a similar state of body hydration (Palmer et al. 2010).

A study by Rivera-Brown et al. (2006a) provides information about hydration practices of girl athletes when both drinks are offered together during the same training session, which is a common practice in the playing field. In that study, the training sessions were led by coaches in a tropical climate (WBGT = 31.9°C) in different sports: tennis, track and field, soccer, and triathlon. Approximately 70% of the girls arrived to training already hypohydrated as shown by USG values pre-exercise >1.020 g/mL. Results showed that while as a group they drank similar amounts of both beverages and maintained the initial state of hydration, there were large inter-individual differences in hydration practices in girls and the need for personalized hydration guidelines. While more than half of the group (58%) maintained or improved their initial state of hydration (Δ body weight = −0.5% to +3%) by drinking both water and sports drink during the session, 35% dehydrated up to 2% while drinking both beverages or water only. Interestingly, of the 18 girls who reported that they never consumed sports drinks during training, 10 of them drank it when it was offered.

In a field study with young female soccer players, Gibson et al. (2012) noted that in a cool outdoor environment, only one of 34 junior elite female soccer players brought to practice a sports drink to ingest during two 90-min training sessions and two-thirds of the group consumed less than 250 mL during the whole session.

10.5.2 PERFORMANCE-ENHANCING BENEFITS OF CARBOHYDRATE–ELECTROLYTE DRINKS

In adults, the majority of field studies that have investigated the performance-enhancing benefits of carbohydrate–electrolyte beverages have been in endurance athletes during continuous prolonged (>60 min) cycling, long distance running, and triathlons. In contrast, in young athletes studies have examined the effects of sports drinks on the improvement or maintenance of skill performance or high intensity endurance capacity in team sports.

Team sports like soccer, basketball, American football, ice hockey, volleyball, and lacrosse are characterized by bouts of high intensity exercise that rely on stored muscle glycogen. Consuming carbohydrate–electrolyte drinks will help replace fluid and glycogen loses and delay fatigue.

Phillips et al. (2010) confirmed the benefits of consuming a 6% carbohydrate–electrolyte drink (5 min before and at 15 min intervals) on increasing high-intensity endurance capacity in young team (males and females) sport players (football, rugby, and field hockey). Time to fatigue was increased by 24% and distance covered by 20% during the run to exhaustion in the exercise session where the players ingested the sports drink (5 mL/kg/min) compared to when they drank a non-carbohydrate placebo.

Soccer is an intermittent high-intensity prolonged sport where about 10% of the total distance covered during a match is in sprinting. Guerra et al. (2004) found that 16-year-old male soccer players who were consuming a 6% carbohydrate–electrolyte drink performed more sprints during the first half of a match compared to players who consumed only a small amount of water during the 15-min intermission. There was a tendency for core temperature to be higher in the no-fluid condition. A limitation of this study is that they did not have a euhydration with water condition. Nevertheless, the study provides field-based evidence that soccer players benefit from consuming carbohydrate–electrolyte drinks.

Dougherty et al. (2006) showed deterioration in performance of basketball skills such as shooting, sprint, lateral movements, and defensive drills with 2% dehydration in 12- to 15-year-old players. When players were given a 6% carbohydrate–electrolyte drink to prevent dehydration, an improvement was observed in shooting performance and on-court sprinting that was not observed when they drank only water. The study clearly showed the benefit of euhydration and ingesting a sports drink versus water on basketball skill performance. However, it is difficult to distinguish pure hydration effects from those associated with carbohydrate intake and the maintenance of blood glucose.

Carvalho et al. (2011) found that consumption of a 6% carbohydrate–electrolyte drink had no effect of improving performance in a set of drills that included free throws, sprints, and defensive skills in basketball when compared to fluid restriction that resulted in a 2% dehydration or *ad libitum* water intake. The only significant effect was that the perceived exertion of the players was lower when drinking the sports drink. According to the available data, sports drinks containing 6%–8% carbohydrates and 18 mmol/L of sodium may be recommended for young athletes who have a need to replenish carbohydrates and electrolytes in combination with water during prolonged vigorous exercise in hot and humid conditions such as cycling, triathlons, and long distance running, and team sports like soccer, basketball, and ice hockey where high sweat rates are expected. Consuming sports drinks may not be necessary for young athletes during exercise at moderate intensity for short periods of time or for obese young athletes who should avoid the calories in sports drinks.

10.6 FLUID INTAKE GUIDELINES FOR YOUNG ATHLETES

10.6.1 GENERAL GUIDELINES

Leading sports and medical organizations have published guidelines to optimize fluid replacement of young athletes during sports training and competition (Table 10.4). Recommendations are provided for hydration before, during, and after exercise, including the composition of carbohydrate–electrolyte drinks. These guidelines provide a general recommendation that can be applied when hydration assessment techniques (*sweat tests*) are not accessible. However, the wide variability in sweat losses and fluid intake patterns of young athletes hinder the real-world application of these guidelines in all sports and environmental conditions.

10.6.2 SWEAT TEST FOR INDIVIDUALIZED FLUID INTAKE RECOMMENDATIONS IN CHILDREN AND ADOLESCENTS

As noted in Table 10.1, sweat rates vary from 0.5 to 2.0 L/h in young athletes. Guidelines based on average sweat rate values are of limited use when providing recommendations to individual athletes who vary in age, maturation, sport specialty, and heat acclimatization level. Providing similar fluid intake guidelines to young athletes in different sports and playing positions may not be appropriate and may lead to dehydration and/or hyperhydration.

The amount, type, and rate of fluid replacement should be determined on an individual basis according to the sweat rate, in different training/competition situations and environmental conditions. The best approach is to develop an individualized hydration plan based on the young athletes sweat rate determined from a *sweat test*, preferably in actual training/competition and climatic conditions. In addition to measuring initial body hydration status and fluid lost in sweat, a sweat sample can be collected with a patch, and the sodium concentration of sweat can be determined for the identification of young athletes with salty sweat. It is also beneficial to include measures of the core temperature response using ingestible sensors and relate it to hydration practices, especially if the exercise is in hot and humid conditions. An individualized hydration plan can be developed that includes the type, amount, and frequency of fluid intake to minimize the risk of dehydration. One or several athletes can be tested at the same time.

In many sports like baseball, cycling, triathlon, American football, and track and field, rehydration opportunities are frequent. In these sports, young athletes can consume smaller volumes repeatedly depending on sweat production in specific environmental conditions. In sports in which rehydration opportunities are limited by the rules of the game or must occur at specific times during competition like soccer, ice hockey, lacrosse, long distance running, and basketball, a drinking strategy should be planned in advance to take advantage of the few opportunities to drink and maximize rehydration within the limitations imposed by the rules.

If the training session is prolonged or very intense, carbohydrate–electrolyte drinks are recommended. Young athletes can become familiar with their pre-exercise

TABLE 10.4

Guidelines to Optimize Fluid Replacement of Young Athletes during Exercise

Reference	(Sports Medicine Australia. Safety guidelines for children in sport and recreation 1997)	(US Soccer Federation-Youth heat and hydration guidelines 2006)	American College of Sports Medicine 2007 (Armstrong et al. 2007)	Rowland, T.W. (Rowland 2011)	American Academy of Pediatrics (American Academy of Pediatrics Policy statement—Climatic heat stress and the exercising child and adolescent 2000; American Academy of Pediatrics Policy Statement—Climatic heat stress and the exercising child and adolescent 2011a).
Before training	*45 min before* *~10 years old:* 150–200 mL *~15 years old:* 300–400 mL	*30 min before* 375–500 mL	*4 h before* 5–7 mL/kg *2 h before* ~3–5 mL/kg	—	Should be well hydrated pre-exercise.
During training	*Every 20 min* *~10 years old:* 75–100 mL *~15 years old:* 150–200 mL	*Every 20 min* ~281 mL/20 min for players >41 kg Avoid fruit juices, carbonated beverages, caffeinated beverages, and energy drinks	Should be customized to prevent >2% dehydration	*Every 15 min* 3.25 mL/kg Choice of fluid should be dictated by taste preference. Some young athletes may need individualized approach to determine fluid intake strategies.	*Every 20 min* *9–12-year olds:* 100–250 mL *>12 years old:* 1.0–1.5 L/h Pre-activity to post-activity body weight changes can provide specific hydration needs.
After training	Liberal until urination	Start drinking immediately and every 20 min for 1 h	~1.5 L/kg body mass lost	480 mL/0.5 kg lost or 4 mL/kg for each hour of exercise.	No guideline.
Sports drinks	—	Replace sodium lost in sweat	Consume beverages containing carbohydrates <8% and electrolytes	—	Include electrolyte-supplemented beverages if exercise ≥1 h or repeated same-day sessions.

body weight and weigh themselves after exercise to confirm that they are drinking enough according to exercise intensity and climatic conditions and replace water and sodium losses.

10.6.3 PREVENTION OF DEHYDRATION DURING EXERCISE

Young athletes should start exercise well hydrated. Water must be readily available at training sites in enough quantities for all players on a team. If exercise is prolonged and/or in a hot and humid environment, sports drinks should also be available. Young athletes should bring their own water bottle to training and coaches must provide frequent fluid breaks (at least every 15 min). Young athletes should be weighed frequently to assure they are drinking enough to replace fluid lost in sweat. The intake of energy drinks, sodas, and fruit juices should be discouraged.

10.7 HYDRATION KNOWLEDGE, ATTITUDES, AND BEHAVIORS

10.7.1 EFFECTS OF EDUCATIONAL INTERVENTIONS

Many young athletes, parents, and coaches are aware of the importance of drinking fluids during exercise to optimize performance especially when exercising outdoors in the heat. However, knowledge does not always translate into proper strategies for adequate fluid intake. Educational interventions are effective to increase awareness of young athletes of their body hydration status pre-, during, and post-exercise so that they drink more to prevent dehydration on a day-to-day basis. Kavouras et al. (2012) implemented a nutritional intervention program emphasizing water consumption in 13- to 14-year-old volleyball and basketball players that attended a summer camp in Greece in warm conditions (28°C), the majority of whom started the camp in a state of hypohydration. The educational program included a talk about the importance of hydration and an orientation about how to use a urine color chart to assess body hydration status and resulted in improvement in hydration status over a 2-day period through *ad libitum* water intake and an increase in endurance exercise performance. However, although there was a significant reduction in the percentage of dehydrated athletes after the intervention, many remained hypohydrated throughout the camp while perceiving that they were drinking enough.

10.7.2 IMPEDIMENTS TO ADEQUATE HYDRATION PRACTICES

Studies show that young athletes understand that drinking fluids during exercise is important for good performance and decreases the possibility of suffering heat illness but they face barriers that hinder the consumption of adequate amounts and types of fluids. Decher et al. (2008) examined the association between hydration status and knowledge regarding hydration, in active young athletes during their stay in a 4-day summer sports camp. Although these athletes understood the importance of hydration and they could recognize when they were doing a good or a bad job hydrating, they arrived to camp in a state of hypohydration which they sustained throughout the camp. A reason for this disparity between knowledge and behavior is the impediments young athletes face when trying to rehydrate: not given enough breaks to drink nor time during breaks, water stations are crowded, and they do not have water bottles or ways to fill them (Decher et al. 2008; McDermott et al. 2009). Identifying impediments to adequate hydration practices and how to overcome them to facilitate fluid intake before, during, and after exercise will help young athletes avoid fluid imbalances.

10.7.3 STRATEGIES TO IMPROVE HYDRATION PRACTICES

The younger athletes depend more on adults to provide the right type and amount of fluid they need during training and competition. It is common to see that when there is fluid available in the training field it is only plain water. Young athletes may benefit from being facilitated different types of fluids before, during, and after practice. Stover et al. (2006) implemented a drinking strategy in high-school American football players who experienced mild fluid deficits during twice per day training sessions over a two-week period in the summer. After one practice session, they provided the athletes with their choice of two bottles of water or sports drink with instructions to consume one between dinner and bedtime and one after waking up in the morning. They observed that after providing fluids and instructions following one practice session athletes arrived to the next training session better hydrated. Another good strategy is to have each athlete bring their own bottle to practice and provide fluids so they can refill their bottles as necessary.

10.8 CHALLENGES FOR FUTURE RESEARCH

Additional research on the effects of dehydration on health and performance of the young athlete as well as on prevention strategies is needed. Particularly lacking is information in relation to physiological responses of young female athletes, both in laboratory settings and in the playing field, which hinders the implementation of strategies geared specifically to address their hydration needs. Future studies should address the following:

- The effects of dehydration on core temperature in the playing field
- Cooling devices/strategies to prevent hyperthermia if dehydration cannot be avoided
- Barriers to drinking and real-life hydration practices in different sports
- Effectiveness of educational programs to reduce fluid imbalances
- Level of dehydration which impairs endurance performance in prolonged duration sports
- Fluid intake variability and its relation to thirst level due to a pre-training fluid deficit
- Sweating, drinking patterns, and core temperature in young female athletes in the playing field
- Strategies to optimize body water recovery after exercise

REFERENCES

American Academy of Pediatrics. 2000. Policy statement—Climatic heat stress and the exercising child and adolescent. *Pediatrics* 106:158–9.

American Academy of Pediatrics. 2011a. Policy statement—Climatic heat stress and the exercising child and adolescent. *Pediatrics* 128:e741–7.

American Academy of Pediatrics. 2011b. Clinical report—sports drinks and energy drinks for children and adolescents: Are they appropriate? *Pediatrics* 127:1182–9.

Aragón-Vargas, L. F., B. Wilk, B. W. Timmons, and O. Bar-Or. 2013. Body weight changes in child and adolescent athletes during a triathlon competition. *Eur J Appl Physiol* 113 (1):233–9.

Armstrong, L. E., D. Casa, M. Millard-Stafford, D. S. Moran, S. W. Pyne, and W. O. Roberts. 2007. American College of Sports Medicine position stand: Exertional heat illness during training and competition. *Med Sci Sports Exerc* 39:566–72.

Armstrong, L. E., C. M. Maresh, C. V. Gabaree, J. R. Hoffman, S. A. Kavouras, R. W. Kenefick, J. W. Castellani, and L. E. Ahlquist. 1997. Thermal and circulatory responses during exercise: Effects of hypohydration, dehydration, and water intake. *J Appl Physiol* 82 (6):2028–35.

Arnaoutis, G., S. A. Kavouras, A. Angelopoulou, C. Skoulariki, S. Bismpikou, S. Mourtakos, and L. S. Sidossis. February 2014. Fluid balance during training in elite young athletes of different sports. *J Strength Cond Res*, DOI:10.1519/JSC.0000000000000400.

Arnaoutis, G., S. A. Kavouras, Y. P. Kotsis, Y. E. Tsekouras, M. Makrillos, and C. N. Bardis. 2013. Ad libitum fluid intake does not prevent dehydration in suboptimally hydrated young soccer players during a training session of a summer camp. *Int J Sport Nutr Exerc Metab* 23 (3):245–51.

Baker, L. B., J. R. Stofan, A. A. Hamilton, and C. A. Horswill. 2009. Comparison of regional patch collection vs. whole body washdown for measuring sweat sodium and potassium loss during exercise. *J Appl Physiol* 107 (3):887–95.

Bardis, C. N., S. A. Kavouras, G. Arnaoutis, D. B. Panagiotakos, and L. S. Sidossis. 2013. Mild dehydration and cycling performance during 5-kilometer hill climbing. *J Athl Train* 48 (6):741–7.

Bar-Or, O., R. Dotan, O. Inbar, A. Rotshtein, and H. Zonder. 1980. Voluntary hypohydration in 10–12-year-old boys. *J Appl Physiol Respir Environ Exerc Physiol* 48:104–8.

Bergeron, M. F. 2003. Heat cramps: Fluid and electrolyte challenges during tennis in the heat. *J Sci Med Sport* 6 (1):19–27.

Bergeron, M. F. 2009. Youth sports in the heat: Recovery and scheduling considerations for tournament play. *Sports Med* 39 (7):513–22.

Bergeron, M. F., M. D. Laird, E. L. Marinik, J. S. Brenner, and J. L. Waller. 2009. Repeated-bout exercise in the heat in young athletes: Physiological strain and perceptual responses. *J Appl Physiol* 106 (2):476–85.

Bergeron, M. F., D. B. McKeag, D. J. Casa, P. M. Clarkson, R. W. Dick, E. R. Eichner, C. A. Horswill, A. C. Luke, F. Mueller, T. A. Munce, W. O. Roberts, and T. W. Rowland. 2005. Youth football: Heat stress and injury risk. *Med Sci Sports Exerc* 37:1421–30.

Bergeron, M. F., K. S. McLeod, and J. F. Coyle. 2007. Core body temperature during competition in the heat: National Boys' 14s Junior Championships. *Br J Sports Med* 41 (11):779–83.

Bergeron, M. F., J. L. Waller, and E. L. Marinik. 2006. Voluntary fluid intake and core temperature responses in adolescent tennis players: Sports beverage versus water. *Br J Sports Med* 40 (5):406–10.

Broad, E. M., L. M. Burke, G. R. Cox, P. Heeley, and M. Riley. 1996. Body weight changes and voluntary fluid intakes during training and competition sessions in team sports. *Int J Sport Nutr* 6 (3):307–20.

Buono, M. J., and A. J. Wall. 2000. Effect of hypohydration on core temperature during exercise in temperate and hot environments. *Pflugers Arch* 440 (3):476–80.

Carvalho, P., B. Oliveira, R. Barros, P. Padrao, P. Moreira, and V. H. Teixeira. 2011. Impact of fluid restriction and ad libitum water intake or an 8% carbohydrate-electrolyte beverage on skill performance of elite adolescent basketball players. *Int J Sport Nutr Exerc Metab* 21 (3):214–21.

Casa, D. J., R. L. Stearns, R. M. Lopez, M. S. Ganio, B. P. McDermott, S. Walker Yeargin, L. M. Yamamoto, S. M. Mazerolle, M. W. Roti, L. E. Armstrong, and C. M. Maresh. 2010. Influence of hydration on physiological function and performance during trail running in the heat. *J Athl Train* 45 (2):147–56.

CDC. 1998. Hyperthermia and dehydration related deaths associated with intentional rapid weight loss in three collegiate wrestlers—North Carolina, Wisconsin, and Michigan, November–December, 1997. *Morb Mortal Wkly Rep* 47:105–8.

Coelho, D. B., E. R. Pereira, E. C. Gomes, L. Coelho, D. D. Soares, and E. Silami-Garcia. 2012. Evaluation of hydration status following soccer matches of different categories. *Rev Bras Cineantropom Desempenho Hum* 14:276–86.

Convertino, V. A., L. E. Armstrong, E. F. Coyle, G. W. Mack, M. N. Sawka, L. C. Senay, Jr., and W. M. Sherman. 1996. American College of Sports Medicine position stand: Exercise and fluid replacement. *Med Sci Sports Exerc* 28 (1):i–vii.

Da Silva, R. P., T. Mundel, A. J. Natali, M. G. Bara Filho, R. C. Alfenas, J. R. Lima, F. G. Belfort, P. R. Lopes, and J. C. Marins. 2012. Pre-game hydration status, sweat loss, and fluid intake in elite Brazilian young male soccer players during competition. *J Sports Sci* 30 (1):37–42.

Decher, N. R., D. J. Casa, S. W. Yeargin, M. S. Ganio, M. L. Levreault, C. L. Dann, C. T. James, M. A. McCaffrey, C. B. Oconnor, and S. W. Brown. 2008. Hydration status, knowledge, and behavior in youths at summer sports camps. *Int J Sports Physiol Perform* 3 (3):262–78.

Dougherty, K. A., L. B. Baker, M. Chow, and W. L. Kenney. 2006. Two percent dehydration impairs and six percent carbohydrate drink improves boys basketball skills. *Med Sci Sports Exerc* 38 (9):1650–8.

Falk, B., O. Bar-Or, and J. D. MacDougall. 1992. Thermoregulatory responses of pre-, mid-, and late-pubertal boys to exercise in dry heat. *Med Sci Sports Exerc* 24 (6):688–94.

FIFA. 2007. *Big Count 2006: Statistical Summary Report.* FIFA Communications Division.

Gibson, J. C., L. A. Stuart-Hill, W. Pethick, and C. A. Gaul. 2012. Hydration status and fluid and sodium balance in elite Canadian junior women's soccer players in a cool environment. *Appl Physiol Nutr Metab* 37 (5):931–7.

Guerra, I., R. Chaves, T. Barros, and J. Tirapegui. 2004. The influence of fluid ingestion on performance of soccer players during a match. *J Sports Sci Med* 3 (4):198–202.

Gutierres, A. P. M., A. J. Natali, J. M. Vianna, V. M. Reis, and J. C. B. Marins. 2011. Dehydration in soccer players after a match in the heat. *Biol. Sport* 28:249–54.

Higham, D. G., G. A. Naughton, L. A. Burt, and X. Shi. 2009. Comparison of fluid balance between competitive swimmers and less active adolescents. *Int J Sport Nutr Exerc Metab* 19 (3):259–74.

Hoffman, J. R., H. Stavsky, and B. Falk. 1995. The effect of water restriction on anaerobic power and vertical jumping height in basketball players. *Int J Sports Med* 16 (4):214–18.

Horswill, C. A., D. H. Passe, J. R. Stofan, and R. Murray. 2005. Adequacy of fluid ingestion in adolescents and adults during moderate intensity exercise. *Pediatr Exerc Sci* 17:41–50.

Inoue, Y., G. Havenith, W. L. Kenney, J. L. Loomis, and E. R. Buskirk. 1999. Exercise-and methylcholine-induced sweating responses in older and younger men: Effect of heat acclimation and aerobic fitness. *Int. J. Biometeorol* 42:210–16.

Institute of Medicine. 2005. *Dietary Reference Intakes for Water, Potassium, Sodium, Chloride, and Sulfate.* Washington, DC: National Academies Press.

Institute of Medicine. 2014. *Nutrition Standards for Foods in Schools: Leading the Way toward Healthier Youth.* Washington, DC: National Academies Press.

Iuliano, S., G. Naughton, G. Collier, and J. Carlson. 1998. Examination of the self-selected fluid intake practices by junior athletes during a simulated duathlon event. *Int J Sport Nutr* 8 (1):10–23.

Kavouras, S. A., G. Arnaoutis, M. Makrillos, C. Garagouni, E. Nikolaou, O. Chira, E. Ellinikaki, and L. S. Sidossis. 2012. Educational intervention on water intake improves hydration status and enhances exercise performance in athletic youth. *Scand J Med Sci Sports* 22 (5):684–9.

Kilding, A. E., H. Tunstall, E. Wraith, M. Good, C. Gammon, and C. Smith. 2009. Sweat rate and sweat electrolyte composition in international female soccer players during game specific training. *Int J Sports Med* 30 (6):443–7.

Kucera, K. L., D. Klossner, B. Colgate, and R. C. Cantu. 2014. *Annual Survey of Football Injury Research 1931–2013.* National Center for Catastrophic Sports Injury, The American Football Coaches Association, The National Collegiate Athletic Association, The National Federation of State High School Associations, and The National Athletic Trainers' Association.

Kutlu, M., and Guler, G. 2006. Assessment of hydration status by urinary analysis of elite junior taekwon-do athletes in preparing for competition. *J Sports Sci* 24 (8):869–73.

Laursen, P. B., R. Suriano, M. J. Quod, H. Lee, C. R. Abbiss, K. Nosaka, D. T. Martin, and D. Bishop. 2006. Core temperature and hydration status during an Ironman triathlon. *Br J Sports Med* 40 (4):320–5.

Lew, C. H., G. Slater, G. Nair, and M. Miller. 2010. Relationship between changes in upon-waking urinary indices of hydration status and body mass in adolescent Singaporean athletes. *Int J Sport Nutr Exerc Metab* 20 (4):330–5.

Logan-Sprenger, H. M., M. S. Palmer, and L. L. Spriet. 2011. Estimated fluid and sodium balance and drink preferences in elite male junior players during an ice hockey game. *Appl Physiol Nutr Metab* 36 (1):145–52.

Mao, I. F., M. L. Chen, and Y. C. Ko. 2001. Electrolyte loss in sweat and iodine deficiency in a hot environment. *Arch Environ Health* 56 (3):271–7.

McDermott, B. P., D. J. Casa, S. W. Yeargin, M. S. Ganio, R. M. Lopez, and E. A. Mooradian. 2009. Hydration status, sweat rates, and rehydration education of youth football campers. *J Sport Rehabil* 18 (4):535–52.

Meyer, F., O. Bar-Or, D. MacDougall, and G. J. F. Heigenhauser. 1992. Sweat electrolyte loss during exercise in the heat: Effects of gender and maturation. *Med Sci Sports Exerc* 24:776–81.

O'Neal, E. K., C. R. Caufield, J. B. Lowe, M. C. Stevenson, B. A. Davis, and L. K. Thigpen. 2014. 24-h fluid kinetics and perception of sweat losses following a 1-h run in a temperate environment. *Nutrients* 6 (1):37–49.

Palmer, M. S., H. M. Logan, and L. L. Spriet. 2010. On-ice sweat rate, voluntary fluid intake, and sodium balance during practice in male junior ice hockey players drinking water or a carbohydrate-electrolyte solution. *Appl Physiol Nutr Metab* 35 (3):328–35.

Palmer, M. S., and L. L. Spriet. 2008. Sweat rate, salt loss, and fluid intake during an intense on-ice practice in elite Canadian male junior hockey players. *Appl Physiol Nutr Metab* 33 (2):263–71.

Perrone, C. A., P. L. Sehl, J. B. Martins, and F. Meyer. 2011. Hydration status and sweating responses of boys playing soccer and futsal. *Med Sport* 15:188–93.

Phillips, S. M., D. Sykes, and N. Gibson. 2014. Hydration status and fluid balance of elite European youth soccer players during consecutive training sessions. *J Sports Sci Med* 13:817–22.

Phillips, S. M., A. P. Turner, S. Gray, M. F. Sanderson, and J. Sproule. 2010. Ingesting a 6% carbohydrate-electrolyte solution improves endurance capacity, but not sprint performance, during intermittent, high-intensity shuttle running in adolescent team games players aged 12–14 years. *Eur J Appl Physiol* 109 (5):811–21.

Rico-Sanz, J., W. R. Frontera, M. A. Rivera, A. Rivera-Brown, P. A. Mole, and C. N. Meredith. 1996. Effects of hyperhydration on total body water, temperature regulation and performance of elite young soccer players in a warm climate. *Int J Sports Med* 17 (2):85–91.

Rivera-Brown, A. M., Y. Cabrera-Dávila, J. Frontera-Cantero, L. E. Berrios, and J. Gonzalez. 2006a. Fluid intake in heat-acclimatized girl athletes when sports drink and water are provided during training. *Med Sci Sports Exerc* 38 (5):S11.

Rivera-Brown, A. M., and R. A. De Felix-Dávila. 2012. Hydration status in adolescent judo athletes before and after training in the heat. *Int J Sports Physiol Perform* 7 (1):39–46.

Rivera-Brown, A. M., R. Gutierrez, J. C. Gutierrez, W. R. Frontera, and O. Bar-Or. 1999. Drink composition, voluntary drinking, and fluid balance in exercising, trained, heat-acclimatized boys. *J Appl Physiol* 86 (1):78–84.

Rivera-Brown, A. M., F. A. Ramirez-Marrero, B. Wilk, and O. Bar-Or. 2008. Voluntary drinking and hydration in trained, heat-acclimatized girls exercising in a hot and humid climate. *Eur J Appl Physiol* 103 (1):109–16.

Rivera-Brown, A. M., T. W. Rowland, F. A. Ramirez-Marrero, G. Santacana, and A. Vann. 2006b. Exercise tolerance in a hot and humid climate in heat-acclimatized girls and women. *Int J Sports Med* 27 (12):943–50.

Rodríguez-Santana, J. R., A. M. Rivera-Brown, W. R. Frontera, M. A. Rivera, P. M. Mayol, and O. Bar-Or. 1995. Effect of drink pattern and solar radiation on thermoregulation and fluid balance during exercise in chronically heat acclimatized children. *Am J Hum Biol* 7:643–50.

Roshan, D. B., M. Hosseinzadeh, and M. Saravi. 2012. The effects of dehydration and rehydration on electrocardiographic and echocardiographic parameters in Greco-Roman wrestlers. *Eur J Sport Sci* 12:49–56.

Rowland, T. 2011. Fluid replacement requirements for child athletes. *Sports Med* 41 (4):279–88.

Shirreffs, S. M., and R. J. Maughan. 2008. Water and salt balance in young male football players in training during the holy month of Ramadan. *J Sports Sci* 26(Suppl 3):S47–54.

Silva, R. P., T. Mundel, A. J. Natali, M. G. Bara Filho, J. R. Lima, R. C. Alfenas, P. R. Lopes, F. G. Belfort, and J. C. Marins. 2011. Fluid balance of elite Brazilian youth soccer players during consecutive days of training. *J Sports Sci* 29 (7):725–32.

Soler, R., M. Echegaray, and M. A. Rivera. 2003. Thermal responses and body fluid balance of competitive male swimmers during a training session. *J Strength Cond Res* 17 (2):362–7.

Stofan, J. R., J. J. Zachwieja, C. A. Horswill, R. Murray, S. A. Anderson, and E. R. Eichner. 2005. Sweat and sodium losses in NCAA football players: A precursor to heat cramps? *Int J Sport Nutr Exerc Metab* 15 (6):641–52.

Stover, E. A., J. Zachwieja, J. Stofan, R. Murray, and C. A. Horswill. 2006. Consistently high urine specific gravity in adolescent American football players and the impact of an acute drinking strategy. *Int J Sports Med* 27 (4):330–5.

Timmons, B. W., O. Bar-Or, and M. C. Riddell. 2003. Oxidation rate of exogenous carbohydrate during exercise is higher in boys than in men. *J Appl Physiol* 94 (1):278–84.

US Soccer Federation. *Youth Heat and Hydration Guidelines*. 2006. http://www.ussoccer.com/.

Volterman, K. A. J. Obeid, B. Wilk, and B. W. Timmons. 2014. Effect of milk consumption on rehydration in youth following exercise in the heat. *Appl Physiol Nutr Metab* 39(11):1257–64.

Wilk, B., and O. Bar-Or. 1996. Effect of drink flavor and NaCL on voluntary drinking and hydration in boys exercising in the heat. *J Appl Physiol* 80 (4):1112–17.

Wilk, B., F. Meyer, O. Bar-Or, and B. W. Timmons. 2014. Mild to moderate hypohydration reduces boys' high-intensity cycling performance in the heat. *Eur J Appl Physiol* 114 (4):707–13.

Wilk, B., A. M. Rivera-Brown, and O. Bar-Or. 2007. Voluntary drinking and hydration in non-acclimatized girls exercising in the heat. *Eur J Appl Physiol* 101 (6):727–34.

Williams, C. A., and J. Blackwell. 2012. Hydration status, fluid intake, and electrolyte losses in youth soccer players. *Int J Sports Physiol Perform* 7 (4):367–74.

Yard, E. E., J. Gilchrist, T. Haileyesus, M. Murphy, C. Collins, N. McIlvain, and R. D. Comstock. 2010. Heat illness among high school athletes—United States, 2005–2009. *J Safety Res* 41 (6):471–4.

Yeargin, S. W., D. J. Casa, D. A. Judelson, B. P. McDermott, M. S. Ganio, E. C. Lee, R. M. Lopez, R. L. Stearns, J. M. Anderson, L. E. Armstrong, W. J. Kraemer, and C. M. Maresh. 2010. Thermoregulatory responses and hydration practices in heat-acclimatized adolescents during preseason high school football. *J Athl Train* 45 (2):136–46.

11 Water Balance and Master Athletes

Zbigniew Szygula

CONTENTS

11.1 INTRODUCTION

The aging of populations is an indisputable fact. According to the World Health Organization (WHO), it is predicted that the proportion of the world's population over the age of 60 will double from about 11% in 2000 to 22% in 2050. In absolute values, this represents an increase from 605 million to 2 billion people over the age of 60 during this period of time. A considerable increase in average life expectancy exceeding 81 years in several countries has also been observed (http://www.who.int/ageing/about/facts/en/).

Age classification varies between countries and elderly people can now be divided into several age groups. According to the WHO, the age ranges are as follows: between 45 and 59—pre-elderly age, between 60 and 74—young elderly/young old, between 75 and 89—old elderly/oldest-old, and above the age of 90—*hoary aged* (http://www.who.int/healthinfo/survey/ageingdefnolder/en/).

Aging is a natural life process and it has its negative effects on many aspects of life. It is an irreversible process but may occur in an optimal manner, which can be defined as healthy aging. However, there are obligatory changes in the body with age, which fit into a so-called physiological course of healthy aging. They include progressive molecular changes in cells and tissues, body composition and body fluids, progressive organ system changes inducing a reduction of reserves in particular systems, that is, systemic changes (e.g., in the circulatory system, respiratory system, hormonal system, and locomotor organs), and in consequence a reduction of organism capacity to adapt to different loads (biological, physical, and psychosocial) (Chester and Rudolph 2011). Due to the aging process, the organism's regulatory abilities decrease and the efficiency of homeostasis mechanisms deteriorate, including the ability to effectively regulate fluid and electrolyte balance, which may predispose to dehydration and impaired exercise tolerance in heat environment (Blatteis 2012; Bongers et al. 2014; Chester and Rudolph 2011; Chodzko-Zajko et al. 2009) or overhydration (Bossingham et al. 2005; Miller 1998).

Progressive declines in fitness, psychic efficiency, social activity, and autonomy are observed with aging. The risk of developing many chronic diseases, including civilization-related diseases such as coronary heart disease, type 2 diabetes, obesity, as well as some neoplastic diseases, degenerative joint diseases, or osteoporosis, also increases with age. Furthermore, the age-related reduction in physical activity, that is, the time being allowed for lifestyle physical activity, and above all the reduction in exercise intensity contribute to acceleration of these negative changes and worsening of the quality of life in the process of aging. Although no physical exercises can stop biological aging, there is a lot of scientific evidence that regular physical activity can minimize negative physiological effects by limiting the development and progression of chronic disease and disabling conditions, especially in highly active and fit people compared to moderately fit or low fit people (i.e., those with normal activity or sedentary lifestyle) (Chodzko-Zajko et al. 2009). The higher the level of physical activity, the greater the expected benefits. For instance, the prevalence of systemic arterial hypertension, hypercholesterolemia, and type 2 diabetes lowers with the increasing number of marathons that have been run (Leppers and Cattagni 2012). There is also rising evidence that regular perticipation in exercise by older adults has psychological and cognitive benefits (Chodzko-Zajko et al. 2009; Tanaka and Seals 2008).

According to numerous scientific reports, master athletes are an ideal model of so-called "successful aging" and an example of the fact that physical activity is one of more effective anti-aging agents. They are fascinating examples of how high physical performance and physiological function can be maintained despite the years that have passed (Leppers and Cattagni 2012; Tanaka and Seals 2008; Weir et al. 2010). Studies from the past years have provided clear evidence that the aging processes and the decline of physiological functions are significantly slowed down by involving sports until old age. Although the practice of sports does not fully prevent a decline in aerobic capacity, it has beneficial effects on the circulatory system and attenuates sarcopenia (Trappe 2007). Regular physical activity may reduce the risk of all-cause mortality by about 25% and increase life expectancy even by 2 years (Jokl et al. 2004). Therefore, it is strongly recommended for older adults to be engaged in regular physical activity and to avoid an inactive lifestyle (Chodzko-Zajko et al. 2009).

Recommendations of various scientific medical societies, as well as promoting physical activity by mass media, have increased the number of old-aged people practicing various kinds of sports at different levels. Many of them train at recreational level; however, there are many competitive athletes (master athletes) who are focused on sports performance with years of training and competition experience. Over the past three decades, there is an increased number of older athletes participating in sports competitions such as marathon running, swimming, cycling, rowing, and wrestling (Lepers and Cattagni 2012; Tanaka and Seals 2008; Trappe 2001).

11.2 MASTER ATHLETES

People who continue physical training above the age of 35 and are oriented towards competition in sport until old age are considered master athletes. Most sports organizations, which bring together master athletes, grant the right to compete only to people aged 35–40 and above. In some sports (golf or baseball), the age of master athletes has shifted from the 4th and even the 5th decade of life (Rosenbloom 2013; Weir et al. 2010).

In competitions organized by the International Masters Games Association (IMGA—www.imga.ch), or by the World Masters Athletics (http://www.world-masters-athletics.org), contestants must be at least 35 years old, whereas people being at least 50 years old are admitted to sports competitions organized by National Senior Games Association (NSGA—www.nsga.com) and the Huntsman World Senior Games (http://www.seniorgames.net). It is not unusual that people over the age of 70 (70+) participate in marathon runs, swims, or triathlons. Recently, Ironman organizers have introduced an age group of 75+ and later 80+, especially for 82-year-old Sister Madonna Buder, *Iron Nun*, who ran and successfully crossed the finish line at the 2012 Ironman Hawaii 70.3, as the oldest woman to ever finish an Ironman Triathlon (http://en.wikipedia.org/wiki/Madonna_Buder—of 27.08.2014; http://www.huffingtonpost.com/2014/07/04/iron-nun-triathlon_n_5558429—of 27.08.2014). Generally, masters competitions are organized in *age categories*, with 5-year age ranges: 35–39, 40–44, 45–49, and so on (Weir et al. 2010), while in sports competitions organized by the Huntsman World Senior Games the age ranges start from 50–54 and end at 90–94, 95–99, 100+ (http://www.seniorgames.net/about/registrationrules).

As mentioned above, the participation of older people in sports competitions has been increasing for several decades. Jokl et al. (2004) have shown that in 1983–1999 the number of participants over the age of 50 (both women and men) in the New York City Marathon increased proportionally more in comparison with younger athletes. More importantly, athletes over the age of 50, and even those of the 70–79 age group improved their running times over these years at a greater rate than young athletes. An increased participation of master runners in the New York City Marathon in 1980–2009 has also been reported by Leppers and Cattagni (2012), with the participation of women increased at a greater extent than that of men. At the same time, a statistically significant shortening of running times in men over the age of 54 and women over the age of 44 has been reported. An increased number of participants over the age of 35 (master runners) and an improved time performance in ultramarathon runs have also been reported by other researchers (Rüst et al. 2014; Zingg et al. 2013, 2014). Apart from these *professionals* training for several years, older males and females (>60 years) who start regular sports training at this age and even decide to participate in sports competitions are seen more often (Trappe 2001).

The purpose of this chapter is to present the effect of aging on body water balance and thermoregulation and to attempt to develop recommendations on hydration and protection against these imbalances in master athletes.

11.3 WATER BALANCE AND THERMOREGULATION

Water is a major component of our bodies. Adequate water content in the human body is of great importance in maintaining homeostasis and health from birth to old age and its volume is the resultant between its intake and output.

In epidemiological studies, a correlation has been observed between low fluid intake and some chronic diseases, including cardiovascular diseases, asthma, diabetic hyperglycemia, constipation, urolithiasis, or even some cancer types, whereas the total lack of access to water in extreme cases may lead to death within a few days (Maughan 2012a). Deficiency of water in the body (hypohydration) may cause disorders with serious consequences, including (but not limited to) thermoregulatory dysfunction during exercise in a hot environment or even during passive body exposure to heat on hot days, leading even to death, particularly in people over the age of 60. It is estimated that a heat wave that happened in Western Europe in August 2003 caused the death of over 30,000 people, primarily elderly people, living alone, suffering from chronic diseases and therefore taking various medicines, which could impair the efficiency of thermoregulatory mechanisms in different ways (Robine et al. 2008).

Dehydration is a frequent geriatric problem, while access to water may be a matter of life or death for elderly people, even in the countries where droughts are only known from mass media. Dehydration at different degrees was observed in 25%–31% of nursing home residents, and it is also a common cause of elderly people hospitalization, with a mortality rate reaching about 45% (Benelam and Wyness 2010; Hodgkinson et al. 2003; Morley et al. 1998; Rolls and Phillips 1990).

11.3.1 CONSEQUENCES OF DEHYDRATION ON PHYSICAL PERFORMANCE

Both acute and chronic dehydration have a negative effect on exercise capacity, which has been extensively described in previous chapters (Chapters 7 and 8) and in many original papers being published in recent years (Casa et al. 2010; Lopez et al. 2011; Naharudin and Yusof 2013) and in review articles (e.g., Cheuvront et al. 2010, 2013; Coyle 2004; Goulet 2012; Maughan 2012b; Murray 2007; and Sawka et al. 2012).

It is generally accepted that dehydration resulting in a 2% reduction in body mass has a negative effect on human exercise capacity (Murray 2007; Sawka et al. 2007). Based on laboratory tests, it is suggested that even 1% dehydration may deteriorate exercise capacity and impair the cardiovascular and thermoregulatory responses, especially when the exercise is performed in a hot environment (Montain and Coyle 1992; Murray 2007; Sawka 1992). Of course, the higher the dehydration, the greater the cardiovascular strain, that is, the increase in heart rate (HR) and the decline in stroke volume (SV) are graded in proportion to the amount of dehydration accured during exercise in a hot environment. The consequence is a decreased blood flow through working muscles and skin and a further rapid development of hyperthermia, being particularly rapid during exercise in a hot environment (Gonzalez-Alonso et al. 2000; Montain and Coyle 1992; Murray 2007; Sawka et al. 2012, see also Chapters 4, 5, 8).

11.3.2 THERMOREGULATION IN ELDERLY PEOPLE

Because humans belong to the group of warm-blooded organisms, a precise regulation of body temperature is crucial for optimal functioning of our body and survival. The maintenance of core body temperature depends on the dynamic equilibrium between heat acquisition by the body and its dissipation (elimination, loss) into the environment. Disturbances in maintaining this equilibrium lead to body temperature changes. Metabolic heat production during physical exercise causes rapid increase in body temperature. Metabolic heat production during physical exercise causes rapid increase in body temperature. A temperature increase is particularly rapid when physical activity is performed in an unfavorable environment—high temperature and humidity. A primary way to eliminate heat from the body during physical exercise is through sweat evaporation, but a good functioning of many organs and systems is equally important for efficient thermoregulation.

There is ample prevailing scientific evidence of impaired thermoregulatory ability and reduced tolerance to heat in older healthy people both during passive heat exposure (Armstrong and Kenney 1993; Dufor and Candas 2007; Miescher and Fortney 1989) and during physical exercise in hot environment (Inbar et al. 2004; Larose et al. 2013a,b,c). However, there are reports (de Viveiros et al. 2012; Kenny et al. 2010a; Wright et al. 2014) where no significant difference has been observed in thermoregulatory responses between older, well-trained people compared to physically active younger persons or others (Kenney 1997; Pandolf et al. 1988) where a slight deterioration has been seen in thermoregulatory functions in individuals over the age of 50.

Among many causes of age-related deterioration in heat dissipation, the following changes are listed:

1. Reduction in the local and whole-body sweat rate (Dufour and Candas 2007; Inbar et al. 2007; Inue 1996; Kenney and Munce 2003), as early as after the age of 40 (Larose et al. 2013a)
2. Reduction in skin blood flow, even by 25%–40%, in older adults both at rest and during exercise in warm environments when compared with younger subjects, under both euhydration and hypohydration (Armstrong and Kenney 1993; Kenney 1997)
3. Decrease in thermal sensitivity, which induces a reduction in sensitivity to heat and later activation of thermoregulatory mechanisms than in younger people (Armstrong and Kenney 1993; Dufour and Candas 2007; Kenny et al. 2010b)
4. Deterioration in the efficiency of the cardiovascular system (Kenney 1997; Kenney and Munce 2003)

Since there is a close correlation between body water content (hydration, in terms of fluid volume and osmolality) and thermoregulation, deterioration in the thermoregulatory capacity may also be affected, apart from the above reasons, by impaired regulation of fluid and electrolyte balance in older people (Kenney 1997).

11.4   WATER BALANCE IN THE PROCESS OF AGING

The maintenance of water–electrolyte homeostasis depends on the balance between the volume of water intake and water output. In healthy individuals, water intake is regulated mostly by thirst, whereas water output is regulated by kidneys through an antidiuretic hormone (ADH), that is, arginine vasopressin (AVP), renin–angiotensin–aldosterone (RAA) system, and atrial natriuretic peptide (ANP).

Water imbalance induces a number of responses, which lead to the adjustment of this disturbance. The water and electrolyte deficiencies are adjusted through a combination of changes in the perception of thirst and through sodium and water retention by kidneys. Water deficiency in the body and increased plasma osmolality (Posm) result in the induction of central osmoreceptors of the thirst center in the hypothalamus, stimulating thirst. There is also vasopressin secretion by the posterior part of the pituitary, which increases the water retention by kidneys. The reduction in plasma volume (PV; and the same of the effective circulating blood volume) results in the induction of volume change sensitive baroreceptors, which, in turn, increase the AVP secretion and the activity of the sympathetic system and the RAA system. These changes lead to sodium and water retention by the kidneys, which contribute to the normalization of circulating blood volume (Benelam and Wyness 2010; Hodgkinson et al. 2003; Horgen et al. 1998; Miller 1998).

Under the conditions of unlimited access to water, the volume of ingested fluids depends on the amount and frequency of consumed food. A sufficient amount of fluids is ingested with food and the kidneys are able to maintain body water balance. In the case that water access is limited or the organism is exposed to a thermal load, or body water is lost due to any disease, the risk of dehydration associated with serious consequences increases in elderly people (Benelam and Wyness 2010; Cowen et al. 2013; Kenney 1997; Kenney and Chiu 2001; Rolls and Phillips 1990).

Due to the aging process, the organism's regulatory abilities decrease, including also the efficiency of homeostasis mechanisms, particularly the capability of physiological adjustments in stress situations, such as exposure to high (and also low) ambient temperature. Elderly people, particularly above the age of 60, are a vulnerable group. They are more prone to dehydration and associated thermal disorders, including heat stroke, on one hand, and to hyperhydration on the other. Therefore, they require a special treatment to assure proper hydration (Benelam and Wyness 2010; Chester and Rudolph 2011; Kenny et al. 2010b; Maughan 2012a; Rosenbloom 2013). It should be noted, however, that some differences in the dynamics of aging are the result of various factors, including genetic ones, causing that the chronological age does not always coincide with the biological (functional) one.

Based on clinical observations, it is believed that elderly healthy people, usually those over the age of 60, are chronically dehydrated (Kenney 1997). According to Mack et al. (1994), they are *hyperosmotic and hypovolemic*, not only because of normal water deficit but also due to a shift in the operating point for control of body fluid volume and composition. However, this is not confirmed by laboratory studies evaluating the fluid balance in elderly healthy people, which reported that only some people were slightly hyperosmotic and hypovolemic (Kenney 1997).

Nevertheless, the risk for developing fluid and water imbalances, and in consequence thermal diseases, increases when the aging is accompanied by diseases such as cardiovascular disease, hypertension, obesity, diabetes mellitus, and respiratory diseases (*morbid aging*), which additionally worsen the adaptability to environmental changes (Kenny et al. 2010b). This situation is even worse in the presence of at least two diseases, most frequently chronic ones in the same person, what we refer to as *comorbidity* or *multimorbidity*. In this case, medicines are being used, sometimes many simultaneously, which may negatively affect fluid and electrolyte balance through different mechanisms and deteriorate body responses to thermal load (Cheshire and Fealey 2008; Kenny et al. 2010b; Schlanger et al. 2010).

The above observations have referred to elderly or old physically inactive patients; thus, the question arises whether healthy, active elderly people are similarly predisposed to developing such disturbances in fluid and electrolyte balance?

11.5 CAUSES OF FLUID–ELECTROLYTE IMBALANCES RELATED TO AGING

Water balance is subject to strict control mechanisms and complex physiological and behavioral interactions which undergo unfavorable changes with human aging, inevitably leading to abnormalities of water homeostasis at several discrete locations along the neurorenal axis responsible for maintaining water and electrolyte balance. There are various changes in the body, which cause elderly people to be more prone to disturbances in fluid and electrolyte balance, mainly sodium balance. As the result of these changes, they face a reduced capacity, both to conserve and to obtain fluids, leading to dehydration and hyperosmolality on one hand, and to excrete water, leading to overhydration and hypoosmolality on the other (Beck 2000; Cowen et al. 2013).

Among the main causes of these age-related changes that may reduce homeostatic reserves of the body and deteriorate the so far precise regulation of fluid and electrolyte balance are (1) changes in body composition, (2) changes in hypothalamic-pituitary regulation of thirst, and (3) deterioration in the functioning of kidneys. (4) Decreased secretion and activity of hormones that regulate fluid and electrolyte balance—AVP (ADH), RAA, and ANP are also listed (Ayus and Arieff 1996; Beck 2000; Benelam and Wyness 2010; Cowen et al. 2013; Kenney 1997; Kenney and Chiu 2001; Luckey and Parsa 2003; Miller 1998; Rolls and Phillips 1990). The result of these unfavorable changes is a reduction in homeostatic mechanisms of the body and disturbance of precise fluid balance regulation when the stress, metabolic, or environmental factors are activated.

11.5.1 CHANGES IN BODY COMPOSITION AND WATER RESERVES

As the organism ages, there is an increase in the content of water-free adipose tissue and a decrease in the total content of water (total body water [TBW]), which is primarily the result of decrease in the muscle mass and intracellular water (Ayus and Arieff 1996; Chumlea et al. 1999; Morley et al. 1998; Steen 1997, see also Chapter 1 for more information on TBW). The percentage of TBW starts to decrease in women and men as early as in mid-life, with the highest decrease observed after the age of 60 (Steen 1997). It is estimated that the average TBW loss between the ages of 20 and 80 is about 4 L in men and 6 L in women (after Chumlea et al. 1999). According to Chumlea et al. (1999), TBW in men decreases from 58% at the age of 18 years to about 46% at the age of 64 and, respectively, from 48% to 43% in women. Bossingham et al. (2005) have noted that TBW was about 47% in healthy women aged 21–46 and about 39% in healthy women aged 70–81. In men, the extent of TBW decrease was lower—from about 53% at the age of 23–43 to about 50% at the age of 63–79. However, neither in women nor in men, the average TBW decrease was statistically significant, probably because of the large age variability within groups.

In cross-sectional studies, Rikkert et al. (1998) have found that TBW decrease from 50% at the age of 26 to 46.8% in women aged 72 and from 60.4% at the age of 26 to 53.4% in men aged 72, and such decreases were statistically significant.

With changes referring to TBW, a decrease in PV has also been observed at rest in elderly people. Davy and Seals (1994) found that absolute total blood volume (BV) was 24% lower in the older (aged 66) than in the young men (aged 25) due to a 21% lower PV and a 28% lower erythrocyte volume. Total BV and PV were also lower in the older than in the young men when expressed relative to body weight, body surface area, or estimated fat-free mass. These authors concluded that total BV decreases with age in healthy men of similar size and physical activity levels (Davy and Seals 1994). When comparing physically active older (over 65) and younger (under 30) men, Mack et al. (1994) observed significantly lower PV in older than in younger men at rest. However, according to Kenney (1997), PV is probably more strongly affected by physical activity and VO_2max than by chronological age.

A reduction in body water reserves in elderly people implies that a given loss in body water volume induces greater changes in PV and osmolality compared to younger ones.

11.5.2 Impairment of Thirst in Elderly People

Thirst is a biological need being manifested as the increased drive to drink, playing a crucial role in controlling body water balance and maintaining the stable volume of water in the body. Thirst perception increases as dehydration level rises and decreases with rehydration. It is the most important factor in the process of rehydration, that is, adjustment of the body's water balance in case of dehydration.

Thirst is stimulated via two routes, through intracellular dehydration and extracellular dehydration, with the increase in Posm being the most important stimulating factor.

Even as small as 1% increase in Posm induces cellular dehydration and stimulation of anterior hypothalamus osomoreceptors (thirst centre) due to an increase in the concentration of osmotically active substances, mainly sodium, in the extracellular fluid (plasma hyperosmolality). This activates the process of searching for water or water-containing products and the act of drinking in other parts of the brain. The water ingested and absorbed from the intestine causes the osmolality to decrease and the thirst to be quenched (Morley et al. 1998; Thornton 2010).

At the same time when thirst develops, and particularly when the access to water is difficult, the activated brain osmoreceptors stimulate hypothalamus neurons to release vasopressin from the posterior pituitary to the blood. The released vasopressin increases the reabsorption of water in the distal tubule of the nephrons through aquaporin-2, decreasing the water output from the body, but simultaneously AVP secretion increases the sensation of thirst (Thornton 2010).

Baroreceptors located in large vessels, activated by decreased BV (hypovolemia) and renal mechanisms, through the RAA system in response to reduced renal perfusion, also participate in the regulation of thirst. However, at least 8%–10% decrease in PV is needed to stimulate thirst as a response in the sensitive baroreceptors (Kenney and Chiu 2001; Morley et al. 1998).

One more mechanism participating in the regulation of thirst should be mentioned here, that is, the oropharyngeal receptors. These receptors send signals to the brain when water is ingested to reduce both thirst and AVP secretion well before a decrease in Posm (Kenney and Chiu 2001; Morley et al. 1998; Thornton 2010).

Thirst, however, is not only a simple consequence of physiological need for water but a complex response to many interrelated stimuli. These factors belong, on one hand, to the physiological control system (homeostatic control) but, on the other hand, there are behavioral factors (non-homeostatic control), such as habit, ritual, drink taste and temperature, body warming or cooling effect, and mouth and throat dryness. These factors may initiate/induce drinking of liquids. On the contrary, the feeling of stomach fullness and discomfort during exercise may result in the abandonment of drinking before body water replenishment occurs. However, it may occur that the effects from higher brain levels can exceed physiological regulatory mechanisms and lead to irresponsible behaviors related to drinking. In such situation, excessive consumption of water or fluids containing only small amounts of electrolytes is observed, which leads to body overhydration and development of hyponatremia with risk consequences, including even death (Maughan 2013).

Scientific literature (Beck 2000; Cowen et al. 2013; Kenney 1997; Kenney and Chiu 2001; Luckey and Parsa 2003; Morley et al. 1998) has documented the presence of an intrinsic defect in the thirst mechanism in elderly people. This may decrease their fluid intake and impair old people to recover from dehydration as effectively as the young ones. Lower sensation of thirst in older men (+65) was reported, for example, by Mack et al. (1994) at any given body fluid status both at rest and after dehydration. In a classic experiment, Phillips and colleagues (1984) showed a difference in the response to 24 h of complete water deprivation in seven healthy active elderly men (aged 67–75) and seven healthy young men (aged 20–31). Although after water deprivation, the older men had a greater level of dehydration (as evidenced by greater increases in Posm and sodium concentration and decrease in body mass), they were less thirsty, and no significant increase in thirst, mouth dryness, or unpleasantness of the taste in the mouth was noted. More importantly, when the subjects had unlimited access to water for 60 min after 24 h of dehydration, as opposed to young men, the older ones drank significantly less water and were not able to restore the blood parameters of pre-deprivation (Phillips et al. 1984). It could be concluded that elderly men experience less thirst perception after water deprivation. Thirst is also reduced in response to heat stress and thermal dehydration in elderly compared to younger groups, and this age-related deficit in thirst may predispose them to fluid balance disturbances (Phillips et al. 1993; Rolls and Phillips 1990).

The mechanism of thirst regulation in elderly people is also less sensitive to exercise-induced dehydration, as demonstrated by Ainslie et al. (2002) during a 10-day very intense mountain hiking when a group of males aged 24 years and another aged 56, walked 10–35 km a day, with elevations of 800–2,540 m. Daily water intake in the older group (2.4 L) was significantly lower than that of the younger (3.5 L), which led them to increased dehydration from the sixth day. Furthermore, the older group had a lower thirst perception compared with the younger group. After 10 days of mountain walking, urine osmolality value in older men increased almost twice from that of baseline, and it was almost two times higher than that of the younger subjects, with a simultaneous decrease in the rating of thirst in the group of older men.

In another study (Miescher and Fortney 1989), older males (61–67 years) rated themselves less thirsty than younger (21–29 years) after thermal dehydration, despite a greater rise in body temperature and more pronounced hemoconcentration. Within 30 min of drinking water, young men had restored PV and Posm, whereas the older ones showed slower responses, restoring Posm after 60 min and PV only after a subsequent 30 min of rehydration.

It should also be noted that impaired thirst reduces the capacity of acclimation to exercise in a hot environment. It has been reported (Zappe et al. 1996) that repeated exercises at high ambient temperature induces a 10% increase in PV (one of the important changes in the process of acclimation) in young men as early as after 4 days, whereas PV expansion is attenuated in older people (aged 67). These authors concluded that the failure to increase PV was caused by a smaller increase in 24 h fluid intake in older men, which averaged 32 mL/kg/day versus 45 mL/kg/day in the younger (aged 24) (Zappe et al. 1996).

11.5.2.1 Causes of Impaired Thirst in Elderly People

The causes of age-related deterioration of thirst in humans are not fully known. Based on literature, the following causes of hypodypsia in elderly people can be listed: (1) occult central nervous system disease (even without any neurological symptoms), (2) impaired sensitivity of thirst center osomoreceptors, (3) diminution of baroreceptor-mediated regulation of thirst in the response to changes in PV, (4) higher osmolar set point for thirst and for control of body fluid volume and composition, (5) deterioration in the function of oropharyngeal receptors, inducing a reduction in the response to stimuli from the oral cavity, such as, for instance, the feeling of throat dryness or a decrease in pleasure associated with drinking, (6) alteration in the endogenous opioid system, leading to a deficit in the opioid-mediated drinking drive in central nervous system, (7) reduction in the activity of RAA system and application of angiotensin-converting enzyme receptor or inhibitor blockers, which leads to a decrease in angiotensin II (AII) level and reduction of thirst, and (8) increase in ANP level, which reduces thirst and inhibits renin and aldosterone secretion (Cowen et al. 2013; Mack et al. 1994; Rolls and Phillips 1990).

Therefore, because of reduced thirst perception and slower body fluid restoration in elderly people, the planning of their participation in training and competitions, particularly on hot days, should be carefully considered.

11.5.3 AGE-RELATED CHANGES IN KIDNEYS

Kidneys are the most important organs that regulate the volume and composition of body fluids, primarily the extracellular fluid. This is achieved through the production of urine in different concentrations of each element. Because liquids are ingested periodically throughout the day, maintenance of extracellular fluid volume is possible through constant renal and neurohormonal control of fluid and electrolyte balance. With excess body water, kidneys are capable to adjust excretion, protecting against overhydration. With water and solutes deficiency, on the contrary, kidneys are able to retain them. Although the functions of kidneys are relatively autonomous, they are affected by the sympathetic part of vegetative nervous system, AVP secreted from the pituitary glands, aldosterone from the adrenal cortex or ANP. Unfortunately, like in other organs, changes related to aging occur also in the kidneys and involve reduction in both anatomical (morphological) and functional reserves. Some of these changes occur as early as after the age of 30 but, although it is a slow process, they are progressive and inevitable. The rate and severity of these changes are affected by many factors and it is not completely clear whether they are typically of the aging process or a result from other conditions, such as discrete changes in the cardiovascular system, increased blood pressure, dyslipidemia, diabetes mellitus, urinary tract infections, or tobacco smoking, as well as the use of antibiotics and non-steroidal anti-inflammatory drugs (NSAIDs; Abdelhafiz et al. 2010; Miller 1998; Zhou et al. 2008).

The changes occurring within kidneys with age can be divided into anatomical and functional ones.

11.5.3.1 Anatomical Changes in Kidneys

In most people between the ages of 30 and 80–85, a 10%–30% decrease in the kidneys size and a 20%–25% reduction in the renal mass are observed, primarily in the cortical layer. In microscopic examination, a hyalinization of blood vessel walls, glomerulosclerosis, a decrease in the number of glomeruli, and tubular atrophy are observed, reducing the number of active nephrons (Abdelhafiz et al. 2010; Beck 2000; Luckey and Parsa 2003; Miller 1998; Schlanger et al. 2010; Zhou et al. 2008). This cortical glomerulosclerosis increases from less than 5% at the age of 40 up to 30% by 80 years old (Abdelhafiz et al. 2010; Schlanger et al. 2010). Unfavorable changes, such as a decrease in effective filtering surface area, decrease in the number of epithelial cells, and an increase in the number of mesangial cells and thickening of the glomerular basement membrane, also occur in other active glomeruli (Miller 1998; Zhou et al. 2008).

Involutional changes in renal tubules refer mainly to the proximal tubules and consist in their shortening and volume reduction, tubular atrophy, and formation of diverticula and cysts. These changes refer to about half of people over the age of 40 (Zhou et al. 2008).

11.5.3.2   Functional Changes in Kidneys

Structural changes in kidneys are accompanied by deterioration in their function, inducing an *inelasticity* in fluid homeostasis (Beck 2000). This means that elderly people have limited abilities to adjust water balance and this defect may occur in challenging situations of dehydration or overhydration, leading to body water and solute imbalances (Beck 2000; Cowen et al. 2013).

According to the *nephrogeriatric giants* concept of Musso (2002), there are many functional changes in the aging kidneys even in healthy people. Three of them are of particular importance to kidney function deterioration, inducing, among others, the impairment of renal concentrating and diluting capacity. These changes include senile hypofiltration, tubular dysfunction, and medullary hypotonicity.

> *Senile hypofiltration*: Glomerular filtration rate (GFR) decreases from the age of 30 by about 0.8–1 ml/min/1.73 m² within a year. This process accelerates after 40 years of age and refers to about 70% of healthy people. Further acceleration of GFR decrease occurs after 65 years of age. A reduction in GFR results from decreased renal perfusion, which decreases from the age of 30 by about 10% per decade (Cowen et al. 2013; Miller 1998; Musso and Oreopoulos 2011; Luckey and Parsa 2003; Zhou et al. 2008). Also, a decrease is observed in the effective renal plasma flow by 10% per decade from 600 ml/min/1.73 m² in youth to 300 ml/min/1.73 m² by the age of 80 years (Musso and Oreopoulos 2011).
>
> *Tubular dysfunction*: A reduction in the function of many renal tubules is observed, primarily impaired distal renal tubular diluting capacity, impaired renal concentrating capacity, impaired sodium conservation, and blunted renal response to circulating vasopressin (Miller 1998). Disturbed renal concentrating and diluting capacity may lead, on one hand, to dehydration but, on the other hand, to overhydration (Miller 1998).
>
> *Medullary hypotonicity*: This change induces a reduction in the effect of vasopressin and, in consequence, a reduction in the water reabsorption and a decrease in urine concentration. This, combined with a decreased sensation of thirst, may lead to severe dehydration (Musso 2002).

11.5.4   Hormonal Regulation of Fluid and Electrolyte Balance in Elderly People

Hormones that participate in the regulation of water balance, such as AVP and cardiac natriuretic peptides (ANP and BNP) as well as RAA system, may also have their secretion and effects altered with age.

Vasopressin (AVP): A decrease in the volume of extracellular fluid is the main stimulus for secretion of vasopressin by the pituitary gland. AVP, also known as an ADH, increases water resorption from renal tubules, decreasing the volume of urine output. It also stimulates thirst. AVP secretion seems to be unchanged with age; moreover, it has been demonstrated that circulating AVP level is increased in older people for a given Posm when compared with younger individuals. This increased AVP secretion may be a result from the lower renal function of older people, as abovementioned (Beck 2000; Cowen et al. 2013; Miller 1998).

Atrial natriuretic peptide (ANP): Expansion of atrial receptors due to increased circulating volume and increased blood inflow to the atria induces an increase in the secretion of this hormone, while a decrease in the atrial filling decreases its secretion. ANP increases glomerular filtration and inhibits sodium and water resorption in renal tubules, increasing natriuresis and diuresis. It also inhibits the activity of RAA system, AVP secretion, and thirst. Even a fivefold increase in baseline ANP level and an excessive response both to sodium chloride infusion and exercise are observed with increasing age (Beck 2000; Miller 1998). Increased ANP secretion may be responsible for age-related renal sodium loss (Miller 1998).

Renin–angiotensin–aldosterone system (RAA): Oligovolemia is the strong stimulus of renin secretion in the granular cells of the juxtagromerular apparatus. Angiotensin I, which is formed under its influence, changes into AII. AII, having an effect on kidneys, stimulates directly sodium reabsorption in renal tubules. It also has a stimulating effect on aldosterone and vasopressin secretion. Aldosterone secreted by the adrenal cortex stimulates sodium and water reabsorption in the distal portion of renal tubules. The activity of this system is decreased with age—renin secretion is reduced in elderly people when compared with younger individuals, like the response of renal tubules to aldosterone. As a consequence, there is a *leakage* of sodium to urine and its wasting (*salt wasting*), being particularly unfavorable in the situations requiring its conservation (Beck 2000; Horgen et al. 1998; Schlanger et al. 2010).

The abovementioned renal changes, together with the neural and hormonal impairments, limit the adaptations to extracellular fluid volume overload (volume expansion) and depletion as well to electrolyte imbalances in older people. Due to impairment in fluid and electrolyte balance, there is an increased risk of hyper- or hypovolemia as well as hyper- or hyponatremia with ageing.

Hypervolemia and hyponatremia: Renal water retention and volume expansion result from deterioration in renal diluting capacity and excess free water excretion from the body. The main cause is a decrease in both renal perfusion and GFR, but also impaired urinary dilution in the distal part of renal tubules and excessive AVP secretion (Beck 2000; Luckey and Parsa 2003; Miller 1998; Taaren et al. 2005), which is additionally increased during physical exercise (Hew-Butler 2010). Due to decreased capacity of extra water excretion, a risk of water overload and life-threatening hyponatremia increases in elderly people. This phenomenon, known as *exertional hyponatremia*, may happen in prolonged type of sports, since it develops in situations of excessive water intake or excessive intake of beverages containing relatively small amounts of sodium compared to that of sweat. Excessive consumption of beverages during physical exercise and exercise-associated hyponatremia is further discussed in Chapter 2 and in other comprehensive reviews (Beltrami et al. 2008; Noakes 2003, 2007; Noakes and Speedy 2007) and original papers (Almond et al. 2005; Black et al. 2014; Hew 2005; Kipps et al. 2011; Wall et al. 2013).

Hypovolemia: The capacity to concentrate urine in renal tubules decreases by about 5% per decade, inducing a decrease in urine relative density from above 1.026 at a young age to even 1.023 after the age of 60 and an excessive loss of water from the body (Beck 2000; Miller 1998). Hypovolemia may be due to a decreased renal sensitivity to AVP observed with age (Beck 2000; Luckey and Parsa 2003; Miller 1998; Taaren et al. 2005) but also due to decreased body sodium content caused by fluid losses. In older people, sodium loss adjustment lasts twice (or thrice) as long than in younger people, at least for two reasons: (1) renin release is blunted in older individuals, which results in inadequate aldosterone response and a *leakage* of sodium into the urine, and (2) overlap between hypovolemia and increased baseline ANP level in older people induces suppression of renin secretion, with subsequent drop in AII and aldosterone levels. The consequence of the above changes is *salt wasting* in situations when sodium saving is required and, as a result, exaggeration of hypovolemia (Beck 2000).

Hypovolemia is also favored by decreased thirst perception in older people as described earlier; therefore, the recommended *drinking to thirst* may prove to be insufficient to replenish the fluid loss during exercise in this population.

11.5.5 Sweating in Elderly People

Sweating, or in fact evaporation of sweat from the skin surface, is the main body protection against overheating. The rate of sweating depends on ambient temperature and body heat production. Increased sweating during exercise occurs almost simultaneously with increased metabolic heat production, while maximum sweat gland recruitment occurs within the first 8 min, also during passive overheating (Kondo et al. 2001, see also Chapter 3).

Regarding age-related changes in sweating, there are controversial reports in literature. It has been shown in many studies that compared to young gender-matched subjects, older (after the age of 60) men and women sweat less, both during passive heat exposure (Armstrong and Kenney 1993; Dufour and Candas 2007; Inoue 1996) and during exercise in the heat (Inbar et al. 2004; Kenney and Munce 2003; Larose et al. 2013a). It is indicated that this results from reduced sweat production by the sweat glands (reduced sweat gland output) with not reduced number of active sweat glands, both after pharmacological stimulation (using cholinergic analogues) and under the effect of physical exercise (Inoue et al. 1999; Kenney 1997; Kenney and

Munce 2003). It is suggested that this results from a reduction in the sweat output per active gland progressing with age (from the age of 33), induced by changes in the sweat glands themselves or/and by a decrease in their sensitivity to stimuli (Inoue et al. 1999). According to Inoue (1996), the sweat gland function decreases over 5 years in healthy active men in their sixth and seventh decade, despite them having similar physical characteristics and exercise habits. The exposure of skin to UV rays and other environmental factors, so frequent during outdoor sport activity, accelerates the development of senile changes in the skin, which may translate into lower performance of sweat glands (Kenney 1997; Kenney and Munce 2003).

There seems to be no difference in sweating rate between middle-aged (54 years old) and younger (28 years old) men, all well trained under hot and moderate conditions and with similar aerobic capacity. The observed decrease in the sweat rate per sweat gland in middle-aged men may be then compensated by a higher number of heat-activated sweat glands (Viveiros et al. 2012). When young sedentary men (aged 24–30) were compared to older (aged 58–74) but physically active men and similar VO_2max, the sweat rate and skin blood flow were similar between the groups. However, when young sedentary men were compared with the old sedentary men, the sweat rate and skin blood flow were significantly reduced in the older ones by 62% and 40%, respectively (Trankersley et al. 1991). These authors concluded that there is no primary effect of aging on heat loss responses but changes are rather associated with an age-related decrease in VO_2max or training adaptation may mask (or attenuate) the body heat loss decline due to aging (Trankersley et al. 1991). According to Kenny and Munce (2003), the sweating rate is strongly determined by the individual VO_2max rather than by chronological age.

Furthermore, it has been demonstrated that the sweat glands may be *trained* by immersion in hot water (Kenney 1997; Kenney and Munce 2003) or during daily aerobic exercises (two exercise bouts at 50% VO_2max for 45 min) in a hot environment (40°C, 30% RH) for 10 days (Lorenzo and Minson 2010).

11.5.6 EFFECT OF MEDICINES ON FLUID AND ELECTROLYTE BALANCE

The above-described situations may be aggravated by some diseases and respective medicines (see Table 11.1) that are taken by elderly people and may, in different ways, affect fluid and electrolyte balance and/or thermal metabolism. NSAIDs, very often used by athletes, may indirectly increase the risk of exertional heat illness (Bergeron 2014). Prostaglandins, being present in kidneys, facilitate the release of renin and decrease sodium reabsorption in the loop and the collecting tubules, partly antagonizing ADH capacity to increase water absorption in the collecting tubules (Kramer et al. 1981; Schlondorff 1993). As prostaglandins are locally released, they mediate natriuretic effects of dopamine and natriuretic peptides (Stichtenoth et al. 2005). NSAIDs, decreasing the synthesis of prostaglandins, may therefore induce a number of changes in the kidney function (Clark et al. 1994), causing disturbances such as hyperkalemia, hyponatremia, and edemas. The pathomechanism of these complications is associated with a decrease in renin and aldosterone release and increased AVP activity (Campbell et al. 1979).

TABLE 11.1

List of Some Drugs/Drug Classes Predisposing to Disturbances in Fluid and Electrolyte Balance and Thermal Balance of the Body

Reduce Sweating	Increase Sweating	Increase Metabolic Heat Production	Affect Cardiovascular System and Circulating Volume	Increase the Risk of Developing Hyponatremia	Increase the Risk of Developing Hypernatremia
Anticholinergic agents	Anticholinesterases	Ephedrine	Digitalis glycosides	Antidepressants—selective serotonin re-uptake inhibitors	Mannitol
Antidepressants—tricyclics	Antidepressants—selective serotonin re-uptake inhibitors	MAO inhibitors	Beta-blockers		Lithium
Antiepileptics		Lithium	Converting enzyme inhibitors	Antidepressants—tricyclics	Loop diuretics
Antihistamines		Thyroid hormones	Diuretics	Carbamazepine	Mannitol
Antihypertensives (Clonidine)	Antidepressants—tricyclics		Adrenolytic drugs	Desmopressin	Vasopressin
Antipsychotics	Antiglaucoma agents		Laxatives	Opiate derivatives	V2 receptor antagonists
Antiemetics	Bladder stimulants			Phenothiazides	
Antivertigo drugs	Opioids			Prostaglandin-synthesis inhibitors	
Bladder antispasmodics	Sialogogues			Thiazides	
Gastric antisecretory drugs (propantheline)					
Muscle relaxants					
Opioids					

Source: Cheshire, W.P., and Freeman, R., *Semin. Neurol.*, 23(4):399–406, 2003; Cheshire, W.P., and Fealey, R.D., *Drug Saf.*, 31(2):109–26, 2008; Schlanger, L.E. et al., *Adv. Chronic Kidney Dis.*, 17(4):308–19, 2010. doi:10.1053/j.ackd.2010.03.008; Tareen, N. et al., *J. Natl. Med. Assoc.*, 97(2):217–24, 2005.

11.6 SUMMARY AND RECOMMENDATIONS FOR MASTER ATHLETES

Master athletes, particularly those over the age of 60, constitute a group of athletes prone to disturbances in fluid and electrolyte balance and thermal reactions, especially when prolonged exercise takes place in a hot and humid environment. Decreased renal concentrating capacity and thirst perception may delay adjustments of fluid and electrolyte imbalances in older athletes, extending the recovery period in case of post-exertional dehydration. This should be considered when planning the next training or competition, particularly on hot days.

On the other hand, decreased renal diluting capacity and free water excretion as well as impaired sodium retention by kidneys may lead to overhydration and hyponatremia in case of excessive consumption of water and other beverages containing small amounts of sodium during prolonged physical exercise.

Age-related structural and functional changes in kidneys, in conjunction with a reduction in the efficiency of neurohormonal regulation, blood pressure and volume, sensation of hunger and thirst, decrease the capacity of efficient and precise adjustment of fluid and electrolyte balance in stressful situations, dehydration, or fluid overload (Beck 2000).

Because of the individual nature of the aging process, responses related to body water content, thirst perception, sweating, as well as cardiovascular and renal adjustments to exertional and thermal stress, may differ, despite the same chronological age. Apart from age-related changes in the body, the alterations in fluid and electrolyte balance and the efficiency of thermoregulatory mechanisms are also affected by factors such as training and fitness level, intensity and duration of physical exercise, ambient thermal conditions, hydration status before exercise, or acclimation to a hot environment. Also, the current health state, use of medicines, and supplements should be taken into consideration.

Therefore, the provision of universal, the same for all, guidelines referring to hydration before, during, and after exercise is pointless; more importantly, it may be even dangerous for the health of older athletes.

Official guidelines referring to ingestion of fluids during extended exercise have been changing, starting from the first guidelines published by the American College of Sports Medicine (ACSM) in 1975 (see also Chapter 7). However, all these guidelines refer to athletes at a young age; instead, there are no guidelines for older participants. In the 2007 ACSM guidelines (Sawka et al. 2007), some general phrases applying to older athletes were included such as: "Thus, older adults should be encouraged to rehydrate during or after exercise, but they should also consider the risks of excess water (i.e., hyponatremia) or sodium ingestion (i.e., hypertension) because they may be slower to excrete both the water and electrolytes."

Practical guidelines which help prevent fluid and electrolyte imbalances and exertional heat illness in older participants are as follows:

1. Athletes should start training/enter competitions being well hydrated. To this end, it is necessary to consume adequate amounts of fluids (see Chapter 7) and control the hydration status by

a. *Monitoring of body weight* by morning fasted weighing, wearing only underwear, on an accurate scale. Daily fluctuations of body weight should be within ±1%.

b. *Monitoring of urine volume and its color* using the urine color chart, but preferably assessment of urine-specific gravity (USG) using a refractometer.

A healthy man excretes daily 1.3–1.6 L and a healthy woman about 1.13 ± 0.42 L of urine, *pale yellow* or *straw colored* (1 or 2 on the urine color chart), and USG should be less than 1.010. Too small volume and dark color of urine, and increased USG, indicate dehydration (Armstrong 2007; Casa et al. 2010).

2. Athletes should control changes in body weight induced by exercise under different thermal conditions. Knowing the magnitude of water losses associated with intensified sweating during exercise under different conditions helps to implement appropriate hydration strategies. In situations of dehydration, the quality of training and impaired performance can be avoided. In situations that participants tend to ingest too much fluid, recommendations can be directed to avoid overhydration. However, because of decreased thirst perception, older athletes cannot rely on thirst sensation but should have own well-prepared strategies of behavior counteracting the excessive dehydration (body weight loss >2%) during exercise and quickly compensate the fluid loss after completing exercise by drinking 1–1.25 L of fluids with addition of salt per each kilogram of lost body mass (Bergeron 2014). These contestants should also eat food containing a large amount of water (fruits and vegetables) and consume beverages with food.

3. When planning the next training, it should be considered that adjustment of fluid and electrolyte imbalances is slower in older people, prolonging the post-exercise recovery period.

4. It is necessary to educate older athletes on how to assess dehydration degree and how to hydrate properly during training and competitions. This education should be extended to the coaching and health personnel. Considering a decreased capability of adjusting fluid and electrolyte balance in older people, advice and guidelines referring to ingestion of fluids during exercise by older athletes should be carefully delivered.

5. Exercise (up to two weeks or more) in a hot and/or humid environment, with progressively increased intensity and duration (from initially 75% of typical training intensity and 50% of typical distance) in order to acclimate is also recommended (Bergeron 2014), especially when competitions are to take place in a hot and humid climate.

REFERENCES

Abdelhafiz, A. H., S. H. M. Brown, A. Bello, and M. El Nahas. 2010. Chronic kidney disease in older people: Physiology, pathology or both? *Nephron Clin. Pract.* 116:c19–24. doi:10.1159/000314545.

ACSM. 1975. Position statement of the American College of Sports Medicine: Prevention of heat injuries during distance running. *Med. Sci. Sports Exerc.* 7:vii–ix.

Ainslie, P. N., I. T. Campbell, K. N. Frayn et al. 2002. Energy balance, metabolism, hydration, and performance during strenuous hill walking: The effect of age. *J. Appl. Physiol.* 93(2):714–23. doi:10.1152/japplphysiol.01249.

Almond, C. S., A. Y. Shin, E. B. Fortescue et al. 2005. Hyponatraemia among runners in the Boston Marathon. *N. Engl. J. Med.* 352:1550–6.

Armstrong, C. G., and W. L. Kenney. 1993. Effects of age and acclimation on responses to passive heat exposure. *J. Appl. Physiol.* 75:2162–7.

Armstrong, L. E. 2007. Assessing hydration status—The elusive gold standard. *J. Am. Coll. Nutr.* 26(5):575S–84.

Ayus, J. C., and Al Arieff. 1996. Abnormalities of water metabolism in the elderly. *Semin. Nephrol.* 16(4):277–88.

Beck, L. H. 2000. The aging kidney: Defending a delicate balance of fluid and electrolytes. *Geriatrics* 55(4):26–8, 31–2.

Beltrami, F. G., T. Hew-Butler, and T. D. Noakes. 2008. Drinking policies and exercise-associated hyponatraemia: Is anyone still promoting overdrinking? *Br. J. Sports Med.* 42:496–501.

Benelam, B., and L. Wyness. 2010. Hydration and health: A review. *Nutr. Bull.* 35:3–25.

Bergeron, M. F. 2014. Heat stress and thermal strain challenges in running. *J. Orthop. Sports Phys. Ther.* 44(10):831–8.

Black, K. E., P. Skidmore, and R. C. Brown. 2014. Fluid balance of cyclists during a 387-km race. *EJSS* 14(S1):S421–8. doi:10.1080/17461391.2012.711860.

Blatteis, C. M. 2012. Age-dependent changes in temperature regulation—A mini review. *Gerontology* 58:289–95. doi:10.1159/000333148.

Bongers, C. C. W. G., T. M. H. Eijsvogels, J. Nyakayiru, M. T. W. Veltmeijer, D. H. J. Thijssen, and M. T. E. Hopman. 2014. Thermoregulation and fluid balance during a 30-km march in 60- versus 80-year-old subjects. *Age* 36:9725. doi:10.1007/s11357-014-9725-1.

Bossingham, M. J., N. S. Carnell, and W. W. Campbel. 2005. Water balance, hydration status, and fat-free mass hydration in younger and older adults. *Am. J. Clin. Nutr.* 81:1342–50.

Campbell, W. B., C. E. Gomez-Sanchez, B. V. Adams, J. M. Schmitz, and H. D. Itskovitz. 1979. Attenuation of angiotensin II- and III-induced aldosterone release by prostaglandin synthesis inhibitors. *J. Clin. Invest.* 64(6):1552–7.

Casa, D. J. 2000. National athletic trainer's association position statement: Fluid replacement for athletes. *J. Athl. Train.* 35:212–4.

Casa, D. J., R. L. Stearns, R. M. Lopez et al. 2010. Influence of hydration on physiological function and performance during trail running in the heat. *J Athl. Train.* 45(2):147–56. doi:10.4085/1062-6050-45.2.147.

Cheshire, W. P., and R. D. Fealey. 2008. Drug-induced hyperhidrosis and hypohidrosis: Incidence, prevention and management. *Drug Safety* 31(2):109–26.

Cheshire, W. P., and R. Freeman. 2003. Disorders of sweating. *Semin. Neurol.* 23(4):399–406.

Chester, J. G., and J. L. Rudolph. 2011. Vital signs in older patients: Age-related changes. *J. Am. Med. Dir. Assoc.* 12(5):337–43. doi:10.1016/j.jamda.2010.04.009.

Cheuvront, S. N, R. W. Kenefick, N. Charkoudian, and M. N. Sawka. 2013. Physiologic basis for understanding quantitative dehydration assessment. *Am. J. Clin. Nutr.* 97:455–62.

Cheuvront, S. N., R. W. Kenefick, S. J. Montain, and M. N. Sawka. 2010. Mechanisms of aerobic performance impairment with heat stress and dehydration. *J. Appl. Physiol.* 109:1989–95. doi:10.1152/japplphysiol.00367.2010.

Chodzko-Zajko, W. J., D. N. Proctor, M. A. Fiatarone Singh et al. 2009. Exercise and physical activity for older adults. ACSM Position Stand. *Med. Sci. Sports Exerc.* 41:1510–30.

Chumlea W. C., S. S. Guo, Ch. M. Zeller, N. V. Reo, and R. M. Siervogel. 1999. Total body water data for white adults 18 to 64 years of age: The Fels longitudinal study. *Kidney Int.* 56, 244–52. doi:10.1046/j.1523-1755.1999.00532.x.

Clark, B. A., R. P. Shannon, R. M. Rosa, and F. H. Epstein. 1994. Increased susceptibility to thiazide-induced hyponatremia in the elderly. *J. Am. Soc. Nephrol.* 5:1106–11.

Cowen, L. E., S. P. Hodak, and J. G. Verbalis. 2013. Age-associated abnormalities of water homeostasis. *Endocrinol. Metab. Clin. North. Am.* 42:349–70.

Coyle, E. F. 2004. Fluid and fuel intake during exercise. *J. Sports Sci.* 22:39–55. doi:10.1080/0264041031000140545.

Davy, K. P., and D. R. Seals. 1994. Total blood volume in healthy young and older men. *J. Appl. Physiol.* 76(5):2059–62.

Dufour, A., and V. Candas. 2007. Ageing and thermal responses during passive heat exposure: Sweating and sensory aspects. *Eur. J. Appl. Physiol.* 100:19–26.

Gonzalez-Alonso, J., R. Mora-Rodriguez, and E. F. Coyle. 2000. Stroke volume during exercise: Interaction of environment and hydration. *Am. J. Physiol. Heart Circ. Physiol.* 278:H321–30.

Goulet, E. D. B. 2012. Dehydration and endurance performance in competitive athletes. *Nutr. Rev.* 70(Suppl 2):S132–6. doi:10.1111/j.1753-4887.2012.00530.x.

Hew, T. D. 2005. Women hydrate more than men during a marathon race: Hyponatremia in the Houston marathon: A report on 60 cases. *Clin. J. Sport Med.* 15(3):148–53.

Hew-Butler, T. D. 2010. Arginine vasopressin, fluid balance and exercise. Is exercise associated hyponatraemia a disorder of arginine vasopressin secretion? *Sports Med.* 40(6):459–79.

Hodgkinson, B., D. Evans, and J. Wood. 2003. Maintaining oral hydration in older adults: A systematic review. *Int. J. Nurs. Pract.* 9:S19–28.

Horgen, L., V. Andrieux, C. Badier et al. 1998. Aging and the renin-angiotensin system. In: *Hydration and Aging*, edited by M. J. Arnaud, R. Baumgartner, J. E. Morley et al., pp. 103–18. Paris: Serdi Publisher.

Inbar, O., N. Morris, Y. Epstein, and G. Gass. 2004. Comparison of thermoregulatory responses to exercise in dry heat among prepubertal boys, young adults and older males. *Exp. Physiol.* 89:691–700.

Inoue, Y. 1996. Longitudinal effects of age on heat-activated sweat gland density and output in healthy active older men. *Eur. J. Appl. Physiol.* 74(1–2):72–7.

Inoue, Y., G. Havenith, W. L. Kenney, J. L. Loomis, and E. R. Buskirk. 1999. Exercise- and methylcholine-induced sweating responses in older and younger men: Effect of heat acclimation and aerobic fitness. *Int. J. Biometeorol.* 42:210–16.

Jokl, P., P. M. Sethi, and A. J. Cooper. 2004. Master's performance in the New York City Marathon 1983–1999. *Br. J. Sports Med.* 38:408–12. doi:10.1136/bjsm.2002.003566.

Kenney, W. L. 1997. Thermoregulation at rest and during exercise in healthy older adults. *Exerc. Sport Sci.* 25:41–76.

Kenney, W. L., and P. Chiu. 2001. Influence of age on thirst and fluid intake. *Med. Sci. Sports Exerc.* 33:1524–32.

Kenney, W. L., and T. A. Munce. 2003. Invited review: Aging and human temperature regulation. *J. Appl. Physiol.* 95:2598–603.

Kenny, G. P., D. Gagnon, L. E. Dorman, S. G. Hardcastle, and O. Jay. 2010a. Heat balance and cumulative heat storage during exercise performed in the heat in physically active younger and middle-aged men. *Eur. J. Appl. Physiol.* 109:81–92. doi:10.1007/s00421-009-1266-4.

Kenny, G. P., J. Yardley, C. Brown, R. J. Sigal, and J. Ollie. 2010b. Heat stress in older individuals and patients with common chronic diseases. *CMAJ* 182(10):1053–60. doi:10.1503/cmaj.081050.

Kipps, C., S. Sharma, and P. D. Tunstall. 2011. The incidence of exercise-associated hyponatraemia in the London marathon. *Br. J. Sports Med.* 45:14–19. doi:10.1136/bjsm.2009.059535.

Kondo, N., S. Manabu, A. Ken, K. Shunsaku, Y. Inoue, and C. G. Crandall. 2001. Function of human eccrine sweat glands during dynamic exercise and passive heat stress. *J. Appl. Physiol.* 90:1877–81.

Kramer, H. J, K. Glänzer, and R. Düsing.1981. Role of prostaglandins in the regulation of renal water excretion. *Kidney Int.* 19:851–9.

Larose, J., P. Boulay, R. J. Sigal, H. E. Wright, and G. P. Kenny. 2013a. Age-related decrements in heat dissipation during physical activity occur as early as the age of 40. *PLoS ONE* 8(12): e83148. doi:10.1371/journal.pone.0083148.

Larose, J., H. E. Wright, R. J. Sigal, P. Boulay, S. Hardcastle, and G. P. Kenny. 2013b. Do older females store more heat than younger females during exercise in the heat? *Med. Sci. Sports Exerc.* 45(12):2265–76. doi:10.1249/MSS.0b013e31829d24cc.

Larose, J., H, E. Wright, J. Stapleton et al. 2013c. Whole body heat loss is reduced in older males during short bouts of intermittent exercise. *Am. J. Physiol. Regul. Integr. Comp. Physiol.* 305:R619–29. doi:10.1152/ajpregu.00157.2013.

Lepers, R., and T. Cattagni. 2012. Do older athletes reach limits in their performance during marathon running? *Age* 34:773–81. doi:10.1007/s11357-011-9271-z.

Lopez, R. M., D. J. Casa, K. A. Jensen et al. 2011. Examining the influence of hydration status on physiological responses and running speed during trail running in the heat with controlled exercise intensity. *J. Strength Cond. Res.* 25(11):2944–54.

Lorenzo, S., and Ch. T. Minson. 2010. Heat acclimation improves cutaneous vascular function and sweating in trained cyclists. *J. Appl. Physiol.* 109:1736–43. doi:10.1152/japplphysiol.00725.2010.

Luckey, A. E., and C. J. Parsa. 2003. Fluid and electrolytes in the aged. *Arch. Surg.* 138:1055–60.

Mack, G. W., Ch. A. Weseman, G. W. Langhans, H. Scherzer, Ch. M. Gillen, and E. R. Nadel. 1994. Body fluid balance in dehydrated healthy older men: Thirst and renal osmoregulation. *J. Appl. Physiol.* 76(4):1615–23.

Maughan, R. 2012a. Hydration, morbidity, and mortality in vulnerable populations. *Nutr. Rev.* 70(Suppl 2):S152–55. doi:10.1111/j.1753-4887.2012.00531.x.

Maughan, R. J. 2012b. Investigating the associations between hydration and exercise performance: Methodology and limitations. *Nutr. Rev.* 70(Suppl. 2):S128–31. doi:10.1111/j.1753-4887.2012.00536.x.

Maughan, R. J. 2013. Water and electrolyte loss and replacement. In: *Sports Nutrition. The Encyclopaedia of Sports Medicine: An IOC Medical Commission Publication, Sports Nutrition*, edited by R. J. Maughan, pp. 174–84. Oxford: John Wiley & Sons.

Miescher, E., and S. M. Fortney. 1989. Responses to dehydration and rehydration during heat exposure in young and older men. *Am. J. Physiol.* 257(5 Pt 2):R1050–6.

Miller, M. 1998. Water metabolism in the elderly in health and disease: Aging changes affecting risk for hypernatremia and hyponatremia. In: *Hydration and Aging*, edited by M. J. Arnaud, R. Baumgartner, J. E. Morley et al., pp. 59–81. Paris: Serdi Publisher.

Montain, S. J., and E. F. Coyle. 1992. Influence of graded dehydration on hyperthermia and cardiovascular drift during exercise. *J. Appl. Physiol.* 73(4):1340–50.

Morley, J. E., D. K. Miller, C. Zdrodowski, B. Gutierrez, and H. M. Perry III. 1998. Fluide intake, hydration and aging. In: *Hydration through Life*, edited by M. J. Arnaud, pp. 107–15. Montrouge, France: John Libbey Eurotext.

Murray, B. 2007. Hydration and physical performance. *J. Am. Coll. Nutr.* 26:542S–8S.

Musso, C. G. 2002. Geriatric nephrology and the "nephrogeriatric giants." *Int. Urol. Nephrol.* 34:255–6.

Musso, C. G., and D. G. Oreopoulos. 2011. Aging and physiological changes of the kidneys including changes in glomerular filtration rate. *Nephron Physiol.* 119(Suppl 1):p1–5. doi:10.1159/000328010.

Naharudin, M. N., and A. Yusof. 2013. Fatigue index and fatigue rate during an anaerobic performance under hypohydrations. *PLoSONE* 8(10):e77290. doi:10.1371/journal. pone.0077290.

Noakes, T. D. 2003. Overconsumption of fluids by athletes. *BMJ* 327:113–14.

Noakes, T. D. 2007. Hydration in the marathon, using thirst to gauge safe fluid replacement. *Sports Med.* 37(4–5):463–6. doi:0112-1642/07/0004-C4A3/S44.95/0.

Noakes, T. D., and D. B Speedy. 2007. Lobbyists for the sports drink industry: An example of the rise of "contrarianism" in modern scientific debate. *Br. J. Sports Med.* 41:107–9.

Pandolf, K. B., B. S. Cadarette, M. N. Sawka et al. 1988. Thermoregulatory responses of middle-aged and young men during dry-heat acclimation. *J. Appl. Physiol.* 65:65–71.

Phillips, P.A., C. I. Johnston, and L. Gray. 1993. Disturbed fluid and electrolyte homoeostasis following dehydration in elderly people. *Age Ageing* 22(1):S26–33.

Phillips, P.A., B. J. Rolls, J. G. Ledingham et al. 1984. Reduced thirst after water deprivation in healthy elderly men. *N. Engl. J. Med.* 311(12):753–9.

Rikkert M. G. M. O., W. H. L. Hoefnagels, and P. Deurenberg. 1998. Age-related changes in body fluid compartments and the assessment of dehydration in old age. In: *Hydration and Aging*, edited by M. J. Arnaud, R. Baumgartner, J. E. Morley et al., pp. 13–32. Paris: Serdi Publisher.

Robine, J.-M., S. L. K. Cheung, S. L. Roy et al. 2008. Death toll exceeded 70,000 in Europe during the summer of 2003. *CR Biol.* 331(2):171–8. doi:10.1016/j.crvi.2007.12.001.

Rolls, B. J., and P. A. Phillips. 1990. Aging and disturbances of thirst and fluid balance. *Nutr. Rev.* 48(3):137–44.

Rosenbloom, C. A. 2013. The aging athlete. In: *Sports Nutrition. The Encyclopaedia of Sports Medicine: An IOC Medical Commission Publication, Sports Nutrition*, edited by R. J. Maughan, pp. 369–91. Oxford: John Wiley & Sons.

Rüst, C. A., T. Rosemann, M. A Zingg, and B. Knechtle. 2014. Age group performances in 100 km and 100 miles ultra-marathons. *Sprigerplus* 3:331. http://www.springerplus. com/content/3/1/331.

Sawka, M. N. 1992. Physiological consequences of hypohydration: Exercise performance and thermoregulation. *Med. Sci. Sports Exerc.* 24:657–70.

Sawka, M. N., L. M. Burke, E. R. Eichner, R. J. Maughan, S. J. Montain, N. S. Stachenfeld. 2007. American College of Sports Medicine position stand: Exercise and fluid replacement. *Med. Sci. Sports Exerc.* 39:377–90.

Sawka, M. N., S. N. Cheuvront, and R. W. Kenefick. 2012. High skin temperature and hypohydration impair aerobic performance. *Exp. Physiol.* 97(3):327–32.

Schlanger, L. E., J. L. Bailey, and J. M. Sands. 2010. Electrolytes in the aging. *Adv. Chronic Kidney Dis.* 17(4):308–19. doi:10.1053/j.ackd.2010.03.008.

Schlondorff, D. 1993. Renal complications of nonsteroidal anti-inflammatory drugs. *Kidney Int.* 44:643–53.

Steen, B. 1997. Body water in the elderly—A review. *J. Nutr. Health Aging* 1(3):142–5.

Stichtenoth, D. O., V. Marhauer, D. Tsikas, F.-M. Gutzki, and J. C. Frölich. 2005. Effects of specific COX-2-inhibition on renin release and renal and systemic prostanoid synthesis in healthy volunteers. *Kidney Int.* 68:2197–207. doi:10.1111/j.1523-1755.2005.00676.x.

Tanaka, H., and D. R. Seals. 2008. Endurance exercise performance in masters athletes: Age-associated changes and underlying physiological mechanisms. *J. Physiol.* 586:55–63. doi:10.1113/jphysiol.2007.141879.

Tankersley, C. G., J. Smolander, W. L. Kenney, and S. M. Fortney. 1991. Sweating and skin blood flow during exercise: Effects of age and maximal oxygen uptake. *J. Appl. Physiol.* 71:236–42.

Tareen, N., D. Martins, G. Nagami, B. Levine, and K. C. Norris. 2005. Sodium disorders in the elderly. *J. Natl. Med. Assoc.* 97(2):217–24.

Thornton, S. N. 2010. Thirst and hydration: Physiology and consequences of dysfunction. *Physiol. Behav.* 100:15–21. doi:10.1016/j.physbeh.2010.02.026.

Trappe, S. 2001. Master athletes. *Int. J. Sport Nutr. Exerc. Metab.* 11:S194–205.

Trappe, S. 2007. Marathon runners, how do they age? *Sports Med.* 37(4–5):302–5. doi:0112-1642/07/0004-0302/$44.95/0.

Viveiros, J. P., F. T. Amorim, M. N. Alves, R. L. Passos, and F. Meyer. 2012. Run performance of middle-aged and young adult runners in the heat. *Int. J. Sports Med.* 33(3):211–17. doi:10.1055/s-0031-1295444.

Wall, B. A, G. Watson, J. J. Peiffer, C. R. Abbiss, R. Siegel, and P. B. Laursen. 2015. Current hydration guidelines are erroneous: Dehydration does not impair exercise performance in the heat. *Br. J. Sports Med.* 49(16):1077–83. doi:10.1136/bjsports-2013-092417.

Weir, P., J. Baker, and S. Horton. 2010. The emergence of masters sport. In: *The Masters Athlete*, edited by J. Baker, S. Horton, and P. Weir, pp. 7–14. London: Routledge.

Wright, H. E., J. Larose, T. M. McLellan, S. G. Hardcastle, P. Boulay, G. P. Kenny. 2014. Moderate-intensity intermittent work in the heat results in similar low-level dehydration in young and older males. *J. Occup. Environ. Hyg.* 11(3):144–53. doi:10.1080/154596 24.2013.817676.

Zappe, D. H., G. W. Bell, H. Swartzentruber, R. F. Wideman, and W. L. Kenney. 1996. Age and regulation of fluid and electrolyte balance during repeated exercise sessions. *Am. J. Physiol.* 270(1 Pt 2):R71–9.

Zhou, X. J., D. Rakheja, X. Yu, R. Saxena, N. D. Vaziri, and F. G. Silva. 2008. The aging kidney. *Kidney Int.* 74:710–20. doi:10.1038/ki.2008.319.

Zingg, M., C. A. Rüst, R. Lepers, T. Rosemann, and B. Knechtle. 2013. Master runners dominate 24-h ultramarathons worldwide—A retrospective data analysis from 1998 to 2011. *Extrem. Physiol. Med.* 2:21. http://www.extremephysiolmed.com/content/2/1/21.

Zingg, M. A., C. A. Rüst, T. Rosemann, R. Lepers, and B. Knechtle. 2014. Runners in their forties dominate ultra-marathons. *Clinics* 69:203–11. doi:10.6061/clinics/2014(03)11.

12 Athletes with Chronic Conditions

Diabetes

Jane E. Yardley and Michael C. Riddell

CONTENTS

12.1 INTRODUCTION

The incidence and prevalence of both type 1 (Diamond Project Group 2006; Vehik and Dabelea 2011) and type 2 (Whiting et al. 2011; Yisahak et al. 2013) diabetes have been increasing around the globe in recent years. According to the International Diabetes Federation, 382 million people worldwide currently have diabetes, with this number expected to rise to 592 million by the year 2035 (International Diabetes Federation 2013). Of those diagnosed with diabetes, roughly 90%–95% of cases will be type 2 diabetes, with the remaining 5%–10% being type 1 diabetes.

With the substantial increase in the number of individuals worldwide with diabetes, there has also been an increase in the amount of research dedicated to diabetes management and prevention. Physical activity has been associated with a multitude of benefits for individuals with both type 1 and type 2 diabetes including higher cardiovascular fitness and vascular health, improved serum cholesterol levels, augmented insulin sensitivity, more favorable body composition, a lower risk of diabetes-related complications, increased life expectancy, and an overall higher quality of life (Duclos et al. 2011; Chimen et al. 2012). For these reasons, it is generally recommended that individuals with diabetes perform at least 150 min of moderate to vigorous aerobic activity weekly (spread over at least three days of the week) in addition to resistance training at least twice per week (Sigal et al. 2004, 2013). While many individuals with diabetes will struggle to perform the recommended minimum, others will go far beyond clinical recommendations and will take part in frequent, high-intensity, high-volume training and competition.

For many trying to become more active, taking part in an *event* such as a charity walk, a run, or a triathlon often provides the required motivation. The term *athlete* has taken on a new meaning in recent decades, as the types of athletic events being offered are catering to a much wider range of abilities than ever before. Running USA remarked that the number of individuals completing running events has increased from just under 5 million in 1990 to over 15 million in 2012 (Running USA 2013). In 2013, participation in non-traditional running events, such as obstacle runs, outnumbered those of traditional running events for the first time, with the majority of the individuals taking part being first-time competitors (Running USA 2014). Likewise, USA Triathlon has seen its membership quadruple between 1999 and 2012. With the increased emphasis on including physical activity as part of the management of diabetes and prevention of its complications, it is highly likely that many individuals with diabetes are included among these newcomers. Athletes with type 1 diabetes can be found in most professional sports organizations (e.g., Jay Cutler and Jay Leeuwenberg [NFL]; Bobby Clark, Nick Boynton, and Cory Conacher [NHL]; Kelli Keuhne, Michelle McGann, and Scott Verplank [LPGA/PGA]; Adam Morrison and Chris Dudley [NBA]) in world championships and Olympics (Gary Hall Jr., swimming; Chris Jarvis, rowing) and in many professional events (Scott Kimball, IndyCar; all members of Team Novo, elite cyclists, triathletes and runners).

It is critical to fully understand the risks and benefits of undertaking physical activity, sport, and competition for individuals with diabetes. Where fluid and electrolytes are concerned, blood glucose management, both in the short term and in the long term, can impact the effectiveness of the kidneys in maintaining appropriate balance within the system. There is also evidence to indicate that sweat production may differ in individuals with diabetes compared to their non-diabetic counterparts, indicating that both thermoregulation and hydration may be impacted differently during intense competition or exercise in the heat. In addition, medications used to treat diabetes, and common complications thereof, may have an impact on hydration and overall kidney function. A complete understanding of all of these factors is essential to ensure the health and safety of athletes with diabetes during prolonged/high-intensity exercise, or activity in the heat.

12.2 TYPE 1 DIABETES

Type 1 diabetes is generally caused by autoimmune destruction of the insulin-producing β-cells in the pancreas. As a result, individuals affected by type 1 diabetes lose the ability to synthesize and secrete their own insulin and are thus dependent on exogenous sources of insulin (through injection or by insulin pump infusion) to control blood glucose levels. Without insulin, the body is unable to take glucose from the circulation for storage or energy production in the tissues. While exogenous insulin allows glucose uptake to take place, the fine-tuned adjustments that occur naturally with a healthy pancreas (i.e., declines in insulin secretion and increases in glucagon release during physical activity to maintain relatively constant blood glucose levels) are very difficult to replicate. Depending on the timing and dosage of injections with respect to the exercise sessions, athletes with type 1 diabetes might find themselves in either a hyperinsulinemic or hypoinsulinemic state. The end result is that they may be prone to large fluctuations in blood glucose, which can have potentially negative outcomes.

As glucose is the main source of fuel for the brain, maintaining adequate blood glucose levels (generally at or above 4 mmol/L or 72 mg/dL) is essential. When blood glucose levels drop below 4 mmol/L, a condition known as *hypoglycemia* (low blood glucose), symptoms such as confusion, vision disturbances, heart palpitations, anxiety, sweating, shakiness, or hunger may start to occur. Should blood glucose levels drop far enough, people with type 1 diabetes can experience loss of consciousness, coma, or even death if carbohydrates are not consumed. Conversely, high blood glucose levels (generally above 10 mmol/L or 180 mg/dL) can lead to thirst, increased urination (and consequently dehydration), fatigue, and headaches in the short term and can contribute to long-term complications such as blindness, cardiovascular disease, and kidney disease. It is therefore of the utmost importance for individuals with type 1 diabetes to carefully monitor their carbohydrate intake, along with their insulin dosage, in order to maintain their blood glucose within safe parameters.

12.2.1 EXERCISE AND BLOOD GLUCOSE LEVELS IN TYPE 1 DIABETES

Exercise is one of the primary factors perturbing blood glucose levels in individuals with type 1 diabetes. Aerobic exercise is associated with rapid uptake of glucose from the bloodstream into the tissues for oxidation and fuel production. This increased glucose utilization by skeletal muscles during exercise must be matched precisely by increased glucose production by the liver or hypo- or hyperglycemia occurs. In individuals without diabetes, aerobic exercise promotes a decrease in insulin secretion and a concomitant rise in glucagon and other counter-regulatory hormones (cortisol, catecholamines, and growth hormone). In individuals with type 1 diabetes, where insulin levels are dependent on the timing, location, and size of insulin dosages either injected or infused into the body prior to exercise, a state of hyperinsulinemia during aerobic exercise is not uncommon (Yardley et al. 2013). This brings with it an accompanying risk of hypoglycemia once exercise is undertaken. To prevent this, common strategies for aerobic exercise include decreasing pre-exercise insulin dosage, and increasing carbohydrate intake before, during, and after exercise (Perkins and Riddell 2006; Riddell and Iscoe 2006; Sigal et al. 2013). Where miscalculation produces too large a decrease in insulin, or excessive carbohydrate intake, the side effect of these strategies is high blood glucose (hyperglycemia). Both of these situations may impact electrolyte balance in the body if they are occurring during competition, although there is evidence to indicate that the effects of high and low blood glucose on plasma electrolytes can be reversed by returning blood glucose levels to a healthy physiological range (Caduff et al. 2011).

12.2.1.1 Hypoglycemia

The term *hypoglycemia* is generally used to describe blood glucose levels that are lower than 4.0 mmol/L (72 mg/dL) (Clayton et al. 2013). Most individuals with type 1 diabetes are extremely vigilant of their blood glucose levels at the onset of exercise and/or competition, due to the fact that hypoglycemia both hinders performance (Kelly et al. 2010) and poses an acute risk to the well-being of the athlete. Blood glucose levels in the day(s) prior to competition and training are equally important, as experiencing a bout of hypoglycemia can blunt counter-regulatory responses to subsequent exercise and increase the risk of blood glucose levels dropping to dangerous levels during activity (Galassetti et al. 2003). In addition, hypoglycemia has been shown to induce a sharp drop in potassium levels, which can persist for several hours after a return to euglycemia (Caduff et al. 2011) (discussed in Section 12.2.3.2). The resulting hypokalemia can negatively impact athletic performance by interfering with heart rhythm and skeletal muscle contraction.

12.2.1.2 Hyperglycemia

The definition of hyperglycemia varies from source to source. In its strictest definition, a blood glucose level higher than the *normal* range (i.e., >approximately 7.0 mmol/L or 126 mg/dL) could be considered hyperglycemia. For the purposes of individuals with diabetes, however, where blood glucose fluctuations outside the normal physiological range occur more frequently, the term *hyperglycemia* is generally used to describe blood glucose levels greater than 11.1 mmol/L (200 mg/dL) (Clement et al. 2004; Canadian Diabetes Association Clinical Guidelines Expert Commmittee 2013). While mild hyperglycemia can result in thirst, blurred vision, fatigue, and headaches, symptoms of severe and/or prolonged hyperglycemia can include vomiting and diarrhea, which can occasionally make it difficult to differentiate it from heat stroke, if training and/or competition is taking place in the heat. Unfortunately, and for a variety of reasons, most individuals with type 1 diabetes are likely to encounter blood glucose levels greater than 11.1 mmol/L (200 mg/dL) on a regular basis (Yardley et al. 2013). The reasons for hyperglycemia related to exercise are discussed in more detail below.

As glucose is a very useful substance in the body, it is one that the kidneys will often try to conserve as they filter the blood. When blood glucose is within the normal physiological range, the vast majority of glucose in the filtrate leaving the glomerulus in the kidney is reabsorbed in the proximal tubule (Sherwood et al. 2013). As a result, the amount of glucose found in urine is generally negligible. The average tubular maximum for glucose reabsorption is around 375 mg/min, although a great deal of inter-individual variability exists with respect to this parameter (Johansen et al. 1984). If an individual has an average glomerular filtration rate (GFR), then the renal threshold for glucose is approximately 300 mg/mL. However, not all nephrons have the same tubular maximum, and as such, glucose may be found in the urine before this threshold is reached (Figure 12.1). This generally starts occurring once blood glucose levels reach 9–10 mmol/L (162–180 mg/dL) (Johansen et al. 1984), but can be lower in some individuals and higher in others. The presence of glucose in the urine, known as *glucosuria*, can lead to substantial water loss through osmotic diuresis (Dhatariya 2008) thereby increasing the risk of dehydration. In individuals with chronic hyperglycemia (as demonstrated by elevated glycosylated hemoglobin [HbA_{1c}] levels), the risk of dehydration is higher: a renal resistance to vasopressin has been demonstrated in type 1 diabetic individuals with poor blood glucose control (McKenna et al. 2000) which can decrease the kidneys' ability to conserve water during exercise.

Hyperglycemia can also affect kidney function with respect to maintaining electrolyte balance. Sodium reabsorption in both the proximal (Turner et al. 1997) and the distal tubule (Simkova et al. 2004) of the kidney can be elevated after as little as

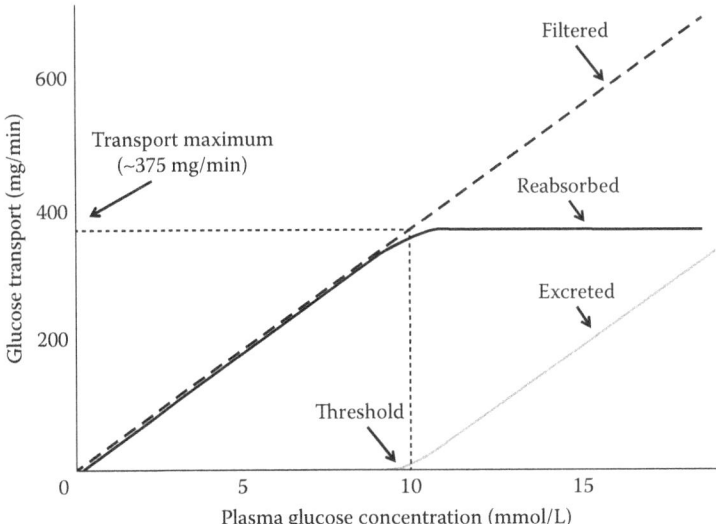

FIGURE 12.1 Approximate values of glucose reabsorption and excretion in normoglycemia and hyperglycemia.

90 min spent with blood glucose levels of 12 mmol/L or higher. With persistent hyperglycemia, the activation of sodium/glucose cotransporters in the proximal tubule is elevated leading to sodium retention (Korner et al. 1994). While kidney function is affected by high blood glucose concentration, if hyperglycemia is short lived (<2 h) it might not have a large impact on plasma concentration of sodium, chloride, and calcium (Caduff et al. 2011). It might, however, lead to an increase in potassium concentration, potentially leading to muscle weakness and cardiac arrhythmia. High blood glucose levels can also increase urinary magnesium excretion (Djurhuus et al. 2000), which could eventually lead to muscle cramping. In extreme cases where hyperglycemia is severe and accompanied by ketoacidosis resulting in diarrhea and/ or vomiting, hypokalemia may result (Arora et al. 2012). As ketoacidosis has several symptoms in common with heat stroke (nausea/vomiting, headache, confusion, and weakness), it is important for athletes with diabetes as well as their coaches/trainers to be able to differentiate between these two.

12.2.1.2.1 Potential Causes of Hyperglycemia in Athletes with Type 1 Diabetes

As it was briefly mentioned above, individuals with type 1 diabetes often need to decrease their insulin intake and/or increase their carbohydrate intake prior to exercise in order to prevent deleterious declines in blood glucose, and consequently hypoglycemia, during and after exercise (Perkins and Riddell 2006; Riddell and Perkins 2006; Yardley et al. 2013). How much the individual decreases their insulin dosage or increases their carbohydrate intake is often largely based on personal experience. It will also depend on the time of day exercise is being performed, the anticipated duration and intensity of the exercise, blood glucose trends throughout the day (i.e., antecedent hypoglycemia can increase the risk of another hypoglycemic event [Cryer 2013]), as well as the timing of previous snacks/meals and related insulin intake prior to exercise. Fear of hypoglycemia is a very common barrier to exercise in individuals with type 1 diabetes (Brazeau et al. 2008) and often leads to excessive caution with respect to exercise-related adjustments. As such, it is not unusual for individuals with type 1 diabetes to start exercise with their blood glucose levels in the hyperglycemic range (Yardley et al. 2013, 2015), in order to prevent encountering hypoglycemic levels before the end of the exercise session (Figure 12.2). This can lead to varying degrees of glucosuria and dehydration depending on the severity and duration of hyperglycemia.

The sympathetic nervous system also has a potent effect on blood glucose levels. Elevated levels of the catecholamines epinephrine and norepinephrine (part of the *fight or flight* response) are associated with enhanced hepatic glycogenolysis and a concomitant increase in blood glucose levels (Mitchell et al. 1988; Sigal et al. 1999). Periods of high emotional stress or illness can trigger sympathetic responses leading to catecholamine release, with the intensity of the stimulus being directly related to the size of the catecholamine response (Wortsman 2002). In individuals without type 1 diabetes, the resultant increase in blood glucose is generally countered by an increase in insulin secretion to maintain blood glucose levels in healthy range, in spite of demands being placed on the body. Where physiological production of insulin is absent, however, individuals with type 1 diabetes may see gradual and sustained increases in blood glucose to the point of hyperglycemia unless a correction

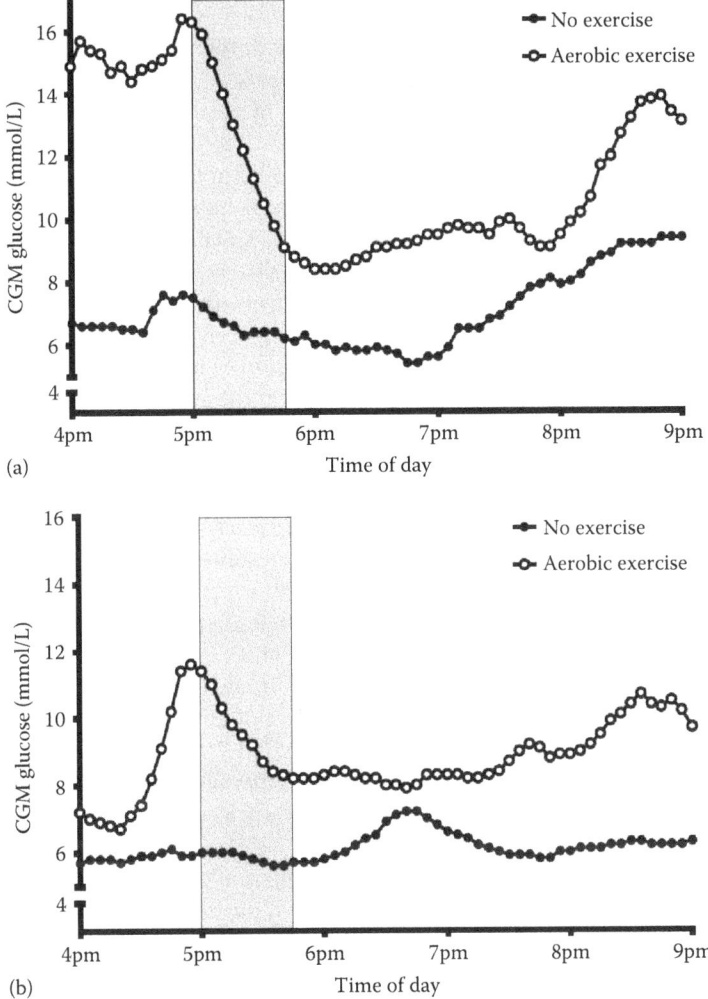

FIGURE 12.2 Continuous glucose monitoring tracings of two patients with type 1 diabetes (one patient per panel, indicated as patient (a) and (b) respectively) on a day where aerobic exercise was performed (open circles) versus a day when no exercise was performed (closed circles). The gray box represents the timing of the 45 min of treadmill running at 60% of peak fitness on the aerobic exercise day. Participant adjustments in insulin dosage before and after exercise resulted in substantially higher blood glucose levels over a 5-h period. (Reprinted from *Canadian Journal of Diabetes*, 37, Yardley, J. et al., Vigorous intensity exercise for glycemic control in patients with type 1 diabetes, pp. 427–432, Copyright 2013, with permission from Elsevier.)

dose of insulin is administered. However, if the stress in question is an anticipated sporting event, a correction dose of insulin may be inadvisable, as this would then increase the risk of hypoglycemia both during and after the event. Athletes with type 1 diabetes will thus often enter competition with elevated blood glucose levels and therefore an elevated requirement for fluid intake to prevent dehydration.

12.2.1.2.2   High-Intensity Exercise, Hyperglycemia, and Dehydration

It has been known for several years that sustained, high-intensity exercise (generally >85% of the individual's peak aerobic capacity, or VO_{2peak}) is associated with increasing blood glucose levels and consequently hyperglycemia in individuals with type 1 diabetes (Mitchell et al. 1988; Sigal et al. 1999). One study showed that a maximal run to exhaustion (which took about 8–10 min for most participants) led to a sustained increase in average blood glucose levels both during exercise and over the 2 h following exercise in individuals with type 1 diabetes (Mitchell et al. 1988). Conversely, blood glucose levels declined immediately post-exercise and returned to normal within 40 min in the non-diabetic controls. Increasing exogenous insulin intake after exercise can help prevent sustained hyperglycemia in athletes with diabetes (Sigal et al. 1994); however, this must be carefully balanced with carbohydrate intake in order to avoid hypoglycemia, especially if the athlete is involved in a competition with repeated performances throughout the day (such as a tournament or a meet).

The effects of high intensity activity on blood glucose levels are slightly less pronounced when bouts of activity are performed intermittently, with aerobic exercise being performed in between. This type of intermittent, high-intensity exercise is the characteristic of many *stop and go* sports such as soccer, rugby, football, and hockey and can also be found in training programs for such sports as running, cycling, swimming, and cross-country skiing. Three groups of researchers have attempted to duplicate these efforts with small studies in the laboratory in order to examine their effects on blood glucose levels in individuals with type 1 diabetes (Guelfi et al. 2005a, b, 2007; Iscoe and Riddell 2011; Campbell et al. 2015). While one study (Iscoe and Riddell 2011) found that blood glucose decreased just as much during intermittent high-intensity activity as during continuous aerobic activity, two studies (Guelfi et al. 2005a; Campbell et al. 2015) found that short, very high intensity bursts of activity such as *stop and go* sports protected against these quick declines in blood glucose levels both during and in the 2 h following exercise. While laboratory-testing sessions may not adequately mimic competition, blood glucose measurements taken on individuals with type 1 diabetes during competitive events have shown a high degree of variability (Miadovnik et al. 2013; Yardley et al. 2013). This would indicate once again that athletes with type 1 diabetes should monitor their blood glucose levels frequently during competition (which can be aided with the use of real-time continuous glucose monitoring) (Riddell and Milliken 2011) in order to avoid any complications that accompany large fluctuations in blood glucose levels. In situations of intense-exercise induced hyperglycemia, patients should monitor for elevated ketone levels, increase carbohydrate-free fluid intake, and administer a conservative insulin corrective dose (i.e., half of the insulin typically administered to normalize glucose concentrations) (Zaharieva and Riddell 2015).

Importantly, fluid and carbohydrate provision during prolonged exercise is critical for safety and performance in type 1 diabetes. Drink concentrations of >6% carbohydrate may help to prevent blood glucose reduction during exercise in type 1 diabetic subjects. However, the increased osmolality of high-carbohydrate drinks may cause gastrointestinal distress (Gisolfi et al. 1998). Supplementation with an 8%–10% carbohydrate drink with electrolytes, ingested before and during exercise, may be an ideal beverage for the maintenance of fluid, electrolytes, and carbohydrate provision in type 1 diabetes (Perrone et al. 2005).

12.2.2 SWEATING ABNORMALITIES IN TYPE 1 DIABETES

Studies examining sweat responses in individuals with type 1 diabetes are few and inconclusive. A handful of studies have examined sweat responses to a passive heat stress test and have generally found that individuals with type 1 diabetes have lower sweat rates in the lower body and higher sweat rates in the upper body than individuals without diabetes (Kennedy et al. 1984; Navarro et al. 1989; Berghoff et al. 2006). For the most part, however, these studies found that sweating impairments were related to diabetic neuropathy as a result of poorly controlled or long-standing type 1 diabetes, and no conclusions regarding overall fluid loss and hydration status were made. Two small studies have examined thermoregulatory function in otherwise healthy, fit, individuals with type 1 diabetes during exercise in the heat (Stapleton et al. 2013; Carter et al. 2014). One study found no difference in sweat rate in individuals with type 1 diabetes compared to non-diabetic individuals matched for age, fitness level, and body surface area during 60 min of moderate cycling at 35°C (Stapleton et al. 2013). A more recent study, however, showed that, at the same temperature, higher levels of heat production/exercise intensity were associated with lower sweat rates in individuals with type 1 diabetes compared to matched controls (Carter et al. 2014). This would imply that the risk of dehydration related to sweating is not increased in individuals with type 1 diabetes; however, the risk of inadequate body cooling and subsequently heat stress are increased. Further studies are necessary to determine the full impact of differences in sweating on hydration and sports performance in individuals with type 1 diabetes.

12.2.3 ELECTROLYTE CONSIDERATIONS FOR ATHLETES WITH TYPE 1 DIABETES

12.2.3.1 Sodium

Type 1 diabetes has been associated with increased sodium reabsorption in the kidneys at rest (Korner et al. 1994; Turner et al. 1997; Djurhuus et al. 2000; Simkova et al. 2004). As insulin stimulates sodium uptake in the distal tubules of the kidneys in individuals without (DeFronzo et al. 1976; Skott et al. 1989; Stenvinkel et al. 1992) and with type 1 diabetes, part of the increased sodium reabsorption may be related to the relative hyperinsulinemic state in which type 1 diabetic individuals often find themselves. Higher GFRs have also been associated with higher rates of water and sodium reabsorption in the proximal tubule of the kidney in individuals with type 1 diabetes compared to control subjects without diabetes (Turner et al. 1997).

Very few studies to date have examined whether rates of sodium and water reabsorption are different between individuals with type 1 diabetes and those without diabetes in the context of physical activity. An observational study of endurance exercise found that antidiuretic hormone levels were increased more through exercise (a half-iron distance triathlon) in individuals with type 1 diabetes compared to athletes without diabetes performing the same event (Koivisto et al. 1992). The authors suggest this may have been due to elevated pre-race blood glucose levels and consequently decreased plasma volume in the diabetic participants. The same study (Koivisto et al. 1992) and one other (Boehncke et al. 2009) reported that renin and aldosterone concentrations immediately post-exercise were similar in athletes with type 1 diabetes compared to athletes without diabetes. This would seem to indicate that physiological mechanisms for the maintenance of fluid and sodium balance are not impaired during endurance exercise in otherwise healthy individuals with type 1 diabetes. It is important to note, however, that these observational studies have very small sample sizes (7–10 diabetic participants) and that blood glucose levels were low to normal in the diabetic participants when these measurements were taken (Koivisto et al. 1992; Boehncke et al. 2009). Considering the impact of high and low blood glucose levels on hydration and/or electrolyte balance in individuals with type 1 diabetes discussed above, it is likely that further studies will need to be performed under dysglycemic conditions in order to understand the full impact of blood glucose levels on fluid and electrolyte homeostasis during exercise.

One small ($n = 7$) study has examined the effect of performing a sprint (130% VO_{2peak}) to exhaustion, and the changes that seven weeks of sprint training (4–10 times 30 s of *all out* sprinting with 3–4 min of rest between intervals) would induce on increases in the sodium–potassium ATPase (and consequently plasma sodium and potassium balance) in the skeletal muscle of individuals with type 1 diabetes compared to individuals without diabetes (Harmer et al. 2006). Resting values of plasma sodium did not differ between groups either before or after training. Mild increases in plasma sodium were found in both groups during exercise, but lower plasma sodium concentration was found in individuals with type 1 diabetes up to 60 min post-exercise (Harmer et al. 2006). This may be an important consideration for athletes with type 1 diabetes performing repeated bouts of very short, high-intensity exercise throughout the day (e.g., track and field, short track speed skating, and gymnastics).

12.2.3.2 Potassium

Disturbances in potassium balance in individuals with diabetes are generally associated with dysglycemia (either high or low blood glucose). Elevated fasting glucose concentration is associated with higher levels of serum potassium in individuals with type 1 diabetes (Pun and Ho 1989); however, where severe hyperglycemia persists and leads to ketoacidosis, the related vomiting and diarrhea can result in hypokalemia (Arora et al. 2012). While there is a paucity of data examining potassium regulation during endurance exercise in individuals with type 1 diabetes, two small studies found that concentrations of two important hormones in sodium and potassium regulation (renin and aldosterone) were found to be similar between diabetic athletes and those without diabetes taking part in an endurance race (Koivisto et al. 1992; Boehncke et al. 2009). These finding may indicate that both sodium and potassium might be handled similarly by the body in the presence of diabetes, provided that blood glucose levels are within the normal physiological range.

One small study has also measured plasma potassium concentrations both at rest and after a single supra-maximal sprint to exhaustion (Harmer et al. 2006). Resting plasma potassium concentration prior to exercise was found to be similar between individuals with type 1 diabetes and controls without diabetes. In response to a single bout of exhaustive exercise, individuals with type 1 diabetes had comparable changes in plasma potassium concentration as non-diabetic controls (Harmer et al. 2006). They also reported that improvements in potassium regulation after seven weeks of sprint training were similar between groups. It was noted, however, that post-exercise increases in plasma potassium concentration were greater in individuals with type 1 diabetes, potentially due to lower plasma insulin and higher plasma glucose levels (Harmer et al. 2006). This, once again, emphasizes the importance of maintaining a healthy range of blood glucose levels before, during, and after high-intensity exercise.

12.2.3.3   Magnesium

Lower plasma and muscle magnesium concentrations have been measured in individuals with type 1 diabetes compared to people without diabetes (Fort and Lifshitz 1986; Sjogren et al. 1986), with magnesium deficiency often being related to poor blood glucose control (McNair et al. 1982; Fort and Lifshitz 1986; Sjogren et al. 1986; Pun and Ho 1989). The lower magnesium levels can likely be attributed to higher levels of urinary excretion of magnesium in individuals with type 1 diabetes compared to non-diabetic controls (Driziene et al. 2005), which can be exacerbated during periods of hyperglycemia (Djurhuus et al. 2000). Exercise can also increase the loss of magnesium through increased sweating (Beller et al. 1975; Costill 1977), and possibly also through an increase in urinary magnesium excretion, particularly when training is intense and/or prolonged (Deuster et al. 1987; Nuviala et al. 1999). It has yet to be determined if these losses are similar in type 1 diabetic athletes compared to their non-diabetic counterparts. Overall, it would be prudent to suggest that athletes with type 1 diabetes should be particularly vigilant with respect to ensuring adequate consumption of foods high in magnesium (e.g., pulses, legumes, green leafy vegetables, and whole grains), as magnesium deficiency has been associated with muscle cramping or spasms, weakness, and neuromuscular dysfunction, all of which could negatively impact athletic performance (Nielsen and Lukaski 2006). Due to the fact that direct measures of magnesium levels in athletes with type 1 diabetes (during training, or pre- and post-competition) are currently lacking, the magnitude of the potential magnesium deficit in type 1 diabetic athletes can only be inferred.

12.2.4   Complications of Type 1 Diabetes: Nephropathy

Diabetic nephropathy is typically defined by macroalbuminuria—defined as a urinary albumin excretion of more than 300 mg in a 24-h collection—or macroalbuminuria and abnormal renal function as represented by an abnormality in serum creatinine, calculated creatinine clearance, or GFR. Clinically, diabetic nephropathy is characterized by a progressive increase in proteinuria and decline in GFR, hypertension, and a high risk of cardiovascular morbidity and mortality. Approximately

30%–50% of patients with diabetes develop some degree of kidney disease over their lifetime (Marshall and Flyvbjerg 2010). Importantly, regular exercise is associated with reduced risk of diabetic nephropathy in patients with type 1 diabetes (Waden et al. 2015). Poor aerobic fitness in adolescents with type 1 diabetes is also associated with early renal dysfunction (Bjornstad et al. 2015). Risk factors for the development of nephropathy in individuals with type 1 diabetes include poor blood glucose control (reflected by sustained elevations in HbA_{1c} levels), dislipidemia, high blood pressure, and diabetes duration (Raile et al. 2007).

In the initial stages of kidney damage, patients may have little or no awareness of the problem until nearly all kidney function is lost. In many patients with type 1 diabetes, GFR can increase (possibly as a result of changes in afferent and efferent arteriolar tone [Sasson and Cherney 2012]). As kidney function gradually declines, filtration rates decrease in tandem, until the kidneys are no longer able to appropriately filter the blood. As kidney disease progresses, there is an increased tendency for fluid retention, hypertension, and hyperkalemia. Other symptoms of kidney disease include loss of sleep, frequent nocturnal urination, poor appetite, upset stomach, weakness, and difficulty concentrating. While it is unlikely that an individual with severe kidney problems would be competing in athletic events, an awareness of possible complications of even mild kidney impairment would be an asset for the athlete and any coaching/medical staff assisting them. Monitoring of kidney function is done by measuring urine protein in a 24-h urine collection, spot protein-to-creatinine ratio and/or albumin-to-creatinine ratio. The main treatments for kidney disease are tight control of blood glucose and blood pressure, typically via angiotensin-converting enzyme (ACE) inhibitors. Once kidneys fail, dialysis is necessary.

12.2.5 COMMON MEDICATIONS FOR INDIVIDUALS WITH TYPE 1 DIABETES

The main glucose regulatory medication that individuals with type 1 diabetes will be using is insulin, which can be administered through injections, or by constant infusion through continuous subcutaneous insulin infusion (i.e., pump therapy). A majority of patients with type 1 diabetes will be using either multiple daily injections or pump therapy, both of which allow for some flexibility in insulin delivery for exercise. As it was mentioned above, insulin can have an effect on electrolyte balance. It is also important to note that the speed with which insulin is absorbed and its onset of action may be increased by both local heating and massage at or near the application site (Freckmann et al. 2012). Other than insulin, there is a range of medications that individuals with type 1 diabetes may be taking. Many of these could impact kidney function, water and/or electrolyte balance, or increase the risk of dehydration due to hyperglycemia and osmotic diuresis. These include a combination of prescription and over the counter medications for everything from hypertension to the common cold.

12.2.5.1 Medications Affecting Fluid and Electrolyte Balance

As hypertension is a common complication related to type 1 diabetes, there is a possibility that ACE inhibitors, angiotensin II receptor blockers (ARB) or some form of diuretic drug have been prescribed to an individual with type 1 diabetes.

ACE inhibitors and ARBs can occasionally cause an increase in potassium levels to the point of hyperkalemia (Rx Files 2013). While many of the symptoms of hyperkalemia are mild in nature, high levels of potassium can impair athletic performance through muscle weakness or cause cardiac arrhythmia, which could be potentially fatal for the athlete. If diuretics have been prescribed to control blood pressure, there can be an increased risk of dehydration due to augmented urine production (Rodenburg et al. 2013). It should also be noted that diuretics can increase the risk of hyponatremia (Rodenburg et al. 2013). Thiazide diuretics, in particular, have been shown to substantially increase the risk of hyponatremia (Rodenburg et al. 2013) by inhibiting sodium chloride reabsorption in the distal tubule of the kidney, creating a situation where sodium loss could exceed water loss (Egom et al. 2011). The risk of thiazide-induced hyponatremia increases with age, is more common in females, and tends to affect those with lower body weight more than those who are heavier (Egom et al. 2011). Hypokalemia can also occur with the use of thiazides due to an increase in potassium secretion in the distal tubule of the kidney (Palmer 2011).

A new class of drug designed to inhibit glucose reabsorption in the kidneys may also theoretically impact hydration status. The sodium/glucose cotransporter 2 (SGLT2) is expressed primarily in the kidneys and is involved in the reabsorption of filtered glucose in the renal tubule. Clinical trials of SGLT2 inhibitors in patients with type 1 diabetes demonstrate a significant clinical effect in decreasing serum glucose, hemoglobin A_{1C}, body weight, and systolic blood pressure (Perkins et al. 2014). SGLT2 inhibition also reduces hyperfiltration in type 1 diabetes (Cherney et al. 2014), which is a risk factor for diabetic kidney disease and vascular dysfunction. It should be noted that SGLT2 antagonists promote increased urine volume, particularly at the onset of use.

12.2.5.2 Medications Causing Hyperglycemia

Several types of medication can also affect hydration indirectly in individuals with type 1 diabetes by causing hyperglycemia and consequently elevating diuresis. The list presented here is by no means exhaustive, but covers some of the main classes of drugs likely to be encountered that could potentially be problematic. Atypical antipsychotic drugs prescribed for a variety of mental health problems are known to decrease insulin sensitivity (Khoza and Barner 2011) and thereby increase the risk of hyperglycemia in individuals with type 1 diabetes. The risk of hyperglycemia is also higher for patients who have been prescribed niacin to control blood cholesterol levels (Elam et al. 2000). In addition, corticosteroids, which are commonly prescribed for ailments such as asthma, arthritis, and rhinitis among others (Rx files 2013), have been associated with hyperglycemia in patients both with and without diabetes (Kwon and Hermayer 2013).

Several over-the-counter decongestants (used for treating colds and/or allergies) have either phenylephrine or pseudoephedrine as their active ingredients. These drugs are generally adrenoceptor agonists that reduce nasal congestion by producing smooth muscle contraction and subsequent vasoconstriction in the nasal blood vessel (Brenner and Stevents 2013). Unfortunately, the side effect of both of these agents is to increase glycogenolysis in the liver in a manner similar to the catecholamines, resulting in an increase in blood glucose levels. An individual with type 1 diabetes taking these medications for several days may find that they are continuously hyperglycemic and thus at high risk of dehydration. Use of such medication should not be expected at elite levels of competition as epinephrine appears on the World Anti-Doping Agency (WADA) prohibited substance list (World Anti-Doping Agency 2014a), and phenylephrine is part of the WADA Monitoring Program (World Anti-Doping Agency 2014b) where testing is performed to prevent the potential misuse of stimulants in sport. Younger and/or less experienced athletes, however, may be unaware of the effects of both of these substances both on their performance and on their blood glucose levels.

12.3 TYPE 2 DIABETES

In contrast with type 1 diabetes, individuals with type 2 diabetes often produce enough insulin, but lack sensitivity to its effects at the cellular level, thereby preventing the removal of glucose from the blood stream into the tissues. In addition, there is evidence to indicate that individuals with type 2 diabetes have higher rates of gluconeogenesis and increased glucose uptake by the kidneys in both the postprandial and postabsorptive states (Mitrakou 2011). As a result, individuals with type 2 diabetes are at a particularly high risk for hyperglycemia, which can be exacerbated by acute stress (Goetsch et al. 1990, 1993, 1994; Konen et al. 1993) and/or very high intensity exercise (Kjaer et al. 1990). Where blood glucose levels are poorly controlled, the kidneys' ability to conserve water (Agha et al. 2004) and maintain electrolyte balance can be affected (Brands and Manhiani 2012).

Because physical activity is known to increase insulin sensitivity (Roberts et al. 2013), it is strongly recommended in the management of type 2 diabetes (Sigal et al. 2004, 2013). Similar to individuals with type 1 diabetes, dehydration secondary to hyperglycemia is likely to be the main problem encountered. In addition, there are also several types of medication that can be prescribed for type 2 diabetes and its related complications that can affect the gastrointestinal tract and impact kidney function, both of which can disturb hydration and electrolyte balance.

12.3.1 Blood Glucose Management during Exercise in Type 2 Diabetes

As fewer individuals with type 2 diabetes require treatment with insulin compared to those with type 1 diabetes, blood glucose management during exercise is often more simple. Some individuals will be managing their blood glucose levels with diet and exercise alone and will have little to consider other than fluid and carbohydrate replacement when performing strenuous or extended periods of activity. Conversely, patients taking insulin or sulfonylureas may require decreases in dosage or increased carbohydrate in order to avoid hypoglycemia. As always, the potential to miscalculate carbohydrate intake and medication dosage exists, with the potential to produce blood glucose levels that exceed the renal threshold for glucose reabsorption and a subsequent increased risk of dehydration. It is also worthy of note that where type 2 diabetes is poorly controlled, chronic hyperglycemia has been shown to impair the renal response to vasopressin, leading to an increased risk of dehydration (Agha et al. 2004).

12.3.2 POTENTIAL CAUSES OF HYPERGLYCEMIA IN TYPE 2 DIABETES

In addition to excessive carbohydrate intake and/or insufficient medication, psychological and physiological factors can increase the risk of hyperglycemia in individuals with type 2 diabetes. It has been suggested that emotional and/or psychological stress can negatively impact blood glucose control (Goetsch et al. 1990, 1993, 1994; Konen et al. 1993). To the best of our knowledge, no studies to date have examined whether stress related to athletic training and/or competition has a similar effect in individuals with type 2 diabetes. It is likely, however, that responses will be extremely individualized. As such, determining an individual's glycemic responses to these stresses should include frequent monitoring of blood glucose prior to and during initial experiences of training and competition.

Similar to individuals with type 1 diabetes, high-intensity exercise results in a counter-regulatory response (particularly increases in epinephrine and glucagon) that increases glucose production to a level that outpaces glucose uptake (Kjaer et al. 1990). This is also accompanied by a brief period of increased insulin resistance post-exercise (Kjaer et al. 1990). The resulting hyperglycemia, however, tends to diminish within an hour of finishing exercise, as insulin secretion will increase in the face of high blood glucose levels, and insulin sensitivity of the tissues is subsequently enhanced for several hours or even days post-exercise (Kjaer et al. 1990; Bordenave et al. 2008; van Dijk et al. 2012). As always, if periods of hyperglycemia occurred prior to exercise and continue for an extended period of time post-exercise, the risk of dehydration can be elevated for the athlete with type 2 diabetes.

12.3.3 COMPLICATIONS OF TYPE 2 DIABETES: NEPHROPATHY

As mentioned above, the progression of kidney disorders is often slow and therefore may remain unnoticed for long periods of time. The risk of damage to the kidneys generally increases with age, disease duration, and worsening blood glucose control. Renal disease, even in its milder forms, is associated with a decrease in GFR, thereby decreasing the kidneys' ability to regulate fluid and electrolyte balance (Kidney Disease Improving Global Outcomes Group 2012). As renal disease progresses, acid–base imbalances become more common, which can be increasingly problematic where high-intensity exercise is performed (Mogensen 2002). However, patients with kidney dysfunction typically have very low exercise tolerance and avoid intense exercise because of self-limitation. Kidney disorders have also been associated with decreases in calcium absorption, leading to negative calcium balance, as well as renal potassium and magnesium wasting (Kidney Disease Improving Global Outcomes Group 2012). Fatigue is often a side effect of damaged kidneys, as erythropoeitin levels may decrease, thereby affecting the oxygen carrying capacity of the blood. Individuals with end-stage renal disease have very low exercise tolerance, low aerobic capacity ($\dot{V}O_{2max}$ < 20 mL/kg min), decreased cardiac output, blunted heart rate response to exercise, anemia, and decreased oxygen extraction (Evans and Forsyth 2004). As such, these individuals need special care when prescribing exercise and should be under close supervision.

Cardiovascular complications are also common in individuals with nephropathy, although little evidence exists that exercise triggers any adverse events (Painter 1988; Moore et al. 1998; Painter et al. 2002, 2003). As special care may be needed in these individuals with advanced disease, including erythropoietin administration (Moore et al. 1998; Painter et al. 2002), supervised exercise programs are recommended. While it is unlikely that individuals with advanced stages of kidney disease will be participating in high-intensity physical activity, it is important to note that even mild impairments in kidney function may have an effect on electrolyte balance.

12.3.4 SWEATING ABNORMALITIES IN TYPE 2 DIABETES

Similar to individuals with type 1 diabetes, very few studies examining sweating responses in individuals with type 2 diabetes exist (Fealey et al. 1989; Petrofsky et al. 2005; Rand et al. 2008; Kenny et al. 2013), with only one of these studies examining sweat responses during moderate physical activity (Kenny et al. 2013). For the most part, studies of individuals with type 2 diabetes have found that sweating responses to local or whole-body heating (Fealey et al. 1989; Petrofsky et al. 2005), as well as electrical (Rand et al. 2008) stimuli are lower in individuals with type 2 diabetes compared to non-diabetic controls. These impairments in sudomotor function may be linked to changes in the sweat glands themselves, especially where participants have either long-standing or poorly controlled diabetes (Fealey et al. 1989; Luo et al. 2011). One study examining a 60-min bout of moderate cycling at 30°C did not find any differences in sweat rate between individuals with type 2 diabetes compared to matched controls, but did note that the measurement was highly variable, thereby making it difficult to find statistically significant differences between the two groups (Kenny et al. 2013). Unfortunately, there is a paucity of studies involving whole-body heat exposure with blood sampling or measures of core temperature to explain how these sweating impairments could potentially impact heat loss mechanisms in the body or overall fluid and electrolyte balance. More information on sweating and skin blood flow in diabetes is described in Chapter 3 (Section 3.3.2) of this book.

12.3.5 CHANGES IN RENAL HANDLING OF WATER AND ELECTROLYTES

12.3.5.1 Sodium

While a great deal of inter-individual variability has been found, some individuals with type 2 diabetes may have glomerular hyperfiltration (Vora et al. 1992; Pruijm et al. 2010) and elevated sodium reabsorption in the proximal tubule of the kidney (Pruijm et al. 2010). These conditions have similarly been found in individuals with abdominal obesity (Strazzullo et al. 2001) and the metabolic syndrome (Strazzullo et al. 2006) and may therefore not be an effect of type 2 diabetes per se. Chronic hyperglycemia, however, will increase the activation of sodium/glucose cotransporters in the proximal tubule and thereby promote sodium retention (Korner et al. 1994). As individuals with type 2 diabetes are also at high risk for hypertension, it is not uncommon for a low sodium diet to be recommended. In such cases, and where medication has been prescribed for hypertension (see Section 12.3.6.1), athletes with type 2 diabetes should be aware of possible sodium deficits related to sweating for long periods of time during training and/or participation in athletic events.

12.3.5.2 Potassium

The importance of maintaining healthy blood glucose levels can also be seen through its effects on potassium balance in individuals with type 2 diabetes. Acute hyperglycemia has been associated with a rise in plasma potassium levels (Nicolis et al. 1981; Rosenstock et al. 1982) in individuals with type 2 diabetes. As it was mentioned above, many of the symptoms of hyperkalemia are mild in nature, but very high levels of potassium can impair athletic performance through muscle weakness, or cause cardiac arrhythmia, with potentially fatal consequences for the athlete.

12.3.5.3 Magnesium

Although the mechanism hasn't been fully elucidated, type 2 diabetes, particularly if poorly controlled, has been associated with magnesium deficiency (Sales and Pedrosa 2006). Hypomagnesemia has been reported in anywhere from 13.5% to 47.7% of outpatients with type 2 diabetes, compared to levels of 2.5%–15% in the general population (Mather et al. 1979; McNair et al. 1982; Ma et al. 1995; Pham et al. 2007). As serum magnesium has been inversely correlated with fasting blood glucose and urinary glucose excretion (Mather et al. 1979; Ma et al. 1995), it has been proposed that tubular reabsorption of magnesium is inhibited in diabetic patients during hyperglycemia (McNair et al. 1982). As mentioned above, periods of intense or prolonged training can also lead to losses of magnesium through increased sweating (Beller et al. 1975; Costill 1977) and urinary magnesium excretion (Deuster et al. 1987; Nuviala et al. 1999). In order to avoid potential muscle cramping, weakness, or neuromuscular dysfunction related to magnesium deficiency, athletes with type 2 diabetes should ensure adequate magnesium intake (Nielsen and Lukaski 2006).

12.3.6   COMMON MEDICATION FOR INDIVIDUALS WITH TYPE 2 DIABETES

Individuals with type 2 diabetes can have a wide range of treatments. Some patients manage to make all of the lifestyle changes necessary at an early enough stage in the disease's progression to be completely free of medication. Conversely, some individuals may be on an array of medications for hypertension, dyslipidemia, and blood glucose control (including insulin and insulin sensitizers). As a result, hydration and electrolyte needs must be evaluated on an individual basis depending on the type 2 diabetic participant's level of blood glucose control and medication regimen.

12.3.6.1   Medications That Can Impact Electrolyte Balance

Several medications prescribed for type 2 diabetes have side effects that may interfere with hydration and electrolyte balance. Metformin, a common oral antidiabetic drug that interferes with gluconeogenesis, has been associated with gastrointestinal upset and diarrhea (Rx Files 2013), which can lead to dehydration and potentially a depletion in potassium levels. Acarbose, another antidiabetic drug, decreases the rate of absorption of glucose from the intestines, but is also associated with a risk of diarrhea and subsequently dehydration when taken at higher dosages (Rx Files 2013).

Pioglitozone, an antihyperglycemic drug often prescribed to individuals with type 2 diabetes, is associated with a dose-related risk of peripheral edema (Rx Files 2013). Similarly, the antidiabetic drugs sitagliptin and saxagliptin have an associated risk of diarrhea and vomiting which can negatively impact hydration and increase the risk of hypokalemia (Rx Files 2013).

Individuals with type 2 diabetes are at an increased risk for hypertension and cardiovascular disease for which several therapeutic drug combinations can be prescribed, which can include diuretics, ACE inhibitors, calcium channel blockers, ARB, and β-blockers (Reboldi et al. 2011). As mentioned above in the discussion of the effects of diuretics in type 1 diabetes, these drugs can cause dehydration, hyponatremia, and hypokalemia (Greenberg 2000; Liamis et al. 2013). Calcium channel blockers may lead to a loss of both sodium and water, although this effect tends to be less evident after the first week of treatment (Leonetti 1994). In addition, ACE inhibitors and ARBs have been associated with hyperkalemia (Raebel 2012). Type 2 diabetic individuals participating in sport and competition who are being treated for hypertension should, therefore, be aware of which medication (or combinations thereof) that they are taking in order to understand which electrolyte imbalances they are most likely to encounter.

12.3.6.2 Medications Causing Hyperglycemia

Most of the medications listed above as increasing the risk of hyperglycemia for individuals with type 1 diabetes also increase the risk of hyperglycemia, and subsequently dehydration, in individuals with type 2 diabetes. Atypical antipsychotics, due to their negative impact on insulin sensitivity (Khoza and Barner 2011), can elevate blood glucose levels in individuals with type 2 diabetes. Similarly, blood glucose levels and hydration should be closely monitored before, during, and after physical activity for individuals with type 2 diabetes who have been prescribed niacin (Elam et al. 2000) and corticosteroids (Kwon and Hermayer 2013). As mentioned above, over the counter decongestants such as phenylephrine and pseudoephedrine can also be problematic in terms of blood glucose control.

12.4 RECOMMENDATIONS FOR ATHLETES WITH DIABETES

In the context of diabetes, it is difficult to provide a *one-size-fits-all* recommendation for fluid and electrolyte intake for athletic performance. If the individual is in otherwise perfect health, has impeccable blood glucose control, and is not taking any additional medication, then the evidence to date would indicate that they should be able to follow the same guidelines as an athlete without diabetes with respect to fluid and electrolyte intake. Otherwise, athletes, coaches, and trainers should be aware that the risk of electrolyte and water imbalances increases in tandem with the degree and duration of hyperglycemia, and of the many physical, physiological, and psychological factors that can impact blood glucose levels throughout a period of training and/or competition. Having knowledge of which medications an athlete takes and their potential side effects with respect to hydration and electrolyte balance will also be necessary in order to tailor fluid and food intake for optimal performance.

REFERENCES

Agha, A., D. Smith, F. Finucane, M. Sherlock, A. Morris, P. Baylis, and C. J. Thompson. 2004. Attenuation of vasopressin-induced antidiuresis in poorly controlled type 2 diabetes. *Am J Physiol Endocrinol Metab* 287: E1100–E1106.

Arora, S., D. Cheng, B. Wyler, and M. Menchine. 2012. Prevalence of hypokalemia in ED patients with diabetic ketoacidosis. *Am J Emerg Med* 30: 481–484.

Beller, G. A., J. T. Maher, L. H. Hartley, D. E. Bass, and W. E. Wacker. 1975. Changes in serum and sweat magnesium levels during work in the heat. *Aviat Space Environ Med* 46: 709–712.

Berghoff, M., S. Kilo, M. J. Hilz, and R. Freeman. 2006. Differential impairment of the sudomotor and nociceptor axon-reflex in diabetic peripheral neuropathy. *Muscle Nerve* 33: 494–499.

Bjornstad, P., M. Cree-Green, A. Baumgartner, D. M. Maahs, D. Z. Cherney, L. Pyle, J. G. Regensteiner, J. E. Reusch, and K. J. Nadeau. 2015. Renal function is associated with peak exercise capacity in adolescents with type 1 diabetes. *Diabetes Care* 38: 126–131.

Boehncke, S., K. Poettgen, C. Maser-Gluth, J. Reusch, W. H. Boehncke, and K. Badenhoop. 2009. Endurance capabilities of triathlon competitors with type 1 diabetes mellitus. *Dtsch Med Wochenschr* 134: 677–682.

Bordenave, S., F. Brandou, J. Manetta, C. Fedou, J. Mercier, and J. F. Brun. 2008. Effects of acute exercise on insulin sensitivity, glucose effectiveness and disposition index in type 2 diabetic patients. *Diabetes Metab* 34: 250–257.

Brands, M. W., and M. M. Manhiani. 2012. Sodium-retaining effect of insulin in diabetes. *Am J Physiol Regul Integr Comp Physiol* 303: R1101–R1109.

Brazeau, A. S., R. Rabasa-Lhoret, I. Strychar, and H. Mircescu. 2008. Barriers to physical activity among patients with type 1 diabetes. *Diabetes Care* 31: 2108–2109.

Brenner, G., and C. Stevents. 2013. Adrenoceptor agonists. In G. Brenner and C. Stevents (Eds.), *Pharmacology*, 4th ed. Philadelphia, PA, Elsevier-Saunders, pp. 69–77.

Caduff, A., H. U. Lutz, L. Heinemann, G. Di Benedetto, M. S. Talary, and S. Theander. 2011. Dynamics of blood electrolytes in repeated hyper- and/or hypoglycaemic events in patients with type 1 diabetes. *Diabetologia* 54: 2678–2689.

Campbell, M. D., D. J. West, S. C. Bain, M. I. Kingsley, P. Foley, L. Kilduff, D. Turner, B. Gray, J. W. Stephens, and R. M. Bracken. 2015. Simulated games activity vs continuous running exercise: A novel comparison of the glycemic and metabolic responses in T1DM patients. *Scand J Med Sci Sports* 25: 216–222.

Canadian Diabetes Association Clinical Guidelines Expert Commmittee. 2013. Canadian Diabetes Association 2013 clinical practice guidelines for the prevention and management of diabetes in Canada. *Can J Diabetes* 37 (Suppl 1): S1–S212.

Carter, M. R., R. McGinn, J. Barrera-Ramirez, R. J. Sigal, and G. P. Kenny. 2014. Impairments in local heat loss in type 1 diabetes during exercise in the heat. *Med Sci Sports Exerc* 46: 2224–2233.

Cherney, D. Z., B. A. Perkins, N. Soleymanlou, M. Maione, V. Lai, A. Lee, N. M. Fagan, H. J. Woerle, O. E. Johansen, U. C. Broedl, and M. von Eynatten. 2014. Renal hemodynamic effect of sodium-glucose cotransport 2 inhibition in patients with type 1 diabetes mellitus. *Circulation* 129: 587–597.

Chimen, M., A. Kennedy, K. Nirantharakumar, T. T. Pang, R. Andrews, and P. Narendran. 2012. What are the health benefits of physical activity in type 1 diabetes mellitus? A literature review. *Diabetologia* 55: 542–551.

Clayton, D., V. Woo, and J.-F. Yale. 2013. Canadian Diabetes Association clinical practice guidelines for the prevention and management of diabetes in Canada: Hypoglycemia. *Can J Diabetes* 37 (Suppl 1): S61–68.

Clement, S., S. S. Braithwaite, M. F. Magee, A. Ahmann, E. P. Smith, R. G. Schafer, I. B. Hirsch, and for the American Diabetes Association Diabetes in Hospitals Writing Committee. 2004. Management of diabetes and hyperglycemia in hospitals. *Diabetes Care* 27: 553–591.

Costill, D. L. 1977. Sweating: Its composition and effects on body fluids. *Ann N Y Acad Sci* 301: 160–174.

Cryer, P. E. 2013. Mechanisms of hypoglycemia-associated autonomic failure in diabetes. *N Engl J Med* 369: 362–372.

DeFronzo, R. A., M. Goldberg, and Z. S. Agus. 1976. The effects of glucose and insulin on renal electrolyte transport. *J Clin Invest* 58: 83–90.

Deuster, P. A., E. Dolev, S. B. Kyle, R. A. Anderson, and E. B. Schoomaker. 1987. Magnesium homeostasis during high-intensity anaerobic exercise in men. *J Appl Physiol (1985)* 62: 545–550.

Dhatariya, K. 2008. People with type 1 diabetes using short acting analogue insulins are less dehydrated than those with using human soluble insulin prior to onset of diabetic keto-acidosis. *Med Hypotheses* 71: 706–708.

Diamond Project Group. 2006. Incidence and trends of childhood type 1 diabetes worldwide 1990–1999. *Diabet Med* 23: 857–866.

Djurhuus, M. S., P. Skott, A. Vaag, O. Hother-Nielsen, P. Andersen, H. H. Parving, and N. A. Klitgaard. 2000. Hyperglycaemia enhances renal magnesium excretion in type 1 diabetic patients. *Scand J Clin Lab Invest* 60: 403–409.

Driziene, Z., D. Stakisaitis, and J. Balsiene. 2005. Magnesium urinary excretion in diabetic adolescents. *Acta Medica (Hradec Kralove)* 48: 157–161.

Duclos, M., M. L. Virally, and S. Dejager. 2011. Exercise in the management of type 2 diabetes mellitus: What are the benefits and how does it work? *Phys Sportsmed* 39: 98–106.

Egom, E. E., D. Chirico and A. L. Clark. 2011. A review of thiazide-induced hyponatraemia. *Clin Med* 11: 448–451.

Elam, M. B., D. B. Hunninghake, K. B. Davis, R. Garg, C. Johnson, D. Egan, J. B. Kostis, D. S. Sheps, and E. A. Brinton. 2000. Effect of niacin on lipid and lipoprotein levels and glycemic control in patients with diabetes and peripheral arterial disease: The ADMIT study: A randomized trial. Arterial Disease Multiple Intervention Trial. *JAMA* 284: 1263–1270.

Evans, N., and E. Forsyth. 2004. End-stage renal disease in people with type 2 diabetes: Systemic manifestations and exercise implications. *Phys Ther* 84: 454–463.

Fealey, R. D., P. A. Low, and J. E. Thomas. 1989. Thermoregulatory sweating abnormalities in diabetes mellitus. *Mayo Clin Proc* 64: 617–628.

Fort, P., and F. Lifshitz. 1986. Magnesium status in children with insulin-dependent diabetes mellitus. *J Am Coll Nutr* 5: 69–78.

Freckmann, G., S. Pleus, C. Haug, G. Bitton, and R. Nagar. 2012. Increasing local blood flow by warming the application site: Beneficial effects on postprandial glycemic excursions. *J Diabetes Sci Technol* 6: 780–785.

Galassetti, P., D. Tate, R. A. Neill, S. Morrey, D. H. Wasserman, and S. N. Davis. 2003. Effect of antecedent hypoglycemia on counterregulatory responses to subsequent euglycemic exercise in type 1 diabetes. *Diabetes* 52: 1761–1769.

Gisolfi, C. V., R. W. Summers, G. P. Lambert, and T. Xia. 1998. Effect of beverage osmolality on intestinal fluid absorption during exercise. *J Appl Physiol (1985)* 85: 1941–1948.

Goetsch, V. L., J. L. Abel, and M. K. Pope. 1994. The effects of stress, mood, and coping on blood glucose in NIDDM: A prospective pilot evaluation. *Behav Res Ther* 32: 503–510.

Goetsch, V. L., B. VanDorsten, L. A. Pbert, I. H. Ullrich, and R. A. Yeater. 1993. Acute effects of laboratory stress on blood glucose in noninsulin-dependent diabetes. *Psychosom Med* 55: 492–496.

Goetsch, V. L., D. J. Wiebe, L. G. Veltum, and B. Van Dorsten. 1990. Stress and blood glucose in type II diabetes mellitus. *Behav Res Ther* 28: 531–537.

Greenberg, A. 2000. Diuretic complications. *Am J Med Sci* 319: 10–24.

Guelfi, K. J., T. W. Jones, and P. A. Fournier. 2005a. The decline in blood glucose levels is less with intermittent high-intensity compared with moderate exercise in individuals with type 1 diabetes. *Diabetes Care* 28: 1289–1294.

Guelfi, K. J., T. W. Jones, and P. A. Fournier. 2005b. Intermittent high-intensity exercise does not increase the risk of early postexercise hypoglycemia in individuals with type 1 diabetes. *Diabetes Care* 28: 416–418.

Guelfi, K. J., N. Ratnam, G. A. Smythe, T. W. Jones, and P. A. Fournier. 2007. Effect of intermittent high-intensity compared with continuous moderate exercise on glucose production and utilization in individuals with type 1 diabetes. *Am J Physiol Endocrinol Metab* 292: E865–E870.

Harmer, A. R., P. A. Ruell, M. J. McKenna, D. J. Chisholm, S. K. Hunter, J. M. Thom, N. R. Morris, and J. R. Flack. 2006. Effects of sprint training on extrarenal potassium regulation with intense exercise in type 1 diabetes. *J Appl Physiol* 100: 26–34.

International Diabetes Federation. 2013. *IDF Diabetes Atlas*, 6th ed. Brussels, Belgium, International Diabetes Federation.

Iscoe, K. E., and M. C. Riddell. 2011. Continuous moderate-intensity exercise with or without intermittent high-intensity work: Effects on acute and late glycaemia in athletes with type 1 diabetes mellitus. *Diabet Med* 28: 824–832.

Johansen, K., P. A. Svendsen, and B. Lorup. 1984. Variations in renal threshold for glucose in type 1 (insulin-dependent) diabetes mellitus. *Diabetologia* 26: 180–182.

Kelly, D., J. K. Hamilton, and M. C. Riddell. 2010. Blood glucose levels and performance in a sports camp for adolescents with type 1 diabetes mellitus: A field study. *Int J Pediatr* 2010.

Kennedy, W. R., M. Sakuta, D. Sutherland, and F. C. Goetz. 1984. Quantitation of the sweating deficiency in diabetes mellitus. *Ann Neurol* 15: 482–488.

Kenny, G. P., J. M. Stapleton, J. E. Yardley, P. Boulay, and R. J. Sigal. 2013. Older adults with type 2 diabetes store more heat during exercise. *Med Sci Sports Exerc* 45: 1906–1914.

Khoza, S., and J. C. Barner. 2011. Glucose dysregulation associated with antidepressant agents: An analysis of 17 published case reports. *Int J Clin Pharm* 33: 484–492.

Kidney Disease Improving Global Outcomes Group KDIGO. 2012. Clinical practice guidelines for the evaluation and management of chronic kidney disease. *Kidney Int* 3: 1–150.

Kjaer, M., C. B. Hollenbeck, B. Frey-Hewitt, H. Galbo, W. Haskell, and G. M. Reaven. 1990. Glucoregulation and hormonal responses to maximal exercise in non-insulin-dependent diabetes. *J Appl Physiol (1985)* 68: 2067–2074.

Koivisto, V. A., T. Sane, F. Fyhrquist, and R. Pelkonen. 1992. Fuel and fluid homeostasis during long-term exercise in healthy subjects and type I diabetic patients. *Diabetes Care* 15: 1736–1741.

Konen, J. C., J. H. Summerson, and M. B. Dignan. 1993. Family function, stress, and locus of control: Relationships to glycemia in adults with diabetes mellitus. *Arch Fam Med* 2: 393–402.

Korner, A., A. C. Eklof, G. Celsi, and A. Aperia. 1994. Increased renal metabolism in diabetes: Mechanism and functional implications. *Diabetes* 43: 629–633.

Kwon, S., and K. L. Hermayer. 2013. Glucocorticoid-induced hyperglycemia. *Am J Med Sci* 345: 274–277.

Leonetti, G. 1994. The effects of calcium antagonists on electrolytes and water balance in hypertensive patients. *J Cardiovasc Pharmacol* 24(Suppl A): S25–29.

Liamis, G., E. M. Rodenburg, A. Hofman, R. Zietse, B. H. Stricker, and E. J. Hoorn. 2013. Electrolyte disorders in community subjects: Prevalence and risk factors. *Am J Med* 126: 256–263.

Luo, K. R., C. C. Chao, Y. T. Chen, C. M. Huang, N. C. Yang, H. W. Kan, S. H. Wang, W. S. Yang, and S. T. Hsieh. 2011. Quantitation of sudomotor innervation in skin biopsies of patients with diabetic neuropathy. *J Neuropathol Exp Neurol* 70: 930–938.

Ma, J., A. R. Folsom, S. L. Melnick, J. H. Eckfeldt, A. R. Sharrett, A. A. Nabulsi, R. G. Hutchinson, and P. A. Metcalf. 1995. Associations of serum and dietary magnesium with cardiovascular disease, hypertension, diabetes, insulin, and carotid arterial wall thickness: The ARIC study. Atherosclerosis Risk in Communities Study. *J Clin Epidemiol* 48: 927–940.

Marshall, S. M., and A. Flyvbjerg. 2010. Diabetic nephropathy. In R. I. G. Holt, C. Cochram, A. Flyvbjerg, and B. J. Goldstein (Eds.), *Textbook of Diabetes*, 4th ed. Chichester, UK, Blackwell Publishing, pp. 599–614.

Mather, H. M., J. A. Nisbet, G. H. Burton, G. J. Poston, J. M. Bland, P. A. Bailey, and T. R. Pilkington. 1979. Hypomagnesaemia in diabetes. *Clin Chim Acta* 95: 235–242.

McKenna, K., A. D. Morris, M. Ryan, R. W. Newton, B. M. Frier, P. H. Baylis, T. Saito, S. Ishikawa, and C. J. Thompson. 2000. Renal resistance to vasopressin in poorly controlled type 1 diabetes mellitus. *Am J Physiol Endocrinol Metab* 279: E155–E160.

McNair, P., M. S. Christensen, C. Christiansen, S. Madsbad, and I. Transbol. 1982. Renal hypomagnesaemia in human diabetes mellitus: Its relation to glucose homeostasis. *Eur J Clin Invest* 12: 81–85.

Miadovnik, L. A., M. C. Riddell, R. J. Gumieniak, C. P. Rowen, D. P. Zaharieva, and V. K. Jamnik. 2013. Effects of sport-specific, intermittent high-intensity exercise on post-exercise heart rate variability and glycemia in young athletes with type 1 diabetes. *Can J Diabetes* 37(Suppl 4): S51.

Mitchell, T. H., G. Abraham, A. Schiffrin, L. A. Leiter, and E. B. Marliss. 1988. Hyperglycemia after intense exercise in IDDM subjects during continuous subcutaneous insulin infusion. *Diabetes Care* 11: 311–317.

Mitrakou, A. 2011. Kidney: Its impact on glucose homeostasis and hormonal regulation. *Diabetes Res Clin Pract* 93 (Suppl 1): S66–S72.

Mogensen, C. E. 2002. Nephropathy: Early. In N. B. Ruderman, J. T. Devlin, S. H. Schneider, and A. M. Kriska (Eds.), *Handbook of Exercise in Diabetes*. Alexandria, VA, American Diabetes Association, pp. 433–449.

Moore, G. E., P. L. Painter, K. R. Brinker, J. Stray-Gundersen, and J. H. Mitchell. 1998. Cardiovascular response to submaximal stationary cycling during hemodialysis. *Am J Kidney Dis* 31: 631–637.

Navarro, X., W. R. Kennedy, and T. J. Fries. 1989. Small nerve fiber dysfunction in diabetic neuropathy. *Muscle Nerve* 12: 498–507.

Nicolis, G. L., T. Kahn, A. Sanchez, and J. L. Gabrilove. 1981. Glucose-induced hyperkalemia in diabetic subjects. *Arch Intern Med* 141: 49–53.

Nielsen, F. H., and H. C. Lukaski. 2006. Update on the relationship between magnesium and exercise. *Magnes Res* 19: 180–189.

Nuviala, R. J., M. G. Lapieza, and E. Bernal. 1999. Magnesium, zinc, and copper status in women involved in different sports. *Int J Sport Nutr* 9: 295–309.

Painter, P. L. 1988. Exercise in end-stage renal disease. *Exerc Sport Sci Rev* 16: 305–339.

Painter, P. L., L. Hector, K. Ray, L. Lynes, S. Dibble, S. M. Paul, S. L. Tomlanovich, and N. L. Ascher. 2002. A randomized trial of exercise training after renal transplantation. *Transplantation* 74: 42–48.

Painter, P. L., L. Hector, K. Ray, L. Lynes, S. M. Paul, M. Dodd, S. L. Tomlanovich, and N. L. Ascher. 2003. Effects of exercise training on coronary heart disease risk factors in renal transplant recipients. *Am J Kidney Dis* 42: 362–369.

Palmer, B. F. 2011. Metabolic complications associated with use of diuretics. *Semin Nephrol* 31: 542–552.

Perkins, B. A., D. Z. Cherney, H. Partridge, N. Soleymanlou, H. Tschirhart, B. Zinman, N. M. Fagan, S. Kaspers, H. J. Woerle, U. C. Broedl, and O. E. Johansen. 2014. Sodium-glucose cotransporter 2 inhibition and glycemic control in type 1 diabetes: Results of an 8-week open-label proof-of-concept trial. *Diabetes Care* 37: 1480–1483.

Perkins, B. A., and M. C. Riddell. 2006. Type 1 diabetes and exercise: Using the insulin pump to maximum advantage. *Can J Diabetes* 30: 71–79.

Perrone, C., O. Laitano, and F. Meyer. 2005. Effect of carbohydrate ingestion on the glycemic response of type 1 diabetic adolescents during exercise. *Diabetes Care* 28: 2537–2538.

Petrofsky, J. S., S. Lee, C. Patterson, M. Cole, and B. Stewart. 2005. Sweat production during global heating and during isometric exercise in people with diabetes. *Med Sci Monit* 11: CR515–521.

Pham, P. C., P. M. Pham, S. V. Pham, J. M. Miller, and P. T. Pham. 2007. Hypomagnesemia in patients with type 2 diabetes. *Clin J Am Soc Nephrol* 2: 366–373.

Pruijm, M., G. Wuerzner, M. Maillard, P. Bovet, C. Renaud, M. Bochud, and M. Burnier. 2010. Glomerular hyperfiltration and increased proximal sodium reabsorption in subjects with type 2 diabetes or impaired fasting glucose in a population of the African region. *Nephrol Dial Transplant* 25: 2225–2231.

Pun, K. K., and P. W. Ho. 1989. Subclinical hyponatremia, hyperkalemia and hypomagnesemia in patients with poorly controlled diabetes mellitus. *Diabetes Res Clin Pract* 7: 163–167.

Raebel, M. A. 2012. Hyperkalemia associated with use of angiotensin-converting enzyme inhibitors and angiotensin receptor blockers. *Cardiovasc Ther* 30: e156–e166.

Raile, K., A. Galler, S. Hofer, A. Herbst, D. Dunstheimer, P. Busch, and R. W. Holl. 2007. Diabetic nephropathy in 27,805 children, adolescents, and adults with type 1 diabetes: Effect of diabetes duration, A1C, hypertension, dyslipidemia, diabetes onset, and sex. *Diabetes Care* 30: 2523–2528.

Rand, S., J. S. Petrofsky, and G. Zimmerman. 2008. Diabetes: Sweat response and heart rate variability during electrical stimuulation in controls and people with diabetes. *J Appl Res* 8: 48–54.

Reboldi, G., G. Gentile, F. Angeli, and P. Verdecchia. 2011. Optimal therapy in hypertensive subjects with diabetes mellitus. *Curr Atheroscler Rep* 13: 176–185.

Riddell, M. C., and K. E. Iscoe. 2006. Physical activity, sport, and pediatric diabetes. *Pediatr Diabetes* 7: 60–70.

Riddell, M. C., and J. Milliken. 2011. Preventing exercise-induced hypoglycemia in type 1 diabetes using real-time continuous glucose monitoring and a new carbohydrate intake algorithm: An observational field study. *Diabetes Technol Ther* 13: 819–825.

Riddell, M. C., and B. A. Perkins. 2006. Type 1 diabetes and vigorous exercise: Applications of exercise physiology to patient management. *Can J Diabetes* 30: 63–71.

Roberts, C. K., J. P. Little, and J. P. Thyfault. 2013. Modification of insulin sensitivity and glycemic control by activity and exercise. *Med Sci Sports Exerc* 45: 1868–1877.

Rodenburg, E. M., E. J. Hoorn, R. Ruiter, J. J. Lous, A. Hofman, A. G. Uitterlinden, B. H. Stricker, and L. E. Visser. 2013. Thiazide-associated hyponatremia: A population-based study. *Am J Kidney Dis* 62: 67–72.

Rosenstock, J., S. A. Loizou, I. E. Brajkovich, K. Mashiter, and G. F. Joplin. 1982. Effect of acute hyperglycaemia on plasma potassium and aldosterone levels in type 2 non-insulin-dependent diabetes. *Diabetologia* 22: 184–187.

Running USA. 2013. *2013 State of the Sport–Part III: U.S. Race Trends*. Retrieved May 29, 2014, from http://www.runningusa.org/state-of-sport-2013-part-III?returnTo=annual-reports.

Running USA. 2014. *2014 State of the Sport–Part I: Non-Traditional Running Events*. Retrieved May 29, 2014, from http://www.runningusa.org/state-of-sport-nontraditional?returnTo=annual-reports.

Rx Files. 2013. *Rx Files Drug Comparison Charts*, 9th ed. B. Jensen and L. D. Regier (Eds.). Saskatoon, SK, Rx Files.

Sales, C. H., and F. Pedrosa Lde. 2006. Magnesium and diabetes mellitus: Their relation. *Clin Nutr* 25: 554–562.

Sasson, A. N., and D. Z. Cherney. 2012. Renal hyperfiltration related to diabetes mellitus and obesity in human disease. *World J Diabetes* 3: 1–6.

Sherwood, L., R. Kell, and C. Ward. 2013. *Human Physiology: From Cells to Systems*, Second Canadian ed. Ontario, Canada, Nelson Education.

Sigal, R. J., M. J. Armstrong, P. Colby, G. P. Kenny, R. C. Plotnikoff, S. M. Reichert, and M. C. Riddell. 2013. Canadian Diabetes Association clinical practice guidelines: Physical activity and diabetes. *Can J Diabetes* 37: S40–44.

Sigal, R. J., S. J. Fisher, J. B. Halter, M. Vranic, and E. B. Marliss. 1999. Glucoregulation during and after intense exercise: Effects of beta-adrenergic blockade in subjects with type 1 diabetes mellitus. *J Clin Endocrinol Metab* 84: 3961–3971.

Sigal, R. J., G. P. Kenny, D. H. Wasserman, and C. Castaneda-Sceppa. 2004. Physical activity/exercise and type 2 diabetes. *Diabetes Care* 27: 2518–2539.

Sigal, R. J., C. Purdon, S. J. Fisher, J. B. Halter, M. Vranic, and E. B. Marliss. 1994. Hyperinsulinemia prevents prolonged hyperglycemia after intense exercise in insulin-dependent diabetic subjects. *J Clin Endocrinol Metab* 79: 1049–1057.

Simkova, R., L. Kazdova, L. Karasova, S. Simek, and T. Pelikanova. 2004. Effect of acute hyperglycaemia on sodium handling and excretion of nitric oxide metabolites, bradykinin, and cGMP in type 1 diabetes mellitus. *Diabet Med* 21: 968–975.

Sjogren, A., C. H. Floren, and A. Nilsson. 1986. Magnesium deficiency in IDDM related to level of glycosylated hemoglobin. *Diabetes* 35: 459–463.

Skott, P., O. Hother-Nielsen, N. E. Bruun, J. Giese, M. D. Nielsen, H. Beck-Nielsen, and H. H. Parving. 1989. Effects of insulin on kidney function and sodium excretion in healthy subjects. *Diabetologia* 32: 694–699.

Stapleton, J. M., J. E. Yardley, P. Boulay, R. J. Sigal, and G. P. Kenny. 2013. Whole-body heat loss during exercise in the heat is not impaired in type 1 diabetes. *Med Sci Sports Exerc* 45: 1656–1664.

Stenvinkel, P., J. Bolinder, and A. Alvestrand. 1992. Effects of insulin on renal haemodynamics and the proximal and distal tubular sodium handling in healthy subjects. *Diabetologia* 35: 1042–1048.

Strazzullo, P., G. Barba, F. P. Cappuccio, A. Siani, M. Trevisan, E. Farinaro, E. Pagano, A. Barbato, R. Iacone, and F. Galletti. 2001. Altered renal sodium handling in men with abdominal adiposity: A link to hypertension. *J Hypertens* 19: 2157–2164.

Strazzullo, P., A. Barbato, F. Galletti, G. Barba, A. Siani, R. Iacone, L. D'Elia, O. Russo, M. Versiero, E. Farinaro, and F. P. Cappuccio. 2006. Abnormalities of renal sodium handling in the metabolic syndrome: Results of the Olivetti Heart Study. *J Hypertens* 24: 1633–1639.

Turner, G., P. Coates, S. Warren, J. S. Woodhead, and J. R. Peters. 1997. Proximal tubular reabsorption of growth hormone and sodium/fluid in normo- and microalbuminuric insulin-dependent diabetes mellitus. *Acta Diabetol* 34: 27–32.

van Dijk, J. W., R. J. Manders, K. Tummers, A. G. Bonomi, C. D. Stehouwer, F. Hartgens, and L. J. van Loon. 2012. Both resistance- and endurance-type exercise reduce the prevalence of hyperglycaemia in individuals with impaired glucose tolerance and in insulin-treated and non-insulin-treated type 2 diabetic patients. *Diabetologia* 55: 1273–1282.

Vehik, K., and D. Dabelea. 2011. The changing epidemiology of type 1 diabetes: Why is it going through the roof? *Diabetes Metab Res Rev* 27: 3–13.

Vora, J. P., J. Dolben, J. D. Dean, D. Thomas, J. D. Williams, D. R. Owens, and J. R. Peters. 1992. Renal hemodynamics in newly presenting non-insulin dependent diabetes mellitus. *Kidney Int* 41: 829–835.

Waden, J., H. K. Tikkanen, C. Forsblom, V. Harjutsalo, L. M. Thorn, M. Saraheimo, N. Tolonen, M. Rosengard-Barlund, D. Gordin, H. O. Tikkanen, P. H., and Groop on behalf of the

FinnDiane Study. 2015. Leisure-time physical activity and development and progression of diabetic nephropathy in type 1 diabetes: The FinnDiane Study. *Diabetologia*. 58: 929–936.

Whiting, D. R., L. Guariguata, C. Weil, and J. Shaw. 2011. IDF diabetes atlas: Global estimates of the prevalence of diabetes for 2011 and 2030. *Diabetes Res Clin Pract* 94: 311–321.

World Anti-Doping Agency. 2014a. *The 2014 Monitoring Program*. Montreal, Canada, World Anti-Doping Agency.

World Anti-Doping Agency. 2014b. *The World Anti-Doping Code: The 2014 Prohibited List, International Standards*. Montreal, Canada, World Anti-Doping Agency.

Wortsman, J. 2002. Role of epinephrine in acute stress. *Endocrinol Metab Clin North Am* 31: 79–106.

Yardley, J., R. Mollard, A. Macintosh, F. Macmillan, B. Wicklow, L. Berard, C. Hurd, S. Marks, and J. McGavock. 2013. Vigorous intensity exercise for glycemic control in patients with type 1 diabetes. *Can J Diabetes* 37: 427–432.

Yardley, J. E., D. P. Zaharieva, C. Jarvis, and M. C. Riddell. 2015. The "ups" and "downs" of a bike race in people with type 1 diabetes: Dramatic differences in strategies and blood glucose responses in the Paris-to-Ancaster Spring Classic. *Can J Diabetes* 39: 105–110.

Yisahak, S. F., J. Beagley, I. R. Hambleton, and K. M. Narayan. 2013. Diabetes in North America and the Caribbean: 2013 update for the IDF Diabetes Atlas. *Diabetes Res Clin Pract*. 103: 223–230.

Zaharieva, D. P., and M. C. Riddell. 2015. Prevention of exercise-associated dysglycemia: A case study-based approach. *Diabetes Spectr* 28: 55–62.

13 Athletes with Chronic Conditions
Obesity

Flavia Meyer, Paulo L. Sehl, and Emily Haymes

CONTENTS

13.1 INTRODUCTION

Most athletes and physically active people are lean but they gain weight and fat sometimes due to the aging process or if no longer in competition. Indeed, there appears to be an increased prevalence of overweight and obesity among athletes or individuals who participate in sports events that follows the overall tendency in the general population. For example, among American football high school athletes approximately 45% are overweight or obese (Malina et al. 2007). Athletes from some sport modalities such as American football sometimes perceive a larger body size as advantageous; however, if it is accompanied by a greater fat mass it becomes detrimental. Increased body fat may impair performance (speed, agility, and endurance), impair thermoregulatory responses, and increase risk of injuries (Kerr et al. 2013). Long-distance or marathon swimmers may also benefit from extra subcutaneous fat for buoyance, body position, or to act as an insulation layer to avoid hypothermia when in cooler water (Holmer and Bergh 1974). However, marathon swimmers may face difficulties in thermoregulation when they shift to dry-land warm environments.

Compared to lean, obese individuals may respond with a greater rise in body temperature under a given heat/exercise stress, and consequently, they may be at a greater risk of exertional heat illness (American Academy of Pediatrics 2011; Bedno et al. 2010; Yard et al. 2010). A survey (Kerr et al. 2013) of American high school sports observed that the risk of exertional heat illness increased by 1.7 times among obese football players. Therefore, increased body fat and obesity is a concern in athletes who wear heavy uniforms and train/compete under heat conditions.

Earlier studies, mostly in non-athletic individuals, have tried to explain impairment of thermoregulatory processes based on intrinsic characteristics, which would be associated with greater adipose tissue, such as geometric characteristics, lower sweat volume, (Bar-Or et al. 1969; Dougherty et al. 2009; Haymes et al. 1975), and lower skin blood flow (Vroman et al. 1983). Recent studies (Leites et al. 2013 and Sehl et al. 2012) have proposed other extrinsic factors that are usually associated with obesity, such as physical inactivity, impaired fitness, and poor heat acclimatization. For example, when exercising in the heat, core temperature increase is lower in aerobically fit young adults compared to their unfit peers (Buono and Sjoholm 1988). This may be due to adaptations in sweating and cutaneous circulation to improve thermoregulation.

This chapter will first provide information regarding the evolution of the studies that compared obese and non-obese individuals from thermoregulatory and sweating outcomes. Then, we will discuss intrinsic and extrinsic factors that could explain different responses in obese individuals. Finally, we will present some recommendations since professionals and athletes should be aware of how much obesity could impair their thermoregulatory and body fluid responses to exercise. Research examining the acute thermoregulatory responses and from a fluid balance angle in obese individuals is lacking.

13.2   OBESITY AND EXERCISE FROM A THERMOREGULATORY AND FLUID BALANCE ANGLE: STUDY ADVANCES

Most studies on obesity and exercise have focused on the energy balance equation and the effect of energy expenditure and adiposity. Less research has looked at acute physiologic response particularly with regard to thermoregulatory responses or the fluid balance equation.

The characteristics of the 8 studies that compared obese and non-obese subjects during exercise that aimed at thermoregulatory outcomes are listed in chronological order in Table 13.1.

In the 1960s, the Human Performance Laboratory at Penn State University began a series of studies examining the effects of heat stress on obese and non-obese/lean subjects during exercise (Bar-Or et al. 1968, 1969; Buskirk et al. 1965). The first study (Buskirk et al. 1965) showed greater skin temperature, total heat body content, and lower evaporative heat loss in obese women walking in the heat. In a subsequent study, Bar-Or et al. (1969) also observed greater increase in mean skin temperatures and lower sweat rates per unit of body surface in obese *versus* lean women while walking (six sessions) at range of conditions from neutral (~21°C) to hot (35°C) environments. Rectal temperature increase was more accentuated in the obese in the thermoneutral, but not in the hottest environments (32.2°C and 35°C). These two earlier studies indicated that under some conditions obese women might respond differently than lean ones when exercising in the heat.

Later, Haymes et al. (1974) evaluated 9- to 11-year-old girls while walking (4.8 km/h, 5% grade) three times 20-min bouts in four sessions that varied from neutral (21.1°C) to hot (32.2°C) conditions. Heavy but not obese (fat = ~25%) and lean (fat = ~14%)

TABLE 13.1

Characteristics and Outcomes of Studies Listed by Chronological Order That Compared Obese and Non-Obese Subjects during Exercise

Study (Year)	N	Subjects: Compared to Non-Obese — Aerobic Fitness $\dot{V}O_{2max}$/Peak	Heat-Acclimation Sessions	Exercise/Climate Protocol — Intensity: Fixed (F) or Relative (%$\dot{V}O_{2peak}$)/Duration (min)	Air Temperature, %RH	Outcome: Compared to Non-Obese — T_{core} Baseline	ΔT_{core} or Final T_{core}	ΔT_{skin}	ΔHR or Final HR	Sweat Rate
Miller and Blyth, 1957	Men: Obese = 14 Non-obese = 14	N/A	Yes	F: walk 4.8 km h⁻¹/45	~50°C, ~30% RH	N/A	=	=	=	=
				F: walk 6.4 km h⁻¹/60	25°C		=	←	=	→
Buskirk et al., 1965	Women: Overweight = 4 Non-obese = 3	N/A	Yes 10 sessions	F: walk 4.8 km h⁻¹ 3–20 min bouts	~46°C dry bulb, ~27°C wet bulb	N/A	←	←	=	←
Bar-Or et al., 1969	Women: Obese = 5 Non-obese = 4	N/A	Partially Only 3 sessions	F: walk 4.8 km h⁻¹ 3–20 min bouts	26°C–29°C	N/A	←	←	←	←
					32°C–35°C		→	←	←	←
Haymes et al., 1974	Girls: Heavy = 7 Non-obese = 5	↓relative to kg	Partially Only 3 sessions	F: walk 4.8 km h⁻¹ 3–20 min bouts	21°C–29°C	N/A	=	→	N/A	=
					32.2°C		→	→	N/A	=
Dougherty et al., 2009	Boys: Obese = 7 Non-obese = 7	↓relative to kg	Yes 6 sessions	R: walk + cycle 30% $\dot{V}O_{2peak}$ 3–20 min bouts	38°C, 50% RH	↑	=	N/A	=	→

(Continued)

TABLE 13.1 (Continued)

Characteristics and Outcomes of Studies Listed by Chronological Order That Compared Obese and Non-Obese Subjects during Exercise

Study (Year)	N	Subjects: Compared to Non-Obese		Exercise/Climate Protocol		Outcome: Compared to Non-Obese				
		Aerobic Fitness VO_{2max}/Peak	Heat-Acclimation Sessions	Intensity: Fixed (F) Or Relative (%VO_{2peak})/Duration (min)	Air Temperature, %RH	T_{core} Baseline	ΔT_{core} or Final T_{core}	ΔT_{skin}	ΔHR or Final HR	Sweat Rate
Sehl et al., 2012	Boys: Obese = 17 Non-obese = 16	= allometric	Naturally acclimatized	R: cycle ~50% VO_{2peak}/30	35°C, 45% RH	=	=	N/A	=	=
Leites et al., 2013	Girls: Obese = 13 Non-obese = 14	= relative to muscle mass	Naturally acclimatized	R: cycle ~55% VO_{2peak}/30	24°C, 50% RH 35°C, 40% RH	↑ ↑	= ↓	N/A N/A	=	=
Adams et al., 2014	Women: Obese = 10 Non-obese = 10	= relative to lean body mass	Not heat-acclimatized	F: cycle 300 W (HP) or 175 W m⁻²/60	40°C, 30% RH	=	=	=	=	=

N/A, not available; =, similar; ↑, greater; ↓, lower.

T_{core}, core temperature; T_{skin}, skin temperature; HR, heart rate (beats min⁻¹); RH, relative humidity; W, watts; HP, heat production.

girls showed similar rectal temperature at lower heat stress (21.1°C–29.4°C) but the lean girls had a greater rectal temperature at the hottest condition. In boys (Haymes et al. 1975), the same exercise-heat protocol showed higher rectal temperatures and heart rates in obese (fat ~31%) than in the lean (fat ~16%) boys, independent of the climate. These above-mentioned studies indicated some gender difference in thermoregulatory responses when comparing obese and lean children.

Subsequent studies (Dougherty et al. 2009; Leites et al. 2013; Sehl et al. 2012) by using cycling instead of walking/running eliminated the possibility of inducing a greater metabolic and heat production just by a greater effort of carrying extra load for the obese subjects (Butte et al. 2007; Falk 1998). In addition, exercise protocols in these studies (Dougherty et al. 2009; Leites et al. 2013; Sehl et al. 2012) attempted to set an individual calculated workload to result in similar relative efforts (i.e., percentage of $\dot{V}O_{2peak}$). Also, the sweating volume has been corrected by body surface area to eliminate differences in relation to body mass.

More recent studies (Adams et al. 2014; Cramer and Jay 2014; Jay et al. 2011) proposed that when comparing groups of different mass and surface area (in this case obese vs. non-obese), the heat/exercise protocol to evaluate thermoregulatory effectiveness (using core temperature change and sweating as parameters) should generate the same metabolic heat production per unit of surface area. They argue that core temperature change and sweating responses are reflections of the amount of heat production rather than a given exercise effort. The rationale is that by submitting groups to a fixed heat production corrected by body mass (W/kg), we may isolate the independent influence of adiposity and avoid experimental bias when evaluating heat dissipation. Therefore, the recent and upcoming studies that adopt such experimental fixed heat production protocol should clarify whether thermoregulatory effectiveness is impaired by adipose tissue in excess.

13.3 FACTORS RELATED TO THERMOREGULATION AND FLUID BALANCE IN OBESE INDIVIDUALS

The mechanisms by which obese, compared to non-obese, individuals may differ on the magnitude of thermoregulatory and fluid balance responses during exercise, particularly in the heat, are summarized in Table 13.2. Factors are most likely interrelated; however, for illustrative purpose we classified them as intrinsic and extrinsic. In this text, intrinsic factors are those independent factors purely related to the greater amount of body fat that may influence heat gain, loss, and transport such as fat-specific heat, geometric, and blood flow aspects. On the other hand, extrinsic factors are those related to lifestyle and environmental factors that usually accompany obesity such as physical activity, fitness level, and acclimatization.

13.3.1 INTRINSIC FACTORS

Initial factors to explain thermoregulatory disadvantages of obese individuals while exercising in the heat were those related to their condition of having a greater amount of adipose tissue and the geometric concern related to their ratio of body surface area to body mass (BSA:BM). The lower thermal conductivity of fat in relation to fat-free

TABLE 13.2
Intrinsic and Extrinsic (Compounded) Factors Related to Thermoregulation and Water Balance during Exercise: Obese Compared to Non-Obese Subjects

Factors		Compared to Non-Obese	Implication
Intrinsic	Fat-specific heat	↓	For the same amount of heat produced, a greater rise in adipose tissue temperature is expected
	Body water content of fat	↓	For the same hypohydration status, water deficit relative to total body water may be greater for the obese
	BSA/BM ratio	↓	Lower rate of heat loss produced by exercise
			Lower rate of heat gain in hot temperatures ($T_{ambient} > T_{skin}$)
	Blood flow	↓	Slower heat transfer from core (muscle) to the periphery (skin)
Extrinsic	Aerobic fitness	↓ (Mostly due to physical inactivity)	For the same exercise/heat stress: Lower sweat rate and capacity for evaporative cooling
	Heat-acclimatization	↓ (Mostly due to lack of heat exposure and/or living in temperate climate)	Increased rise of T_{core}
	Metabolic cost for locomotion (walking/running)	↑	Greater relative effort (%VO_{2max}) Increased rise of T_{core} Lower tolerance to exercise, especially when aerobic unfit and not heat-acclimatized

BSA, body surface area; BM, body mass; T_{core}, core temperature; T_{skin}, skin temperature; $T_{ambient}$, ambient temperature; ↓, lower than in non-obese; ↑, greater than in non-obese.

mass (0.4 vs. 0.82 kcal g^{-1} C^{-1}) implies that the magnitude of temperature increase of fat is higher under a given heat production.

Overall, obese subjects present a lower body surface area in relation to body mass. Unless in situations when ambient temperature is greater than skin temperature and less surface will be available to gain heat, the lower BSA:BM ratio may limit body heat release from the skin to the ambient (Bar-Or and Rowland 2004; Haymes et al. 1974). In addition, the subcutaneous adipose tissue may aggravate by acting as a thermic isolator (Savastano et al. 2009).

A lower blood flow that may be present in obese individuals is another factor, which could slower heat transfer from core (muscle) to the periphery (skin). A study by Vroman et al. (1983) showed that peripheral blood flow of obese (fat ~32%) young men while cycling at intensities from 30% to 70% $\dot{V}O_{2peak}$ in the heat (dry bulb = 38°C, wet bulb = 20°C) or thermoneutral (dry bulb = 22°C, wet bulb = 14°C) conditions was about 27% lower compared to their lean (fat ~16%) counterparts. In this study, however, esophageal and skin temperatures were similar between groups. Another study (Karpoff et al. 2009) showed that obese prepubertal boys had a lower blood flow compared with lean in response to a dynamic knee-extensor exercise. A challenge for future studies is to evaluate skin blood flow during the exercise protocol in order to clarify whether it is slower in overweighed/obese individuals and therefore one possible mechanism that could impair central heat transfer to periphery.

13.3.2 Extrinsic Factors

Physical aerobic training affects sweating and therefore thermoregulation and fluid balance. A study (Buono and Sjoholm 1988) showed that endurance trained men and women had a greater sweating rate as compared to their untrained counterparts. Sato and Sato (1983) suggested that training induces sweat gland hypertrophy and an increase in the cholinergic sensitivity that stimulates sweating. But the modality and training environment may still affect sweating. For example, endurance swimmers and runners may not have the same degree of sweating adaptations, as training and competing in the water help dissipate body heat production through convection and conduction. Henkin et al. (2010) showed that when young adult swimmers cycled (on land) at similar relative intensity warm conditions, they do not sweat as much as runners during a 30-min period (0.45 vs. 0.75 L, respectively) even when correcting for body mass in mL per kg (~6 vs. 10, respectively). Heat exposure while living under heat stress, or moving from cold to warm conditions, also produces an increase in sweating rate (Armstrong and Maresh 1991). Therefore, adjustments in volumes of fluid intake should not only be considered according to the sport modality and fitness level, but also along the training season or timing of heat exposure.

Because physical training may be accompanied by heat acclimatization, it is difficult to separate their effects. Heat acclimatization is a series of adaptations that occur in response to a natural exposure to warm environment and acclimation is the same process, but it occurs when adaptations happen in controlled sessions of exposure to heat or heat combined with exercise over a course of 7–14 days. These adaptive responses are aimed to optimize body heat loss, and therefore, they are accompanied by sweat rate. The heat acclimatization process, not necessarily accompanied by physical training, can also affect sweating by anticipating its onset and increasing the volume (Cheung and McLeland 1988; Gisolfi and Robinson 1969; Havenith and van Middendorp 1990).

Sweat sodium and chloride concentration seems to decrease with acclimatization perhaps due to an increase in the aldosterone level that absorbs sodium in the sweat gland duct. Thus, the combination of aerobic training and heat acclimatization could probably optimize these sweat adaptations. There is some indication that acclimation process may be slower in obese compared to non-obese boys. Dougherty et al. (2009) observed that obese boys did not increase as much their sweat rate or had attenuated their increase in core temperature as much as non-obese boys after a acclimation process that consisted of six exercise sessions in the heat. However, a confounder factor could have been that the obese started the acclimation process at a lower aerobic fitness.

Two studies (Leites et al. 2013 and Sehl et al. 2012) attempted to compare sweating and thermoregulatory responses between obese and non-obese children that were

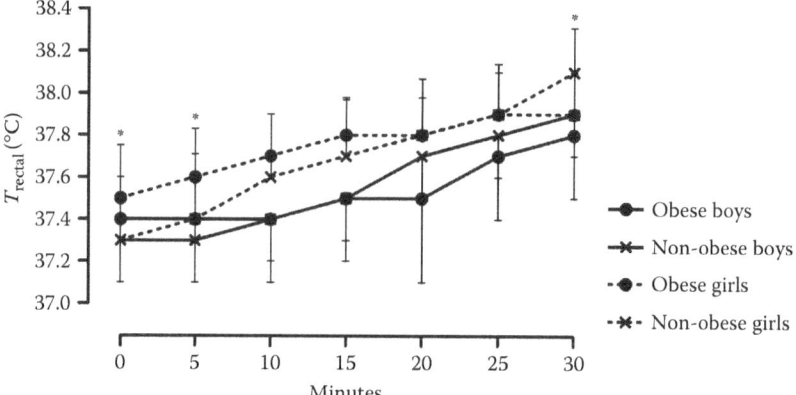

FIGURE 13.1 Rectal temperature response during exercise in obese ($n = 17$) and non-obese ($n = 16$) pubescent (12- to 15-year-old) boys; and obese ($n = 13$) and non-obese ($n = 14$) pre-pubescent (9-year-old) girls. Exercise consisted of 30 min cycling at 50%–55% $\dot{V}O_{2peak}$ in the heat (35°C, 40% relative humidity). *$p < .05$. (Data from Sehl, P.L. et al., *Int. J. Sports Med.*, 33, 497–501, 2012; Leites, G.T. et al., *J. Pediatr.*, 162, 1054–1060, 2013.)

physically active, equally fit, and naturally, heat acclimatized as experiments occurred during summer months in South of Brazil. Boys and girls cycled 30 min at 50%–55% $\dot{V}O_{2peak}$ in the heat (35°C, 40% relative humidity). None of the studies observed difference in sweating rate corrected by body surface area between obese and non-obese boys (Sehl et al. 2012) or girls (Leites et al. 2013). As illustrated in Figure 13.1 from data redrawn from these two studies, in boys the rectal temperature from the start to the end of cycling was similar between the obese and non-obese. In girls, the rectal temperature at the start was higher in the obese, but during exercise the increase was greater among the non-obese girls and actually surpassed that of the obese girls at the end of cycling.

A greater discomfort and perceived exertion of obese individuals while exercising in the heat may indirectly assume that they are at thermoregulatory disadvantage. The obese boys, compared to non-obese, in the study of Sehl et al. (2012) reported a greater heat sensation during the whole 30-min cycling and a greater perceived effort (Borg Scale) in the final 5 min. Obese and non-obese boys presented similar aerobic fitness and heat acclimatization and responded with similar rectal temperature while exercising in the heat. Therefore, a higher perceived effort and fatigue in individuals with greater adiposity levels while exercising in the heat should be considered and may impair motivation and adhesion to training programs.

13.4 FINAL REMARKS AND RECOMMENDATIONS

In summary, research limitation in matching to obese and non-obese individuals by their aerobic fitness and acclimatization may be to limit the knowledge on how much the abovementioned intrinsic factors is impaired by the greater adiposity level. However—independent of the underlying mechanism—current data indicate that extra attention and recommendations should be given to athletes and physically active people who are obese or have greater body fat while training/competing under heat conditions and are depended on their evaporative sweating and thermoregulatory efficiency. The amount and type of clothing worn by the obese individual will affect the transfer of body heat from the skin to the external environment. In warm and hot environments, lightweight clothing will facilitate heat transfer from the skin to the air and evaporation of sweat. Adequate fluid replacement during exercise and recovery is also critical to attenuate body temperature increase in warm environments.

Thermoregulatory impairments and body fluid balance during exercise in individuals with increased adiposity should be identified to avoid any disadvantage on performance and health. Athletic trainers, healthcare professionals, family members, and the athlete should recognize situations when increased body fat represents an extra challenge and causes thermoregulatory risks. Examples of such situations are as follows:

- Athletes who gain body fat and intensify training in order to increase energy expenditure and reduce adiposity. This situation may be aggravated at the start of a training season during hot and humid weather, when athletes are not yet heat acclimated. Another serious misconception is to make athletes sweat heavily to induce dehydration and give a false impression of body (fat) mass lost. On the contrary, dehydration should precipitate fatigue and stop exercise earlier in addition to increased risk of exertional heat illness.
- Long-distance swimmers who shift training to dry land and hot environments.
- Heavy American football players who train or compete under heat stress.
- Prolonged athletic events, such a marathon, when we observe a heterogeneous group, including overweight and obese runners.

REFERENCES

Adams, J.D., M.S. Ganio, J.M. Burchfield et al. 2015. Effects of obesity on body temperature in otherwise-healthy females when controlling hydration and heat production during exercise in the heat. *Eur. J. Appl. Physiol.* 115:167–76.

American Academy of Pediatrics. 2011. Policy statement—Climatic heat stress and exercising children and adolescents. *Pediatr.* 128:1–7.

Armstrong, L.E., C.M. Maresh. 1991. The induction and decay of heat acclimatization in trained athletes. *Sports Med.* 12:302–12.

Bar-Or, O., H.M. Lundegren, E.R. Buskirk et al. 1969. Heat tolerance of exercising obese and lean women. *J. Appl. Physiol.* 26:403–9.

Bar-Or, O., L.I. Magnusson, E.R. Buskirk et al. 1968. Distribution of heat-activated sweat glands in obese and lean men and women. *Human Biol.* 40:235–48.

Bar-Or, O., T.W. Rowland. 2004. Climate, body fluids and the exercising child. In: Bar-Or, O., Rowland, T.W., eds. *Pediatric Exercise Medicine: From Physiologic Principles to Health Care Application.* Human Kinetics Publishing, pp. 69–101.

Bedno, S.A., Y. Li, W. Han et al. 2010. Exertional heat illness among overweight U.S. Army recruits in basic training. *Aviat Space Environ. Med.* 81:107–11.

Buono, M.J., N.T. Sjoholm. 1988. Effect of physical training on peripheral sweat production. *J. Appl. Physiol.* 65:811–14.

Buskirk, E.R., H.M. Lundegren, L.I. Magnusson et al. 1965. Heat acclimatization pattern in obese and lean individuals. *Ann. N Y Acad. Sci.* 131:637–53.

Butte, N.F., M.R. Puyau, F.A. Vohra et al. 2007. Body size, body composition, and metabolic profile explain higher energy expenditure in overweight children. *J. Nutr.* 137:2660–7.

Cheung, S.S., T.M. McLellan. 1988. Heat acclimation, aerobic fitness, and hydration effects on tolerance during uncompensable heat stress. *J. Appl. Physiol. (1985).* 84:1731–9.

Cramer, M.N., O. Jay. 2014. Selecting the correct exercise intensity for unbiased comparisons of thermoregulatory responses between groups of different mass and surface area. *J. Appl. Physiol.* 116(9):1123–32.

Dougherty, K.A., M. Chow, W.L. Kenney. 2009. Responses of lean and obese boys to repeated summer exercise in the heat bouts. *Med. Sci. Sports Exerc.* 41:279–89.

Dougherty, K.A., M. Chow, W.L. Kenney. 2010. Critical environmental limits for exercising heat-acclimated lean and obese boys. *Eur. J. Appl. Physiol.* 108:779–89.

Falk, B. 1998. Effects of thermal stress during rest and exercise in the pediatric population. *Sports Med.* 25:221–40.

Gisolfi, C., S. Robinson. 1969. Relations between physical training, acclimatization, and heat tolerance. *J. Appl. Physiol.* 26:530–34.

Havenith, G., H. van Middendorp. 1990. The relative influence of physical fitness, acclimatization state, anthropometric measures and gender on individual reactions to heat stress. *Eur. J. Appl. Physiol. Occup. Physiol.* 61:419–27.

Haymes, E.M., E.R. Buskirk, J.L. Rodgson et al. 1974. Heat tolerance of exercising lean and heavy prepubertal girls. *J. Appl. Physiol.* 36:566–71.

Haymes, E.M., R.J. McCormick, E.R. Buskirk. 1975. Heat tolerance of exercising lean and obese prepubertal boys. *J. Appl. Physiol.* 39:457–61.

Henkin, S.D., P.L. Sehl, F. Meyer. 2010. Sweat rate and electrolyte concentration in swimmers, runners, and nonathletes. *Int. J. Sports Physiol. Perform.* 5:359–66.

Holmér, I., U. Bergh. 1974. Metabolic and thermal responses to swimming in water at varying temperature. *J. Appl. Physiol.* 35:702–5.

Jay, O., A.R. Bain, T.M. Deren et al. 2011 Large differences in peak oxygen uptake do not independently alter changes in core temperature and sweating during exercise. *Am. J. Physiol. Regul. Integr. Comp. Physiol.* 301:R832–41.

Karpoff, L., A. Vinet, I. Schuster et al. 2009. Abnormal vascular reactivity at rest and exercise in obese boys. *Eur. J. Clin. Invest.* 39:94–102.

Kerr, Z.Y., D.J. Casa, S.W. Marshall et al. 2013. Epidemiology of exertional heat illness among U.S. High School Athletes. *Am. J. Prev. Med.* 44:8–14.

Leites, G.T., P.L. Sehl, G.S. Cunha et al. 2013. Responses of obese and lean girls exercising under heat and thermoneutral conditions. *J. Pediatr.* 162:1054–60.

Malina, R.M., P.J. Morano, M. Barron et al. 2007. Overweight and obesity among youth participants in American football. *J. Pediatr.* 151:378–82.

Miller, A.T., C.S. Blyth. 1958. Lack of insulating effect of body fat during exposure to internal and external heat loads. *J. Appl. Physiol.* 12:17–19.

Sato K., F. Sato. 1983. Individual variations in structure and function of human eccrine sweat gland. *Am. J. Physiol. Regul. Integr. Comp. Physiol.* 245:203–8.

Savastano, D.M., A.M. Gorbach, H.S. Eden et al. 2009. Adiposity and human regional body temperature. *Am. J. Clin. Nutr.* 90:1124–31.

Sehl, P.L., G.T. Leites, J.B. Martins et al. 2012. Responses of obese and non-obese boys cycling in the heat. *Int. J. Sports Med.* 33:497–501.

Veltmeijer, M.T.W., T.M.H. Eijsvogels, D.H.J. Thijssen et al. 2010. Obesity and the risk of water and electrolyte imbalances during prolonged exercise. *Med. Sci. Sports Exerc.* 42:111.

Vroman, N.B., E.R. Buskirk, J.L. Rodgson. 1983. Cardiac output and skin flow in lean and obese individuals during exercise in the heat. *J. Appl. Physiol.* 55:69–74.

Yard E.E., J. Gilchrist, T. Haileyesus et al. 2010. Heat illness among high school athletes—United States, 2005–2009. *J. Safety Res.* 41:471–4.

14 Athletes with Chronic Conditions
Hypertension

François Carré

CONTENTS

14.1 INTRODUCTION

Systemic hypertension is the most common cardiovascular condition in adults and even if it is generally asymptomatic, it is one of the key risk factors for cardiovascular disease (Mancia et al. 2013). Although its prevalence is reported to be lower in athletes than in the general population, it is also the most common cardiovascular risk detected during sport pre-participation evaluation (Leddy et al. 2009; Lehmann et al. 1990). Thus in athletes, as in general population, proven hypertension must be treated to prevent its deleterious health effects.

Focusing on the particularities of the athletic population, this chapter reviews the acute and chronic responses of physical exercise on blood pressure (BP) levels, evaluation, and treatment. It also pointed out some specific and practical issues regarding the interaction between some common prescribed medications, diet, and exercise on body fluid balance and performance.

14.2 ACUTE RESPONSES OF SYSTEMIC BLOOD PRESSURE DURING PHYSICAL EXERCISE

The mean BP that is the controlled variable of the systemic circulation is the product of cardiac output and peripheral resistance. During intense physical effort, the measurement of BP is technically difficult, especially on the treadmill. In addition, the commonly used conventional exercise protocols used have been proposed to detect coronary artery disease and not to assess the adaptation of BP in the effort. Indeed, the recommended exercise test duration that is 12–15 min requires successive increased steps of short stages (1–3 min), so described as progressive and maximal load until exhaustion. That is far from the sport duration on the field. These protocols do not take into account the inertia adaptation of the peripheral resistance that is the major parameter of the BP adaptation. Thus, the protocols used in medical laboratory to study BP adaptation to exercise are far removed from constraints of sport participation on the field (Carré 2013).

The two types of physical exercise, dynamic and static, present different BP responses. The dynamic exercise is characterized by alternating periods of muscular contraction–relaxation. During dynamic exercise, cardiac output increases and peripheral resistance decreased. This decrease appears after 3–5 min of exercise and is proportional to the intensity of the exercise. Cardiac output increases proportionally more than the decrease in peripheral resistance. Thus, systolic BP must rise during exercise linearly with the exercise intensity. The diastolic BP varies little. After exercise, the return to pre-exercise BP values needs 6–10 min, and due to the persistent vasodilation, both systolic and diastolic BP values remain frequently below the baseline BP values for several hours (Carré 2013; Haliwill et al. 2013; Taylor-Tolbert et al. 2000). During static exercise, the muscle contracts against a stable force and does not change its length. The exercise is usually of short duration, with little increase in cardiac output and no significant vasodilation. Thus, both systolic and diastolic BP increase more than during dynamic exercise, especially in case of the Valsalva maneuver often associated. When effort stops, the sudden fall in BP explains the post-exertional syncope sometimes reported in weightlifters (Carré 2103).

Finally, most sport combines the dynamic and static components. During these mixed sports, the rise in BP is intermediate to what is observed in pure dynamic or static exercise. For example, BP increase is more accentuated during riding than during running, because of the greater vasodilation during running.

The BP increase level observed is proportional to the muscle mass used and to the percentage of maximal voluntary force requested by the static part of exercise (Carré 2013).

Considering other parameters of continuous and stable exercise, the intensity plays a role on the acute BP adaptations during exercise. After an initial rise, the systolic BP decreases progressively with the duration of the exercise of constant intensity because of the increase in vasodilation. The decrease in BP is more accentuated when the ambient temperature is high (>15°C) due to the decrease in blood volume explained by dehydration (Carré 2013; Vu Le et al. 2008). Finally, the characteristics of the exercising person such as age, gender, level of training, pathology, and treatment are also involved in these adaptations. Most common sports have been classified according to their dynamic and static components (see Table 14.1).

What are the highest BP values that can be accepted during physical exertion? There is no current good answer to this frequently asked question. We described before the limits of laboratory exercise BP study. It is therefore logical to ask whether measuring BP during a maximum effort level is justified in an athlete

TABLE 14.1
Classification of Sports in Accordance with Their Dynamic and Static Components

	A. Low Dynamic	B. Moderate Dynamic	C. High Dynamic
I. Low static	Bowling	Fencing	Badminton
	Cricket	Table tennis	Race walking
	Golf	Tennis (doubles)	Running (marathon)
	Riflery	Volleyball	Cross country skiing (classic)
		Baseball[a]/softball[a]	Squash[a]
		Field events (jumping)	Basketball[a]
		Figure skating[a]	Biathlon
		Lacrosse[a]	Ice hockey[a]
		Running (sprint)	Field hockey[a]
			Rugby[a]
			Soccer[a]
			Cross country skiing (skating)
			Running (mid/long)
			Swimming
			Tennis (single)
			Team handball[a]
			Boxing[a]
			Canoeing, kayaking
			Cycling[a,b]
			Decathlon
			Rowing
			Speed skating
			Triathlon[a,b]
II. Moderate static	Auto racing[a,b]		
	Diving[b]		
	Equestrian[a,b]		
	Motorcycling[a,b]		
	Gymnastics[a]		
	Karate/Judo[a]		
	Sailing		
	Archering		
III. High static	Bobsledding[a,b]	Bodybuilding[a,b]	
	Field events (throwing)	Downhill skiing[a,b]	
	Luge[a,b]	Wrestling[a]	
	Rock climbing[a,b]	Snowboarding[a,b]	
	Waterskiing[a,b]		
	Weight lifting[a]		
	Windsurfing[a,b]		

Source: Pelliccia, A. et al., *Eur. Heart J.,* 26, 1422–45, 2005.

Low, moderate, and high dynamic components correspond, respectively, to <40%, 40%–70%, and >70% of maximal oxygen uptake.

Low, moderate, and high static components correspond, respectively, to <20%, 20%–50%, and >50% of maximal voluntary force.

[a] Risk of bodily collision.

[b] Danger if syncope occurs.

normotensive at rest and asymptomatic who presents normal BP adaptations during the first stages of exercise.

Thus, how to interpret the BP measured during exercise? In the absence of scientific data validated by large studies, one can propose the following approach. It seems more appropriate to rather analyze the kinetic of BP changes during exercise than to look only to the last value of BP obtained. We must therefore pay attention to sudden and often exponential rise of BP in the first steps of exercise and of course to any associated clinical or electrocardiographic symptom. Of course, any associated diseases must also be considered for interpretation of BP responses to exercise. Concerning BP adaptations on cycloergometer, age of the subject and maximum power developed are the two factors that most influence the maximal exercise BP values. The formula: systolic BP (mm Hg) = 147 + 0.334 × max power (Watts) + 0.31 × age (years) that has the merit of holding account of the two main factors that influence exercise BP values can be used as proposed by Rost et al. (1987). Several systolic BP upper limit values according to age are proposed (300 mm Hg if < 40 years, >280 mm Hg between 40 and 50 years, >260 mm Hg between 51 and 60 years and >250 mm Hg >60 years) for general normotensive population (Douard et al. 1994; Palatini 1988). In the general population, the maximal value proposed for hypertension induced by exercise is systolic BP >210 mmHg for males and >190 mmHg for females (Lauer et al. 1992). These different *limits* are based on observation and not on scientific validation. Thus, systemic exercise hypertension is not currently defined. Regarding diastolic hypertension, the BP values >130 mm Hg during exercise are exceptionally rare in this population and deserve more caution.

In summary, the systemic hypertension is defined only in accordance with validated standards on resting BP measurements (Kjeldsen et al. 2014; Mancia et al. 2013). Consequently, a normotensive athlete at rest remains normotensive whatever are the values of his BP measured at the end of exercise.

The true prognostic value of exercise BP measurement in normotensive athlete is still debated. In this population, a significant positive correlation between maximal exercise systolic BP and left ventricular mass has been reported (Lauer et al. 1992). Similarly, it was proposed that for resting normotensive people the risk to present later systemic hypertension or acute cardiovascular events is increased in case of high BP values during dynamic exercise (Laukkanen et al. 2012; Schultz et al. 2013). Abnormal BP following maximal exercise could also be an indication for the risk to develop future hypertension (Turmel et al. 2012). However, all these correlations are relatively weak and directed mainly to individuals who have high arterial stiffness linked to aging and to other cardiovascular risk factors. Indeed, such higher BP responses induced by exercise are observed mainly in poorly physically trained subjects with low physical capacity and those with metabolic syndrome and/ or elevated chronic inflammatory state (Thanassoulis et al. 2012). Thus, the clinical relevance of these data on the individual level for healthy trained subjects can be debated.

Considering the athletic population, few studies investigated the exercise-induced arterial hypertension. The systemic hypertension may be preceded by a state of pre-hypertension during which stress induced by physical exercise may unmask a trend toward higher BP (Turmel et al. 2012). One study (Dlin et al. 1984) showed that in normotensive water-polo players, the risk to develop hypertension was 10 times higher in athletes with an exaggerated systolic BP response to exercise than in those with normal exercise BP responses. A more recent study (Turmel et al. 2012) showed that endurance athletes with an exaggerated BP response to exercise showed higher systolic BP values on 24-h ambulatory BP monitoring, although the BP parameters remain within normal values. The authors propose that the exaggerated BP response to exercise may imply an abnormal hemodynamic response, explained by slight lipids metabolism disturbance and/or altered autonomic nervous system (Turmel et al. 2012). In triathletes, it has been observed that BP values at the anaerobic threshold were higher in men with greater left ventricular mass (>220g) and an echocardiographic concentric myocardial remodeling pattern (Leischik et al. 2014).

14.3 CHRONIC EFFECTS OF PHYSICAL ACTIVITY ON SYSTEMIC BLOOD PRESSURE

In the general population, higher fitness level is associated with a lower prevalence of hypertension and low fitness is inversely associated with the risk of developing hypertension at baseline. These associations are reported regardless of demographic characteristics and common hypertension risk factors (Juraschek et al. 2014). Globally, the overall prevalence of hypertension in physically active people is approximately 50% lower than in the general population independently of other risk factors (Lehmann et al. 1990).

Cross-sectional as well as longitudinal studies have shown that regular physical training reduces BP and all meta-analysis, despite their methodological limits, concluded that regular endurance training was associated with a decrease in resting BP (Johnson et al. 2014). This beneficial effect is more pronounced in hypertensive than in normotensive subjects. Globally, the magnitude of decreases reported are between 2.4 and 3.4 mm Hg for systolic BP and between 3.1 and 4.7 mm Hg for diastolic BP (Cornelissen et al. 2013a; Fagard 2005; Pescatello et al. 2004). The decrease in BP was more accentuated during the daytime, and this level of BP reduction is close to that reported with recent monotherapies in stage 1 hypertensive individuals (Cornelissen et al. 2013a) (see classification of hypertension in Section 14.4). The effectiveness of the training is similar between men and women, being more pronounced in Caucasians and Asians than among blacks but it seems to be less efficient after 50 years of age (Cornelissen et al. 2013b; Juraschek et al. 2014).

Resistance training and isometric studies have examined smaller samples of subjects (Cornelissen et al. 2011). However, a significant decrease in BP (systolic BP 6.8 mm and diastolic BP 4 mm) is reported (Carlson et al. 2014). No difference between the training circuits and classical training was observed (Millar et al. 2014).

It is a consensus among official societies and academies to recommend lifestyle and dietary changes as the first management of uncomplicated mild hypertension. These changes will combine the practice of regular moderate exercise and a balanced diet without sodium restriction to reach and maintain a good weight balance (Fagard 2005). For moderate (stage 1) hypertension, adequate (low weights and high repetition regimens) resistance training may be allowed but at best in combination with physical endurance activity (Cornelissen et al. 2011; Millar et al. 2014).

In the case of more severe hypertension, such lifestyle changes must be individually elaborated and in combination with pharmacological treatment to achieve a BP of less than 140/90 mm Hg at rest. Even for patients with resistant hypertension, who do not respond to medication, physical exercise can be adapted in addition to the optimal pharmacological treatment (Ribeiro et al. 2015).

The effects of chronic physical training on resting BP are not fully elucidated and many mechanisms have been proposed: (1) indirect effects related to responses in other cardiovascular risk factors such as increased whole-body insulin sensitivity, reduced circulating noradrenaline level, and reduced body weight and, improvement of aortic stiffness by reduced resting heart rate (Fagard 2005; Juraschek et al. 2014); (2) direct effects on vascular resistance by increased endothelial production of nitric oxide synthase and decreased aortic stiffness (Cornelissen et al. 2013a; Millar et al. 2014); and (3) genetic factor because about 25% of hypertensive subjects do not respond to chronic physical exercise (Juraschek et al. 2014).

Although the prevalence of hypertension is reported to be lower in athletes (less than 10% in elite athletes), compared to that of the general population, it must be underlined that systemic hypertension is also the most common cause for exclusion, at least temporary, for competitive sport participation (Fagard 2007). This is an important question because several data indicate that BP level in early adulthood is associated with risk of cardiovascular disease later in life. Thus, identification of populations at risk for early adult hypertension is important (Lewington et al. 2002; Gray et al. 2011).

In hypertensive athletes, physical training is regular but usually of moderate intensity. However, in order to perform at a highest level, they must practice repetitive bouts of high-intensity strength or endurance exercise that may cause BP rise, possibly due to increased sympathetic activity (Traschel et al. 2015). Thus, the effects of this type of training on later development of high resting BP in predisposed athletes may be questioned (Traschel et al. 2015).

BP responses have recently been studied in some groups of athletes of different modalities. In collegiate American football players, an increase in resting systolic and diastolic BP postseason has been reported in more than 50% of players, especially in linemen while such increase was not observed in a group of rowers (Weiner et al. 2013). In this study, lineman field position, intra-season weight gain, and family history of hypertension were the strongest independent predictors of postseason BP (Weiner et al. 2013). It is noteworthy that a higher prevalence of hypertension is reported in retired National Football League players than in age-matched population and that deaths among linemen were mainly attributable to hypertensive and ischemic heart disease (Castle et al. 2009). In endurance athletes, males show significantly higher BP compared to females, despite a comparable amount of training and performance level (Traschel et al. 2015). In male European professional football players, the reported prevalence of high BP was low and independent of ethnicity (Berge et al. 2013a, b). These recent studies showed that systemic hypertension in athletes is rare, although a positive linear relationship was observed between systolic BP level, left ventricle mass concentric remodeling and left atrium size (Berge et al. 2013b; Traschel et al. 2015; Weiner et al. 2013). However, all cardiac morphological values were still within normal limits. Higher and concerning BP values were observed only in systolic BP, and it was associated with a reduced arterial compliance (Berge et al. 2013b). Because athletes' heart patterns are classically explained by increased volume load, it has been proposed that the cardiac remodeling observed in athletes with higher BP is the result of the combination of pressure and volume overload (Berge et al. 2013b). Further studies are needed to confirm if these BP adaptations will be associated with a future systemic hypertension and if they participate to the increased risk of atrial fibrillation reported in master athletes (Traschel et al. 2015).

14.4 DEFINITION AND SPECIFICITIES OF HYPERTENSION IN ATHLETES

All the guidelines propose the same criteria to define hypertension, based on at least two measurements of clinic (or ambulatory) seated BP. The classification is as follows: stage 1 hypertension when systolic BP and/or diastolic BP, respectively, $\geq 140/90$ mm Hg, and stage 2 hypertension as ≥ 160 and/or ≥ 110 mm Hg. To our knowledge, there is no specific recommendation for BP measurement in adult athletes. In children and teenagers, because of their variability in growth we use BP charts depending on their height. Normal value is defined as BP <90th percentile, and BP between 90 and 95th is defined as pre-hypertension (Leddy et al. 2009). The normal values proposed for ambulatory or home BP monitoring must be those used for athletes (Kjeldsen et al. 2014).

All guidelines say that hypertension concerns only abnormal resting BP values. Thus, currently in a normotensive athlete at rest, BP values obtained during exercise can't be used for the diagnosis of hypertension or for any treatment.

Globally, the conventional BP measurement method is used in athletes; however, extra care should be taken. The size of the cuff should be adapted to the arm's muscle mass of the athlete (large cuff when mid-arm circumference >33 cm and small cuff when <23 cm). Given the acute effects of exercise, BP measurement must be made at least 24 h after a training session. During the tapering periods preceding the competition period, the resting BP level may be somewhat increased by the level of adrenergic stimulation. Thus, this period must be avoided for BP measurement. If still in doubt, it is better to check the BP during a detraining period or if needed discontinue it for few days. Any drug often used by athletes which might increase BP level, such as, for example, nonsteroidal anti-inflammatory drugs, caffeine, herbal, or other supplements, should be avoided. Finally, the observation of a recent and unexplained systemic hypertension in a young athlete may be an indication of prohibited use of substances and it needs investigation (Achar et al. 2010). Because of the prevalence of the white coat effect in athletes, elevated BP office measurements should be confirmed by ambulatory BP measurement (Liddy et al. 2009).

The BP should be monitored irrespective of the level of physical training. Indeed, almost 80% of adolescent athletes with exaggerated resting BP during preseason medical evaluation were found to eventually develop sustained hypertension (Leddy et al. 2009; Weiner et al. 2013). The large body size, the frequent use of nonsteroidal anti-inflammatory drugs, and a family history of hypertension are also causing factors to investigate (Leddy et al. 2009).

In athletes, diastolic BP is usually normal or low. Classically, the hypervolemia and increased stroke volume due to endurance training are compensated by the associated bradycardia, and thus, resting cardiac output is not altered. However, mainly in young athletes, an exaggerated pulse pressure amplification that occurs as a result of progressive decrease in arterial diameter from the heart to the arm and the corresponding increase in arterial impedance can explain the increase in systolic BP sometimes in the ranges of pre-hypertension or of isolated systolic hypertension (Hulsen et al. 2006). The moderate (+20 mm Hg) increase in central aortic pressure reported in athletes may reflect their greater stroke volume (McEniery et al. 2005). But in athletes at greater risk, the central (aortic) systolic BP is much lower than the brachial, and central BP contour (pressure wave) patterns are normal (Wilkinson et al. 2001). This specific BP response has been called the spurious systolic hypertension (Mahmud et al. 2003). The long-term significance of this condition is not well known and because central aortic BP is not routinely measured, we need further studies (Nijdam et al. 2008).

Masked hypertension that is defined as a normal office BP in the presence of an elevated BP during the ambulatory 24-h BP measurement presents nearly similar risk of cardiovascular morbidity as classical hypertension (Trachsel et al. 2015). A recent prospective study, in middle-aged endurance competitive athletes, reported a 38% prevalence of masked hypertension. This prevalence was similar to that of professional football players (Berge et al. 2013a) but higher than in general population (<20%). This prevalence may be partly increased by the definition criterion of 24-h or daytime BP used. As discussed before, masked hypertension was associated with a lower diastolic function and a higher left ventricular mass/volume ratio, indicating predominantly concentric myocardial remodeling.

From the results of this study (Trachsel et al. 2015), it could be recommended to use ambulatory BP measurement in endurance athletes presenting with no optimal office BP (≤120/80 mm Hg).

14.5 MANAGEMENT, TREATMENT, AND RECOMMENDATIONS FOR A HYPERTENSIVE ATHLETE

As described before, the systemic hypertension must first be confirmed. In approximately 95% of cases, the hypertension is essential (primary). However, clinical exam must eliminate some other causes of secondary hypertension. Thyroid gland palpation, cardiac and abdominal auscultation are indicated. Palpation of the femoral pulses must eliminate a dissymmetry between arm and legs pulse that is in favor of coarctation of the aorta (Fagard et al. 2005; Leddy et al. 2009).

Then, the overall CV risk must be estimated by searching for other risk factors, target organ damage, and concomitant diseases or accompanying clinical conditions. The classical biological routine tests recommended for untrained hypertensive patients must be performed as standard electrocardiography. Extended evaluation may be necessary based on the results from these classical investigations (Fagard et al. 2005).

Hypertensive individuals who participate in sports activities first need to obtain a good equilibrium of resting BP with adapted drugs (discussed later in this section). Then sport's participation depends on the hypertension state (see Table 14.2).

In hypertensive athletes, the need of other tests such as echocardiography and exercise test depends on the patient's risk and on the characteristics of the sport (see Table 14.3). Hypertensive patients with symptoms must be exhaustively evaluated before giving clearance for sport and in all cases an annual cardiologic evaluation is requested (Fagard et al. 2005).

Although the hypertensive patient is usually asymptomatic, the uncontrolled hypertension increases the work myocardium at rest and during exercise. In hypertensive well-trained athletes, even in the absence of structural or functional heart damage, the high BP value during exercise, together with autonomic dysfunction, leads to decrease in exercise capacity (Mazic et al. 2015).

As with any hypertensive patient, the goal of treatment is to maintain resting BP <140/90 mm Hg. For the athlete, however, we should attempt to offer a treatment that

---

TABLE 14.2
Sport Participation According to Hypertension Level

Competition

No restriction if:
Low-risk hypertension, well controlled and asymptomatic
Partial restrictions:
Hypertension with moderate risk = all sports except IIIC[a]
Hypertension with very high risk = limited to IA, IB[a]
Secondary hypertension = indications adapted to the cause
Polycystic kidney disease or aortic coarctation = no collision sports

Leisure Sports

High-intensity leisure sport: same as for competition
Low- or moderate-intensity leisure sport: no restriction

Source: Fagard, R.H. et al., *Eur. J. Cardiovasc. Prev. Rehab.*, 12, 326–31, 2005; Pelliccia, A. et al., *Eur. Heart J.*, 26, 1422–45, 2005.
Note: In all cases, hypertension must be controlled by treatment.
[a] Classification of sport from Table 14.1.

TABLE 14.3

Type and Frequency of Evaluation in Hypertensive Individuals According to the Risk Level and Sport Situation (Competition and Leisure)

Intensity of Sport	ECG	Exercise Test	Echocardiography	Evaluation
Competition	Yes	Yes	Yes	Yearly
				6 months
				if high risk
Leisure				
≥60% VO$_2$max	Yes	Yes	Yes (?)	Yearly
40%–60% VO$_2$max	Yes	Yes	Yes	Yearly
Low or moderate risk	Yes	Yes	Yes (?)	Yearly
High risk	Yes	No	No	Yearly
<40% VO$_2$max				

Source: Fagard, R.H. et al., *Eur. J. Cardiovasc. Prev. Rehab.*, 12, 326–31, 2005; Pelliccia, A. et al., *Eur. Heart J.*, 26, 1422–45, 2005.

will safely have the least restrictive scheme for performance and, in case of competition, will not be in the list of substances banned by the World Anti-Doping Agency.

No antihypertensive medication significantly limits the anaerobic performance. Renin–angiotensin system inhibitors and calcium channel blockers do not limit aerobic or anaerobic performance (Fagard et al. 2005; Leddy et al. 2009). Thus, they can be suggested as first-line options for hypertensive athletes.

As described in Chapter 11 (Table 11.1), some medications used for hypertension may slightly affect sweating and fluid volume and electrolyte balance. Diuretics, for example, which increase the risk of dehydration and potassium loss, may impair performance in prolonged exercise (Armstrong et al. 1987). These undesirable effects could be explained by the hemodynamic instability and by substrate turnover and oxidation changes caused by the potential effect of dehydration induced by diuretics (Roy et al. 2000a). However, the use of low doses of diuretic, such as hydrochlorothiazide, and generally when association with other drugs, has little effect on limiting athletic performance (Leonetti et al. 1989). No deleterious effect on cardiac instability is reported (Roy et al. 2000b). Another example is the beta-blockers, which are very effective to limit BP increase during exercise; however, they can compromise heat dissipation by reducing skin blood flow. The effects on hydration depend on the drug used. The non-selective drugs can increase the heat storage during exercise because of their limiting effect on hydration. The selective beta-blockers do not seem to present this limit (Gordon et al. 1987; Pescatello et al. 1990). In highly trained athletes, they limit maximal aerobic performance and endurance through their effects on ventilation and on carbohydrate metabolism during exercise. But their detrimental effect on performance is not significant in other athletes (Brion et al. 2000). When beta-blockers are prescribed, the use of cardioselective molecules with vasodilator effect is more indicated. The physician must be aware that beta-blockers are prohibited in competition by some federations such as shooting and auto racing.

Hypertensive athletes, as well as physically active individuals, should discuss with physicians and healthcare staff aspects related to their treatment including not only the medication but also those related to their training and diet regimes. Those hypertensive athletes on typical sodium-restricted diets and involved in prolonged sport modalities that induce high sweat loss should consider the need of adding sodium in their hydration beverage during exercise. This is to avoid hyponatremia (Chapter 3), that more often occur when endurance athletes ingest large amounts of sodium-free or low sodium fluids. Sodium concentration of sports drinks is comparatively lower to that of human sweat, especially in endurance athletes that tend to have greater rates of diluted (less sodium) sweat (Henkin et al. 2010). Therefore, there is apparently no clinical reason to contraindicate adequate volume intake of sport drinks during prolonged exercise in hypertensive athletes.

Finally, in hypertensive athletes, as in other hypertensive individuals, the effectiveness of treatment depends on good compliance. Thus, in asymptomatic athletes the physician must choose a treatment appropriate to the sport practice. Thus, for competitive sports, we must focus on the renin–angiotensin system inhibitors or calcium channel blockers. If their effectiveness is insufficient, beta-blockers can be used after checking that they are not prohibited by the relevant sports federation. For leisure sport practice, all drugs may be prescribed. We must also educate the athletes, reminding them to adjust their salt intake and medication especially in case of prolonged exercise, which may induce dehydration and vasodilation with marked orthostatic hypotension.

REFERENCES

Achar, S., A. Rostamian, and S.M. Narayan. 2010. Cardiac and metabolic effects of anabolic-androgenic steroid abuse on lipids, blood pressure, left ventricular dimensions, and rhythm. *Am. J. Cardiol.* 106:893–901.

Armstrong L.E., D.L. Costill, and W.J. Fink. 1985. Influence of diuretic-induced dehydration on competitive running performance. *Med. Sci. Sports Exerc.* 17:456–61.

Berge, H. M., T. E. Andersen, E. E. Solberg, and K. Steine. 2013a. High ambulatory blood pressure in male professional football players. *Br. J. Sports Med.* 47:521–5.

Berge, H. M., G. F. Gjerdalen, T. E. Andersen, E. E. Solberg, and K. Steine. 2013b. Blood pressure in professional football players in Norway. *J. Hypertens.* 31:672–9.

Brion, R, F. Carré F, J. C. Verdier et al. 2000. Comparative effects of bisoprolol and nitrendipine on exercise capacity in hypertensive patients with regular physical activity. *J. Cardiovasc. Pharmacol.* 35:78–83.

Carlson, D. J., G. Dieberg, N. C. Hess, P. J. Millar, and N. A. Smart. 2014. Isometric exercise training for blood pressure management: A systematic review and meta-analysis. *Mayo Clin. Proc.* 89:327–34.

Carré F. 2013. Acute cardiovascular adaptations to exercise. In: *Cardiologie du Sport*, ed. F Carré, pp. 1–31. De Boeck Supérieur Bruxelles, Belgium.

Cornelissen, V. A., R. Buys, and N. A. Smart. 2013a. Endurance exercise beneficially affects ambulatory blood pressure: A systematic review and meta-analysis. *J. Hypertens.* 31:639–48.

Cornelissen, V. A., R. H. Fagard, E. Coeckelberghs, and L. Vanhees. 2011. Impact of resistance training on blood pressure and other cardiovascular risk factors: A meta-analysis of randomized, controlled trials. *Hypertension* 58:950–8.

Cornelissen, V. A., and N. A. Smart. 2013b. Exercise training for blood pressure: A systematic review and meta-analysis. *J. Am. Heart Assoc.* 2(1):e004473. doi: 10.1161/JAHA.112.004473.

Dlin, R. A., R. Dotan, O. Inbar et al. 1984. Exaggerated systolic blood pressure response to exercise in a water polo team. *Med. Sci. Sports Exerc.* 16:294–8.

Douard, H., C. Vuillemin, P. Bordier et al. 1994. Normal blood pressure profiles during exercise according to age, sex and protocols. *Arch. Mal. Coeur. Vaiss.* 87:311–18.

Douglas, P. S. 1989. Cardiac considerations in the triathlete. *Med. Sci. Sports Exerc.* 21(Suppl 5):S214–18.

Fagard, R. H. 2005. Effects of exercise, diet and their combination on blood pressure. *J. Hum. Hypertens.* 19(Suppl 3):S20–4.

Fagard, R. H. 2007. Athletes with systemic hypertension. *Cardiol. Clin.* 25:441–8.

Fagard, R. H., H. H. Bjornstad, M. Borjesson et al. 2005. ESC Study group of Sports Cardiology Recommendations for participation in leisure-time physical activities and competitive sport for patients with hypertension. *Eur. J. Cardiovasc. Prev. Rehab.* 12:326–31.

Gordon, N. F., J. P. van Rensburg, H. M. et al. 1987. Effect of beta-adrenoceptor blockade and calcium antagonism, alone and in combination, on thermoregulation during prolonged exercise. *Int J Sports Med.* 8:1–5.

Gray, L., I. M. Lee, H. D. Sesso et al. 2011. Blood pressure in early adulthood, hypertension in middle age, and future cardiovascular disease mortality: HAHS (Harvard Alumni Health Study). *J. Am. Coll. Cardiol.* 58:2396–403.

Halliwill, J. R., T. M. Buck, A. N. Lacewell et al. 2013. Postexercise hypotension and sustained postexercise vasodilatation: What happens after we exercise? *Exp. Physiol.* 98:7–18.

Henkin, S. D., P. L. Sehl, and F. Meyer. 2010. Sweat rate and electrolyte concentration in swimmers, runners, and nonathletes. *Int. J. Sports Physiol. Perform.* 5:359–66.

Hulsen, H. T., M. E. Nijdam, W. J. Bos et al. 2006. Spurious systolic hypertension in young adults: Prevalence of high brachial systolic blood pressure and low central pressure and its determinants. *J. Hypertens.* 24:1027–32.

Johnson, B. T., H. V. MacDonald, M. L. Bruneau Jr. et al. 2014. Methodological quality of meta-analyses on the blood pressure response to exercise: A review. *J. Hypertens.* 32:706–23.

Karpinos, A. R., C. L. Roumie, H. Nian et al. 2013. High prevalence of hypertension among collegiate football athletes. *Circ. Cardiovasc. Qual. Outcomes* 6:716–23.

Kjeldsen, S., R. D. Feldman, L. Lisheng et al. 2014. Updated nationally and international hypertension guidelines: A review of current recommendations. *Drugs* 74:2003–51.

Lauer, M., D. Levy, K. Anderson, and J. Plehn. 1992. Is there a relationship between exercise systolic blood pressure response and left ventricular mass? *Ann. Intern. Med.* 116:203–12.

Laukkanen, J. A., and S. Kurl. 2012. Blood pressure responses during exercise testing-is up best for prognosis? *Ann. Med.* 44:218–24.

Leddy, J. J., and J. Izo. 2009. Hypertension in athletes. *J. Hypertens. (Greenwich)* 11:226–33.

Lehmann, M., H. Durr, H. Merkelbach et al. 1990. Hypertension and sports activities: Institutional experience. *Clin. Cardiol.* 13:197–208.

Leischik, R., N. Spelsberg, H. Niggemann et al. 2014. Exercise-induced arterial hypertension— An independent factor for hypertrophy and a ticking clock for cardiac fatigue or atrial fibrillation in athletes? *F1000Research* 3:105. doi:10.12688/f1000research.4001.1.

Leonetti, G., C. Mazzola, C. Pasotti et al. 1989. Antihypertensive efficacy and influence on physical activity of three different treatments in elderly hypertensive patients. *J. Hypertens.* 7:S304–5.

Lewington, S., R. Clarke, N. Qizilbash et al. 2002. Age-specific relevance of usual blood pressure to vascular mortality: A meta-analysis of individual data for one million adults in 61 prospective studies. *Lancet* 360:1903–13.

Mahmud, A., and J. Feely. 2003. Spurious systolic hypertension of youth: Fit young men with elastic arteries. *Am. J. Hypertens.* 16:229–32.

Mancia, G., R. Fagard, and K. Narkiewicz. 2013. ESH/ESC guidelines for the management of arterial hypertension: The task force for the management of arterial hypertension of the European Society of Hypertension (ESH) and of the European Society of Cardiology (ESC). *Eur. Heart J.* 34(28):2159–219.

Mazic, S., J. S. Lazic, M. Dekleva et al. 2015. The impact of elevated blood pressure on exercise capacity in elite athletes. *Int. J. Cardiol.* 180:171–7.

McEniery, C. M., Yasmin, S. Wallace et al. 2005. Increased stroke volume and aortic stiffness contribute to isolated systolic hypertension in young adults. *Hypertension* 46:221–6.

Millar, P. J., C. L. McGowan, V. A. Cornelissen et al. 2014. Evidence for the role of isometric exercise training in reducing blood pressure: Potential mechanisms and future directions. *Sports Med.* 44:345–56.

Nijdam, M. E., Y. Plantinga, H. T. Hulsen et al. 2008. Pulse pressure amplification and risk of cardiovascular disease. *Am. J. Hypertens.* 21:388–92.

Palatini, P. 1988. Blood pressure behaviour during physical activity. *Sports Med.* 5:353–74.

Pelliccia, A., R. Fagard, H. H. Bjornstad et al. 2005. Recommendations for sports participation in athletes with cardiovascular disease. *Eur. Heart J.* 26:1422–45.

Pescatello, L. S., B. A. Franklin, R. Fagard et al. 2004. American College of Sports Medicine position stand. Exercise and hypertension. *Med. Sci. Sports Exerc.* 36:533–53.

Pescatello, L. S., G. W. Mack, C. N. Leach Jr., and E. R. Nadel. 1990. Thermoregulation in mildly hypertensive men during beta-adrenergic blockade. *Med. Sci. Sports Exerc.* 22:222–8.

Ribeiro, F., R. Costa, and J. Mesquita-Bastos. 2015. Exercise training in the management of patients with resistant hypertension. *World. J. Cardiol.* 7:47–51.

Rost, R., and H. Heck. 1987. Exercise hypertension-significance from the viewpoint of sports. *Herz* 12:125–33.

Roy, B. D., H. J. Green, and M. Burnett. 2000a. Prolonged exercise after diuretic-induced hypohydration: Effects on substrate turnover and oxidation. *Am. J. Physiol. Endocrinol. Metab.* 279:E1383–90.

Roy, B. D., H. J. Green, and M. E. Burnett. 2000b. Prolonged exercise following diuretic-induced hypohydration: Effects on cardiovascular and thermal strain. *Can. J. Physiol. Phamacol.* 78:541–7.

Schultz, M. G., P. Otahal, V. J. Cleland et al. 2013. Exercise-induced hypertension, cardiovascular events, and mortality in patients undergoing exercise stress testing: A systematic review and meta-analysis. *Am. J. Hypertens.* 26:357–66.

Taylor-Tolbert, N. S., D. R. Dengel et al. 2000. Ambulatory blood pressure after acute exercise in older men with essential hypertension. *Am. J. Hypertens.* 13(1, Pt 1):44–51.

Thanassoulis, G., A. Lyass, E. J. Benjamin et al. 2012. Relations of exercise blood pressure response to cardiovascular risk factors and vascular function in the Framingham Heart Study. *Circulation* 125:2836–43.

Trachsel, L. D., F. Carlen, N. Brugger et al. 2015. Masked hypertension and cardiac remodeling in middle-age endurance athletes. *J. Hypertens.* 33. doi:10.1097/HJH.0000000000000506000–000.

Turmel, J., V. Bougault, L. P. Boulet, and P. Poirier. 2012. Exaggerated blood pressure response to exercise in athletes: Dysmetabolism or altered autonomic nervous system. *Blood Press. Monit.* 17:184–92.

Vu, L., T. Mitiku, G. Sungar et al. 2008. The blood pressure response to dynamic exercise testing: A sytematic review. *Prog. Cardiovasc. Dis.* 51:135–60.

Weiner, R. B., F. Wang, S. K. Isaacs et al. 2013. Blood pressure and left ventricular hypertrophy during American-style football participation. *Circulation* 128:524–31.

Wilkinson, I. B., S. S. Franklin, I. R. Hall et al. 2001. Pressure amplification explains why pulse pressure is unrelated to risk in young subjects. *Hypertension* 38:1461–6.

15 Athletes with Chronic Renal Diseases

Dilip R. Patel and Vimal M.S. Raj

CONTENTS

15.1 INTRODUCTION

Chronic kidney disease (CKD) is generally defined based on kidney damage for more than three months associated with structural or functions abnormalities (Table 15.1) (Hogg et al. 2003, Levey et al. 2003). The glomerular filtration rate (GFR) is the best indicator of kidney function and is the basis for stratification of CKD presented in Table 15.2. Other related terms used in the description of CKD are summarized in Table 15.3. CKD affects approximately 20 million persons in the United States (Collins et al. 2011). Approximately half-million persons have end-stage renal disease (ESRD). In 2008 in the United States, 17,413 kidney transplants were performed and 165,639 functional transplants by the end of the year.

The most common causes of CKD in adults are diabetes mellitus and hypertension. Other causes include glomerular diseases, drug toxicity, and trauma. Common causes of CKD in children are congenital kidney diseases (e.g., renal hypoplasia,

TABLE 15.1
Criteria for CKD Definition by National Kidney Foundation

1. Kidney damage for ≥3 months, as defined by structural or functional abnormalities of the kidney, with or without decreased glomerular filtration rate (GFR), manifest by either pathological abnormalities or markers of kidney damage, including abnormalities in the composition of the blood or urine, or abnormalities in imaging tests
2. GFRe <60 mL/min/1.73 m² for ≥3 months, with or without kidney damage

Source: K/DOQI Clinical Practice Guidelines for Chronic Kidney Disease: Evaluation, Classification, and Stratification. Kidney Disease Outcome Quality Initiative. 2002. *Am J Kidney Dis.* 39:S1–S246.

TABLE 15.2
Stages of CKD

Stage	Description	EGFR (mL/min/1.73 m²)
I	Kidney damage with normal or ↑ eGFR	≥90
II	Kidney damage with mild ↓ eGFR	60–89
III	Moderate ↓ eGFR	30–59
IV	Severe ↓ eGFR	15–29
V	Kidney failure	<15 (or dialysis)

eGFR, estimated glomerular filtration rate calculated on the basis of serum creatinine level, age, and values assigned for sex and race.

dysplasia, obstructive uropathy, and polycystic kidney disease), acquired diseases (e.g., glomerulonephritis), inherited diseases (e.g., Alport syndrome and polycystic kidney disease), and metabolic disease (e.g., cystinosis) (Sreedharan et al. 2011). Main considerations in planning exercise programs for persons with CKD are decreased exercise tolerance, comorbid conditions (such as diabetes, hypertension, and atherosclerotic heart disease), complications (such as decreased bone mineral density), and the potential for damage to the access sites and the transplant kidney. Time burden of participating in exercise program is also an important consideration.

For the purpose of our discussion, the term *physical activity* refers to any bodily movement produced by the contraction of skeletal muscle that increases energy expenditure above a basal level. The term *exercise* (exercise training) refers to a subcategory of physical activity that is planned, structured, repetitive, and purposive in the sense that the improvement or maintenance of one or more components of physical fitness is the objective. Sport is a type of physical activity with added social, psychological, and competitive dimensions.

TABLE 15.3
Terms Related to CKD

Kidney damage	Structural or functional abnormalities of the kidney, includes abnormal results on blood and urine tests or imaging studies.
Kidney failure	Either (1) a level of GFR 15 mL/min/1.73 m^2 which is accompanied in most cases by signs and symptoms of uremia or (2) a need for initiation of kidney replacement therapy (dialysis or transplantation) for treatment for complications of decreased GFR which would otherwise increase the risk of mortality or morbidity.
End-stage renal disease	End-stage renal disease (ESRD) is used as an administrative term in the United States based on the conditions for payment for healthcare by Medicare ESRD program. It generally refers to patients with renal failure (stage V chronic kidney disease) and those treated by dialysis or transplantation irrespective of the level of GFR.

Source: Hogg, R.J. et al., *Pediatrics,* 111(6), 1416–21, 2003; Levey, A.S. et al., *Ann. Intern. Med.,* 139, 137–47, 2003.

15.2 RENAL PHYSIOLOGY DURING EXERCISE

During exercise, renal blood flow (RBF) decreases (Johnson et al. 2015, Drew et al. 2013, Painter et al. 2009). The magnitude of decrease in RBF is linearly correlated with the intensity of physical activity (Muller et al. 2013, Johanson and Painter, 2012). Baseline hydration status affects the change in RBF with exercise, with initial hyperhydration resulting in no significant decrease, whereas a significant decrease seen in a dehydrated state (Otani et al. 2013, Painter and Marcus 2013, Strachenfeld et al. 1996, Francesconi et al. 1983). During exercise, the blood is preferentially shunted to the exercising skeletal muscles. The GFR initially tends to increase with light intensity of exercise and decreases with moderate to vigorous intensity of exercise. Effect of exercise on the GFR in relation to hydration status is similar to that seen on RBF. Renal tubular function is also altered during exercise, mainly affected by the levels of angiotensin, renin, aldosterone, vasopressin, atrial natriuretic protein, and prostaglandins, all of which increase linearly with the intensity of the exercise; these changes affect the excretion of sodium and chloride, and urinary output helping in maintaining adequate volume (Wade et al. 1984, Wade and Claybaugh 1980, Fenzl et al. 2013).

Hyperfiltration injury to glomeruli is the most significant final common pathway for progression of CKD. In cases where nephrons are lost because of kidney disease, the remaining nephrons undergo structural and functional adaptations to maintain adequate glomerular filtration. In the early stages, this compensatory hypertrophy of nephrons is able to transiently maintain the renal function; however, the increased blood flow results in elevated hydrostatic pressure and damages the capillary walls. Subsequent to this microalbuminuria, overt nephropathy develops. The combined effect of hyperfiltration and direct toxic effects of leaked urinary protein on renal tubules results in progressive loss of kidney function. In individuals who have a solitary kidney or only one functioning kidney, similar changes occur over long term. Uncontrolled systemic hypertension and metabolic components such as elevated serum phosphorus levels and uric acid levels also contribute to further renal damage.

15.3 EXERCISE TOLERANCE IN CKD

Maximal oxygen uptake (VO_{2max}), the greatest amount of oxygen consumed during exercise, expressed as milliliters of oxygen consumed per kilogram of body weight per minute, is a generally accepted measure of aerobic or cardiovascular fitness (Pescatello et al. 2013). Persons with CKD have a much lower VO_{2max} when compared to healthy persons (Clapp et al. 2012, Van den ham et al. 2005, Pattaragarn et al. 2004, Painter et al. 2007). Exercise capacity is much lower in persons with CKD, especially stage IV and stage V, dialysis, and kidney transplant (Weaver et al. 2008). The main causes of decreased exercise tolerance in persons with renal failure and ESRD are anemia, changes in the cardiac function, changes in skeletal muscle function, and physical inactivity.

15.3.1 ANEMIA

Oxygen carrying capacity of the blood to the exercising muscles is decreased in chronic anemia resulting in a decreased VO_{2max} (Leikis et al. 2006). Progressive loss of nephrons in CKD results in decreased production of erythropoietin and anemia. The availability of recombinant human erythropoietin (rHuEPO) has made a significant difference in treating anemia of CKD. However, Pattaragarn et al. (2004) in a prospective study of adolescents on dialysis showed significant lowering of VO_{2max} (28.5 ± 10.5 mL/kg/min) when compared to healthy adolescents (49.1 ± 5.6 mL/kg/min) even after correction of anemia. This and other studies suggest that in addition to anemia, other factors contribute to decreased exercise tolerance in persons with CKD.

15.3.2 CARDIAC CHANGES

In adults with CKD, comorbid conditions such as systemic hypertension and coronary artery disease contribute to decreased exercise tolerance. Abnormalities in both cardiac structure and function have also been documented in children and adolescents with CKD. A study with 33 participants (aged 10–20) with chronic renal insufficiency reported a significant increase in left ventricular mass (LVM) index in chronic renal insufficiency patients when compared to that in healthy controls (Mistnefes et al. 2004). The increase in LVM index was more accentuated in patients on dialysis when compared to those who were on pre-dialysis ESRD. Patients who were on pre-dialysis showed concentric hypertrophy of ventricles, whereas those on dialysis showed eccentric hypertrophy. The indices of diastolic dysfunction and left ventricular compliance were significantly worse in CKD patients when compared to those in controls. The increased LVM and increased sympathetic activity contribute to the diastolic dysfunction, both in children and in adults with CKD

(Petras et al. 2013, Rump et al. 2000). Studies have shown that alteration in cardiac structure and function occur early in the course of CKD, and progressively worsening diastolic dysfunction is associated with a declining VO_{2max} (Colan et al. 1987).

15.3.3 Skeletal Muscle Changes

Poor exercise tolerance and premature exercise termination in persons with CKD secondary to muscle fatigue has been documented in several studies (Eijsermans et al. 2004, Tenbrock et al. 2000). Multiple factors including inadequate nutrition, protein loss, catabolic state, and physical inactivity contribute to skeletal muscle wasting in CKD. The restriction on protein intake to preserve renal function results in reduced synthesis of muscle protein in persons with CKD (Adey et al. 2000). Accelerated degradation of muscle proteins by increased expression of caspase 3 and activation of ubiquitin–proteasome system has been shown in animal models with renal failure (Du et al. 2004, Wang and Mitch, 2013). CKD is also a pro-inflammatory state with documented increase in several inflammatory markers like interleukin-6 and tumor necrosis factor-alpha, which also contributes to skeletal muscle wasting and atrophy (Barton 2001, Becker et al. 2005). Patients with ESRD demonstrate resistance to anabolic factors namely insulin and growth factor. A decreased anabolic to catabolic ratio further contributes to muscle wasting (Becker et al. 2005). A quantitative reduction in mitochondrial content in CKD patients could also contribute to poor exercise tolerance (Yokoi and Yanagita 2014).

Alayh et al. (2008) in a cross-sectional study of 22 children on peritoneal dialysis showed significant decrease in physical performance (6-min walk distance and gait speed) when compared to healthy controls. Quadriceps strength was significantly lower in children with CKD than in control participants. Increased fat depositions in muscles have been shown in children with CKD. Chronic muscle wasting and inactivity also result in reduced oxygen extraction in the muscles.

15.3.4 Physical Inactivity

Persons with CKD are generally physically inactive, often with significantly restricted day-to-day activities. Three weeks of bed rest and restricted activity decreased VO_{2max} by 26% (Heiwe and Jacobson 2011). Physical inactivity in persons with CKD by itself can lead to muscle atrophy, which, in addition to the cardiac and skeletal muscle changes, further contributes to decreased exercise tolerance.

15.4 EFFECTS OF EXERCISE IN CKD

Most studies on exercise training in persons with CKD have focused on the effect of regular aerobic exercise training on maximum oxygen uptake (Malagoni et al. 2008, Cheema and Singh 2005, Ouzouni et al. 2009). In these studies, the exercise regimens have included outpatient supervised aerobic exercises during non-dialysis days, cycling during hemodialysis, and home exercise programs without supervision (Wang et al. 2009, Moinuddin and Leehey 2008, Koudi et al. 2004, Depaul et al. 2002, Deligiannis et al. 1999). Most of these studies are done in persons on hemodialysis, including exercises such as cycling, jogging, and flexibility with a regimen as follows: frequency of 3–5 days per week, moderate intensity, each session between 30 and 60 min. Almost all the studies showed an increase in VO_{2peak} anywhere between 20% and 40% indicating a significant improvement from pre-training levels but still a modest increase when compared to healthy controls. Very few studies were performed in children and adolescents. Goldstein and Montgomery (2009) reported on the effects of twice-weekly exercise regimen during hemodialysis in children older than 8 years and showed improvement in 6-min walk distance and lower extremity strength after 3 months of training.

Studies looking at the effects of regular resistance training in persons with CKD are few with no studies specifically done in children (Castaneda et al. 2001, Headley et al. 2002, Momeni et al. 2014, Kouidi et al. 1998). Castaneda et al. (2001) showed that resistance training improved muscle mass and strength in a randomized controlled trial of 26 adult patients with moderate renal insufficiency prior to dialysis. The findings of this and other similar studies show that resistance training can negate the effects of muscle catabolism due to low protein diet and protein loss.

Headley et al. (2002), in a study of 10 adults on hemodialysis, showed that 12 weeks of resistance training improved their muscle strength and functional capacity. Variables measured included peak torque of quadriceps, 6-min walk test, gait speed, percentage of body fat, and time to complete 10 repetitions of the sit-to-stand-to-sit test. Study participants showed a significant improvement in all the variables after 12 weeks except for percentage of body fat. Other studies in adults with CKD have also shown that regular resistance exercises are well tolerated and resulted in improved muscle strength, local muscular endurance, muscle function, overall ability for daily activities, and slowed down wasting of muscles.

15.4.1 CARDIOVASCULAR STRUCTURE AND FUNCTION

Diastolic dysfunction and alterations in heart morphology in ESRD significantly contribute to exercise intolerance. Deligiannis et al. (1999) in a study in adults with ESRD showed beneficial left ventricular structural and functional adaptations after aerobic exercise training. In the study, participants had significant improvements in cardiac systolic function with increase in ejection fraction, stroke volume, and cardiac output both at rest and at submaximal exercise. These effects were seen after six months of supervised outpatient training aerobic sessions three times per week. Echocardiography showed an increase in left ventricular end diastolic dimension and a decrease in end systolic volume, findings consistent with improved myocardial contractility. An increase in left ventricular posterior wall thickness and interventricular septum thickness was also noted. Similar improvement in left ventricular ejection fraction and decrease in pulmonary artery and systemic vascular resistance were noted in patients doing exercise training for 30 min a day, 3 days a week for even shorter period of 3 months (Momeni et al. 2014).

Although regular aerobic exercise training in persons with CKD has been shown to result in favorable structural adaptations of the heart, improved cardiac function, and improved exercise tolerance, no long-term follow-up studies on survival data are available (Johanson et al. 2012, Painter and Roshanravan 2013, Booher and Smith 2001).

15.4.2 SKELETAL MUSCLES

Muscle fatigue is the most common reason for premature termination of exercise in persons with CKD. Regular resistance training has been shown to improve muscle strength, endurance, and function in persons with CKD. Kouidi et al. (1998) looked at the effects of six months of intensive resistance training on adults on hemodialysis. Study results showed significant improvement in muscle structure and function. Results of muscle biopsy (vastus lateralis) after six months of training showed a significant increase in skeletal muscle fiber cross-sectional area, formation of new muscle fibers, and regeneration of previously degenerated skeletal muscle fibers. In addition, participants showed improved VO_{2max}, exercise time, and isokinetic muscle strength of lower limbs.

Both aerobic and resistance exercise training have been shown to improve VO_{2max} and exercise tolerance in persons with CKD. Because decreased survival in persons with CKD is associated with a progressive decrease in VO_{2max}, regular exercises that improve VO_{2max} are especially important (Siestama et al. 2004).

15.4.3 QUALITY OF LIFE

Main domains of health-related quality of life (HRQoL) generally measured include physical functioning (self-care, daily activities, and employment), emotional well-being (anxiety, depression, and hostility), and social functioning. Persons with CKD report limitations of daily activities such as walking, grasping, kneeling, and house work. They also report emotional difficulties including depression, anxiety, phobia, irritability, and hysteria (Koufaki et al. 2013). Employment status is a significant factor that relates to how persons with CKD perceive their quality of life. With appropriate adjustments, most persons with CKD are able continue to work and most students continue to attend school and college (Smart et al. 2011). Several studies in adults have reported the beneficial effects of exercise on both physical and emotional components of the measures of quality of life (Kolewaski et al. 2005, Levendoglu et al. 2004, Kouidi 2004, Mercer et al. 2004, Tawney et al. 2000).

Adolescents encounter a variety of biological and emotional challenges as they develop their abstract thinking and transition into adulthood. Any chronic disease process is going to add to the challenges that they already face. Loss of school days and difficulty in socializing can play a big part in affecting their quality of life. A lack of adherence to the treatment regimen is one of the most significant factors for transplant failures in adolescents.

15.5 CONSIDERATIONS FOR PHYSICAL ACTIVITY

Before starting an exercise program, medical evaluation and exercise testing are recommended. Exercise programs should include both aerobic and resistance training and must be individualized on the basis of specific goals, circumstances, and needs. Although various regimens of exercise programs have been used in CKD patients, beneficial effect is noted in all study groups (Konstantinidou et al. 2001, Shalom et al. 1984, Heiwe and Jacobson 2014, Deligiannis 2004a, Deligiannis 2004b, Gould et al. 2014).

TABLE 15.4
Considerations for Exercise Recommendations in Persons with CKD

Medical evaluation and exercise testing	• Generally recommended before starting exercise program.
	• Assess cardiovascular risks.
	• Review effects of medications on exerciser response.
	• Exercise is contraindicated in CKD patients with a systolic blood pressure >200 mm Hg or a diastolic blood pressure >110 mm Hg, electrolyte abnormalities, recent myocardial infarction, or recent changes in the electrocardiogram.
Exercise programs	• Exercise programs should include appropriate warm-up, cool down, and flexibility exercises.
	• Start at low-intensity, short duration sessions.
	• Resistance exercise should start at low resistance loads (i.e., 3 repetitions maximum or higher, e.g., 10–12 repetitions maximum).
Peritoneal dialysis	• Avoid breath holding during exercise.
	• Persons on continuous ambulatory peritoneal dialysis should try to exercise with fluid in the abdomen; if not tolerated, exercise with empty abdomen which is generally better tolerated.
	• Ensure adequate dressing of the catheter site.
Hemodialysis	• May perform exercise during dialysis session or non-dialysis days.
	• Can exercise arm with vascular access as long as weight is not directly borne on the access site itself.
	• Heart rate is not a reliable indicator for intensity; use rate of perceived exertion scale.
	• Exercise during first half of dialysis session is preferred to avoid hypotensive episodes.

Source: Painter, P., and Krasnoff, J.B., End stage metabolic disease: Chronic kidney disease and liver failure. In: Durstine, J.L., Moore, G.E., Painter, P.L., and Roberts, S.O., eds. *ACSM's Exercise Management for Persons with Chronic Diseases and Disabilities*, 3rd ed. Human Kinetics, Champaign, IL, 2009, pp. 175–81; Pescatello, L.S. et al., eds., *ACSM's Guidelines for Exercise Testing and Prescription*, 9th ed., Lippincott Williams and Wilkins, Philadelphia, PA, 2013.

The current recommendations of the American College of Sports Medicine for exercise in persons with CKD are summarized in Tables 15.4 and 15.5.

The most common risks associated with exercise are musculoskeletal injuries and cardiovascular adverse events. Most of the data on exercise-related risks are derived from studies on healthy population typically engaged in high-intensity training (Copley and Lindberg, 1999, Foster and Porcari, 2001). Persons with CKD have metabolic bone disease documented by decreased bone mineral density, vitamin D deficiency, and secondary hyperparathyroidism, all these being risk factors for bone fractures and poor healing after a skeletal injury. Poorly conditioned muscles and tendons also make them more prone to injury.

Most adults with CKD also have diabetes mellitus, systemic hypertension, and atherosclerotic heart disease, increasing their risk for exercise-associated adverse

TABLE 15.5

American College of Sports Medicine: Exercise Prescription for Persons with CKD

	Aerobic Exercise	Resistance Exercise
Frequency	3–5 days week⁻¹	2–3 days week⁻¹
Intensity	Moderate intensity (i.e., 40% to <60% VO_2R, RPE 11–13 on a scale of 6–20)	60% to 75% 1-RM
Time	20–40 min. day⁻¹ of continuous aerobic activity; however, if this duration cannot be tolerated, 10-min bouts of intermittent exercise to accumulate 20–60 min day⁻¹.	One set of 10–15 repetitions. Multiple sets may be done depending on patient tolerance and time.
Type	Exercises such as walking and cycling	Choose 8–10 different exercises to work the major muscle groups. Use machines or free weights.

Source: Pescatello, L.S. et al., eds., *ACSM's Guidelines for Exercise Testing and Prescription*, 9th ed., Lippincott Williams and Wilkins, Philadelphia, PA, 2013.

VO_2R, oxygen uptake reserve; RPE, rate of perceived exertion; RM, repetition maximum.

cardiovascular events including death (Mottram et al. 2004, Miyachi et al. 2004). This underscores the importance of proper screening and monitoring and the need for a graded exercise program in which the intensity of exercise is gradually increased. Available data also suggest that once conditioning is achieved, more intense exercise poses less risk.

Trauma to the transplanted kidney is always a concern while participating in certain sports and leisure time activities. Trauma to the kidneys from participation in sports is rare. Most cases are reported in contact or collision sports. There is no consensus on sport participation recommendations for persons with solitary kidney (Bernhardt et al. 2010, Bernard, 2009, Wu and Gaines 2007, Psooy 2006, McAleer et al. 2002) though current recommendation is to give a qualified yes after a thorough physical examination and discussion with the parents on the possible adverse outcomes. In contrast to native solitary kidney as most transplanted kidneys are placed extra peritoneally in the abdomen, risk of injury is relatively higher, and hence, contact sports are not recommended. Surveys indicate that most physicians recommended their patients with solitary kidney to avoid contact sports (Sharp et al. 2002). An analysis of the pediatric trauma cases from the National Pediatric Trauma Registry revealed that high-grade renal injuries requiring a nephrectomy are associated with sledding, skiing, and rollerblading (Johnson et al. 2005). There were no reported nephrectomies for injuries resulting from contact sports. Even though playing American football resulted in more renal injuries, all the injuries were low grade with no grade 3 injuries. Most kidney injuries resulted from motor vehicle accidents and bicycle accidents. Gerstenbluth et al. (2002) in their report on 68 children with blunt renal injury showed that bicycle riding was the main cause for blunt renal trauma.

The current sport-preparticipation guidelines recommend that athletes with one functioning kidney or a solitary kidney may be allowed to participate in contact or collision sports and other similar leisure time activities after explanation of all the risks (Johnson et al. 2005, Kennedy and Siegel 1995). Contact or collision sports are not recommended for persons who have a solitary kidney that is pelvic or iliac, is multicystic, has hydronephrosis, or shows ureteropelvic junction abnormalities. Persons with a transplant have successfully participated in many sports in national and international events. These sports events include track and field, swimming, table tennis, tennis, badminton, bowling, volleyball, golf, bicycling, and road races.

Concerns about exit site contamination of catheter for peritoneal dialysis, especially in children and adolescents who participate in swimming, have been raised, but with proper dressing, the risk is minimal (Plante et al. 1990). Similarly, the vascular access site and arteriovenous fistula should be protected (Kennedy et al. 1995).

15.5.1 PROTEINURIA

Proteinuria is prevalent in school age children. In normal adults, 24-h urine protein measurement greater than 150 mg/day or 80 mg/L is considered abnormal (Johnson et al. 2015). In children, urine protein excretion is calibrated to their body surface area, and values greater than 100 mg/m^2/day are considered abnormal (Patel et al. 2005, Greydanus and Torres 2008). As 24-h urine protein measurement can be tedious, spot urine protein to creatinine ratio has been widely used and normal value in children greater than 2 years of age and adults is less than 0.2.

Incidence of proteinuria in general population by dipstick can be as high as 10% though on repeat screening persistent proteinuria is present in only about 0.1%. Incidence of proteinuria tends to increase with age, peaks during adolescence, and is more common in girls (Collins et al. 2011, Sreedharan et al. 2011). The prevalence of proteinuria in athletes though is much higher and has been attributed to the change in renal hemodynamics during exercise. Sports-related proteinuria in non-pathological conditions of the kidney is usually transient and benign. Values return to normal range usually after 4 h of rest, and persistent proteinuria 48 h after exercise should be thoroughly evaluated for any pathological causes. Prevalence of proteinuria in athletes correlates more to exercise intensity than to the duration of exercise. In patients with preexisting conditions of the kidney like diabetic nephropathy or nephrotic syndrome, moderate to severe exercise is shown to worsen proteinuria in most of the patients though some of the participants did not show changes in the amount of protein excretion.

Orthostatic proteinuria is not uncommon during the adolescent period (Patel et al. 2005, Greydanus et al. 2008). This is a benign proteinuria as well with long-term follow-up studies showing no significant renal consequences. The etiology is not clearly understood, and compression of left renal vein altering glomerular hemodynamics has been suggested. Diagnosis can be made by a split 24-h urine collection but more easily by a spot urine collection for Pr/Cr ratio collected first thing in the morning. In this, the first morning spot urine Pr/Cr will be less than 0.2 after being supine for a few hours but will be higher after being up for a few hours. In 24-h urine collection in cases of orthostatic proteinuria, the total amount of protein will be usually less than 1 g/day.

Proteinuria is usually detected in an asymptomatic athlete during a routine sports physical examination. The screening method by dipstick detects albumin excretion, and levels of 1+ and above are considered abnormal. Rescreening for proteinuria by a spot first morning urine specimen after the athlete stays off exercise for 2–3 days will help in identifying the benign and transient proteinuria. Athletes with persistent proteinuria should be evaluated further.

15.5.2 Hematuria

Hematuria can be classified into macroscopic when the blood is visible to the naked eye or microscopic which may be detected on a dipstick screening. Microscopic hematuria is confirmed by detection of RBC usually greater than 5–10 high power field on spun urine sediment under a high power microscope (Patel et al. 2005, Greydanus et al. 2008). Absence of RBCs with a positive dipstick for hematuria can occur with myoglobinuria and hemoglobinuria.

The reported incidence of macroscopic hematuria is about 1 per 1000 (Patel et al. 2005). For microscopic hematuria, incidence varies with the definition used for diagnosis. It is estimated at 1%–2% in school age children in consecutive two or more urine samples and as high as up to 13% in some series (Patel et al. 2005, Greydanus et al. 2008). Microscopic hematuria is much more prevalent in athletes when compared to the general population. The exact prevalence of microscopic hematuria in athletic population is not known. Exercise-related hematuria is a benign and transient process with no long-term kidney problems. Reassuring the athlete and the family is very important, as the sight of blood in urine can be disturbing. Hematuria will clear in majority of patients when urinalysis is repeated 48–72 h after being off exercise. It is important to stress the need for good hydration during exercise and to avoid any nephrotoxic medication. Persistent microscopic hematuria or gross hematuria in athletes needs to be evaluated further.

15.5.3 Protein Intake in Athletes

The long-term effect on kidneys of excessive protein intake by athletes is a subject of many investigations. The recommended dietary allowance for protein is 0.75–0.8 g/kg/day for general population without CKD. The current recommendation of protein intake for athletes by the American College of Sports Medicine is 1.2–1.7 g/kg/day. This is roughly twice the recommended dietary allowance recommended in general population to account for new muscle protein synthesis and to aid in repair of tissue damage that occurs with high-intensity exercise (Raj et al. 2012).

Current methods of determining the required protein intake is based on nitrogen balance studies. This method takes into account the entire protein intake and measures the nitrogen losses including that of urine, feces, sweat, hair, semen, and skin. The amount of protein required to give a positive balance is considered the optimal intake. Disadvantages of using this method to calculate recommended dietary allowance is the fact that these studies are done on sedentary population as a whole and will not meet the requirements of an athlete who does intensive training. Protein requirement will also vary with the kind of exercise with athletes involved in resistance training requiring much more protein than those involved in aerobic training.

Optimal results including muscle mass and sports performance for an athlete are based not only on the amount of protein ingestion but also on the kind of exercise training, the quality of protein ingested, and the timing of their intake. Also, muscle mass as a unit will adapt with exercise such that in chronically trained individuals, lesser amounts of protein ingested will be sufficient to produce the desirable effect. Whey protein is a superior protein when compared to soy or casein, and whey plus casein gives the greatest gain in lean body mass long-term when compared to whey plus amino acids alone (Tang et al. 2009, Kerksick et al. 2006).

Protein ingestion just prior to or following exercise has the greatest benefit on muscle protein synthesis (Tipton et al. 2001), and on muscle hypertrophy acutely and over time (Esnarck et al. 2001, Cribb and Hayes 2006). Though there is no deleterious effect of ingested protein on a healthy kidney, the same cannot be said for CKDs (Pecoits-Filho 2007). Several studies have shown that the decline in GFR will be faster with an excess protein load. Some studies have pointed toward the association between excess protein intake and kidney stones especially in genetically predisposed individuals (Robertson et al. 2002, Reddy et al. 2002, Nguyen et al. 2001).

15.6 FLUID BALANCE IN CKD

Fluid balance in the human body is closely maintained by the kidneys through variable excretion of solute and water in relationship to their intakes and losses by the lungs, skin, and gastrointestinal tract. The main mechanism for fluid loss by urine is based on the concentrating ability of urinary solutes by the kidney. In a normal functioning adult kidney, the maximal urinary concentrating ability is found to be at 1200 mOsm/kg H_2O. The obligatory urine loss (minimal amount of urine volume needed to get rid of the urinary solutes) is given by the formula (Wang et al. 2013)

$$\text{Obligatory } V \text{ (ml)} = \frac{\text{Daily osmolar excretion (mOsm)}}{\text{maximal urine osmolaity (U osm max)}}$$

Hence, in a patient with normal kidney function with maximal urine concentrating ability, the urine volume needed to get rid of a urinary solute concentration of 400 mOsm could be as low as 300 mL in 24 h. However, as GFR falls and CKD progresses, the urine concentrating ability will go down as well. Most patients with advanced CKD (stage IV and above) have urine osmolality similar to plasma (iso-osmolar). Considering a urine concentrating ability of only 300 mOsm, the obligatory urine volume will be increased to 1.3 L indicating the need for increased fluid intake in CKD.

Arginine vasopressin (AVP) is a peptide hormone that plays a main role in water balance in the body. Increase in plasma osmolality and low intravascular volume are the main stimuli for increase in AVP production. Rat models with 5/6 nephrectomy have shown that decreasing AVP by increasing water intake has shown to slow the progression of decline in GFR and CKD (Bouby et al. 1990, Suguira et al. 1999).

Suggested mechanisms contributing to the progression of CKD by AVP include increasing renin synthesis, increasing sodium and chloride reabsorption in the thick ascending limb, and stimulating mesangial cell proliferation. All these tend to promote glomerular hypertension and hypertension, known factors involved in glomerular sclerosis.

Prospective population studies (Clark et al. 2011, Strippoli et al. 2011) have shown that higher urine volumes and increased water intake were associated with preserved renal function. Inadequate water intake in the setting of intense physical activity could predispose to CKD as shown by Pereza et al. (2012). Studies done in autosomal dominant polycytic kidney disease using tolvaptan as a vasopressin antagonist has reduced cyst growth and slowed the decline of GFR (Higashihara et al. 2011).

No specific recommendation for the amount of fluid intake for CKD is available in the literature. However, based on available evidence, it can be mentioned that patients with early stages of CKD (GFR > 60 mL/min/1.73 m^2) will benefit from increased fluid intake especially with strenuous physical activity. For an average adult, the recommendation is to reach at least 3 L of urine output in 24 h. This will be particularly beneficial in subjects with cystic kidney disease and tubulointerstitial diseases who have a reduced urinary concentrating ability. In patients with more advanced CKD, the amount of fluid intake has to be individualized based on their urine output. Other dietary changes including restricting salt and protein intake to the recommended dietary allowance will also be helpful in slowing the progression of CKD.

15.7 SUMMARY

CKD affects 20 million Americans. The most common causes of CKD in adults are diabetes mellitus and hypertension. The main reasons for poor exercise tolerance in persons with CKD are anemia, cardiovascular changes, change in skeletal muscles, and physical inactivity. Regular aerobic and resistance exercise improve exercise tolerance, cardiovascular function, muscle function, ability to engage in daily physical activities, and overall quality of life of persons with CKD. A medical evaluation and exercise testing are recommended before starting an exercise program. Both aerobic and resistance exercises are recommended. Aerobic exercises such as cycling and walking are recommended for 3–5 days per week at moderate intensity. Resistance exercises using free weights, resistance bands, or machines are recommended for 2–3 days per week at low to moderate resistance with higher repetitions, typically one set of 10–15 repetitions. For athletes with a solitary kidney, although the risks of significant injury to the kidney in contact or collision sports are minimal, there is no consensus recommendation, and each case should be individualized with full explanation of all the potential risks. In most athletes, exercise-related microscopic hematuria and proteinuria is generally benign and transient. Ingestion of excess protein does not seem to result in adverse impact on the kidneys; however, in those with any renal disease appropriate modification is recommended as part of the overall management.

REFERENCES

Adey, D., R. Kumar, J.T. McCarthy et al. 2000. Reduced synthesis of muscle proteins in chronic renal failure. *Am J Physiol Endocrinol Metab.* 278:E219–25.

Alayh, G., O. Ozkaya, K. Bek, A. Calmasur et al. 2008. Physical function, muscle strength and muscle mass in children on peritoneal dialysis. *Pediatr Nephrol.* 23:639–44.

Barton, B.E. 2001. IL-6 like cytokines and cancer cachexia: Consequences of chronic inflammation. *Immunol Res.* 23:41–58.

Becker, B., F. Kronenberg, J.T. Kielstein et al. 2005. Renal insulin resistance syndrome, adiponectin and cardiovascular events in patients with kidney disease: The mild and moderate kidney disease study. *J Am Soc Nephrol.* 16:1091–8.

Bernard, J.J. 2009. Renal trauma: Evaluation, management, and return to play. *Curr Sports Med Rep.* 8(2):98–103.

Bernhardt, D.T., W.O. Roberts, eds. 2010. *Pre-Participation Physical Evaluation*, 4th ed. Elk Grove Village: AAP.

Booher, M.A., B.W. Smith. 2001. Physiological effects of exercise on the cardiopulmonary system. *Clin Sports Med.* 22:1–21.

Bouby, N., S. Bachmann, D. Bichet et al. 1990. Effect of water intake on the progression of chronic renal failure in the 5/6 nephrectomized rat. *Am J Physiol.* 258: F973–9.

Castaneda, C., P.L. Gardon, K.L. Uhlin et al. 2001. Resistance training to counteract the catabolism of a low-protein diet in patients with chronic renal insufficiency. *Ann Intern Med.* 135:965–76.

Cheema, B.S, M.A. Singh, 2005. Exercise training in patients receiving maintenance hemodialysis: As systematic review of clinical trials. *Am J Nephrol.* 25:352–64.

Clapp, E.L., A. Bevington, A.C. Smith et al. 2012. Exercise for children with chronic kidney disease and end stage renal disease. *Pediatr Nephrol.* 27(2):165–72.

Clark, W.F., J.M. Sontrop, J.J. Macnab et al. 2011. Urine volume and change in estimated GFR in a community-based cohort study. *Clin J Am Soc Nephrol.* 6:2634–41.

Colan, S.D., S.P. Sanders, J.R. Ingelfinder et al. 1987. Left ventricular mechanics and contractile state in children and adolescents with end stage renal disease: Effect of dialysis and renal transplantation. *J Am Coll Cardiol.* 10:1085–94.

Collins, A.J., R.N. Foley, C. Herzog et al. 2011. US renal data system 2010 annual data report. *Am J Kidney Diseases* 57(A8):e1–526.

Copley, J., J. Lindberg. 1999. The risks of exercise. *Adv Ren Replace Ther.* 6(2):165–71.

Cribb, P.J., A. Hayes. 2006. Effects of supplement timing and resistance exercise on skeletal muscle hypertrophy. *Med Sci Sports Exerc.* 38:1918–25.

Deligiannis, A. 2004a. Exercise rehabilitation and skeletal muscle benefits in hemodialysis patients. *Clin Nephrol.* 61(Suppl 1):S46–50.

Deligiannis, A. 2004b. Cardiac adaptations following exercise training in hemodialysis patients. *Clin Nephrol.* 61(Suppl 1):S39–45.

Deligiannis, A., E. Koidi, E. Tassolas et al. 1999. Cardiac response to physical training in hemodialysis patients: An echocardiographic study at rest and during exercise. *Int J Cardiol.* 70:253–66.

DePaul, V., J. Moreland, T. Eager et al. 2002. The effectiveness of aerobic and muscle strength training in patients receiving hemodialysis and EPO: A randomized controlled trial. *Am J Kidney Dis.* 40(6):1219–29.

Drew, R.C., M.D. Muller, C.A. Blaha et al. 2013. Renal vasoconstriction is augmented during exercise in patients with peripheral arterial disease. *Physiol Rep.* 1(6):e00154.

Du, J., X. Wang, C. Miereles et al. 2004. Activation of caspase-3 is an initial step triggering accelerated muscle proteolysis in catabolic conditions. *J Clin Invest.* 113:115–23.

Eijsermans, R.M., D.G. Creemers, P.J. Helders et al. 2004. Motor performance, exercise tolerance and health related quality of life in children on dialysis. *Pediatr Nephrol.* 19:1262–6.

Esmarck, B., J.L. Andersen, S. Olsen et al. 2001. Timing of post-exercise protein intake is important for muscle hypertrophy with resistance training in elderly humans. *J Physiol.* 535 (Pt 1):301–11.

Francesconi, R.P., M.N. Sawka, K.B. Pandolf et al. 1983. Hypohydration and heat acclimation: Plasma renin and aldosterone during exercise. *J Appl Physiol.* 55:1790–4.

Fenzl, M., W. Schnizer, N. Aebli et al. 2013. Release of ANP and fat oxidation in overweight persons during aerobic exercise in water. *Int J Sports Med.* 34(9):795–9.

Foster, C., J. Porcari. 2001. The risks of exercise training. *J Cardiopulm Rehabil.* 21:347–52.

Gerstenbluth, R., P. Spirnak, J. Elder et al. 2002. Sports participation and high grade renal injuries in children. *J Urol.* 168:2575–8.

Goldstein, S.L., L.R. Montgomery. 2009. A pilot study of twice weekly exercise during hemodialysis in children. *Pediatr Nephrol.* 24:833–9.

Gould, D.W., M.P. Graham-Brown, E.L. Watson et al. 2014. Physiological benefits of exercise in pre-dialysis chronic kidney disease. *Nephrology (Carlton).* 19:519–27.

Greydanus, D.E., A.D. Torres. 2008. Disorders of the kidneys. In: Patel, D.R., Greydanus, D.E., Baker, R., eds. *Pediatric Practice: Sports Medicine.* New York: McGraw Hill Medical, pp. 130–43.

Headley, S., M. Germain, P. Mailoux et al. 2002. Resistance training improves strength and functional measures in patients with end stage renal disease. *Am J Kidney Dis.* 40(2):355–64.

Heiwe, S., S.H. Jacobson. 2011. Exercise training in adults with chronic kidney disease. *Cochrane Database Syst Rev* (10):CD003236.

Heiwe, S., S.H. Jacobson. 2014. Exercise training in adults with CKD: A systematic review and meta-analysis. *Am J Kidney Dis.* 64:383–93.

Higashihara, E., V.E. Torres, A.B. Chapman et al. 2011. Tolvaptan in autosomal dominant polycystic kidney disease: Three years' experience. *Clin J Am Soc Nephrol.* 6:2499–507.

Hogg, R.J., S. Furth, K.V. Lemley et al. 2003. National Kidney Foundation's kidney disease outcomes quality initiative clinical practice guidelines for chronic kidney disease in children and adolescents: Evaluation, classification, and stratification. *Pediatrics* 111(6):1416–21.

Johanson, K.L., P. Painter. 2012. Exercise in individuals with CKD. *Am J Kidney Dis.* 59(1):126–34.

Johnson, B., C. Christensen, S. DiRusso et al. 2005. A need for reevaluation of sports participation recommendations for children with a solitary kidney. *J Urol.* 174:686–9.

Johnson, R.J., ed. 2015. *Comprehensive Clinical Nephrology*, 5th ed. Philadelphia, PA: Elsevier Saunders.

Kennedy, T.L., N.J. Siegel. 1995. Chronic renal disease. In: Goldberg, B., ed. *Sports and Exercise for Children with Chronic Health Conditions.* Champaign, IL: Human Kinetics, pp. 265–78.

Kerksick, C.M., C.J. Rasmussen, S.L. Lancaster et al. 2006. The effects of protein and amino acid supplementation on performance and training adaptations during ten weeks of resistance training. *J Strength Cond Res.* 20:643–53.

Kolewaski, C.D., M.C. Mullally, T.L. Parsons et al. 2005. Quality of life and exercise rehabilitation in end stage renal disease. *CANNT J.* 15(4):22–9.

Konstantinidou, E., G. Koukouvou, E. Kouidi et al. 2001. Exercise renal rehabilitation: Comparison of three exercise programs. *J Rehabil Med.* 34:40–5.

Koudi, E., D. Grekas, A. Deligiannis et al. 2004. Outcomes of long-term exercise training in dialysis patients: Comparison of two training programs. *Clin Nephrol.* 61(Suppl 1): S31–8.

Koufaki, P., S.A. Greenwood, I.C. Macdougall et al. 2013. Exercise therapy in individuals with chronic kidney disease: A systematic review and synthesis of the research evidence. *Annu Rev Nurs Res.* 31:235–75.

Kouidi, E. 2004. Health-related quality of life in end stage renal disease patients: The effects of renal rehabilitation. *Clin Nephrol.* 61(Suppl 1):S60–71.

Kouidi, E., M. Albani, K. Natsis et al. 1998. The effects of exercise training on muscle atrophy in hemodialysis patients. *Nephrol Dial Transplant.* 13:685–99.

Leikis, M.J., M.J. McKenna, A.C. Petersen et al. 2006. Exercise performance falls over time in patients with chronic kidney disease despite maintenance of hemoglobin concentration. *Clin J Am Soc Nephrol.* 1(3):488–95.

Levendoğlu, F., L. Altintepe, N. Okudan et al. 2004. A twelve-week exercise program improves the psychological status, quality of life, and work capacity in hemodialysis patients. *J Nephrol.* 17(6):826–32.

Levey, A.S., J. Coresh, E. Balk et al. 2003. National Kidney Foundation practice guidelines for chronic kidney disease: Evaluation, classification, and stratification. *Ann Intern Med.* 139:137–47.

Malagoni, A.M., L. Catizone, S. Mandini et al. 2008. Acute and long-term effects of an exercise program for dialysis patients prescribed in hospital and performed at home. *J Nephrol.* 21(6):871–8.

McAleer, I.M., G.W. Kaplan, B.E. LoSasso et al. 2002. Renal and testis injuries in team sports. *J Urol.* 168(4 Pt 2):1805–7.

Mercer, T.H., P. Koufaki, P.F. Naish et al. 2004. Nutritional status, functional capacity and exercise rehabilitation in end stage renal disease. *Clin Nephrol.* 61(Suppl 1):S54–9.

Mitsnefes, M.M., T.R. Kimball, W.L. Border et al., 2004. Impaired left ventricular diastolic function in children with chronic renal failure. *Kidney Int.* 65:1461–6.

Miyachi, M., H. Kawano, J. Sugawara et al. 2004. Unfavorable effects of resistance training on central arterial compliance: A randomized intervention study. *Circulation* 110:2858–63.

Moinuddin, I., D.J. Leehey. 2008. A comparison of aerobic exercise and resistance training in patients with and without chronic kidney disease. *Adv Chronic Kidney Dis.* 15(10):83–96.

Momeni, A., A. Nematolahi, M. Nasr et al. 2014. Effect of intradialytic exercise on echocardiographic findings in hemodialysis patients. *Iran J Kidney Dis.* 8(3):207–11.

Mottram, P.M., B. Haluska, S. Yuda et al. 2004. Patients with hypertensive response to exercise have impaired systolic function without diastolic dysfunction or left ventricular hypertrophy. *J Am Col Cardiol.* 43:848–53.

Muller, M.D., R.C. Drew, J. Cui et al. 2013. Effect of oxidative stress on sympathetic and renal vascular responses to ischemic exercise. *Physiol Rep.* Aug;1(3):e00047.

Nguyen, Q.V., A. Kälin, U. Drouve et al. 2001. Sensitivity to meat protein intake and hyperoxaluria in idiopathic calcium stone formers. *Kidney Int.* 59(6):2273–81.

Otani, H., M. Kaya, J. Tsujita et al. 2013. Effect of the volume of fluid ingested on urine concentrating ability during prolonged heavy exercise in a hot environment. *J Sports Sci Med.* 12(1):197–204.

Ouzouni, S., E. Kouidi, A. Sioulis et al. 2009. Effects of intradialytic exercise training on health-related quality of life indices in haemodialysis patients. *Clin Rehabil.* 23(1):53–63.

Painter, P., J.B. Krasnoff. 2009. End stage metabolic disease: Chronic kidney disease and liver failure, In: Durstine, J.L., Moore, G.E., Painter, P.L., and Roberts, S.O., eds. *ACSM's Exercise Management for Persons with Chronic Diseases and Disabilities*, 3rd ed. Champaign, IL: Human Kinetics, pp. 175–81.

Painter, P., J. Krasnoff, R. Mathias et al. 2007. Exercise capacity and physical fitness in pediatric dialysis and kidney transplant patients. *Pediatr Nephrol.* 22:1030–9.

Painter, P., and R.L. Marcus. 2013. Assessing physical function and physical activity in patients with CKD. *Clin J Am Soc Nephrol.* 8(5):861–72.

Painter, P., and B. Roshanravan. 2013. The association of physical activity and physical function with clinical outcomes in adults with chronic kidney disease. *Curr Opin Nephrol Hyperten.* 22(6):615–23.

Patel, D.R., A.D. Torres, D.E Greydanus et al. 2005. Kidneys and sports. *Adolesc Med Clin.* 16(1):111–19.

Pattaragarn, A., B.A. Warday, R.J. Sabath et al. 2004. Exercise capacity in pediatric patients with end stage renal disease. *Perit Dial Int.* 24:274–80.

Pecoits-Filho, R. 2007. Dietary protein intake and kidney disease in western diet. *Nephrol.* 155:102–12.

Peraza, S., C. Wesseling, A. Aragon et al. 2012. Decreased kidney function among agricultural workers in El Salvador. *Am J Kidney Dis.* 59:531–40.

Pescatello, L.S., R. Arena, D. Riebe et al., eds. 2013. *ACSM's Guidelines for Exercise Testing and Prescription*, 9th ed. Philadelphia, PA: Lippincott Williams and Wilkins.

Petras, D., K. Koutroutsos, A. Kordalis et al. 2013. The role of sympathetic nervous system in the progression of chronic kidney disease in the era of catheter based sympathetic renal denervation. *Curr Clin Pharmacol.* 8(3):197–205.

Plante, B., M. Amadei, E. Herbert et al. 1990. Tegaderm dressings for peritoneal dialysis and gastrojejunostomy catheters in children. *Adv Perit Dial.* 6:279–80.

Psooy, K. 2006. Sports and the solitary kidney: How to counsel parents. *Can J Urol.* 13(3):3120–6.

Raj, V., K. Sturgeon, D.R. Patel et al. 2012. Protein intake in athletes: A review. *Int J Disabili Hum Dev.* 11(3):191–7.

Reddy, S.T., C. Wang, K. Sakhaee et al. 2002. Effect of low carbohydrate high-protein diets on acid- base balance, stone forming propensity and calcium metabolism. *Am J Kidney Dis.* 40(2):265–74.

Robertson, W.G., P.J. Heyburn, M. Peacock et al. 1979. The effect of high animal protein intake on the risk of calcium stone-formation in the urinary tract. *Clin Sci (Lond).* 57(3):285–8.

Rump, L.C., K. Amann, S. Orth et al. 2000. Sympathetic overactivity in renal disease: A window to understand progression and cardiovascular complications of uremia. *Nephrol Dial Transplant.* 15:1735–8.

Shalom, R., J. Blumenthal, R. Williams et al. 1984. Feasibility and benefits of exercise training in patients on maintenance dialysis. *Kidney Int.* 25:958–63.

Sharp, D., J. Ross, R. Kay et al. 2002. Attitudes of pediatric urologists regarding sports participation by children with a solitary kidney. *J Urol.* 168:1811–15.

Sietsema, K.E., A. Amato, S.G. Adler et al. 2004. Exercise capacity as a predictor of survival among ambulatory patients with end stage renal disease. *Kidney Int.* 65:719–24.

Smart, N., M. Steele. 2011. Exercise training in haemodialysis patients: A systematic review and meta-analysis. *Nephrology.* 16(7):626–32.

Sreedharan, R., E.D. Anver. 2011. Chronic kidney disease. In: Kliegman, R.M., Stanton, B., Geme, J., Schor, N.F., and Behrman, R.E., eds. *Nelson Textbook of Pediatrics*, 19th ed. Philadelphia, PA: Elsevier, pp. 1822–5.

Stachenfeld, N.S., G.W. Gleim, P.M. Zabetakis et al. 1996. Fluid balance and renal response following dehydrating exercise in well-trained men and women. *Eur J Appl Physiol.* 72:468–77.

Strippoli, G.F., J.C. Craig, E. Rochtchina et al. 2011. Fluid and nutrient intake and risk of chronic kidney disease. *Nephrology (Carlton).* 16:326–34.

Sugiura, T., A. Yamauchi, H. Kitamura et al. 1999. High water intake ameliorates tubulointerstitial injury in rats with subtotal nephrectomy: Possible role of TGF-beta. *Kidney Int.* 55:1800–10.

Tang, J.E., D.R. Moore, G.W. Kujbida et al. 2009. Ingestion of whey hydrosylate, casein or soy protein isolate: Effects on mixed muscle protein synthesis at rest and following resistance exercise in young men. *J Appl Physiol.* 107:987–92.

Tawney, K.W., P. Tawney, G. Hladik et al. 2000. The life readiness program: A physical rehabilitation program for patients on hemodialysis. *Am J Kidney Dis.* 36(3):581–91.

Tenbrock, K., S. Kruppa, E. Mokov et al. 2000. Analysis of muscle strength and bone structure in children with renal disease. *Pediatr Nephrol.* 14:669–72.

Tipton, K.D., B.B. Rasmussen, S.L. Miller et al. 2001. Timing of amino acid-carbohydrate ingestion alters anabolic response of muscle to resistance exercise. *Am J Physiol Endocrinol Metab.* 281:E197.

Van den Ham, E.C., J.P. Kooman, A.M. Schols et al. 2005. Similarities in skeletal muscle strength and exercise capacity between renal transplant and hemodialysis patients. *Am J Transplant.* 5:1957–65.

Wade, C.E. 1984. Response, regulation, and actions of vasopressin during exercise: A review. *Med Sci Sports Exerc.* 16:506–11.

Wade, C.E., J.R. Claybaugh. 1980. Plasma renin activity, vasopressin concentration and urinary excretory responses to exercise in men. *J Appl Physiol.* 49:930–6.

Wang, C.J., J.J. Grantham, J.B. Wetmore et al. 2013. The medicinal use of water in renal disease. *Kidney Int.* 84(1):45–53.

Wang, X.H., J. Du, J.D. Klein et al. 2009. Exercise ameliorates chronic kidney disease-induced defects in muscle protein metabolism and progenitor cell function. *Kidney Int.* 76(7):751–9.

Wang, X.H., W.E. Mitch. 2013. Muscle wasting from kidney failure-a model for catabolic conditions. *Int J Biochem Cell Biol.* 45(10):2230–8.

Weaver, D.J., Jr., T.R. Kimball, T. Knilans et al. 2008. Decreased maximal aerobic capacity in pediatric chronic kidney disease. *J Am Soc Nephrol.* 19:624–30.

Wu, H.Y., B.A. Gaines. 2007. Dirt bikes and all terrain vehicles: The real threat to pediatric kidneys. *J Urol* 178(4 Pt 2):1672–4.

Yokoi, H., Yanagita, M. 2014. Decrease of muscle volume in chronic kidney disease: The role of mitochondria in skeletal muscle. *Kidney Int.* 85(6):1258–60.

16 Practical Considerations for Fluid Replacement for Athletes with a Spinal Cord Injury

Victoria Goosey-Tolfrey, Thomas Paulson, and Terri Graham-Paulson

CONTENTS

16.1 INTRODUCTION—PARALYMPIC SPORT FOR INDIVIDUALS WITH SPINAL CORD INJURIES

The popularity of paralympic sport has been growing steadily over the years, since its first introduction at the Stoke Mandeville games in 1952. This is evident by the increasing number of disciplines; the number of summer sports for which medals were awarded has increased from six in 1952 to above twenty that will be seen at the

333

Rio Paralympic Games in 2016. Since the 1988 Paralympic Games (held in Seoul, Korea), it is notable that the games have been hosted by the same city, in the same year, and at the same venues as the Olympic competitions. The selection of Athens, Greece, (2004) and Beijing, China, (2008) as sites for the Paralympic Games and the forthcoming games in Rio, Brazil, (2016) poses an environmental challenge for competitors with the potential for heat-related illnesses, dehydration, and impaired performance.

The evaluation of fluid balance, hydration, and sports performance for healthy able-bodied (AB) athletes is an important aspect of this book, and the scientific literature is well versed regarding the problems of exercise in the heat (Coris et al. 2004), the effects of dehydration (Maughan 1991; Sawka and Greenleaf 1992), and the benefits of acclimatization (Nielsen 1994), particularly in endurance events. However, the Paralympic Games include athletes with a variety of impairments which produce additional problems when striving to maintain euhydration and regulate temperature. There are currently 17 summer paralympic sports that involve competition for individuals with a spinal cord injury (SCI), across a range of both indoor and outdoor sporting arenas (Table 16.1, which represents the different events by the risk of heat illness). It has been known for many years that athletes with an SCI are less-effective thermoregulators than AB athletes in hot environments (Price 2006), and yet limited scientific research regarding optimal fluid replacement strategies is available for this group of athletes. For example, fluid intake for AB athletes is based on the fact that fluid losses should be replaced during exercise, preferably at a rate equal to sweat rate (Sawka et al. 2007). Guidance on fluid replacement is difficult to determine for athletes with an SCI, where the individual's fluid loss will depend not only on the environmental conditions but also on the level and completeness of the SCI and available sweating capacity (Guttmann et al. 1958). Moreover, since the physiological consequences of an SCI result in a decreased lean body mass due to muscle atrophy, leading to a lower resting metabolic rate, and in turn, a further reduction in energy expenditure (EE), AB fluid replacement strategies are almost

---

TABLE 16.1

Needs Analysis of the *Heat Stress* Risk Associated with Paralympic Sporting Competition Involving Athletes with an SCI

Heat Stress Level	Event	Outdoor/Indoor
High risk (Level 3)	Athletics, equestrian, paratriathlon, road cycling, wheelchair tennis	Outdoor
	Wheelchair rugby	Indoor
Intermediate risk (Level 2)	Paracanoe, rowing, sailing	Outdoor
	Swimming, table tennis, wheelchair basketball, wheelchair fencing	Indoor
Low risk (Level 1)	Archery boccia, powerlifting	Outdoor
	Shooting, sitting volleyball	Indoor

Source: Webborn, A.D.J., *Br. J. Ther. Rehabil.*, 3(8), 429–35, 1996.

certainly not directly transferable to athletes with an SCI. Thus, the purpose of this chapter is to describe the physiological and anatomical consequences of an SCI, with the main aim to provide some practical information for coaches and practitioners to fully understand the hydration needs of athletes with an SCI so that they can apply their knowledge to a condition perhaps beyond their primary area of expertise.

16.1.1 Epidemiology of Spinal Cord Injury

Traumatic spinal injuries occur when direct or indirect forces applied to the vertebral column lead to damage of the spinal cord, with motor vehicle collisions, falls, violence, and sports activities among the leading causes (Lee et al. 2013). A bimodal distribution in injury prevalence is present with regard to age, with a first peak in young adults aged between 15 and 29 years and a smaller, but growing, second peak over the age of 60. Predicted global estimates suggest an incidence of 200,000 injuries per annum worldwide (Lee et al. 2013) with around 40,000 people living with an SCI in the United Kingdom (Webborn and Goosey-Tolfrey 2008).

Whether an SCI is traumatic or non-traumatic, the resultant deficits in motor, sensory, and autonomic functions are dependent on the level and the pattern of spinal lesion, mainly in the transverse plane. Remaining sensory (dermatome) and motor (myotome) function is examined according to international guidelines for neurological classification (American Spinal Injuries Association—ASIA) to identify lesion level and completeness (Kirschblum et al. 2011), see Figure 16.1. Injury level is described as the most caudal (*away from the head*) spinal segment with remaining function. *Tetraplegia* refers to impairment or loss of motor and/or sensory function in the cervical segments of the spinal cord and results in some impairment of the arms as well as typically the trunk, legs, and pelvic organs (Kirschblum et al. 2011). *Paraplegia* refers to impairment or loss of motor function in the thoracic, lumbar, or sacral segments of the spinal cord, with lesion level-dependent losses in trunk, leg, and pelvic function (Kirschblum et al. 2011). A complete SCI results in the complete loss of both sensory and motor communication between the brain and tissue innervated below the lesion level (classified as ASIA grade A). In contrast, the neurology of incomplete injuries is complex and can differ greatly from one case to another (classified as ASIA grade B–D) (Kirschblum et al. 2011). Damage to the neurons of the anterior ventral roots may lead to a loss of motor function but maintenance of sensory function, whereas the converse may apply with damage to neurons of posterior dorsal roots.

16.1.2 Anatomy of the Spinal Cord and Nervous System

Enclosed within the vertebral column, the spinal cord is the major conduit through which motor and sensory information is relayed between the brain and the body (Kirschblum et al. 2011). The spinal cord provides a key structure of the central nervous system (CNS), organizing both conscious and unconscious motor and sensory function. In total, 31 pairs of spinal nerves (8 cervical, 12 thoracic, 5 lumbar, 5 sacral, and 1 coccygeal) act as the bridge between the CNS and the motor and sensory components of the peripheral nervous system. As shown in Figure 16.2, the

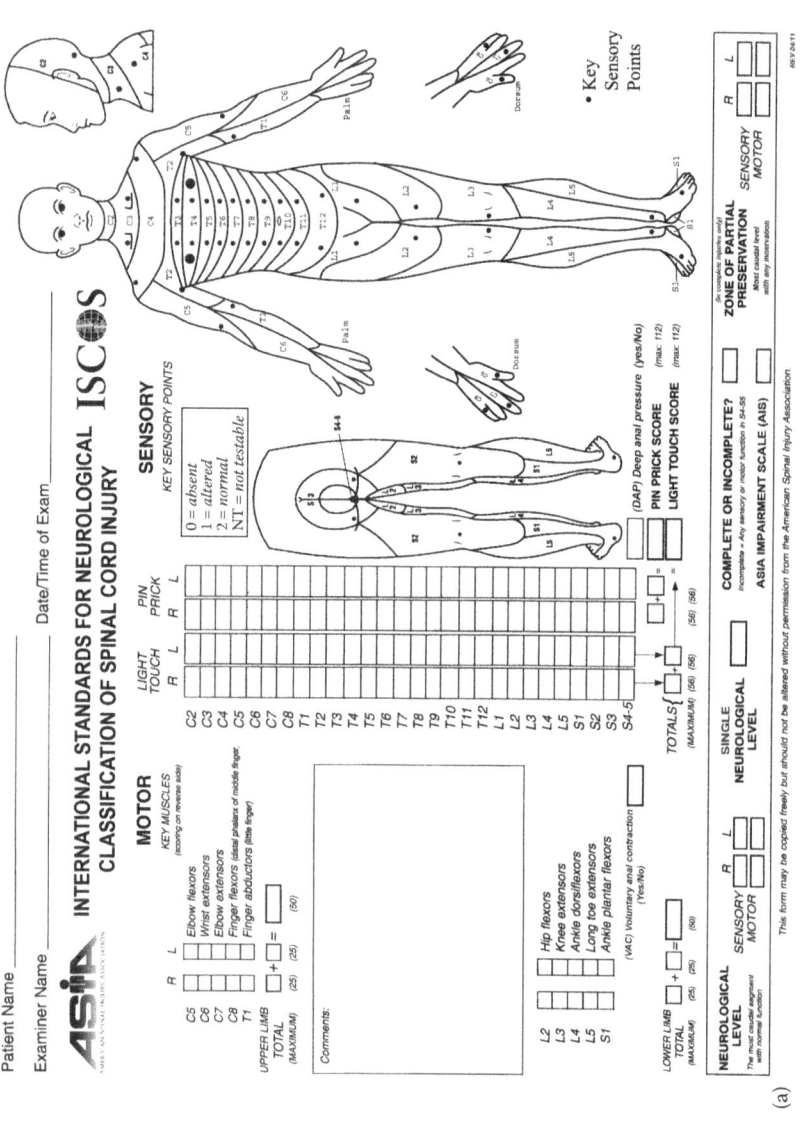

FIGURE 16.1 (a) Assessment sheet. American Spinal Injury Association: International Standards (ASIA) for Neurological Classification of Spinal Cord Injury, revised 2013, Atlanta, GA. Reprinted 2013 (permission granted from ASIA). *(Continued)*

Muscle function grading

0 = Total paralysis

1 = Palpable or visible contraction

2 = Active movement, full range of motion (ROM) with gravity eliminated

3 = Active movement, full ROM against gravity

4 = Active movement, full ROM against gravity and moderate resistance in a muscle specific position.

5 = (Normal) active movement, full ROM against gravity and full resistance in a muscle specific position expected from an otherwise unimpaired peson.

5* = (Normal) active movement, full ROM against gravity and sufficient resistance to be considered normal if identified inhibiting factors (i.e., pain, disuse) were not present.

NT = Not testable (i.e., due to immobilization, severe pain such that the patient cannot be graded, amputation of limb, or contracture of >50% of the range of motion).

ASIA Impairment Scale (AIS)

☐ A = Complete. No sensory or motor function is preserved in the sacral segments S4-S5.

☐ B = Sensory incomplete. Sensory but not motor function is preserved below the neurological level and includes the sacral segments S4-S5 (light touch, pin prick at S4-S5; or deep anal pressure [DAP]), AND no motor function is preserved more than three levels below the motor level on either side of the body.

☐ C = Motor incomplete. Motor function is preserved below the neurological level,** and more than half of key muscle functions below the single neurological level of injury (NLI) have a muscle grade less than 3 (Grades 0–2).

☐ D = Motor incomplete. Motor function is preserved below the neurological level,** and at least half (half or more) of key muscle functions below the NLI have a muscle grade ≥ 3.

E = Normal. If sensation and motor function as tested with the ISNCSCI are graded as normal in all segments, and the patient had prior deficits, then the AIS grade is E. Someone without an initial SCI does not receive an AIS grade.

**For an individual to receive a grade of C or D, that is, motor incomplete status, they must have either (1) voluntary anal sphincter contraction or (2) sacral sensory sparing with sparing of motor function more than three levels below the motor level for that side of the body. The standards at this time allows even non-key muscle function more than 3 levels below the motor level to be used in determining motor incomplete status (AIS B vs. C).

NOTE: When assessing the extent of motor sparing below the level for distinguishing between AIS B and C, the motor level on each side is used; whereas to differentiate between AIS C and D (based on proportion of key muscle functions with strength grade 3 or greater) the single neurological level is used.

(b)

Steps in classification

The following order is recommended in determining the classification of individuals with SCI.

1. Determine sensory levels for right and left sides.

2. Determine motor levels for right and left sides.
   Note: in regions where there is no myotome to test, the motor level is presumed to be the same as the sensory level, if testable motor function above that level is also normal.

3. Determine the single neurological level.
   This is the lowest segment where motor and sensory function is normal on both sides, and is the most cephalad of the sensory and motor levels determined in steps 1 and 2.

4. Determine whether the injury is Complete or Incomplete. (i.e., absence or presence of sacral sparing)
   If voluntary anal contraction = No AND all S4-S5 sensory scores = 0 AND deep anal pressure = No, then injury is COMPLETE. Otherwise, injury is incomplete.

5. Determine ASIA Impairment Scale (AIS) Grade:
   Is injury *Complete?*   If YES, AIS = A and can record ZPP
   (lowest dermatome or myotome on each side with some preservation)

   NO

   Is injury
   motor *incomplete?*   If NO, AIS = B
   (Yes = voluntary anal contraction OR motor function more than three levels below the motor level on a given side, if the patient has sensory incomplete classification)

   YES

   Are *at least half* of the key muscles below the single *neurological* level graded 3 or better?

   NO   YES

   AIS = C   AIS = D

   If sensation and motor function is normal in all segments, AIS = E
   Note: AIS E is used in follow-up testing when an individual with a documented SCI has recovered normal function. If at initial testing no deficits are found, the individual is neurologically intact; the ASIA Impairment Scale does not apply.

FIGURE 16.1 (Continued) (b) Guidelines for assessment. American Spinal Injury Association: International Standards (ASIA) for Neurological Classification of Spinal Cord Injury, revised 2013, Atlanta, GA. Reprinted 2013 (permission granted from ASIA).

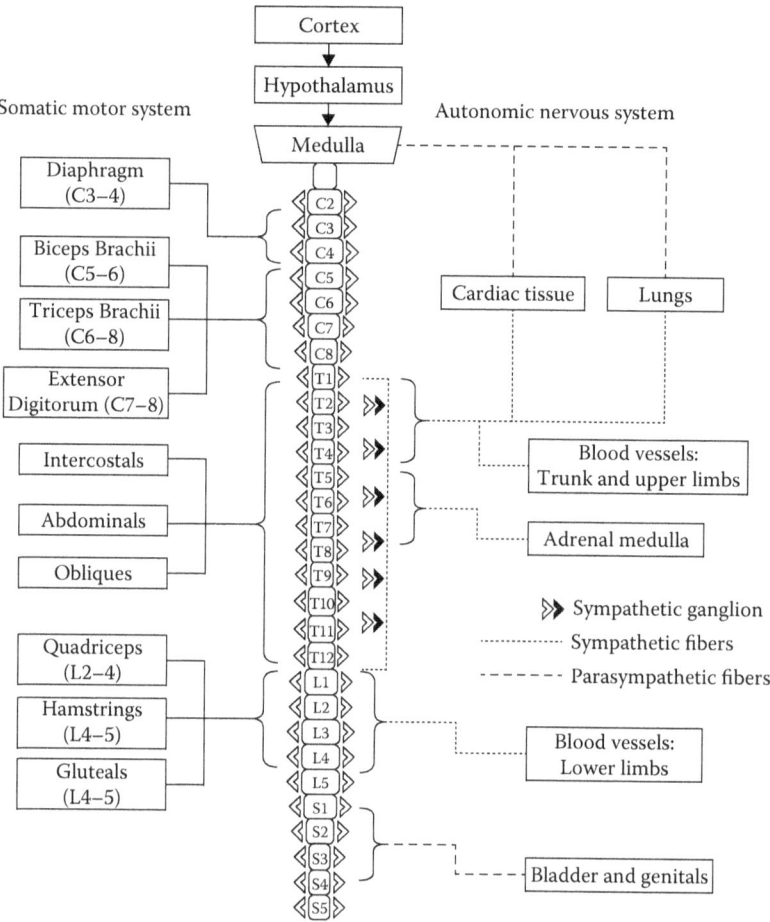

FIGURE 16.2 A schematic representation of the somatic and autonomic nervous systems. (Modified from Krassioukov, A., *Respir. Physiol. Neurobiol.*, 169, 157–64, 2009.)

innervation of target organs is organized in a segmental fashion. The cervical nerves supply the muscles of the upper limbs (e.g., C5–C6 innervates biceps brachii [elbow flexor], C6–C8 innervates triceps brachii [elbow extensor]) and the lumbar and sacral nerves supply lower limbs (e.g., L2–L4 innervates quadriceps [knee extensors]) (Marieb and Hoehn 2007).

While a large portion of motor function is under conscious control, the human body is highly sensitive to challenges against the stability of its internal environment. The autonomic nervous system, with its parasympathetic and sympathetic arms, exists as a division of the peripheral nervous system providing involuntary innervation of smooth muscle (blood vessels), cardiac muscle, and secretory glands (Marieb and Hoehn 2007). Normal autonomic function provides a constant balancing act between activity-supporting sympathetic tone and rest-and-digest parasympathetic tone. With the exception of the sacral outflow to the bladder and genitals

(Figure 16.2), neurons of the parasympathetic nervous system are located in the motor-cranial nerve of the brain stem and synapse directly to the target organ, predominantly via the vagus nerve. Among its many functions, parasympathetic tone decreases heart rate (HR) while promoting energy storage and digestion. In contrast, the gray matter of the thoracic and upper lumbar spinal cord between T1–L2 contains sympathetic nerve fibers within the lateral gray horns. Preganglionic sympathetic fibers enter the adjoining paravertebral sympathetic ganglion (Figure 16.2). Subsequent postganglionic fibers act directly via the neurotransmitter noradrenaline or indirectly via stimulating release of the catecholamines adrenaline and noradrenaline from the adrenal medulla (Marieb and Hoehn 2007). Sympathetic tone elevates HR and breathing rate, controls skeletal muscle blood flow via vasodilation and vasoconstriction of blood vessels in active and inactive muscles, and initiates the secretion of adrenaline and noradrenaline from the adrenal medulla (Figure 16.2).

16.2 PHYSIOLOGICAL CONSEQUENCES OF SPINAL CORD INJURY

Injury to the spinal cord presents a complex, lesion-level-dependent challenge to respiratory, cardiovascular, autonomic, and skeletal muscle function. Due to the increasing loss of functional muscle mass and autonomic control, higher levels of injury result in impaired cardiovascular function and oxygen demand both at rest and during exercise (Haisma et al. 2006; Leicht et al. 2012). The physiological consequences of SCI therefore have a large impact on an individual's physical capacity and EE experienced during exercise.

16.2.1 CARDIOVASCULAR AND AUTONOMIC CONTROL

At rest, and in response to exercise and orthostatic challenge, the redistribution of blood following an SCI is impaired due to the lack of sympathetic vasoconstriction in inactive tissue below the lesion level (Jacobs et al. 2002). The *pooling* of blood in the lower limbs and inactive venous muscle pump reduce ventricular refilling and therefore stroke volume during exercise (Jacobs et al. 2002). Cardiac output ($\dot{Q}$) is maintained by elevations in resting and submaximal HR in individuals with paraplegia (Hopman et al. 1993). Methods to increase ventricular filling, including exercising in a supine position, have been shown to increase stroke volume at submaximal exercise intensities (Hopman et al. 1998). In individuals with tetraplegia, the redistribution of blood and ability to elevate $\dot{Q}$ is further limited due to the loss of autonomic control of vessels in the abdominal bed and cardiac tissue (Thijssen et al. 2009). A spinal cord lesion above T5 results in the loss of sympathetic outflow to the heart and peak HR's (HR_{peak}) of around 100–140 b min^{-1} (Valent et al. 2007). Partial cardio-acceleration is maintained by the withdrawal of parasympathetic tone.

Depressed plasma concentrations of adrenaline and noradrenaline are observed at rest and during exercise (Schmid et al. 1998a, 1998b) following the loss of sympathetic outflow to adrenal glands at T6. This absent exercise-induced systemic vasoconstriction and cardiac stimulation further limits the ability to augment $\dot{Q}$ during exercise (Hopman et al. 1993). This hypokinetic circulation results in a decreased left ventricular mass (~25% less than non-SCI) and smaller left ventricular volume in both untrained (De Groot et al. 2006) and trained (West et al. 2012) individuals with a cervical level SCI compared to non-SCI controls.

The loss of autonomic function following high thoracic and cervical level injury presents two distinct challenges to health and exercise performance, namely, *orthostatic hypotension* and *autonomic dysreflexia*. Vasomotor centers in the medulla reflexively control the cardiovascular system by adjusting sympathetic and parasympathetic outflow to the heart and peripheral vasculature in order to maintain blood pressure (Krassioukov 2009). The loss of sympathetic nervous system outflow in tetraplegia results in bradycardia and chronic hypotension induced via a constant state of vasodilation. Postural change to an upright position results in an unmodulated drop in blood pressure called *orthostatic hypotension*. Extreme hypotension is transient following an SCI, resolving within a few weeks of injury due to a compensatory reduction in parasympathetic outflow and return of sympathetic reflexes.

Autonomic dysreflexia is a potentially life-threatening bout of extreme, uncontrolled hypertension resulting from severe vasoconstriction and cardiac stimulation. Incidences occur when a noxious stimulus below the lesion level triggers an excessive sympathetic response due to the loss of inhibition descending from the medulla and/or hypersensitivity of sympathetic neurons (Krassioukov and Claydon 2006). Common causes include bladder or rectal distension, and failure to remove the stimulus can lead to renal failure, cerebrovascular accidents, and in extreme cases, even death (Krassioukov and Claydon 2006).

16.2.2 Physical Capacity

Physical capacity is the ability of the musculoskeletal, cardiovascular, and respiratory systems to undertake a level of physical activity. Peak anaerobic power output and muscular strength show a similar lesion-level-dependent relationship to cardiovascular and respiratory function (Dallmeijer et al. 1996). This is an important factor as most daily activities during wheelchair propulsion, including negotiating curbs or a slope, require substantial anaerobic power (Hutzler 1998; Janssen et al. 2002). Subsequently, individuals with tetraplegia exhibit higher physical strain during tasks of daily living, including transferring to bed and entering a car, than those with paraplegia (thoracic level injuries) (Janssen et al. 1994). For the purposes of this succinct chapter, the reader is directed to Janssen et al. (2002) to find normative physical capacity values for which an aerobic capacity of >1.19 L min^{-1} and >2.31 L min^{-1} suggests excellent physical capacity for persons with tetraplegia and paraplegia, respectively.

Wheelchair sport adds another layer of complexity, and athletes differ with respect to their sports classification (e.g., for the purposes of this chapter, level and completeness of SCI), which consequently influences physical capacity and EE. A rapid decline in muscle fiber cross-sectional area is observed in the lower limbs following paralysis, with a direct relationship reported between the extent of muscle atrophy and injury duration (Castro et al. 1999). Subsequent reductions in resting and daily EE are directly proportional to the volume of lean body mass lost. Therefore,

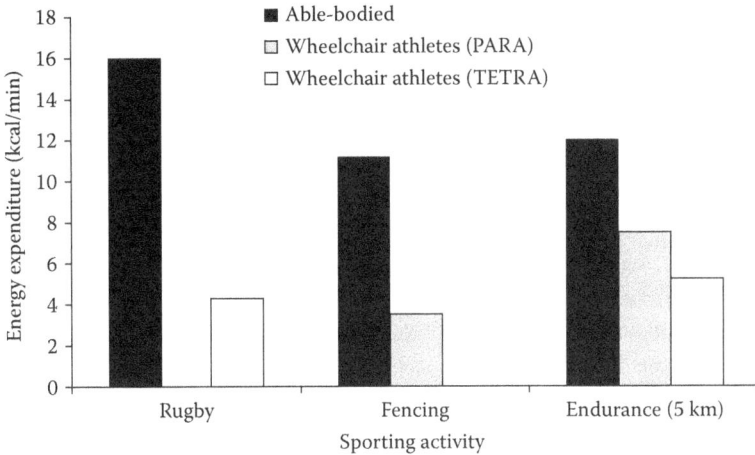

FIGURE 16.3 A comparison between the reported EE values for AB athletes and the equivalent wheelchair-based sports for persons with an SCI. PARA—athletes with paraplegia, TETRA—athletes with tetraplegia. (Data from Coutts, A. et al., *J. Sports Sci.,* 21, 97–103, 2003; Abel, T. et al., *Spinal Cord.,* 46, 785–90, 2008; McArdle, W.D. et al., *Exercise Physiology: Energy, Nutrition and Human Performance,* 2nd edition, Lea and Febiger, Philadelphia, PA, 642–49, 1986; Bernardi, M. et al., *Int. J. Sports Cardiol.,* 5, 58–61, 1998; Lakomy, H.K. et al., *Br. J. Sport Med.,* 21, 130–33, 1987; Ramsbottom, R. et al., *Br. J. Sport Med.* 21, 9–13, 1987.)

using AB values for EE may overestimate the needs of athletes with an SCI due to the reduced muscle mass and type of activity completed as shown in Figure 16.3 for which the reader is referred to the review of Price (2010).

16.2.3 GASTRIC EMPTYING

The ability of the digestive system to absorb and digest nutrients is a key factor to consider when developing any hydration strategy. In a clinical setting, gastric emptying of both liquid and solid meals can be prolonged in individuals with an SCI (Williams et al. 2012), and this may be more prominent in females and those with a high lesion level (Kao et al. 1999). Hence, individuals with tetraplegia may be more susceptible to delayed gastric emptying and commonly reported gastrointestinal (GI) symptoms such as dysphagia, nausea, bloating, and early satiety. Gastric emptying times were reported to be 10.6 ± 7.2 h in subjects with an SCI (complete and incomplete paraplegia and tetraplegia) compared to 3.5 ± 1.0 h in control AB subjects (Williams et al. 2012). This prolonged gastric emptying time may have implications on the timing, and composition of liquid and/or solid food athletes should ingest before and during exercise or competition.

GI symptoms in athletes can have a detrimental effect on sports performance (De Oliveira and Jeukendrup 2013) and given the demands and postural requirements of many wheelchair sports, this is an important consideration for athletes with an SCI. A recent review (de Oliveira et al. 2014) suggests that moderate intensity exercise does not have a significant impact on GI tract motility in AB individuals but gastric emptying in particular may be inhibited at higher exercise intensities. A number of disability sports played by individuals with an SCI (wheelchair basketball, rugby, tennis, and ice sledge hockey) are intermittent in nature, and some evidence suggests that gastric emptying may also be delayed during this type of exercise (Leiper et al. 2001a, b).

Given the variable demands placed on an athlete during different wheeled sports (e.g., endurance vs. intermittent; contact vs. non-contact) and the various roles an athlete can play within a team, there are a number of seating positions and abdominal binding options which an athlete can adopt (Goosey-Tolfrey 2010). It has been noted that the posture assumed by AB cyclists can increase the prevalence of upper GI symptoms, possibly due to increased pressure on the abdomen (De Oliveira and Jeukendrup 2013). In wheelchair racing, a similar and sometimes more extreme aerodynamic posture is commonly used to help optimize performance (Figure 16.4). It is therefore important to consider the impact an athlete's position may have on their GI motility when recommending drinking and fueling practices. Training the gut through the use of these positions during training sessions can help to reduce the incidence of GI symptoms in those who are susceptible. It should also be highlighted that GI discomfort, especially during intense exercise can be exacerbated when an individual is hypohydrated (Rehrer et al. 1990). All athletes, especially individuals with an SCI, should therefore take care to remain hydrated during exercise and competition.

Evidence shows that nutritional supplement use is common among disabled athletes and that sports drinks are one of the top three supplements being used by this population (Graham-Paulson et al. 2015). There are no disability-specific guidelines regarding the type, amount, or frequency, which these products should be ingested and therefore we look to the AB literature once again. When utilizing isotonic sports drinks for hydration and fueling, lowering the carbohydrate concentration (hypotonic: compared to an isotonic ~6% solution) does not appear to improve gastric emptying or intestinal water absorption in AB participants (Rogers et al. 2005). However, higher carbohydrate concentrations and therefore higher osmolalities (>500 mOsm/L) are likely to delay gastric emptying and could cause unwanted symptoms (De Oliveira et al. 2014) such as cramping, bloating, and diarrhea. Many unwanted GI symptoms can be prevented or at least managed through appropriate fluid and fuel choices.

FIGURE 16.4 Posture of an athlete with an SCI competing in wheelchair racing and hand cycling. (Courtesy of Alexandra Shill.)

16.2.4 THERMOREGULATION AND SWEAT RATES

Surprisingly, given that many Paralympic Games have taken place in hot climates (Sydney, Athens, and Beijing) and international qualification events for wheelchair athletes are worldwide, little information is available on either fluid replacement strategies or thermoregulatory responses during wheelchair propulsion in the heat (Goosey-Tolfrey et al. 2008). It is well documented that athletes with an SCI have a (1) reduced sympathetic input to the thermoregulatory center, (2) loss of sweating capacity, and (3) loss of vasomotor control below the level of the spinal lesion (Price 2006). Since the redistribution of blood and sweating are two major thermoregulatory effectors, persons with an SCI have compromised thermoregulation and are at a greater risk of heat illness than AB individuals (Price 2006). Further specific discussion on sweating and cutaneous circulation in SCI is covered in Chapter 3 (Section 3.3.3). As with physical capacity, the extent of the thermoregulatory impairment is proportional to the level and completeness of the lesion level. Athletes with tetraplegia will demonstrate little or no sweating response due to the spinal lesion being above the sympathetic outflow (cervical region), placing them at a greater risk of a heat-related illness. A high level of thermal strain may be apparent not only in outdoor sports such as athletics/wheelchair tennis but also in wheelchair rugby which although is played indoors involves high intensity intermittent efforts (Rhodes et al. 2015) (see Table 16.1). Even during steady-state arm cranking (60% $\dot{V}O_{2peak}$ for 60 min at ~21.5°C), Price and Campbell (1999) found a continuous increase in core temperature (T_{core}), for an athlete with tetraplegia, in contrast to a plateau experienced by AB athletes and athletes with paraplegia. While the athlete with tetraplegia did not experience high thermal strain in these conditions, the continuous rise in T_{core} shows that thermal balance was not achieved. Early work reported by Hopman (1994) has suggested that when arm cranking in a hot environment, there is a linear relationship between sweat produced and the heat liberated for both AB and persons with an SCI (718, 425, 335, and 176 mL) between T10–T12, T7–T9, and T2–T6, respectively. Evidently, the majority of literature will suggest that for persons with an SCI above T6 that a balance between heat production and heat dissipation during exercise will not be met (Price 2006). Given the differences in physical strain between arm-crank and wheelchair ergometry, it is important to understand the fluid intakes and sweat rates during wheelchair propulsive exercise representing various sporting scenarios as shown in Table 16.2 which will be discussed in the Section 16.3.

Athletes with an SCI can take part in many different sports and a number of these require them to wear heavy and/or protective clothing during competition such as wheelchair fencers. Athletes taking part in these sports must be aware of the influence clothing can have on thermoregulation and sweat losses. In athletes who already have thermoregulatory issues, clothing can further hamper the loss of heat during exercise (Havenith 1999). The potential increase in body temperature and therefore sweat losses should not be underestimated and where possible should be monitored in order to produce an individual hydration strategy for these athletes (Table 16.2).

TABLE 16.2

Fluid Intakes and Sweat Rates during Intermittent Sprint Performance (ISP), Training and Steady-State Wheelchair Propulsion for Wheelchair Athletes in a Variety of Environmental Conditions and Cooling Interventions for Wheelchair Athletes

Reference	Participants/Sport	Disability—Level/Completeness of SCI	Duration/Intensity/Modality	Intervention	Fluids Provided	Pre-BM (kg)	Post-BM (kg)	Fluid Intake (mL/h)	Sweat Rate (mL/h)	Actual Change in BM (with Intake of Fluids) (%)	kg
Goosey-Tolfrey et al. (2008)	8 male/female tennis players	Single leg amputation to C6/C7 SCI	60 min ISP (WERG) at 30.4°C ± 0.6°C	Localized head/neck cooling garments	Water ad libitum at five predetermined intervals	66.7 (17.3)	66.7 (17.4)	700 (393)	687 (409)	0.00 (0.59)	0.00 (0.40)
				No cooling		66.5 (17.3)	67.1 (17.4)[b]	1198 (657)[b]	648 (360)	0.85 (0.62)[b]	0.60 (0.40)
Diaper and Goosey-Tolfrey (2009)	1 female tennis player	SCI: L1 incom	60 min ISP (WERG) at 30.4°C ±0.6°C	Localized head/neck cooling garments	Water ad libitum at five predetermined intervals	47.6	47.2	904	1304	-0.84	-0.40
				No cooling		47.2	48.1	1949	1049	1.91	0.90
Black et al. (2013)	12 male wheelchair rugby players	SCI: C5–6 com to C7–8 incom	On-court wheelchair rugby training sessions	Training sessions in the morning (am) and afternoon (pm)	Own sports drink ad libitum	76.8 (18.2)	–	250 (110)	170 (210)	0.41 (0.65)	–
			Morning 157 min 23.0°C ± 1.4°C Afternoon 106 min 27.5°C ± 0.7°C			–	–	520 (300)	160 (340)	0.69 (1.24)	–

(Continued)

TABLE 16.2 (Continued)

Fluid Intakes and Sweat Rates during Intermittent Sprint Performance (ISP), Training and Steady-State Wheelchair Propulsion for Wheelchair Athletes in a Variety of Environmental Conditions and Cooling Interventions for Wheelchair Athletes

Reference	Participants/ Sport	Disability— Level/ Completeness of SCI	Duration/ Intensity/ Modality	Intervention	Fluids Provided	Pre-BM (kg)	Post-BM (kg)	Fluid Intake (mL/h)	Sweat Rate (mL/h)	Actual Change in BM (with Intake of Fluids) (%)	kg
Goosey-Tolfrey et al. (2014)	6 male wheelchair rugby players	SCI[a]: C5–6 com to C7–8 incom	On-court wheelchair rugby training session 60 min 21°C–22°C	Training session	Own sports drink *ad libitum*	85.2 (13.2)	86.2 (13.4)	1025 (351)	32 (443)	1.15 (0.30)	1.00 (0.37)
Price and Campbell (2003)	28 wheelchair athletes	SCI; 8 TP, 10 HP, 10 LP	60-min steady-state wheelchair exercise at 60% $\dot{V}O_{2peak}$ at 31°C	None	Water *ad libitum*			TP 764 (342) HP 472 (252) LP 381 (251)	TP 310 (420) HP 710 (530) LP 690 (530)	TP 0.45 (0.60) HP −0.24 (0.38) LP −0.31 (0.44)	–

Note: Values are means (±SD) for each trial.

a Calculations excluding amputee athlete, BM, body mass; WERG, wheelchair ergometer; ISP, intermittent sprint protocol; TP, tetraplegia; HP, high level paraplegia; LP, low-level paraplegia.

b Sig difference between conditions of the study.

16.3   FLUID REPLACEMENT

16.3.1   DRINKING PRACTICES IN ATHLETES WITH A SPINAL CORD INJURY

As with AB individuals and athletes, fluid consumption in SCI populations can vary greatly. In a clinical population, daily fluid intakes appear to differ slightly between acute (2600 ± 800 mL/day) and chronic patients with an SCI (3100 ± 1200 mL/day) (Perret and Stoffel-Kurt 2011). Nonetheless, they appear to be in line with the recommendations for young men and women (aged 19–30 years) (Institute of Medicine 2005). Fluid intakes in an athletic population also appear to be adequate with 23 wheelchair basketball players reporting average daily intakes of 2300 mL over a 7-day period (Goosey-Tolfrey and Crosland 2010). Sufficient daily fluid intakes are also deemed important, often in combination with the consumption of fiber, to help prevent constipation in some individuals with an SCI. Persons with an SCI are at increased risk of developing urinary tract infections (UTIs) due to a number of structural and physiological factors such as high-pressure voiding, bladder over-distention, and their method of bladder drainage. The consumption of cranberry juice is widespread among individuals with an SCI given the belief that it can help to prevent and relieve symptoms of UTIs (Goosey-Tolfrey and Crosland 2010; Graham-Paulson et al. 2015). For those with an indwelling catheter, it is important to maintain a steady fluid intake across the day to ensure hydration and to help *flush* through the catheter to help prevent the risk UTIs.

In line with Hopman's (1994) sweat rate data, fluid intakes also appear to differ during exercise in persons with high and low lesion levels during wheelchair propulsive exercise (Table 16.2). Athletes with tetraplegia consumed greater fluid volumes (764 ± 342 mL/h) during exercise (60 min at 60% $\dot{V}O_{2peak}$ at 31°C) than those with high- and low-level paraplegia (472 ± 252 and 381 ± 251 mL/h, respectively) (Price and Campbell 2003). For the athlete with a higher lesion level, the inability to thermoregulate effectively, as described earlier, results in a continual increase in core body temperature and consequently a greater degree of thermal strain. The consumption of cool/cold fluids has been shown to help attenuate the thermal discomfort experienced during exercise (Lee and Shirreffs 2007). However, the combination of reduced sweating capacity and fluid loss, and high fluid intakes in athletes with tetraplegia can result in the athlete gaining weight during exercise. Table 16.2 shows elite athletes with an SCI gaining up to 1.9% body mass during only 1 h of exercise. An increase in body weight during exercise may be detrimental to performance because it will increase the rolling resistance of the wheelchair-user combination.

Other ways to help reduce thermal strain include external cooling methods such as ice vests, fans, water sprays, and hand and foot cooling techniques (Griggs et al. 2015). These methods appear to improve the athlete's thermal comfort and ratings of perceived exertion (Webborn et al. 2005; Goosey-Tolfrey et al. 2008), yet there is limited evidence on any gains in performance and the effectiveness of one technique over the other with reducing core body temperature (Griggs et al. 2015; Trbovich et al. 2014). Nevertheless, Goosey-Tolfrey et al. (2008) highlighted a 42% reduction in fluid intake when cooling methods (localized to the head and neck) were used compared to no cooling (Table 16.2). This observation of reduced fluid consumption, when using a cooling method, may be a greater issue during prolonged events such as outdoor wheelchair tennis and cycling (road). Both the athlete and the practitioner should therefore be aware of this possible change in drinking behavior alongside the use of cooling methods and should ideally practice the planned drinking strategy with the cooling method to ensure hypohydration is prevented.

The act of gaining substantial weight via fluid consumption during exercise can be dangerous in extreme cases where hyponatremia (a serum sodium concentration <135 mmol/L) can result. This is not a common occurrence in athletes with an SCI; however, it is more likely to happen if an athlete hydrates with water only as this dilutes the plasma sodium concentration. As previously mentioned, excess fluid consumption can result in a distended bladder which can lead to autonomic dysreflexia. In athletes with a high-level SCI, autonomic dysreflexia occurs spontaneously but it can also be triggered deliberately to help improve performance, known as *boosting*. This reflex is a health risk and so the International Paralympic Committee (IPC) forbids athletes to compete in this state (systolic blood pressure ≥180 mmHg). Any deliberate attempt to induce autonomic dysreflexia will result in disqualification and possible further action by the IPC Legal and Ethical Committee. More information can be found in the IPC position statement on autonomic dysreflexia and boosting (IPC 2014). The most important point for consideration by practitioners is that if autonomic dysreflexia occurs, it is essential that prompt action is taken to remove the cause. Potential causes include over distention of the bladder, a blocked catheter, constipation, an UTI, bladder stones, pressure sores, and any condition that would have produced pain prior to the injury. If an athlete appears prone to autonomic dysreflexia, they should seek further medical advice.

Hyperhydration and consequent weight gain can also influence performance, especially in sports such as wheelchair rugby and tennis where acceleration and the ability to repeatedly overcome inertia (made harder following an increase in body mass) are key contributors to performance. The majority of the AB literature focuses on the effects of hypohydration and practitioner recommendations tend to reflect this. However, it is clear that the effects of hyperhydration must also be considered in an SCI population before recommending hydration strategies.

16.3.2 MAKING FLUID CHOICES IN ATHLETES WITH A SPINAL CORD INJURY

16.3.2.1 What?

Water should remain the fluid of choice on a daily basis and during short-duration, low-intensity exercise sessions. Recommended volumes will vary based on the individual athlete's ability to sweat. There is very little evidence regarding the use of sports drinks by athletes with an SCI (Spendiff and Campbell 2005). However, there is no reason to believe they would not be useful during long-duration, high-intensity exercise in which glycogen stores are challenged or when an athlete is unable to consume appropriate and sufficient fuel prior to exercise. The EEs of various sports were provided in the introduction and highlight the low values reported during selected wheeled sports. Some athletes with an SCI may avoid the use of sports drinks because they are fearful of consuming more energy than they expend, which could ultimately contribute to weight gain. During light sessions or for those with a weight loss goal, this viewpoint may be understandable. However, for elite athletes participating in heavy and/or prolonged exercise, some athletes may need to be educated on the advantages of exogenous carbohydrate delivery to enable a high-quality training session and to aid performance in competition.

As mentioned earlier, it is important to consider the composition of a drink. As with any AB athlete, this will vary depending on the sporting situation, the environment, the duration and intensity of the session, their drink preferences, as well as the level and completeness of the SCI, which will provide an indication of EE. A lack of SCI-specific research in this area limits the evidence-based recommendations that a practitioner can provide. However, until this evidence is available, practitioners should continue to use AB guidelines as a starting point and use their knowledge and any available monitoring data to adapt their recommendations. The carbohydrate content of the drink should reflect the energy demand placed on the athlete, their ability to consume carbohydrate in the lead up to a session, and the ability of their GI system to tolerate it.

Given reduced sweat rates and therefore fluid losses of individuals with an SCI, electrolyte losses may be small but the inclusion of electrolytes may still be important to aid the retention of fluid, especially in those with an indwelling catheter, and consequently to help reduce the risk of hyponatremia. Where possible, the analysis of an athlete's sweat composition (via absorbent sweat patches) allows the practitioner to gain a greater understanding about their electrolyte losses. This will therefore help to inform the electrolyte concentration required in a drink.

The temperature of the fluid may also be important as it can encourage/discourage drinking. The current position stand of the American College of Sports Medicine (Sawka et al. 2007) suggests that fluids between 15°C and 21°C are preferred but this can vary between individuals, cultures, and environmental conditions. To help reduce the thermal discomfort experienced by athletes with an SCI, cool/cold fluids may be desired and may encourage drinking. Warm/hot fluids may also stimulate drinking in cool/cold environments such as in cold water events, cold and/or wet outdoor endurance events, or in-between competition bouts.

The inclusion of protein in a post-exercise recovery drink is a common practice among athletes to help stimulate skeletal muscle protein synthesis (Koopman et al. 2007). A recent survey indicates that this practice is also common among athletes with an impairment (Graham-Paulson et al. 2015). At present, there is very little evidence regarding the influence of post-exercise protein on muscle protein synthesis in athletes with an SCI. However, Kressler et al. (2014) recently reported improvements in anaerobic fatigue resistance and aerobic capacity in individuals with tetraplegia following immediate (36–37 g) protein supplementation during a 6-month circuit training program compared to a delayed protein feeding. Yet there was no influence of protein supplementation on upper body strength. Given the dearth of evidence, practitioners should continue with the assumption that an athlete with an SCI has similar protein requirements to those of AB athletes, and hence, this macronutrient should be considered for use in a post-exercise drink in certain situations.

16.3.2.2   How Much?

The amount of fluid an athlete requires on a daily basis will be extremely individual and should be based on their bladder and bowel routine, their training schedule, and consequently their sweat losses. Athletes employing indwelling catheterization or penile sheath urinary drainage typically intake frequent, low volumes of fluid to flush the catheter through and lower the risk of urinary tract complications. Those employing intermittent catheterization techniques (4–5 times per day) are required to closely monitor fluid intake to ensure bladder routines fit around training and competition schedules. Ideally, the amount of fluid consumed should reflect a number of pre- and post-exercise weighing sessions to enable an individual drinking strategy. When providing recommendations for fluid intake, it is important to consider a number of factors: access to toilet facilities, the athlete's habitual bladder control mechanism, and the use of *voluntary dehydration* when traveling. Access to toilet facilities may be limited in some situations such as on the mountain for disability sit-skiers, when an individual is out on the road for a long cycling (road) race or training session, and when a tennis player has his/her racquet strapped to their hand during a match. When traveling, especially flying long-haul or without assistance, some individuals may choose to restrict fluid intake to prevent the hassle of using the toilet facilities. Remember that dehydration increases the risk of UTIs and so where possible safer/healthier alternatives should be encouraged. For individuals who normally use intermittent catheterization as a means of bladder control, an indwelling catheter could be a good option during the travel period. Self-catheterization during travel and when abroad should be performed with care as the risk of infection can be increased. Individuals should aim to start their travel period in a hydrated state and ideally remain so throughout. However, if the individual does choose to dehydrate to any degree, the most important thing is to put a hydration strategy in place upon arrival at their destination. This will hopefully ensure they rehydrate effectively and are euhydrated in time to train or compete. It is also important to consider those with an ileostomy or any GI incompetence; large fluid boluses should be discouraged because they may cause dumping of fluid.

16.3.2.3   When?

As with all athletes, individuals should aim to start exercise in a hydrated state. To achieve this, athletes should drink regularly throughout the day. Sipping little and often is the simplest advice for athletes. Athletes should try not to *catch up* at night as this may result in a disturbed night's sleep. During exercise, athletes should aim to prevent fluid losses of more than 2% body mass to prevent any influence of hypohydration on performance. On the other hand, athletes should also aim to limit weight gain during exercise (hyperhydration) as this may increase the risk of hyponatremia and autonomic dysreflexia in athletes with an SCI.

The timing and frequency of fluid intake may be dependent on the sports rules and/or availability of fluids. A number of events in which athletes with an SCI can compete allow for drinking; sports such as wheelchair tennis and rugby have scheduled breaks, and some sports such as shooting, boccia, and fencing allow for drinks between multiple short bouts/games. Hydration strategies for disability endurance-based sports such as wheelchair racing or cycling (road) may differ slightly to the AB equivalent of the sports. Athletes often make use of portable drinks containers such as Camelbaks on their chairs/handcycles to prevent lost time due to slowing down to consume and/or collect a drink.

As with any athlete, it is recommended that athletes with an SCI practice their hydration strategy in a training situation prior to use in competition. This practice allows an athlete to assess any effects on their normal bladder and bowel routine, to become accustomed to the carbohydrate and electrolyte content of a drink and to understand how much they require.

16.3.3 Implications of Monitoring Fluid Balance in Athletes with a Spinal Cord Injury

Monitoring hydration status can be an extremely useful tool in educating an athlete on the strengths and weaknesses of their current practices. Monitoring can also allow the development of individual drinking strategies to help improve sports performance in different situations. Aside from the common symptoms experienced due to hypohydration (e.g., thirst, a dry mouth, headache, little or no urination, and/or irritability and confusion), many practitioners and athletes wish to determine hydration status in a more quantitative manner.

Body mass can be used as an indication of hydration status for athletes with an SCI as long as wheelchair accessible platform or seated scales are available. However, it requires an athlete to know their normal body mass, which can be problematic given limited availability of scales and/or a prolonged colonic transit time of up to 80 h (Goosey-Tolfrey et al. 2014). However, the use of pre- and post-exercise body mass is a valid tool to assess acute body water losses/gains during different training sessions and in varying environments with an aim of providing session and athlete-specific hydration strategies. There are a few things the practitioner should consider when using pre- and post-exercise weighing with an SCI population: (1) weighing the athlete in minimal clothing can be hard as the athlete may be strapped into a chair wearing gloves or they may not wish to have their catheter bag on show, (2) measuring urine output requires you to reweigh the athlete and chair pre- and post-visiting the toilet (easier than asking them to collect it) and you should be sure to check for accidental adding of weight (bottles, gloves, strapping, etc.), and (3) for heavy sweaters (i.e., paraplegics) be aware that sweat may also be absorbed into the seating material of the wheelchair (this is likely to be minimal).

Blood and urine indices are further quantitative methods of assessing hydration status, and again, both are feasible in athletes with an SCI. Blood plasma markers include osmolality, sodium concentration, hematocrit and haemoglobin changes, and the concentration of hormones that help to regulate body fluids. These measures can be expensive, require analytical expertise, and often require tight methodological control; therefore, they are not as accessible to the practitioner as urinary markers. The most commonly used vein for blood sampling is the antecubital fossa (located at the bend in the elbow). Accessing this site may produce some unnecessary discomfort at the elbow joint which is repeatedly flexed and extended to enable the athlete to propel their wheelchair or handcycle, and so on, and hence, this method should be considered carefully.

Normal urinary markers of hydration used in an AB population are suitable for use with athletes with an SCI. The easiest and most accessible tool to use when monitoring hydration status on a daily basis is urine color. For those who use urinary catheterization, acute changes in hydration status are harder to establish via urine color given the accumulation of urine in the bag. One must also consider the influence of nutritional supplements and medications (more common in this population) on urine color, which can reduce the accuracy of this measure. It is also important to be aware that discoloration or a *cloudy* urine sample may be an indicator of a UTI and may limit the use of urinary markers of hydration. In such a case, the player should be referred to their team doctor or general practitioner for further tests.

When using urine-specific gravity or urine osmolality for a more accurate and quantifiable measure, the practicalities of urine collection for athletes with an SCI should be considered. When asking for a first morning sample, it must be made clear that their catheter bags should be emptied prior to providing this first morning void. For those with limited hand function, they may require assistance with this process or require larger collection pots. Following the monitoring of an athlete's hydration status, their SCI lesion level and completeness, and the individual athlete's bladder and bowel management routine must be considered when providing fluid recommendations.

16.4   PRACTITIONER CHECKLIST

The proceeding summary provides a practical overview of key considerations when implementing and assessing fluid replacement strategies in athletes with an SCI. This final section provides take home messages for practitioners working directly with athletes with an SCI in an applied setting.

- Close collaboration with medical practitioners and physicians is vital to understand an individual's level and completeness of injury as well as commonly employed bladder and bowel management routines (e.g., indwelling or intermittent catheterization).
- Morning urine collections for the assessment of hydration status in athletes with tetraplegia may require assistance; decoloration of urine may represent presence of an underlying UTI and may affect measures of osmolality or urine-specific gravity. Collection must be made from the first morning void and not from urine collected in a night-bag.
- Sweat rates should be assessed on an individual basis and in ambient temperatures reflective of training and competition environments. Where a normal thermoregulatory response is present, traditional methods for determining sweat volume and composition should be employed. For athletes with a thermoregulatory impairment, pre- and post-exercise body mass weighing is an effective practice to indicate whether a state of euhydration is being maintained during a session. Catheters should be emptied prior to weighing and weight loss due to urinary output should be accounted for by weighing and reweighing after each void.

- In athletes with impaired thermoregulation, rates of fluid consumption may be increased in order to offset increases in thermal strain when no cooling strategies are implemented. Attention should be given to the potential increased risk of autonomic dysreflexia via bladder distention and repeated rapid voiding. Conversely, if cooling methods are employed, lower perceptions of thirst and thermal strain may result in inadequate fluid replacement practices. Therefore, cooling methods and drinking strategies should be practiced in combination.
- The composition of fluid replacement drinks will vary depending on the demands of training and competition. Athletes with tetraplegia exhibit lower EEs in training than those with paraplegia, and therefore, nutrient needs should be managed on an athlete-to-athlete basis.
- Wheelchair seating position and the sports rules and/or fluid availability (endurance event or intermittent team sport) directly impact hydration behaviors with respect to access to toilet facilities, GI discomfort, and opportunity for fluid intake. Sport-specific practices must therefore be considered.

ACKNOWLEDGMENTS

We are grateful for the advice provided by Lee Stutely and the proof reading by Katy Griggs.

REFERENCES

Abel, T., P. Platen, S. Rojas Vega, S. Schneider, and H. K. Strüder. 2008. Energy expenditure in ball games for wheelchair users. *Spinal Cord.* 46(12):785–90.

Bernardi, M., I. Canale, F. Felici, and P. Marchetti. 1998. Field evaluation of the energy cost of different wheelchair sports. *Int. J. Sports Cardiol.* 5:58–61.

Black, K. E., J. Huxford, T. Perry, and R. C. Brown. 2013. Fluid and sodium balance of elite wheelchair rugby players. *Int. J. Sport Nutr. Exer. Metabol.* 23(2):110–18.

Castro, M. J., D. F. Apple, R. S. Staron, G. E. Campos, and G. A. Dudley. 1999. Influence of complete spinal cord injury on skeletal muscle within 6 months of injury. *J. Appl. Phys.* 86(1):350–8.

Coris, E. E., A. M. Ramirez, and D. J. Van Durme. 2004. Heat illness in athletes. *Sports Med.* 34(1):9–16.

Coutts, A., P. Reaburn, and G. Abt. 2003. Heart rate, blood lactate concentration and estimated energy expenditure in a semi-professional rugby league team during a match: A case study. *J. Sports Sci.* 21(2):97–103.

Dallmeijer, A., M. Hopman, H. van As, and L. H. van der Woude. 1996. Physical capacity and physical strain in persons with tetraplegia: The role of sport activity. *Spinal Cord.* 34(12):729–35.

De Groot, P. C., A. van Dijk, E. Dijk, and M.T. Hopman. 2006. Preserved cardiac function n after chronic spinal cord injury. *Arch. Phys. Med. Rehabil.* 87(9):1195–200.

De Oliveira, E. P., R. C. Burini, and A. E. Jeakendrup. 2014. Gastrointestinal complaints during exercise: Prevalence, aetiology, and nutritional recommendations. *Sports Med.* 44(S1):79–85.

De Oliveria, E. P., and A. Jeukendrup. 2013. Nutritional recommendations to avoid gastrointestinal complaints during exercise. *Sports Science Exchange*. 26:1–4.

Diaper, N., and V. L. Goosey-Tolfrey. 2009. A physiological case study of a paralympic wheelchair tennis player: Reflective practise. *J. Sport Sci. Med*. 8:300–7.

Goosey-Tolfrey, V. L. ed. 2010. *Wheelchair Sport*. Champaign, IL: Human Kinetics. pp: 1–232.

Goosey-Tolfrey, V. L., and J. Crosland. 2010. Nutritional practices of competitive British wheelchair games players. *Adapt. Phys. Act. Q*. 27:47–59.

Goosey-Tolfrey, V. L., N. Diaper, J. Crosland, and K. Tolfrey. 2008. Fluid intake during wheelchair exercise in the heat: Effects of localized cooling garments. *Int. J. Sports Physiol. Perform*. 3:145–56.

Goosey-Tolfrey, V. L., M. J. Price, and J. Krempien. 2014. Chapter 4: Spinal cord injured athletes. In: Broad, E. ed. *Sports Nutrition for Athletes with a Disability*. Oxford, UK: Taylor & Francis. pp: 67–90.

Graham-Paulson, T., C. Perret, B. S. Smith, J. Crosland, and V. L. Goosey-Tolfrey. 2015. Nutritional supplement habits of athletes with an impairment and their sources of information. *Int. J. Sports Nutr. Exer. Metab*. (in press).

Griggs, K. E., M. J. Price, and V. L. Goosey-Tolfrey. 2015. Cooling athletes with a spinal cord injury. *Sports Med*. 45(1):9–21.

Guttmann L., J. Silver, and C. H. Wyndham. 1958. Thermoregulation in spinal man. *J. Physiol*. 142(3):406–19.

Haisma, J., L. Woude van der, H. Stam, M. Bergen, T. Sluis, and J. Bussmann. 2006. Physical capacity in wheelchair-dependent persons with a spinal cord injury: A critical review of the literature. *Spinal Cord*. 44(11):642–52.

Havenith, G. 1999. Heat balance when wearing protective clothing. *Ann. Occup. Hyg*. 43:289–96.

Hopman, M. T., M. Monroe, C. Duek, W. T. Phillips, and J. S. Skinner. 1998. Blood redistribution and circulatory responses to submaximal arm exercise in persons with spinal cord injury. *Scand. J. Rehabil. Med*. 30(3):167–74.

Hopman, M. T., M. Pistorius, I. C. Kamerbeek, and R. A. Binkhorst. 1993. Cardiac output in paraplegic subjects at high exercise intensities. *Eur. J. Appl. Physiol*. 66(6):531–5.

Hopman, M. T. E. 1994. Circulatory responses during arm exercise in individuals with paraplegia. *Int. J. Sports Med*. 15:126–31.

Hutzler, Y. 1998. Anaerobic fitness testing of wheelchair users. *Sports Med*. 25(2):101–13.

Institute of Medicine. Water. 2005. *Dietary Reference Intakes for Water, Sodium, Chloride, Potassium and Sulfate*. Washington, DC: National Academy Press. pp: 73–185.

IPC. 2014. *International Paralympic Committee Anti-Doping Guide* (December 2011). Accessed http://www.paralympic.org/sites/default/files/document/140122160211761_2012_ipc+anti+doping+code_dec2011v_final.pdf.

Jacobs, P. L., E. T. Mahoney, A. Robbins, and M. S. Nash. 2002. Hypokinetic circulation in persons with paraplegia. *Med. Sci. Sport Exer*. 34(9):1401–7.

Janssen, T. W., A. J. Dallmeijer, D. H. Veeger, and L. H. van der Woude. 2002. Normative values and determinants of physical capacity in individuals with spinal cord injury. *J. Rehabil. Res. Dev*. 39(1):29–39.

Janssen, T. W., C. A. van Oers, H. E. Veeger, A. P. Hollander, L. H. van der Woude, and R. H. Rozendal. 1994. Relationship between physical strain during standardised ADL tasks and physical capacity in men with spinal cord injuries. *Paraplegia* 32(12):844–59.

Kao, C., Y. Ho, S. Changlai, and H. Ding. 1999. Gastric emptying in spinal cord injury patients. *Dig. Dis. Sci*. 44:1512–15.

Kirschblum, S. C., S. P.Burns, F. Biering-Sorensen et al. 2011. International standards for neurological classification of spinal cord injury. *J. Spinal Cord Med*. 34(6):535–46.

Koopman, R., W. H. M. Saris, A. J. M. Wagenmakers, and L. J. C. van Loon. 2007. Nutritional interventions to promote post-exercise muscle protein synthesis. *Sports Med*. 37:895–906.

Krassioukov, A. 2009. Autonomic function following cervical spinal cord injury. *Respir. Physiol. Neurobiol.* 169(2):157–64.

Krassioukov, A., and V. E. Claydon. 2006. The clinical problems in cardiovascular control following spinal cord injury: An overview. *Prog. Brain Res.* 152:223–9.

Kressler, J., P. A. Burns, L. Betancourt, and M. Nash. 2014. Circuit training and protein supplementation in persons with chronic tetraplegia. *Med. Sci. Sport Exer.* 46:1277–84.

Lakomy, H. K., I. Campbell, and C. Williams. 1987. Treadmill performance and selected physiological characteristics of wheelchair athletes. *Br. J. Sport Med.* 21(3):130–3.

Lee, B., R. Cripps, M. Fitzharris, and P. Wing. 2013. The global map for traumatic spinal cord injury epidemiology: Update 2011, global incidence rate. *Spinal Cord.* 52(2):110–16.

Lee, J. K. W., and S. Shirreffs. 2007. The influence of drink temperature on thermoregulatory responses during prolonged exercise in a moderate environment. *J. Sports Sci.* 25:975–85.

Leicht, C., N. Bishop, and V. L. Goosey-Tolfrey. 2012. Submaximal exercise responses in tetraplegic, paraplegic and non-spinal cord injured elite wheelchair athletes. *Scand. J. Med. Sci. Sport* 22(6):729–36.

Leiper, J. B., N. Broad, and R. J. Maughan. 2001a. Effect of intermittent high-intensity exercise on gastric emptying in man. *Med. Sci. Sport Exerc.* 33:1270–8.

Leiper, J. B., A. S. Prentice, C. Wrightson, and R. J. Maughan. 2001b. Gastric emptying of a carbohydrate-electrolyte drink during a soccer match. *Med. Sci. Sport Exer.* 33:1932–8.

Marieb, E. N., and K. Hoehn. 2007. *Human Anatomy and Physiology, 7th edition.* San Francisco, CA: Pearson Benjamin Cummings. 490–527, 532–51.

Maughan R. J. 1991. Fluid and electrolyte loss and replacement in exercise. *J Sports Sci.* 9(1):117–42.

McArdle, W. D., F. I. Katch, and V. L. Katch. 1986. *Exercise Physiology: Energy, Nutrition and Human Performance,* 2nd edition. Philadelphia, PA: Lea and Febiger. pp: 642–9.

Nielsen, B. 1994. Heat stress and acclimation. *Ergonomics.* 37(1):49–58.

Perret, C., and N. Stoffel-Kurt. 2011. Comparison of nutritional intake between individuals with acute and chronic spinal cord injury. *J. Spinal Cord. Med.* 34(6):569–75.

Price, M. 2010. Energy expenditure and metabolism during exercise in persons with a spinal cord injury. *Sports Med.* 40(8):681–96.

Price, M. J. 2006. Thermoregulation during exercise in individuals with spinal cord injuries. *Sports Med.* 36(10):863–79.

Price, M. J., and I. G. Campbell. 1999. Thermoregulatory responses of spinal cord injured and able-bodied athletes to prolonged upper body exercise and recovery. *Spinal Cord.* 37(11):772–9.

Price, M. J., and I. G. Campbell. 2003. Effects of spinal cord lesion level upon thermoregulation during exercise in the heat. *Med. Sci. Sport Exer.* 35:1100–7.

Ramsbottom, R., M. G. Nute, and C. Williams. 1987. Determinants of five kilometre running performance in active men and women. *Br. J. Sport Med.* 21(2):9–13.

Rehrer, N. J., E. J. Beckers, F. Brouns, T. Hoor, and W. H. M. Saris. 1990. Effects of dehydration in gastric emptying and gastrointestinal distress while running. *Med. Sci. Sport Exer.* 22:790–5.

Rhodes, J., B. S. Mason, B. Perrat, M. J. Smith, L. A. Malone, and V. L. Goosey-Tolfrey. 2015. Activity profiles of elite wheelchair rugby players during competition. *Int. J. Sports Physiol. Perform.* 10(3):318–24.

Rogers, J., R. W. Summers, and G. P. Lambert. 2005. Gastric emptying and intestinal absorption of a low-carbohydrate sport drink during exercise. *Int. J. Sport Nutr. Exerc. Metab.* 15:220–35.

Sawka, M. N., L. M. Burke, E. R. Eichner, R. J. Maughan, S. J. Montain, and N. S. Stachenfeld. 2007. American College of Sports Medicine position stand: Exercise and fluid replacement. *Med. Sci. Sport Exer.* 39:377–90.

Sawka, M. N., and J. E. Greenleaf. 1992. Current concepts concerning thirst, dehydration, and fluid replacement: Overview. *Med. Sci. Sport Exer.* 24(6):643–4.

Schmid, A., M. Huonker, J. M. Barturen et al. 1998a. Catecholamines, heart rate, and oxygen uptake during exercise in persons with spinal cord injury. *J. Appl. Physiol.* 85(2):635–41.

Schmid, A., M. Huonker, F. Stahl et al. 1998b. Free plasma catecholamines in spinal cord injured persons with different injury levels at rest and during exercise. *J. Autonom. Nerv. Syst.* 68(1):96–100.

Spendiff, O., and I. G. Campbell. 2005. Influence of pre-exercise glucose ingestion of two concentrations on paraplegic athletes. *J. Sports Sci.* 23(1):21–30.

Thijssen, D., S. Steendijk, and M. Hopman. 2009. Blood redistribution during exercise in subjects with spinal cord injury and controls. *Med. Sci. Sport Exer.* 41(6):1249–54.

Trbovich, M., C. Ortega, J. Schroeder, and M. Frederickson. 2014. Effect of a cooling vest on core temperature in athletes with and without spinal cord injury. *Top Spinal Cord Inj. Rehabil.* 20:70–80.

Valent, L. J., A. Dallmeijer, H. Houdijk, J. Slootman, T. Janssen, and A. Hollander. 2007. The individual relationship between heart rate and oxygen uptake in people with a tetraplegia during exercise. *Spinal Cord.* 45(1):104–11.

Webborn, A. D. J. 1996. Heat-related problems for the Paralympic Games, Atlanta 1996. *Br. J. Ther. Rehabil.* 3(8):429–35.

Webborn, N., and V. Goosey-Tolfrey. 2008. Spinal cord injury In: Buckley, J. eds. *Exercise Physiology in Special Populations*. London: Elsevier. pp: 309–34.

Webborn, N., M. J. Price, P. C. Castle, and V. L. Goosey-Tolfrey. 2005. Effects of two cooling strategies on thermoregulatory responses of tetraplegic athletes during repeated intermittent exercise in the heat. *J. Appl. Physiol.* 98:2101–7.

West, C. R., I. G. Campbell, R. E. Shave, and L. M. Romer. 2012. Resting cardiopulmonary function in paralympic athletes with cervical spinal cord injury. *Med. Sci. Sport Exer.* 44(2):323–9.

Williams, R. S., W. A. Bauman, A. M. Spungen et al. 2012. SmartPill technology provides safe and effective assessment of gastrointestinal function in persons with spinal cord injury. *Spinal Cord.* 50:81–4.

17 Athletes with Chronic Conditions
Sickle Cell Trait

Philippe Connes

CONTENTS

17.1 DEFINITION

Sickle cell anemia (SCA) is a genetic disease characterized by a single nucleotide mutation (adenine → thymine) that leads to the presence of sickle hemoglobin (HbS) resulting from the substitution of valine for glutamic acid at the sixth position of the β-globin chain. The hydrophobic residue of valine associates with other hydrophobic residues causing HbS molecules to aggregate, forming fibrous precipitates when hemoglobin is deoxygenated. This phenomenon is called *HbS polymerization* and is responsible for the characteristic shape change termed *sickling* of red blood cells (RBCs). Sickle RBCs are rigid and do not easily flow through the microcirculation, causing frequent vaso-occlusive episodes in affected patients. Recurrent HbS polymerization leads to numerous RBC and systemic physiological abnormalities with variable phenotypic severity (Embury 1986).

Sickle cell trait (SCT) is the heterozygous form of SCA, which is marked by the presence of both HbS (less than 50%) and normal Hb (HbA). SCT prevalence can reach 8%–10% for African-Americans and 10% in the Caribbean Islands.

17.2 SCT, VASCULAR DYSFUNCTION, AND BIOLOGICAL ABNORMALITIES

SCT is usually considered as an asymptomatic benign condition, compared to SCA, and reportedly does not affect growth and mental development. However, it has been suggested that SCT carriers could increase the risk for developing hyposthenuria, hematuria, and renal medullary carcinoma (Connes et al. 2008). More recently, epidemiological data on large U.S. cohorts demonstrated that SCT increases the risks for venous thromboembolism (Austin et al. 2007; Folsom et al. 2015), ischemic stroke (Caughey et al. 2014), and glomerulopathy (Key et al. 2015; Naik et al. 2014) in African-American individuals. These observations suggest that vascular function could be impaired in SCT (Key et al. 2015). Moreover, several studies reported increased blood viscosity (Tripette et al. 2009), slightly decreased RBC deformability (Tripette et al. 2010a), enhanced oxidative stress (Faes et al. 2012), and inflammation (Aufradet et al. 2010) in SCT carriers compared to controls at baseline, which could promote micro- and macrocirculatory disorders. Even if a direct causal relationship between SCT and the observed medical complications is difficult to prove (Goldsmith et al. 2012), evidences suggest that the medical status of this population should not be dismissed.

17.3   SCT AND EXERCISE PERFORMANCE

Of note, SCT is not a limiting factor for sport participation, even at the highest level (Le Gallais et al. 1991; Thiriet et al. 1994). Few laboratory studies have investigated the ability of SCT carriers to perform in brief and intense exercise and it seems they could be able to reach higher performance than non-SCT carriers (Hue et al. 2002). These experimental data have been confirmed by large epidemiological studies done in Ivory Coast and in the French West Indies (Connes et al. 2008; Le Gallais et al. 1991). The underlying mechanisms are unknown but a recent study reported that SCT carriers are characterized by a higher cross-sectional surface area of the IIb fibers, known to be involved in explosive works (Vincent et al. 2010).

Works done on anaerobic lactic power/capacity did not report large difference between SCT carriers and control group. Nevertheless, it seems that the ability of SCT carriers to repeat short anaerobic exercise with limited recovery is lower than in non-SCT carriers (Connes et al. 2006b).

Previous laboratory and epidemiological studies have suggested that aerobic physical fitness could be decreased in SCT carriers compared to controls (Freund et al. 1995; Le Gallais et al. 1991). However, comparisons of VO_2max, maximal aerobic power, ventilatory thresholds, lactic thresholds or VO_2-on kinetics between SCT carriers and subjects with no hemoglobinopathy demonstrated no difference between the two populations (Marlin et al. 2007, 2008; Sara et al. 2006). The conclusions of these laboratory works do not support a lower aerobic physical fitness in SCT carriers. Nevertheless, one study reported higher magnitude of the VO_2 slow component during a prolonged exercise at 70% maximal aerobic power in SCT carriers compared with a control group (Connes et al. 2006a), suggesting that exercise tolerance of SCT carriers during intense prolonged aerobic effort could be lower than for subjects with no SCT (Connes et al. 2006a). Recent data obtained on muscle biopsies coming from SCT carriers and non-carriers demonstrated lower cytochrome C oxidase (Cox) activity in muscle fibers from SCT carriers that could partly support the hypothetical lower aerobic capacity during prolonged endurance exercise in this population (Vincent et al. 2010).

17.4　EXERCISE-RELATED COMPLICATIONS AND ROLE OF HYDRATION

17.4.1 EPIDEMIOLOGICAL DATA

One of the most debated issue for SCT carriers concerns the true existence or not of a greater risk to die unexpectedly in response to exercise compared to the general population (Connes et al. 2007; Goldsmith et al. 2012; Key et al. 2015; O'Connor et al. 2012). If it is usually admitted that SCT carriers are at risk for developing complications, such as splenic syndrome, when exercising in altitude, the debate is always running concerning exercising at sea level.

A greater attention was devoted to this association following the publication of a large epidemiological study that retrospectively evaluated the risk of exercise-related death in U.S. Armed Forces recruits over a 5-year period (1977–1981) (Kark et al. 1987). The relative risk of exercise-related death explained by preexisting disease (largely silent heart disease) was 2.3 for SCT, which was not statistically significant. In contrast, the relative risk of exercise-related death unexplained by preexisting disease was 28 for SCT subjects (95% CI, 9–100; $p < .001$). The timing of exercise-related complications in SCT has generally been linked to the immediate stress of exercise. About 50% of the deaths resulted from heat illness from overexertion, while the remaining cases were classified as *idiopathic*. Poor physical conditioning of recruits and dehydration appeared to increase the risk for adverse events during exercise (Kark et al. 1987, Kark and Ward 1994; O'Connor et al. 2012). Harmon and colleagues (2012) recently reviewed the causes of all cases of sudden death in student athletes from the National Collegiate Athletic Association (NCAA) in the United States from January 2004 through December 2008. Seventy-two deaths occurred in American football players due to trauma unrelated to sports activities or medical causes. The overall risk of exertional death from any cause was 5.2 times higher in African-American players with or without SCT compared with that in non-African-American football players. The calculated relative risk of exertional death in all Division I football players (i.e., all ethnicities) with SCT was 1:827, which represented a 37-fold higher relative risk than in athletes without SCT. It has been suggested that the repetition of very intense exercise with incomplete recovery during football training sessions could result in massive sickling and rhabdomyolysis leading to death in affected individuals (Eichner 2011).

17.4.2　HYPOTHETICAL MECHANISMS

Abnormal blood rheology, impaired nitric oxidative bioavailability, enhanced oxidative stress, inflammation, and loss of vascular reactivity are among the factors involved in the pathophysiology of vaso-occlusive like events in SCA.

Indeed, several studies investigated the biological responses to various exercises in small groups of SCT carriers, trying to identify abnormalities that could be responsible for the exercise-related complications occurring, sometimes, in this population. On the whole, it has been shown that acute intense exercise in SCT carriers may (1) cause mild RBC sickling (Bergeron et al. 2004), (2) increase blood viscosity at higher level than controls (Connes et al. 2006c; Tripette et al. 2010c, 2013), and (3) cause higher oxidative stress and inflammatory responses in SCT carriers than in non-SCT carriers (Aufradet et al. 2010; Faes et al. 2012; Monchanin et al. 2007; Tripette et al. 2010b). However, fortunately, these abnormalities seem insufficient to trigger frequent exercise-related complications in SCT carriers. It seems that additional factors are needed such as (1) exercising in extreme heat and humidity exposure, (2) severe dehydration, (3) exercising at high altitude, (4) exercise-induced asthma, and (5) pre-exercise fatigue due to illness (viral infection for example) or lack of sleep and poor physical fitness (Key et al. 2015; Tripette et al. 2013).

17.4.3 ROLE OF ADEQUATE HYDRATION

Exercising in the heat and dehydration seem to play an important role in the occurrence of exercise-related complications in SCT carriers (Hedreville et al. 2009; Kark et al. 1994). Few works focused on the strategies to normalize, or at least to reduce, the biological alterations described in SCT carriers. Bergeron et al. (2004) previously demonstrated that 45 min of brisk walking in hot condition favored RBC sickling in SCT carriers but promoting active hydration prevented it. More recently, Tripette et al. (2010c) and Diaw et al. (2014) compared the hemorheological responses of SCT carriers and control individuals during a cycling test and a soccer game, respectively, with and without *ad libitum* hydration. The two studies clearly observed blood hyperviscosity in SCT carriers before exercising. Blood viscosity moderately increased with exercise in controls and hydration had very limited effect. In contrast, blood viscosity exhibited a very large rise in dehydrated SCT carriers but *ad libitum* hydration was able to limit the rise and to normalize the blood viscosity at the same level as control individuals. These findings clearly demonstrate the major role played by dehydration on the worsening of the biological profile of SCT carriers. Indeed, adequate hydration, by normalizing the biological abnormalities of SCT carriers, should help in preventing exercise-related complications in this population (O'Connor et al. 2012).

17.5 SUMMARY

SCT is generally considered as a benign condition. However, recent epidemiological data demonstrated an increased risk for several complications such as glomerulopathy, thromboembolism, and stroke. In addition, while exercise performances are almost similar in SCT carriers and non-carriers; the risk for exercise-related idiopathic sudden death is increased in the former population. Several biological abnormalities could be involved in these complications and promoting adequate and active hydration could be an easy way to normalize them.

REFERENCES

Aufradet, E., G. Monchanin, S. Oyonno-Engelle et al. 2010. Habitual physical activity and endothelial activation in sickle cell trait carriers. *Med. Sci. Sports Exer.* 42:1987–94.
Austin, H., N. S. Key, J. M. Benson et al. 2007. Sickle cell trait and the risk of venous thromboembolism among blacks. *Blood* 110:908–12.

Bergeron, M. F., J. G. Cannon, E. L. Hall, and A. Kutlar. 2004. Erythrocyte sickling during exercise and thermal stress. *Clin. J. Sport Med.* 14:354–6.

Caughey, M. C., L. R. Loehr, N. S. Key et al. 2014. Sickle cell trait and incident ischemic stroke in the Atherosclerosis Risk in Communities study. *Stroke* 45:2863–7.

Connes, P., M. D. Hardy-Dessources, and O. Hue. 2007. Counterpoint: Sickle cell trait should not be considered asymptomatic and as a benign condition during physical activity. *J. Appl. Physiol.* 103:2138–40; discussion 2140–1.

Connes, P., G. Monchanin, S. Perrey et al. 2006a. Oxygen uptake kinetics during heavy submaximal exercise: Effect of sickle cell trait with or without alpha-thalassemia. *Int. J. Sports Med.* 27:517–25.

Connes, P., S. Racinais, F. Sara et al. 2006b. Does the pattern of repeated sprint ability differ between sickle cell trait carriers and healthy subjects? *Int. J. Sports Med.* 27:937–42.

Connes, P., H. Reid, M. D. Hardy-Dessources, E. Morrison, and O. Hue. 2008. Physiological responses of sickle cell trait carriers during exercise. *Sports Med.* 38:931–46.

Connes, P., F. Sara, M. D. Hardy-Dessources et al. 2006c. Effects of short supramaximal exercise on hemorheology in sickle cell trait carriers. *Eur. J. Appl. Physiol.* 97:143–50.

Diaw, M., A. Samb, S. Diop et al. 2014. Effects of hydration and water deprivation on blood viscosity during a soccer game in sickle cell trait carriers. *Br. J. Sports Med.* 48:326–31.

Eichner, E. R. 2011. Sickle cell considerations in athletes. *Clin. Sports Med.* 30:537–49.

Embury, S. H. 1986. The clinical pathophysiology of sickle cell disease. *Annu Rev. Med.* 37:361–76.

Faes, C., C. Martin, E. N. Chirico et al. 2012. Effect of alpha-thalassaemia on exercise-induced oxidative stress in sickle cell trait. *Acta Physiol. (Oxf)* 205:541–50.

Folsom, A. R., W. Tang, N. S. Roetker et al. 2015. Prospective study of sickle cell trait and venous thromboembolism incidence. *J. Thromb. Haemost.* 13:2–9.

Freund, H., J. Lonsdorfer, S. Oyono-Enguelle et al. 1995. Lactate exchange and removal abilities in sickle cell trait carriers during and after incremental exercise. *Int. J. Sports Med.* 16:428–34.

Goldsmith, J. C., V. L. Bonham, C. H. Joiner et al. 2012. Framing the research agenda for sickle cell trait: Building on the current understanding of clinical events and their potential implications. *Am. J. Hematol.* 87:340–6.

Harmon, K. G., J. A. Drezner, D. Klossner, and I. M. Asif. 2012. Sickle cell trait associated with a RR of death of 37 times in National Collegiate Athletic Association football athletes: A database with 2 million athlete-years as the denominator. *Br. J. Sports Med.* 46:325–30.

Hedreville, M., P. Connes, M. Romana et al. 2009. Central retinal vein occlusion in a sickle cell trait carrier after a cycling race. *Med. Sci. Sports Exer.* 41:14–18.

Hue, O., M. E. Julan, S. Blonc et al. 2002. Alactic anaerobic performance in subjects with sickle cell trait and hemoglobin AA. *Int. J. Sports Med.* 23:174–7.

Kark, J. A., D. M. Posey, H. R. Schumacher, and C. J. Ruehle. 1987. Sickle-cell trait as a risk factor for sudden death in physical training. *N. Engl. J. Med.* 317:781–7.

Kark, J. A., and F. T. Ward. 1994. Exercise and hemoglobin S. *Semin. Hematol.* 31:181–225.

Key, N. S., P. Connes, and V. K. Derebail. 2015. Negative health implications of sickle cell trait in high income countries: From the football field to the laboratory. *Br. J. Haematol.* 170:5–14.

Le Gallais, D., C. Prefaut, C. Dulat, J. Macabies, and J. Lonsdorfer. 1991. Sickle cell trait in Ivory Coast athletic champions, 1956–1989. *Int. J. Sports Med.* 12:509–10.

Marlin, L., P. Connes, S. Antoine-Jonville et al. 2008. Cardiorespiratory responses during three repeated incremental exercise tests in sickle cell trait carriers. *Eur. J. Appl. Physiol.* 102:181–7.

Marlin, L., F. Sara, S. Antoine-Jonville et al. 2007. Ventilatory and lactic thresholds in subjects with sickle cell trait. *Int. J. Sports Med.* 28:916–20.

Monchanin, G., L. D. Serpero, P. Connes et al. 2007. Effects of a progressive and maximal exercise on plasma levels of adhesion molecules in athletes with sickle cell trait with or without alpha-thalassemia. *J. Appl. Physiol.* 102:169–73.

Naik, R. P., V. K. Derebail, M. E. Grams et al. 2014. Association of sickle cell trait with chronic kidney disease and albuminuria in African Americans. *JAMA* 312:2115–25.

O'Connor, F. G., M. F. Bergeron, J. Cantrell et al. 2012. ACSM and CHAMP Summit on sickle cell trait: Mitigating risks for warfighters and athletes. *Med. Sci. Sports Exer.* 44:2045–56.

Sara, F., P. Connes, O. Hue et al. 2006. Faster lactate transport across red blood cell membrane in sickle cell trait carriers. *J. Appl. Physiol.* 100:437–42.

Thiriet, P., J. Y. Le Hesran, D. Wouassi et al. 1994. Sickle cell trait performance in a prolonged race at high altitude. *Med. Sci. Sports Exer.* 26:914–18.

Tripette, J., T. Alexy, M. D. Hardy-Dessources et al. 2009. Red blood cell aggregation, aggregate strength and oxygen transport potential of blood are abnormal in both homozygous sickle cell anemia and sickle-hemoglobin C disease. *Haematologica* 94:1060–5.

Tripette, J., P. Connes, E. Beltan et al. 2010a. Red blood cell deformability and aggregation, cell adhesion molecules, oxidative stress and nitric oxide markers after a short term, submaximal, exercise in sickle cell trait carriers. *Clin. Hemorheol. Microcirc.* 45:39–52.

Tripette, J., P. Connes, M. Hedreville et al. 2010b. Patterns of exercise-related inflammatory response in sickle cell trait carriers. *Br. J. Sports Med.* 44:232–7.

Tripette, J., M. D. Hardy-Dessources, M. Romana et al. 2013. Exercise-related complications in sickle cell trait. *Clin. Hemorheol. Microcirc.* 55:29–37.

Tripette, J., G. Loko, A. Samb et al. 2010c. Effects of hydration and dehydration on blood rheology in sickle cell trait carriers during exercise. *Am. J. Physiol. Heart Circ. Physiol.* 299:H908–14.

Vincent, L., L. Feasson, S. Oyono-Enguelle et al. 2010. Skeletal muscle structural and energetic characteristics in subjects with sickle cell trait, alpha-thalassemia, or dual hemoglobinopathy. *J. Appl. Physiol.* 109:728–34.

Section IV

Recommendations

18 Water Replacement before, during, and after Exercise
How Much Is Enough?

Ronald J. Maughan and Susan M. Shirreffs

CONTENTS

18.1 INTRODUCTION AND RATIONALE

Sport embraces an enormous diversity of activities and each places different demands upon the individual. Whatever the intensity and duration of the activity, however, there is the potential for hydration status, and therefore for drinking behavior, to affect the performance capacity. While the literature on this topic is mixed, there can be no doubt that hypohydration, if sufficiently severe, will have an adverse effect on all types of activity. The more relevant question is whether the levels of water loss normally encountered in sporting contexts are sufficient to impair performance or to compromise health. Following on from this are questions about how much should be consumed, when it should be consumed, and what drinks might be most effective. Abstinence from fluid intake will have measurable effects on performance over a timescale that may range from an hour or less in hard exercise in a warm environment to a few days for individuals at rest in a cool environment. Some activities will be more affected than others, and the effect on performance will also be influenced by the environment, by the clothing worn, by the training and acclimation status of the individual, and by other factors too.

The extent of fluid and electrolyte losses during exercise will obviously have a major influence on the need for replacement during exercise, and the loss will depend on the volume of sweat lost and on the composition of the sweat. This in turn will depend on the size and surface area of the individual, the intensity and duration of the exercise, the training and acclimation status of the individual, the clothing worn, and the environmental conditions. Sweat losses also vary greatly between individuals, making predictions of loss, and therefore of the need for replacement, difficult. Typical values for sweating rate are often reported in the literature, but the range of values is very large, even when the obvious factors are accounted for. In a number of studies of relatively homogeneous groups of young male football (soccer) players, for example, all doing the same training at the same time (and therefore in the same environment), sweat losses varied from less than 1 L to more than 3 L (Maughan et al. 2010). The main electrolytes lost in sweat are sodium (at concentrations that typically range from about 20 to 80 mmol/L) and chloride (20–60 mmol/L) with a range of other electrolytes present at much lower concentrations (Shirreffs and Maughan 1997). With such a large individual variability, formulating general guidelines for replacement strategies is therefore almost impossible, and individualization of rehydration strategies is essential (Maughan and Shirreffs 2008).

The need for fluid intake during exercise will be dictated by many factors, including the immediate effects of the exercise and the environment but also the pre-exercise hydration status. Fluid intake in the hours before exercise is therefore a critical consideration, especially when there is a possibility that hypohydration may exist due to prior exercise or because of deliberate hypohydration to achieve a required body mass target. The role of drinks as a vehicle for the provision of energy substrates during exercise, primarily carbohydrate, is also important when the exercise stress results in depletion of the endogenous glycogen stores or when these are inadequate at the onset of exercise. Effects other than those on hydration status must also be considered. The temperature of ingested drinks will affect body heat content, with implications for thermoregulatory responses and for performance, especially in conditions of heat stress (Lee et al. 2008a, b). The act of drinking itself, or the conscious response to the withholding of drinks, will also induce effects on mood, expectation, and other psychological states that may affect performance. Finally, restoration of water and salt balance is a critical aspect of the recovery process after exercise.

Against this, background of the positive effects of drinking and maintaining an adequate hydration status is the issue of overdrinking and the risk of hyponatremia. Although rare, the consequences are serious and, in extreme cases, occasionally fatal. This condition was first recognized in men undertaking occupational tasks such as coal mining that involved many hours of hard physical work in hot environments (Haldane 1923, 1929). Much later, it was reported in slower participants in endurance road races and triathlons (Noakes et al. 1985), but has now been seen to occur in other sports, for example, American football (Dimeff 2006). There is a real need, therefore, to give careful consideration to the amount of fluid that should be consumed before, during, and after exercise and to the composition of that fluid.

In assessing the evidence, it must be remembered that performance effects that are highly meaningful to the athlete are often far below the sensitivity of the laboratory tests of performance that are commonly used (Hopkins 2001). There has been much debate about the appropriateness of the tests used where the intention is to apply results to a competitive environment. Laboratory tests differ in several important aspects from real competition: tests may involve measurement of time to fatigue at a constant speed or power output, a simulated time trial where a fixed distance or a fixed amount of work must be completed as fast as possible, or a combination of the two where a period of exercise at a fixed power output is followed by a simulated time trial. None of these can truly reflect what happens in a race, where tactical considerations come into play. The argument that time trial tests are ecologically valid while time to fatigue tests are not is often advanced and is used to dismiss the results of studies that have used time-to-fatigue tests (Goulet 2013). This argument is, however, fundamentally flawed: for the serious athlete, the time to fatigue test more closely resembles the race situation than does a laboratory time trial. Although some continue to voice concerns over a lack of ecological validity, the challenge for most runners or cyclists is to maintain contact with the leading group: if dropped, the chances of winning are almost nil, and the point at which contact with the leaders is lost is in effect the point of fatigue. The ability of inexperienced individuals to maximize performance by choosing an optimum pacing strategy is also highly questionable.

A key factor to consider when selecting an appropriate exercise test is its sensitivity and the smallest worthwhile effect that can be reliably detected (Currell and Jeukendrup 2008). Data from our research group report a relatively consistent performance (coefficient of variation = 6%) in time-to-fatigue tests (Maughan et al. 1989), and more recent reports have highlighted similar errors of measurement when changes in performance are normalized across a series of tests (Hinckson and Hopkins 2005). Although this variability may seem large when compared with the margin between success and defeat in elite sport, Amann et al. (2008) demonstrated that time-to-fatigue and time trial protocols display a similar sensitivity to the effects of hypoxia and hyperoxia on performance and suggest that this finding will extend to other factors influencing performance. This is brought about by larger effects on performance in response to an intervention with constant power tests than are typically observed in time trial protocols, and this greater effect compensates for the larger test–retest variability, resulting in a signal-to-noise ratio that is very similar to that seen with time trial protocols (Amann et al. 2008; Currell and Jeukendrup 2008). It must nevertheless be recognized that there are some differences between the competitive environment and exercise in a laboratory on a rowing or cycle ergometer or on a treadmill. In the absence of forward movement, convective heat loss may be dramatically reduced and local humidity rises sharply, reducing evaporative heat loss. Laboratory studies therefore generally impose a much higher heat stress on subjects than do field trials.

It is also the case that most of the available information is derived from recreationally active individuals, many or whom are not competitive athletes and some of whom have done little or no training. The factors that limit performance are likely to be different in the highly trained athlete than in the less well-trained, so it is also likely that the responses to interventions such as fluid provision are also different. Whether hydration status and fluid ingestion are more or less important factors in these distinct populations is, however, not known.

18.2　PRE-EXERCISE HYDRATION

The evidence on the effects of pre-exercise hydration status and exercise performance is mixed, perhaps because of the various experimental protocols that have been applied. These have involved exercise performance tests ranging from the assessment of isometric or dynamic muscle strength through short duration cycling or running tests to endurance performance. In some cases, these exercise tests have been performed in a cool or temperate environment, while others have been undertaken under conditions of heat stress. Hypohydration has been induced prior to the exercise test by heat exposure, exercise, or a combination of both: other studies have used a period of fluid restriction or incomplete recovery from prior dehydration, and some have induced hypohydration by the administration of diuretic agents. In some cases, the degree of hypohydration has been modest (1%–3% of body mass), while in others it has been more severe (6%–8% of body mass). Pre-exercise hyperhydration has been induced by water ingestion or by administration of saline or glycerol solutions. Given all the possible permutations of experimental conditions, it is perhaps unsurprising that there is no simple answer to the question of whether or how pre-exercise hydration status affects performance.

Armstrong et al. (1985) induced hypohydration by administration of a diuretic agent prior to participation in simulated distances over distances of 1500, 5,000, and 10,000 m: they found that performance was dramatically impaired in the hypohydrated state relative to performance in races where the subjects were fully hydrated. Not all studies have found performance impairments, though. Stewart et al. (2014) showed that pre-exercise hypohydration equivalent to a 4% body mass loss had no effect on performance in a 5-km cycling time trial lasting about 7 min that was performed in a temperate environment.

In spite of the potential for negative effects on performance, it is clear that significant numbers of athletes are hypohydrated when they begin training (Maughan et al. 2005) or competition (Maughan et al. 2007). In the case of morning training sessions, many of these players simply eschewed breakfast, meaning that they had drunk nothing since the evening before. In matches played later in the day, there seems to be little reason for players not to be well hydrated when arriving at the stadium. Education programs aimed at athletes, coaches, and support staff have generally focused on issues related to drinking during training and competition, but there is perhaps a need for greater attention to fluid intake in the hours before competition.

It has been suggested that if dehydration is one of the major factors contributing to fatigue in prolonged exercise, then increasing body water content prior to exercise should, by analogy with glycogen loading, improve performance. Drinking large volumes of plain water provokes a diuretic response, but some degree of temporary hyperhydration appears to result when drinks with high (100 mmol/L or more) sodium concentrations are ingested. Sims et al. (2007a) have shown that pre-exercise ingestion of a high-sodium beverage (164 mmol/L Na^+ in a volume of 10 mL/kg body mass) increased plasma volume before exercise and resulted in reduced thermoregulatory and perceived strain of men during exercise in the heat, with a consequent improvement in exercise capacity. They subsequently reported very similar effects in women using the same procedures (Sims et al. 2007b). An alternative strategy that has been attempted with the aim of inducing an expansion of the blood volume prior to exercise is to add glycerol to ingested fluids. Glycerol, as further discussed in Chapter 20, exerts an osmotic action: although its distribution in the body water compartments is variable, glycerol will expand the extracellular space and some of the water ingested with the glycerol will be retained, at least for a few hours (Riedesel et al. 1987; Freund et al. 1995). The elevated osmolality of the extracellular space will, however, result in some degree of intracellular dehydration, and the implications of this are unknown. It might be expected, however, that the raised plasma osmolality will have negative consequences for thermoregulatory capacity. The evidence for increased water retention and improved exercise performance after administration of glycerol and water prior to prolonged exercise is, however, fairly convincing (Goulet et al. 2007; Van Rosendal and Coombes 2012). Athletes liable to testing under the regulations of the World Anti-Doping Agency must be aware of the prohibition on the use of glycerol that was introduced in 2010 (Koehler et al. 2014).

A concern with pre-exercise hyperhydration is that it may result in reductions in plasma sodium concentrations before starting exercise and therefore increase the risk of dilutional hyponatremia if large volumes of fluid are consumed during subsequent exercise (Sawka et al. 2007). Hyponatremia is further discussed below, but it should be noted that most of the reports on exercise-associated hyponatremia (EAH) did not assess pre-exercise sodium status. Where it was measured, this was most often done at the pre-race registration 2–3 days before the event rather than in the immediate pre-exercise period. While those reported cases were therefore undoubtedly associated with exercise, it is less clear that they were caused by the exercise or by behaviors specifically associated with exercise.

18.3   FLUID INGESTION DURING EXERCISE: HOW MUCH IS ENOUGH?

There has been considerable controversy in recent years about the amount of fluid that it is appropriate to drink during exercise and the advice that should be given to athletes and to participants in various exercise activities. It was recognized many years ago that dehydration, if sufficiently severe, would impair performance. As a consequence, it was also recognized that opportunities to drink should be made available during endurance events and that participants should be encouraged to make use of these facilities. Nonetheless even as late as the 1970s, drinks stations were provided at only four points during marathon races and most runners consumed little during the race, perhaps in part because drinking was often seen as a sign of weakness, but also because of the physical difficulty in drinking from an open cup while running at speed. With the growing popularity of mass participation events in the 1980s, the picture changed: race organizers provided large amounts of fluid every mile (1.6 km) and participants, many of whom were totally unprepared for the challenge of running 42.2 km, were encouraged to drink copiously. Many were happy to stop and linger at drinks stations, and an unforeseen consequence was that a small number of individuals drink far in excess of their requirements, leading to overhydration and EAH (Noakes et al. 1985). At that time, guidelines on drinking were directed to serious athletes who were generally well trained and experienced in race participation. These guidelines recognized both the limited availability of drinks in most endurance events and the difficulties in drinking during intense exercise: they therefore encouraged athletes to drink *as much as tolerable* up to a rate equal to that lost from sweating (Convertino et al. 1996). Subsequent guidelines were more conservative, recognizing the potential problem of overhydration and telling participants in endurance events that they should aim to prevent excessive (>2% body weight loss from water deficit) dehydration, but again emphasizing that they should not drink so much as to gain weight (Sawka et al. 2007).

In assessing the evidence on hyponatremia in sport, it is important to remember that the normal range is defined as that embracing 95% of the normal population values. In other words, 5% of the normal population will be outside the reference range: if the values are normally distributed—as they should be—then we would expect that 2.5% of the population will return a serum sodium concentration of less than the lower limit of the normal range and that 2.5% will be above the normal range. This has real implications for the interpretation of the literature in this area, especially as most studies of EAH have not collected pre-event blood samples. We also do not know for sure whether the normal range of values for athletes is that same as that for non-athletes with whom their values are compared. It is almost certainly the case too that mild hyponatremia that develops slowly will be less harmful than the same level of hyponatremia that develops rapidly. A serum sodium concentration of 132 mmol/L for the individual whose habitual value is 132 mmol/L is very different from the individual whose normal value is 145 mmol/L but who has experienced an acute fall to 132 mmol/L. We need to know the normal baseline value rather than trying to overinterpret single values.

Concerns over the possibility of excessive fluid intake leading to the development of hyponatremia, with potentially serious sequelae, led to the proposal that, rather than encouraging participants to follow a planned drinking strategy, they should instead be told simply to drink according to the dictates of thirst (Noakes 2007). In a recent investigation, however, Armstrong et al. (2014) pointed out that instructing athletes to drink according to the sensation of thirst is different from the complex behavior of drinking *ad libitum*. To evaluate this, they compared two groups of trained cyclists: participants in one group were told to drink only when thirsty, while those in the other group consumed fluid *ad libitum* (i.e., whenever and in whatever volume they desired). There were no differences between groups on the day of a 164-km race for total exercise time, urinary indices of hydration status, body mass change or fluid intake, but those instructed to drink *ad libitum* drank more on the day after the race. The authors concluded that specific instructions to drink to thirst are unnecessary and may distract athletes by requiring ongoing evaluation of thirst sensations. In a recent review, Goulet (2012) concluded that "In athletes whose thirst sensation is untrustworthy or when external factors such as psychological stress or repeated food intake may blunt thirst sensation, it is recommended to program fluid intake to maintain exercise-induced body weight loss around 2% to 3%."

The fear of development of a potentially fatal hyponatremia has led to a general consensus that athletes should never drink so much that they gain weight. This ignores the situation of the athlete who begins exercise in a hypohydrated state and who might benefit, but if proper pre-exercise hydration strategies are adopted this situation should seldom arise.

It is not unusual for some athletes and coaches to deliberately restrict fluid intake in training in the belief that the body will adapt to a lack of fluid and will therefore cope better with dehydration in competition. Deliberate restriction of fluid intake has long been a common practice in sport and added to this the use of purgatives, laxatives, and *sweating liquors* was also common among early athletes to promote weight loss (Thom 1813; Shearman 1888). Shearman (1888) did, though, express some doubts when he wrote that "Sweating, meat-eating and purging constituted the old system of training. It was chiefly applied to men of the lower classes, used to coarse food, and with no highly-organised nervous system. Such a method could not be beneficial to an amateur, who takes up athletic sports as a recreation, and not as a business." Practices in many sports have changed substantially, but particularly in the weight category or weight-sensitive sports, where acute, and sometimes severe, dehydration is a normal part of pre-competition preparation for the majority of competitors, chronic hypohydration is not uncommon (Wilson et al. 2012). It is perhaps also likely that those who persist in participation in weight-category sports are those who are most resistant to the effects of dehydration. Those who cannot tolerate the weight-making process are unlikely to be successful and are therefore more likely to discontinue participation in the sport.

Even when fluids are readily available, it is common for athletes in weight-category sports to drink little or nothing during training, perhaps in the belief that they will adapt to the dehydrated state (Rivera Brown et al. 2012). There is very limited evidence of any physiological adaptations to dehydration, although those who frequently restrict fluid intake may learn to complain less about the symptoms that accompany it (Fleming and James 2014). Some recent evidence, though, does suggest that the adaptations induced by short-term (5-day) heat acclimation may be more pronounced if dehydration is allowed to develop (Garrett et al. 2012). This seemed to be a consequence of the greater elevation of core temperature that occurs when exercising in the heat in a hypohydrated state and the consequently greater cardiovascular adaptations that ensue, including the extent of the increase in vascular volume (Fan et al. 2008; Ikegawa et al. 2011). A follow-up study, however, confirmed these responses but suggested that they are not due entirely to effects on body temperature (Garrett et al. 2014). Emerging evidence suggests that the adaptations that occur in response to even a short period of heat acclimation may promote enhanced endurance performance in a temperate environment (Lorenzo et al. 2010).

18.4   POST-EXERCISE REHYDRATION AND RECOVERY

The need for rehydration as part of the post-exercise recovery process will depend largely on the extent of water and salt losses during the exercise period and on the time available before the next exercise. Where losses are small and a prolonged period (24 h or more) is available, no special effort is needed to achieve an adequate hydration status; eating and drinking normally will achieve this. Large losses that must be replaced over a short period of time, however, will require a more aggressive approach to restoring fluid balance.

The important aspects of volume replenishment are the ingestion of an adequate quantity of fluid and replacement of electrolyte losses. The need for ingestion of fluid is obvious, and it is common to recommend that athletes drink 1 L of fluid for each kg of mass lost during training. The volume of fluid consumed should, however, exceed that lost in sweat—perhaps by as much as 50%—if effective rehydration is to be achieved (Shirreffs et al. 1996). This stems from the need to allow for ongoing urine and other losses during the recovery period. It is also now known that restoration of euhydration (a state of fluid balance) after exercise-induced sweat losses will not be achieved without replacement of the electrolytes lost in sweat (Maughan et al. 1994). These electrolytes (especially sodium) prevent the drop in plasma sodium concentration and plasma osmolality that would occur with the acute ingestion of large volumes of plain water. If plasma osmolality and sodium concentration fall, a diuresis is initiated, leading to the prompt loss of a significant fraction of the ingested fluid (Maughan and Leiper 1995). The thirst mechanism is also inhibited by these changes, leading to the termination of drinking before a volume sufficient to meet the body's needs has been ingested (Nose et al. 1988). Wong et al. (1998) showed that a prescribed fluid intake was more effective than an *ad libitum* drinking schedule in restoring exercise capacity when applied during the 4-h recovery period between two endurance exercise bouts. Even though the total fluid intake was the same on both trials, a greater volume was ingested in the early stages of recovery with the prescribed drinking schedule.

The addition of salt to drinks can prevent or at least reduce the changes in plasma composition that follow the ingestion of large volumes of plain water, thereby reducing the urinary loss and maintaining the drive to drink. If food is eaten together with water, the electrolytes present in the food eaten may achieve this, depending on the foods chosen, but there are many situations, especially when the recovery time is short, when athletes may not be willing or able to tolerate solid food, or where the amount and type of food eaten may not supply an adequate amount of sodium (Maughan et al. 1996). The presence of food or the ingestion of drinks with a high energy content will also slow the rate of gastric emptying, delaying water absorption and avoiding large excursions of plasma osmolality, which may result in a less pronounced diuresis (Evans et al. 2009a). If no food is consumed, the drinks consumed during the recovery period must provide sufficient electrolytes to replace the sweat losses and should also perhaps contain carbohydrate and protein to address other aspects of the recovery process (Maughan and Burke 2011). Drinks with a very high solute content will, however, have a high osmolality and may result in a temporary net secretion of water into the small intestine. This will be reabsorbed in time but has the effect of causing a temporary reduction in the total body water pool that may be large enough to have physiological effects (Evans et al. 2009b).

Because the electrolyte concentration of sweat varies so greatly between individuals, it is not possible to produce general recommendations that will meet the needs of all athletes in all situations. Nonetheless, the average sweat sodium concentration is about 50 mmol/L, and it is recognized that the amount of sodium that must be replaced to maintain volume homeostasis approximates to that lost (Shirreffs and Maughan 1998). Ingesting more is probably not harmful for most active individuals, but there is a need to be cautious about chronic high salt intakes that are in excess of requirements. The ingestion of salt tablets is seldom necessary, but may be contemplated when sweat losses are exceptionally high or if the intake of food is limited.

18.5 CONCLUSION

Hydration is an important issue for everyone who engages in physical activity and sport, whether at the elite level or simply for enjoyment and for the health benefits that ensue. In recent years, the simple act of drinking before, during, or after participation in exercise has become the subject of heated debate and controversy. It should remain simple. Dehydration, if sufficiently severe, will impair all aspects of physical and cognitive performance: it will make exercise feel harder and it will reduce performance. Those engaging in exercise should ensure that they begin exercise with an adequate hydration status and, where the situation requires, should drink sufficient but not too much during exercise. Rehydration practices after exercise should be appropriate to the situation.

REFERENCES

Amann, M., W. G. Hopkins, and S. M. Marcora. 2008. Similar sensitivity of time to exhaustion and time-trial time to changes in endurance. *Med. Sci. Sports Exer.* 40:574–8.

Armstrong, L. E., D. L. Costill, and W. J. Fink. 1985. Influence of diuretic-induced dehydration on competitive running performance. *Med. Sci. Sports Exer.* 17:456–61.

Armstrong, L. E., E. C. Johnson, L. J. Kunces et al. 2014. Drinking to thirst versus drinking ad libitum during road cycling. *J. Athl. Train.* 49:624–31.

Convertino, V. A., L. E. Armstrong, E. F. Coyle et al. 1996. American College of Sports Medicine position stand: Exercise and fluid replacement. *Med. Sci. Sports Exer.* 28:i–vii.

Currell, K., and A. E. Jeukendrup. 2008. Validity, reliability and sensitivity of measures of sporting performance. *Sports Med.* 38:297–316.

Dimeff, R. J. 2006. Seizure disorder in a professional American football player. *Curr. Sports Med. Rep.* 5:173–6.

Evans, G. H., S. M. Shirreffs, and R. J. Maughan. 2009a. Post-exercise rehydration in man: The effects of osmolality and carbohydrate content of ingested drinks. *Nutrition* 25:905–13.

Evans, G. H., S. M. Shirreffs, and R. J. Maughan. 2009b. Acute effects of ingesting glucose solutions on blood and plasma volume. *Br. J. Nutr.* 101:1503–8.

Fan, J.-L., J. D. Cotter, R. A. I. Lucas et al. 2008. Human cardiorespiratory and cerebrovascular function during severe passive hyperthermia: Effects of mild hypohydration. *J. Appl. Physiol.* 105:433–45.

Fleming, J., and L. J. James. 2014. Repeated familiarisation with hypohydration attenuates the performance decrement caused by hypohydration during treadmill running. *Appl. Physiol. Nutr. Metab.* 39:124–9.

Freund, B. J., S. J. Montain, A. J. Young et al. 1995. Glycerol hyperhydration: Hormonal, renal and vascular fluid responses. *J. Appl. Physiol.* 79:2069–77.

Garrett, A. T., R. Creasy, and N. J. Rehrer. 2012. Effectiveness of short-term heat acclimation for highly trained athletes. *Eur J. Appl. Physiol.* 112:1827–37.

Garrett, A. T., N. G. Goosens, N. J. Rehrer et al. 2014. Short-term heat acclimation is effective and may be enhanced rather than impaired by dehydration. *Am. J. Hum. Biol.* 26:311–20.

Goulet, E. D. B. 2012. Dehydration and endurance performance in competitive athletes. *Nutr. Rev.* 70(Suppl 2):S132–6.

Goulet, E. D. B. 2013. Effect of exercise-induced dehydration on endurance performance: Evaluating the impact of exercise protocols on outcomes using a meta-analytic procedure. *Br. J. Sports Med.* 47:679–86.

Goulet, E. D. B., M. Aubertin-Leheudre, G. E. Plante, and I. J. Dionne. 2007. A meta-analysis of the effects of glycerol-induced hyperhydration on fluid retention and endurance performance. *Int. J. Sport Nutr. Exer. Metab.* 17:391–410.

Haldane, J. S. 1923. Water Poisoning. *Br. Med. J.*, 986.

Haldane, J. S. 1929. Salt depletion by sweating. *Br. Med. J.*, 469.

Hinckson, E. A. and W. G. Hopkins. 2005. Reliability of time to exhaustion analyzed with critical-power and log–log modeling. *Med. Sci. Sports Exer.* 37:696–701.

Hopkins, W. G. 2001. Clinical vs statistical significance. *Sportscience* 5:1–2.

Ikegawa, S., J. Kamijo, K. Okazaki et al. 2011. Effects of hypohydration on thermoregulation during exercise before and after 5-day aerobic training in a warm environment in young men. *J. Appl. Physiol.* 110:972–80.

Koehler, K., H. Braun, M. de Marees et al. 2014. Glycerol administration before endurance exercise: Metabolism, urinary glycerol excretion and effects on doping-relevant blood parameters. *Drug Test. Anal.* 6:202–9.

Lee, J. K. W., S. M. Shirreffs, and R. J. Maughan. 2008a. The influence of serial feeding of drinks at different temperatures on thermoregulatory responses during prolonged exercise. *J. Sport Sci.* 26:583–90.

Lee, J. K. W., S. M. Shirreffs, and R. J. Maughan. 2008b. Cold drink ingestion improves exercise endurance capacity in the heat. *Med. Sci. Sport Exer.* 40:1637–44.

Lorenzo, S., J. R. Halliwill, M. N. Sawka, and C. T. Minson. 2010. Heat acclimation improves exercise performance. *J. Appl. Physiol.* 109:1140–7.

Maughan, R. J., and L. M. Burke. 2011. Practical nutritional recommendations for the athlete. In *Sports Nutrition: More Than Just Calories—Triggers for Adaptation,* ed. R. J. Maughan, and L. M. Burke, 131–49. Nestlé Nutr Inst Workshop Ser. vol 69, Basel, Switzerland: Karger AG.

Maughan, R. J., C. E. Fenn, and J. B. Leiper. 1989. Effects of fluid, electrolyte and substrate ingestion on endurance capacity. *Eur. J. Appl. Physiol. Occup. Physiol.* 58:481–6.

Maughan, R. J., and J. B. Leiper. 1995. Effects of sodium content of ingested fluids on post-exercise rehydration in man. *Eur J. Appl. Physiol.* 71:311–19.

Maughan, R, J., J. B. Leiper, and S. M. Shirreffs. 1996. Restoration of fluid balance after exercise-induced dehydration: Effects of food and fluid intake. *Eur. J. Appl. Physiol.* 73:317–25.

Maughan, R. J., J. H. Owen, S. M. Shirreffs, and J. B. Leiper. 1994. Post-exercise rehydration in man: Effects of electrolyte addition to ingested fluids. *Eur. J. Appl. Physiol.* 69: 209–15.

Maughan, R. J., and S. M. Shirreffs. 2008. Development of individual hydration strategies for athletes. *Int. J. Sport Nutr. Exer. Metab.* 18:457–72.

Maughan, R. J., S. M. Shirreffs, S. J. Merson, and C. A. Horswill. 2005. Fluid and electrolyte balance in elite male football (soccer) players training in a cool environment. *J. Sports Sci.* 23:73–9.

Maughan, R. J., S. M. Shirreffs, K. T. Ozgünen et al. 2010. Living, training and playing in the heat: Challenges to the football player and strategies for coping with environmental extremes. *Scand. J. Med. Sci. Sport.* 20(Suppl 3):117–24.

Maughan, R. J., P. Watson, G. H. Evans, N. Broad, and S. M. Shirreffs. 2007. Water balance and salt losses in competitive football. *Int. J. Sport Nutr. Exer. Metab.* 17:583–94.

Noakes, T. D. 2007. Hydration in the marathon—Using thirst to gauge safe fluid replacement. *Sports Med.* 37, 463–6.

Noakes, T. D., N. Goodwin, B. L. Rayner, T. Branken, and R. K. Taylor. 1985. Water intoxication: A possible complication during endurance exercise. *Med. Sci. Sports Exer.* 17:370–5.

Nose, H., G. W. Mack, X. Shi, and E. R. Nadel. 1988. Role of osmolality and plasma volume during rehydration in humans. *J. Appl. Physiol.* 65:325–31.

Riedesel, M. L., D. L. Allen, G. T. Peake, and K. Al-Qattan. 1987. Hyperhydration with glycerol solutions. *J. Appl. Physiol.* 63:2262–8.

Rivera-Brown, A. M., and R. A. De Felix-Davila. 2012. Hydration status in adolescent judo athletes before and after training in the heat. *Int. J. Sport Physiol. Perf.* 7:39–46.

Sawka, M. N., L. M. Burke, E. R. Eichner, R. J. Maughan, S. J. Montain, and N. S. Stachenfeld. 2007. Exercise and fluid replacement. *Med. Sci. Sports Exer.* 39:377–90.

Shearman, M. 1888. *Athletics and Football.* London: Longmans, Green.

Shirreffs, S. M., and R. J. Maughan. 1997. Whole body sweat collection in man: An improved method with some preliminary data on electrolyte composition. *J. Appl. Physiol.* 82:336–41.

Shirreffs, S. M., and R. J. Maughan. 1998. Volume repletion following exercise-induced volume depletion in man: Replacement of water and sodium losses. *Am. J. Physiol.* 43:F868–75.

Shirreffs, S. M., A. J. Taylor, J. B. Leiper, and R. J. Maughan. 1996. Post-exercise rehydration in man: Effects of volume consumed and sodium content of ingested fluids. *Med. Sci. Sports Exer.* 28:1260–71.

Sims, S. T., N. J. Rehrer, M. L. Bell, and J. D. Cotter. 2007b. Preexercise sodium loading aids fluid balance and endurance for women exercising in the heat. *J. Appl. Physiol.* 103:534–41.

Sims, S. T., L. van Vliet, J. D. Cotter, and N. J. Rehrer. 2007a. Sodium loading aids fluid balance and reduces physiological strain of trained men exercising in the heat. *Med. Sci. Sports Exer.* 39:123–30.

Stewart, C. J., D. G. Whyte, J. Cannon, J. Wickham, and F. E. Marino. 2014. Exercise-induced dehydration does not alter time trial or neuromuscular performance. *Int. J. Sports Med.* 35:725–30.

Thom, W. 1813. *Pedestrianism.* Aberdeen, Scotland: Chalmers.

Van Rosendal, S. P., and J. S. Coombes. 2012. Glycerol use in hyperhydration and rehydration: Scientific update. *Med. Sport Sci.* 59:104–12.

Wilson, G., N. Chester, and M. Eubank. 2012. An alternative dietary strategy to make weight while improving mood, decreasing body fat, and not dehydrating: A case study of a professional jockey. *Int. J. Sport Nutr. Exer. Metab.* 22:225–31.

Wong, S. H., C. Williams, M. Simpson et al. 1998. Influence of fluid intake pattern on short-term recovery from prolonged, submaximal running and subsequent exercise capacity. *J. Sports Sci.* 16:143–52.

19 Plain Water or Carbohydrate–Electrolyte Beverages

Lindsay B. Baker, Kelly A. Barnes,
and John R. Stofan

CONTENTS

19.1 INTRODUCTION

There is not a one-size-fits-all type of answer to the question of whether athletes should drink water or carbohydrate–electrolyte beverages. The optimal composition of a fluid replacement beverage during/after exercise depends on several factors, including the duration and intensity of exercise, environmental conditions, and rate of sweat water and electrolyte losses, as well as the athlete's personal preferences and goals. The primary goal for many competitive athletes is to enhance performance or mitigate fatigue during training/competition. Fatigue during prolonged exercise can be associated with dehydration and/or energy (carbohydrate) depletion, and beverage composition can have a substantial impact on both processes. In this chapter, we discuss the role that water, carbohydrate, electrolytes, and flavor play in fluid replacement and energy provision for athletes during/after exercise. Based on these scientific principles as well as some practical considerations, we conclude this chapter with recommendations related to beverage composition choices for athletes. Recommendations regarding timing and volume of fluid intake are discussed in Chapter 18 by Maughan and Shirreffs.

19.2 OBJECTIVES OF A FLUID REPLACEMENT BEVERAGE FOR ATHLETES

The main objective of drinking a beverage during exercise is to replace fluids lost from sweating. Fluid replacement is a process that involves thirst and voluntary fluid intake, gastric emptying and intestinal absorption of fluid, distribution within the body fluid compartments, and whole-body fluid retention. Another potential objective of a fluid-replacement beverage (unrelated to hydration) is to provide nutrients or compounds to enhance exercise performance, such as carbohydrate for energy provision. In this section, we discuss the scientific principles and practical ramifications of fluid replacement and energy provision during exercise (see also Baker and Jeukendrup 2014; Coombes and Hamilton 2000; Murray and Stofan 2001).

19.2.1 THIRST AND VOLUNTARY FLUID INTAKE

In a laboratory, the type and amount of fluid intake is usually controlled to determine the effect of beverage formulation on factors such as fluid absorption, rehydration, or performance. However, in *real life*, the athlete's drinking pattern and their choice of beverage is often decided on an *ad libitum* or voluntary basis. Voluntary fluid intake is dictated in part by thirst (both physiological and perceived), but also by many other factors such as social influences, cultural preferences, and beverage-related factors (e.g., availability and organoleptic properties) (Adolph et al. 1954; see Passe 2001 for review). As discussed in Sections 19.3.2 and 19.3.3, the inclusion of sodium and/or flavoring in a fluid replacement beverage can impact thirst and/or voluntary fluid intake (Rothstein et al. 1947).

19.2.2 Gastric Emptying and Intestinal Absorption

Fluid absorption through the stomach is very limited; thus, ingested fluid must be emptied from the stomach and delivered to the lumen of the small intestine before it can enter the circulation. The rates of gastric emptying and fluid absorption are impacted by several factors, including various types of stressors. For example, ≥3% dehydration has been shown to impair gastric emptying (Neufer et al. 1989; Rehrer et al. 1990; Van Nieuwenhoven et al. 2000). Exercise intensity >~70%–75% VO$_2$max decreases gastric emptying and intestinal water absorption (Costill and Saltin 1974; Fordtran and Saltin 1967). Further discussion on the factors affecting gastrointestinal responses during exercise is provided in Chapter 6.

In this chapter, the discussion will primarily focus on the effects of beverage composition, particularly energy (carbohydrate) content and osmolality, which can significantly impact the rate at which gastric emptying and intestinal absorption of water occur. In addition, intestinal absorption is an important rate limiting step for energy provision (i.e., carbohydrate oxidation) during exercise.

19.2.3 Fluid Distribution

Once fluid is absorbed into circulation, it is distributed to the intracellular and extracellular fluid compartments according to osmotic gradients. The most osmotically active ions in the intracellular and extracellular fluid compartments are potassium and sodium, respectively. The maintenance of extracellular volume, particularly plasma volume, is important for cardiovascular and thermoregulatory function. Sodium ingestion plays a critical role in the maintenance and/or restoration of plasma volume when drinking to replace sweat fluid losses, especially rehydration after exercise (Maughan et al. 1994, 1997; Nose et al. 1988).

19.2.4 Physiologic and Performance Effects

The physiologic and performance effects of ingesting water to maintain body water balance are discussed in detail in this book (Chapters 4 through 9) and elsewhere (Cheuvront and Kenefick 2014; Sawka et al. 2007). In brief, drinking water to avoid a significant body mass deficit can prevent the deleterious effects of dehydration on cardiovascular and thermoregulatory function. Performance during activities greatly dependent upon the cardiovascular system, such as endurance exercise, is more likely to be impaired by dehydration than strength or anaerobic exercise (Cheuvront and Kenefick 2014). Although some individuals may be more or less sensitive to dehydration, the level associated with performance degradations approximates >2% decrease in body mass (Sawka et al. 2007; Shirreffs and Sawka 2011) and further body water deficits can result in further deteriorations in performance (Baker et al. 2007; McConell et al. 1997; Montain and Coyle 1992; Sawka et al. 1985). The likelihood and magnitude of performance impairment with dehydration is also dependent upon the environmental conditions. When heat stress is combined with dehydration, there is greater cardiovascular strain resulting from the competition between the central and peripheral circulation for limited blood volume. In short, hyperthermia (particularly hot skin) exacerbates the performance decrement for a given level of dehydration (>2% body mass deficit), such that the percentage decrement in aerobic time trial performance declines linearly by ~1.3% for each 1°C rise in skin temperature (starting at a skin temperature of ~27°C–29°C) (see Cheuvront and Kenefick 2014 for review).

Fatigue during prolonged exercise is often associated with muscle glycogen depletion, diminished total carbohydrate oxidation, and/or reduced blood glucose concentrations. It is well established that provision of carbohydrate during exercise can improve endurance performance (Burke et al. 2011; Cermak and van Loon 2013; Stellingwerff and Cox 2014). In addition, carbohydrate intake has been shown to improve sustained high-intensity (Below et al. 1995; Jeukendrup et al. 1997) and intermittent high-intensity exercise capacity (Phillips et al. 2011) as well as sport-specific skills (Currell et al. 2009; Dougherty et al. 2006). However, documentation of the performance-enhancing effects of carbohydrate ingestion on skill is less consistent when compared with findings on endurance outcomes, and more work is needed to establish the potential underlying mechanisms (Baker et al. 2014; Russell and Kingsley 2014).

19.2.5 POST-EXERCISE REHYDRATION (FLUID RETENTION)

After exercise in which the athlete has incurred a body mass deficit, the goal is to drink to replace sweat losses and restore body fluid balance (Sawka et al. 2007; Shirreffs and Sawka 2011). Electrolytes, specifically sodium, play a critical role in the completeness of rehydration. The increase in blood sodium concentration and osmolality with sodium ingestion stimulates renal water reabsorption (Maughan and Leiper 1995). In addition, slowing the appearance of the ingested fluid into the circulation can help attenuate diuresis during rehydration. This can be achieved by a gradual drinking pattern (Kovacs et al. 2002; Wong et al. 1998) or decreasing the rate of gastric emptying of the ingested beverage by increasing the energy density (e.g., carbohydrate concentration) of the drink (Evans et al. 2009a, 2011).

19.3 COMPONENTS OF CARBOHYDRATE–ELECTROLYTE BEVERAGES

Table 19.1 lists the composition of water and carbohydrate–electrolyte beverages, as well as other common beverages for comparative purposes. Carbohydrate–electrolyte beverages, or sports drinks, are formulated for consumption before, during, and/or after physical activity. They are generally composed of 6%–8% carbohydrate, low-moderate amounts of the primary electrolytes lost in sweat (i.e., sodium, chloride, and potassium), and flavor (usually fruit-related). Carbohydrate–electrolyte beverages are typically formulated with the types and amounts of carbohydrates (or combinations of carbohydrates) that are rapidly digested and absorbed to facilitate delivery of fluid and carbohydrate to the body. The following sections discuss the roles that carbohydrates, electrolytes, and flavor play in the fluid replacement process (see Table 19.2 for a summary) (see also Baker and Jeukendrup 2014; Coombes and Hamilton 2000; Murray and Stofan 2001).

TABLE 19.1

Composition of Water, Sports Drinks, and Other Common Beverages

	Carbohydrate (% Weight/Volume)	Sodium (mmol/L)	Potassium (mmol/L)
Water (plain, bottled)	0	0	0
Sports drinks	Moderate (~6–8)	Low to moderate (~15–35)	Low (~2–10)
Oral rehydration solutions	Low (~1–3)	High (~45–75)	Moderate (~20)
Juice	High (~10–15)	Low (~0–5)	Low to high (~15–50)
Regular sodas	High (~10–13)	Low (~3–10)	Low (~0)
White milk	Moderate (~5–6)	Moderate (~20–25)	Moderate to high (~35–45)
Chocolate milk	Moderate to high (~8–11)	Moderate (~30–35)	Moderate to high (~35–45)
Coffee	Low (~0–2)	Low (~0–5)	Low to moderate (~10–20)
Tea	Low to high (~0–10)	Low (~1–20)	Low (~0–10)
Energy drinks	High (~11–13)	Low (~5–20)	Low (~0–1)
Beer	Low (~1–4)	Low (~1–5)	Low (~0–10)

The ranges for each category include a variety of more common types and brands of beverages consumed (primarily US-based). White milk includes fat-free, 2% fat, and whole fat. Chocolate milk is low fat. Coffee includes black and with two tablespoons of cream and/or one packet of sugar per cup. Tea includes self-brewed, bottled, sweetened, and unsweetened varieties.

19.3.1 ROLE OF CARBOHYDRATE

19.3.1.1 Voluntary Fluid Intake

Lightly sweetened beverages help improve beverage palatability (hedonic ratings such as liking and acceptability) and can increase voluntary drinking compared with water (Passe et al. 2000; Rolls 1994; Szlyk et al. 1989; Wilk and Bar-Or 1996). However, based on anecdotal accounts the desire for or palatability of a sweet beverage may decline during very prolonged exercise. The amount and type of carbohydrate impact the sweetness level of a beverage. Compared with sucrose (table sugar, relative sweetness rating of 100), solutions that contain crystalline fructose (rating of 180) or high fructose corn syrup (rating 105–130) have higher relative sweetness, whereas glucose (rating of 50–70) and maltose (rating 50) taste less sweet. Maltodextrin, especially those of a relatively long chain length of glucose units, and starch have very low sweetness levels (BeMiller 1992; Murray and Stofan 2001).

TABLE 19.2
Summary of the Effects of Beverage Carbohydrate–Electrolyte Composition and Other Beverage Characteristics on Fluid Replacement in Athletes

Beverage Attribute	Thirst, Beverage Palatability, and Voluntary Fluid Intake	Gastric Emptying of Fluid	Intestinal Fluid Absorption	Fluid Distribution	Fluid Retention (Post-Exercise)
			Phase of the Fluid Replacement Process		
Carbohydrate amount	Lightly sweetened beverages are generally preferred by athletes and are associated with increased fluid intake compared with water; however, anecdotally the desire for a sweet beverage declines during very prolonged exercise.	Increases in beverage CHO increase energy density, which generally decreases gastric emptying rate; for example, ≥8% CHO solution has been found to slow gastric emptying rate and increase GI discomfort compared with water.	Some CHO (~1%) is important to stimulate water absorption via the SGLT1; however, increasing the amount of CHO in a beverage increases osmolality and intestinal fluid absorption is inversely related to beverage osmolality.	N/A	6%–12% CHO may improve fluid retention; however, more work is needed to better understand mechanisms.
Carbohydrate type	Fructose, HFCS, and sucrose are very sweet; maltodextrins, especially those of long chain lengths, have very low sweetness; maltose and glucose have intermediate sweetness.	Multiple transportable CHO increase gastric emptying rate of beverages with >8% CHO.	Different types of CHO contribute differently to beverage osmolality (e.g., mono- and disaccharides > maltodextrin and starch). Multiple transportable CHO (i.e., utilizing both SGLT1 and GLUT5) increase the rate of intestinal absorption of solutes and water.	N/A	N/A

(Continued)

TABLE 19.2 (Continued)

Summary of the Effects of Beverage Carbohydrate–Electrolyte Composition and Other Beverage Characteristics on Fluid Replacement in Athletes

Beverage Attribute	Thirst, Beverage Palatability, and Voluntary Fluid Intake	Gastric Emptying of Fluid	Intestinal Fluid Absorption	Fluid Distribution	Fluid Retention (Post-Exercise)
			Phase of the Fluid Replacement Process		
Sodium	Low to moderate [Na⁺] improve beverage palatability and stimulate physiological thirst and fluid intake; however, these responses decrease with high beverage [Na⁺] (≥50 mmol/L).	Limited data available; however, one study reported no effect of beverage Na⁺ on gastric emptying rate.	Na⁺ is important for SGLT1 transport of CHO and water flux across enterocytes; however, beverage Na⁺ plays only a minor role since Na⁺ is readily available through intestinal secretions.	Na⁺ is the primary extracellular cation; Na⁺ ingestion promotes fluid retention in the vascular space (i.e., plasma volume maintenance).	Significant inverse relation between beverage [Na⁺] and urine output volume.
Flavor	Flavored beverages are generally preferred by athletes and are associated with increased fluid intake compared with water.	N/A	N/A	N/A	N/A

See text for supporting references.

CHO, carbohydrate; GI, gastrointestinal; GLUT5, intestinal transporter for fructose; HFCS, high fructose corn syrup; Na⁺, sodium; N/A, not applicable or not enough information available; SGLT1, sodium–glucose link transporter.

19.3.1.2   Gastric Emptying and Intestinal Absorption

Carbohydrate–electrolyte beverages intended for consumption during exercise are formulated to promote rapid delivery of fluid and carbohydrate. The type and amount of beverage carbohydrate have a significant effect on fluid/substrate delivery by influencing the rates of gastric emptying and intestinal absorption. The presence of some carbohydrate (~1%) is important to stimulate water absorption via the sodium–glucose link transporter (SGLT1) (Gisolfi 1994). However, in general, beverages with high energy density (i.e., carbohydrate concentration) result in a slower rate of gastric emptying (Brouns et al. 1995; Lambert et al. 2012; Murray 1987). For example, beverages with ≥8% carbohydrate solution have been found to decrease gastric emptying rate and increase gastrointestinal discomfort compared with water (Murray et al. 1999; Rehrer et al. 1989, 1992; Shi et al. 2004). Furthermore, there is a significant inverse relation between beverage osmolality and intestinal fluid absorption (Shi and Passe 2010; Shi et al. 1995). However, formulating drinks with multiple transportable carbohydrates (i.e., utilizing both SGLT1 and GLUT5 transporters) can increase gastric emptying rate (Shi et al. 2000; Jeukendrup and Moseley 2010) and intestinal absorption of beverages with >8% carbohydrate (Gisolfi et al. 1998; Jeukendrup 2010; Shi et al. 1994, 1995). Thus, as discussed in more detail in subsequent sections (19.3.1.3 and 19.5), multiple transportable carbohydrates are an important consideration when high rates of carbohydrate (>60 g/h) delivery are needed (e.g., during prolonged exercise >2–3 h). For more information, the reader is referred to recent review papers, which discuss in more detail the effect of carbohydrate amount and type on intestinal fluid and carbohydrate absorption (Baker and Jeukendrup 2014; Jeukendrup 2010; Shi and Passe 2010).

19.3.1.3   Physiologic and Performance Effects

Several reviews have discussed the ergogenic effects of carbohydrate ingestion during exercise in detail (Cermak and van Loon 2013; Coombes and Hamilton 2000; Jeukendrup 2004; Karelis et al. 2010). In brief, carbohydrate delivery is important because muscle glycogen and blood glucose are the primary substrates for the contracting muscle during exercise (Romijn et al. 1993). Maintaining carbohydrate oxidation through carbohydrate feeding has been shown to be an effective strategy to delay fatigue (Coyle et al. 1983). During prolonged exercise (>2 h) there seem to be dose–response relations among carbohydrate intake, carbohydrate oxidation, and improved performance (Smith et al. 2010, 2013; Vandenbogaerde and Hopkins 2011). In a large multicenter study, Smith et al. (2013) tested the effects of ingesting 0–120 g carbohydrate/h during a 2-h constant load ride on subsequent 20-km cycling time trial performance. The ingestion of carbohydrate significantly improved performance in a curvilinear dose-dependent manner, and the greatest performance enhancement was seen with an ingestion rate of 68–88 g carbohydrate/h (Smith et al. 2013).

Exogenous carbohydrate oxidation is mainly limited by the intestinal absorption of carbohydrates. In general, carbohydrates can be divided into two categories: faster oxidized (up to about 60 g/h) and slower oxidized (up to about 40 g/h). The faster oxidized carbohydrates include glucose, maltose, sucrose, maltodextrins, and amylopectin starches. The slower oxidized carbohydrates include fructose, galactose, isomaltulose, trehalose, and insoluble (amylose) starches (see Jeukendrup 2010 for review).

Commercially available carbohydrate–electrolyte beverages typically contain glucose, sucrose, fructose (in relatively small amounts), and/or maltodextrins (Coombes and Hamilton 2000; Murray and Stofan 2001).

During prolonged exercise (i.e., >2 h), the benefits of carbohydrate are mainly metabolic in nature. However, the underlying mechanism for the ergogenic effect of carbohydrate ingestion during high-intensity, shorter duration exercise may be related more so to the central nervous system. Rinsing the mouth with a carbohydrate-containing drink has been shown to improve performance (Jeukendrup and Chambers 2010) and activate regions of the brain associated with reward and motor control (Chambers et al. 2009). Thus, it is thought that carbohydrate may influence the brain via carbohydrate-sensitive receptors in the mouth. However, more work regarding this potential mechanism is clearly needed to identify carbohydrate receptors in the oral cavity and better understand the brain regions involved. The reader is referred to review papers for a more detailed discussion on the topic of carbohydrate mouth rinse (Jeukendrup 2013; Silva et al. 2014).

19.3.1.4 Post-Exercise Rehydration (Fluid Retention)

Recent research indicates that the carbohydrate concentration of a fluid replacement beverage may have an impact on fluid retention during post-exercise rehydration. Carbohydrate solutions ranging from 6% to 12% have been shown to promote greater fluid retention compared with electrolyte-matched, carbohydrate-free placebo solutions (Evans et al. 2009b; Kamijo et al. 2012; Osterberg et al. 2010). Highly concentrated carbohydrate solutions (e.g., 10%–12%) could affect fluid retention by delaying gastric emptying or intestinal absorption (from the increased energy density and/or osmolality of the higher carbohydrate beverages), which would effectively delay the appearance of fluid in the circulation (Evans et al. 2009b; Osterberg et al. 2010). The delayed absorption of fluid and/or the higher plasma osmolality elicited by the carbohydrate drink would attenuate renal water excretion. Other proposed mechanisms, particularly regarding the fluid retention benefits of less concentrated carbohydrate solutions (e.g., 6%), include an insulin-mediated increase in renal sodium and water reabsorption (Kamijo et al. 2012). However, more work is needed to elucidate the exact mechanisms involved and their relative contributions to enhanced fluid retention.

19.3.2 Role of Electrolytes

19.3.2.1 Voluntary Fluid Intake

The presence of low to moderate amounts of sodium improves beverage palatability and stimulates the physiological drive to drink. Consumption of plain water decreases plasma osmolality and sodium concentration, which reduce the drive to drink, oftentimes before body water volume has been fully restored (Nose et al. 1988; Sawka et al. 2007; Takamata et al. 1994; Wilk and Bar-Or 1996). However, palatability and voluntary fluid intake also decrease with high beverage sodium concentrations (≥50 mmol/L) (Passe 2001; Wemple et al. 1997). In these instances, other anions (such as citrate) could be used in place of chloride (the typical anion with sodium) to decrease beverage saltiness (Shirreffs 2003).

19.3.2.2 Gastric Emptying and Intestinal Absorption

There is limited data available; however, there seems to be no effect of beverage sodium concentration on the rate of gastric emptying of ingested fluids (Gisolfi et al. 2001). Regarding intestinal fluid absorption, it is well established that sodium is important for SGLT1 transport of substrate and water flux across enterocytes. However, beverage sodium has been found to play only a minor role in intestinal fluid absorption, likely because sodium is readily available through intestinal secretions (Gisolfi et al. 1995, 2001; Murray and Stofan 2001). Very limited information is available on the effects of other electrolytes on gastric emptying and intestinal fluid absorption. One study reported faster intestinal absorption when the conjugate anion of sodium was chloride compared with bicarbonate or sulfate (Fordtran 1975); however, more research is needed for other anion comparisons.

19.3.2.3 Physiologic and Performance Effects

Sodium is the predominant electrolyte lost in sweat; average sweat sodium concentration is ~40 mmol/L, but ranges from ~15 to ~90 mmol/L (Baker et al. 2009). Potassium is also lost in sweat, but in much smaller amounts (~3–5 mmol/L) compared with sodium (Baker et al. 2009). The presence of sodium and potassium in a fluid replacement beverage helps replace sweat electrolyte losses. However, there are currently not enough data to elucidate exactly how much of the sweat electrolyte loss should be replaced during exercise. Acute sweat sodium loss does not typically have deleterious effects except in situations where losses are especially large (>3–4 g) (Coyle 2004; Shirreffs and Sawka 2011). There is insufficient evidence to date to suggest that sweat potassium loss and/or the replacement of those losses during exercise impacts physiological function and/or performance (Leiper 2001; Murray and Stofan 2001; Shirreffs and Sawka 2011).

19.3.2.4 Post-Exercise Rehydration (Fluid Retention)

The sodium concentration of a fluid replacement beverage has a significant impact on fluid retention and thus the completeness of rehydration after exercise. In fact, there is a significant direct linear relation between beverage sodium concentration and fluid retention (Maughan and Leiper 1995). The increase in serum sodium concentration and osmolality with sodium ingestion stimulates renal water reabsorption and thus decreases urine output. In addition, because sodium is the primary extracellular cation, ingesting a beverage with sodium can promote fluid retention in the vascular space (i.e., plasma volume maintenance) (Gonzalez-Alonso et al. 1992; Maughan and Leiper 1995; Shirreffs et al. 2007; Wemple et al. 1997).

19.3.3 Role of Flavor

19.3.3.1 Voluntary Fluid Intake

Flavored beverages are generally preferred by athletes and thus associated with increased *ad libitum* fluid intake. This has been demonstrated in a range of athletes, including male and female as well as youth and adults (Minehan et al. 2002; Passe 2001; Passe et al. 2000, 2004; Rivera-Brown et al. 1999; Rolls 1994).

It is also important to note that the athletes' perception and acceptability of beverage flavor differs from rest to exercise. For example, in one study subjects rated the acceptability of two flavored sports drinks and water prior to exercise (Passe et al. 2000). As expected, the subjects consumed greater quantities of their *most acceptable* flavor of sports drink than water and the *least acceptable* sports drink flavor at rest and during the first 75 min of exercise. However, at 90 and 180 min of exercise, subjects' hedonic ratings of the *least acceptable* flavor dramatically improved compared to the ratings taken at rest (from a *liking* rating of ~4 to ~7 on a 9-point scale). Furthermore, total consumption of the *least acceptable* flavor of sports beverage was greater than water (Passe et al. 2000).

In addition to changes in flavor perception, the perception of other organoleptic properties of beverages, such as sweetness, tartness, mouthfeel, and aftertaste, is also modified by physical activity (Murray and Stofan 2001). Thus, sports drinks are formulated with consideration of these changes in sensory parameters to optimize palatability during exercise. Other factors such as time of day (Birch et al. 1984), previous experience (Pliner 1982), cultural background (Rozin and Vollmecke 1986), and environment (Sohar et al. 1962) can also influence beverage palatability and voluntary drinking behavior.

19.4 PRACTICAL CONSIDERATIONS

19.4.1 MEETING HYDRATION AND FUEL DEMANDS DURING EXERCISE

There are two main nutrition-related factors that impact an athlete's performance: (1) fluid replacement (hydration) and (2) carbohydrate provision (fuel). It is important to keep in mind that the goals of fluid replacement and carbohydrate provision are usually not mutually exclusive. Oftentimes when there is a need for fluid intake to replace heavy sweat losses, there is also a need for carbohydrate intake to support energy demands (e.g., prolonged or intermittent high-intensity exercise). When both fluid replacement and energy provision are important, the goal is to deliver carbohydrate to the body without impeding gastric emptying and intestinal absorption of water. In short, carbohydrate intake should be balanced with consideration for fluid replacement needs.

In some instances, either fluid replacement or energy intake may take top priority. For example, there may be individuals (e.g., heavy sweaters) or situations (e.g., exercise in a hot environment) where replacing sweat losses is of higher priority and thus water intake is the primary aim. On the other hand, when sweat losses are low (e.g., exercise in very cool weather) or high rates of carbohydrate delivery are necessary (e.g., >2 h of moderate to high-intensity exercise), more concentrated beverages or solid/semisolid forms of carbohydrate may be warranted. The reader is referred to papers that discuss the idea of potential trade-offs between carbohydrate and fluid ingestion during exercise (Coyle and Montain 1992a, 1992b).

To meet fuel needs, carbohydrate can be consumed in a liquid or solid/semisolid form. Liquid forms of carbohydrate provide an option to replace fluid and sweat electrolyte losses in addition to meeting fuel needs. However, it has been shown in endurance studies that solid (bar) or semisolid (gel or chew) carbohydrate is oxidized as effectively as a carbohydrate solution, provided that sufficient water is also consumed (Pfeiffer et al. 2010a, 2010b).

19.4.2 INDIVIDUAL PREFERENCES AND STRATEGIES

It is clear that the optimal nutrition/hydration strategy is personalized rather than based on one-size-fits-all recommendations (Jeukendrup 2014; Maughan and Shirreffs 2008, 2010). For example, an appropriate drinking strategy during exercise takes into account pre-exercise hydration status as well as sweat water and electrolyte losses, which are largely dependent upon genetics, body mass, exercise intensity, exercise duration, and environmental conditions (Maughan and Shirreffs 2008, 2010; Sawka et al. 2007). In addition, recommendations for carbohydrate intake during exercise depend on the athlete's exercise duration and absolute exercise intensity (Jeukendrup 2014). Given the many factors governing hydration and carbohydrate needs, it follows that the recommendations vary from athlete to athlete and even within individuals between training sessions/competitions. For more detail on personalized nutrition for endurance athletes, the reader is referred to a recent review (Jeukendrup 2014).

Research has shown that athletes' actual fluid and/or carbohydrate intake rates during exercise are highly variable and in some cases fall short of recommendations (Baker et al. 2014; Garth and Burke 2013; Pfeiffer et al. 2012). There are several possible reasons why some athletes fail to meet recommendations, some of which may be related to practicality. Thus, it is important to promote a hydration and carbohydrate intake strategy that is flexible and relatively easy to follow. This includes offering a variety of options for the source of fluid (water and/or carbohydrate–electrolyte beverages) and carbohydrate (carbohydrate–electrolyte beverages and/or semisolid or solid foods), with consideration for personal preferences (flavor, sweetness, and liquid/solid form preferences) and fluid/food availability (e.g., on a race course or other competition setting).

Personalized strategies should also take into account an individual's gut tolerance for different types and amounts of fluid and carbohydrate intake. In some circumstances, gastrointestinal discomfort may deter athletes from consuming sufficient fluid or carbohydrate during exercise. Importantly, there is some evidence that training the gut can be an effective strategy to reduce gastrointestinal discomfort when aiming to increase fluid (Lambert et al. 2008) and carbohydrate (Cox et al. 2010; De Oliveira et al. 2014) intake during exercise, which further demonstrates that athletes should practice their competition nutrition strategy in training. For a more in-depth discussion on the gut and nutritional strategies to reduce the risk of gastrointestinal problems, the reader is referred to a recent review (De Oliveira et al. 2014).

19.5 RECOMMENDATIONS

Table 19.3 summarizes the current recommendations for water, carbohydrate, and electrolyte intake for before, during, and after exercise. These guidelines (from Burke et al. 2011; Sawka et al. 2007) can be used to help athletes determine when water and/or carbohydrate beverages are needed to meet their hydration and fuel demands.

TABLE 19.3

Water, Carbohydrate, and Electrolyte Intake Recommendations for Athletes

	Description	Water Ingestion	Carbohydrate Ingestion	Electrolyte Ingestion
Before exercise	Before competition or training lasting >1 h	Slowly drink water and/or other fluid (e.g., ~5–7 mL/kg) ≥4 h before exercise. If no urine is produced, or urine is dark or highly concentrated, slowly drink more fluid (e.g., ~3–5 mL/kg) ~2 h before exercise.	1–4 g CHO/kg body mass consumed 1–4 h before activity from drinks, gels, bars, or food	Na⁺ from fluids (20–50 mmol/L) or solids (e.g., Na⁺-containing foods or snacks)
During exercise	Competition or training lasting 1.0–2.5 h	Drink water and/or 6%–8% carbohydrate– electrolyte beverage to prevent >2% body mass deficit, especially in warm-hot environments	30–60 g/h from fluids (e.g., 6%–8% carbohydrate solution) or solids (e.g., gels, bars)	Na⁺ and K⁺ from fluids (20–30 mmol Na⁺/L and 3–5 mmol K⁺/L) or solids (e.g., gels and bars)
After exercise	When there is <8–12 h recovery between two competition/ training sessions	Drink ~1.5 L of water and/or other fluid for each 1 kg of body mass deficit	1.0–1.2 g CHO/ kg body mass/h for first 4 h from drinks, gels, bars, or food	Na⁺ from fluids (20–50 mmol/L) or solids (e.g., Na⁺-containing foods or snacks)

Sources: Burke, L.M. et al., *J. Sports Sci.,* 29(Suppl 1), S17–27, 2011; Sawka, M.N. et al., *Med. Sci. Sports. Exer.* 39(2), 377–90, 2007.

CHO, carbohydrate; K⁺, potassium; Na⁺, sodium.

Further guidelines (from Jeukendrup 2014) specific to the amount and type of carbohydrate recommended for various exercise durations are in Figure 19.1. It is important to reiterate that the recommended intake rates of water, carbohydrate, and electrolytes are not one-size-fits-all across all athletes or even static for a given individual athlete, but instead shift according to changes in training goals, exercise intensity and duration, and environmental conditions.

The recommendations for water, carbohydrate, and electrolyte intake are mostly drawn from studies involving endurance exercise such as cycling and running.

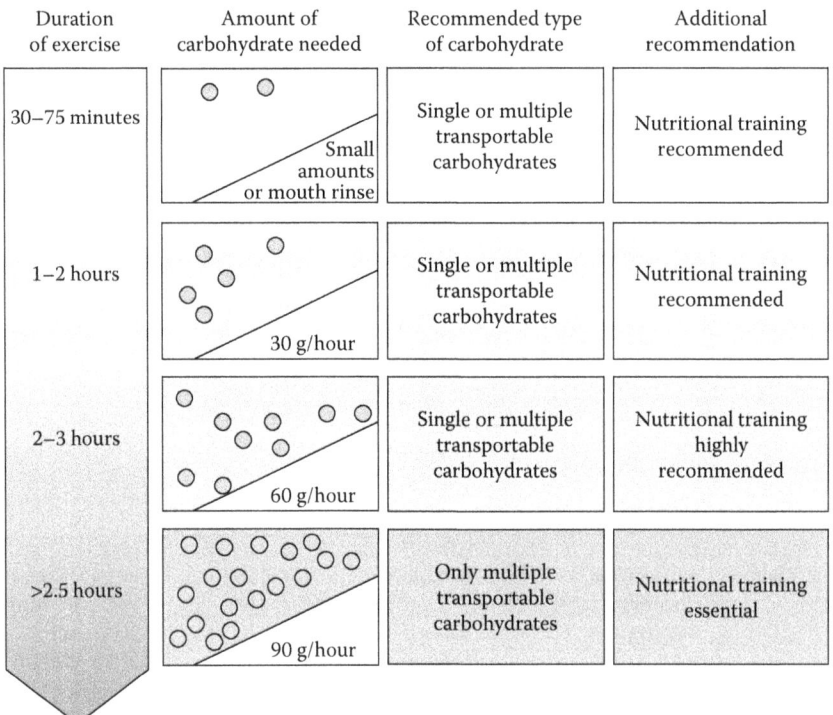

Duration of exercise	Amount of carbohydrate needed	Recommended type of carbohydrate	Additional recommendation
30–75 minutes	Small amounts or mouth rinse	Single or multiple transportable carbohydrates	Nutritional training recommended
1–2 hours	30 g/hour	Single or multiple transportable carbohydrates	Nutritional training recommended
2–3 hours	60 g/hour	Single or multiple transportable carbohydrates	Nutritional training highly recommended
>2.5 hours	90 g/hour	Only multiple transportable carbohydrates	Nutritional training essential

FIGURE 19.1 The new carbohydrate intake guidelines. Carbohydrate intake recommendations during exercise depend on the duration of exercise. In general, carbohydrate intake recommendations increase with increasing duration. The type of carbohydrate may also vary as well as recommendations for nutritional training. These recommendations are for well-trained athletes. Aspiring athletes (exercising at lower absolute exercise intensities) may need to adjust these recommendations downward. (Data from Jeukendrup, A., *Sports Med.*, 44:S25–33, 2014.)

Historically, research specific to intermittent high-intensity exercise or skill-based sports (e.g., team sports) has been sparse, but is gaining increased attention more recently. Emerging research suggests that sport-specific skills and endurance capacity in team sports (such as soccer and basketball) can be improved by the ingestion of fluid (Baker et al. 2007; Edwards et al. 2007) and carbohydrate (Currell et al. 2009; Dougherty et al. 2006; Phillips et al. 2012). More detailed reviews and recommendations specific to carbohydrate and team sports can be found elsewhere (Holway and Spriet 2011; Mujika and Burke 2010; Phillips et al. 2011). The impact of hydration status on intermittent high-intensity sports performance is covered in Chapter 8 of this book. Nevertheless, the effects of fluid and carbohydrate ingestion on team sports performance have been inconsistent and many gaps still exist in the literature. Therefore, more work is needed to determine the effects and underlying mechanisms of carbohydrate and fluid ingestion on performance outcomes related to team sports.

19.6 CONCLUSION

Athletes should take into account several factors when deciding the optimal fluid replacement beverage to meet their individual needs. The most important factors to consider are the athlete's hydration and fuel needs, and these depend on timing (before, during, or after exercise), exercise duration and intensity, sweating rate, electrolyte losses, and environmental conditions. Water is likely a sufficient fluid replacement drink when fuel demands are low, such as during short duration (<1 h) and/or low intensity exercise. However, during/after prolonged or intermittent high-intensity exercise, carbohydrate ingestion can enhance performance and sodium can aid rehydration. Depending on the athlete's preference, hydration and carbohydrate provision can come in the form of either a carbohydrate–electrolyte solution or water in combination with a solid/semisolid form of carbohydrate and electrolytes.

DISCLOSURE

The views expressed in this chapter are those of the authors and do not necessarily reflect the position or policy of PepsiCo, Inc.

REFERENCES

Adolph, E. F., J. P. Barker, and P. A. Hoy. 1954. Multiple factors in thirst. *Am. J. Physiol.* 178(3):538–62.

Baker, L. B., K. A. Dougherty, M. Chow, and W. L. Kenney. 2007. Progressive dehydration causes a progressive decline in basketball skill performance. *Med. Sci. Sports Exer.* 39(7):1114–23. doi:10.1249/mss.0b013e3180574b02.

Baker, L. B., L. E. Heaton, R. P. Nuccio, and K. W. Stein. 2014. Dietitian-observed macronutrient intakes of young skill and team-sport athletes: Adequacy of pre, during, and post-exercise nutrition. *Int. J. Sport. Nutr. Exer. Metab.* 24:166–76.

Baker, L. B., and A. E. Jeukendrup. 2014. Optimal composition of fluid-replacement beverages. *Compr. Physiol.* 4(2):575–620. doi:10.1002/cphy.c130014.

Baker, L. B., R. P. Nuccio, and A. E. Jeukendrup. 2014. Acute effects of dietary constituents on motor skill and cognitive performance in athletes. *Nutr. Rev.* 72(12):790–802. doi:10.1111/nure.12157.

Baker, L. B., J. R. Stofan, A. A. Hamilton, and C. A. Horswill. 2009. Comparison of regional patch collection vs. whole body washdown for measuring sweat sodium and potassium loss during exercise. *J. Appl. Physiol.* 107(3):887–95. doi:10.1152/japplphysiol.00197.2009.

Below, P. R., R. Mora-Rodriguez, J. Gonzalez-Alonso, and E. F. Coyle. 1995. Fluid and carbohydrate ingestion independently improve performance during 1 h of intense exercise. *Med. Sci. Sports. Exer.* 27(2):200–10.

BeMiller, J. N. 1992. Sucrose. In *Encyclopedia of Food Science and Technology*, ed. Y. H. Hui. pp. 2441–42. New York: John Wiley & Sons.

Birch, L. L., J. Billman, and S. S. Richards. 1984. Time of day influences food acceptability. *Appetite* 5(2):109–16.

Brouns, F., J. Senden, E. J. Beckers, and W. H. Saris. 1995. Osmolarity does not affect the gastric emptying rate of oral rehydration solutions. *J. Parenter. Enteral. Nutr.* 19(5):403–6.

Burke, L. M., J. A. Hawley, S. H. Wong, and A. E. Jeukendrup. 2011. Carbohydrates for training and competition. *J. Sports Sci.* 29(Suppl 1):S17–27. doi:10.1080/02640414.2011.585473.

Cermak, N. M., and L. J. van Loon. 2013. The use of carbohydrates during exercise as an ergogenic aid. *Sports Med.* 43(11):1139–55. doi:10.1007/s40279-013-0079-0.

Chambers, E. S., M. W. Bridge, and D. A. Jones. 2009. Carbohydrate sensing in the human mouth: Effects on exercise performance and brain activity. *J. Physiol.* 587(Pt 8):1779–94. doi:10.1113/jphysiol.2008.164285.

Cheuvront, S. N., and R. W. Kenefick. 2014. Dehydration: Physiology, assessment, and performance effects. *Compr. Physiol.* 4(1):257–85. doi:10.1002/cphy.c130017.

Coombes, J. S., and K. L. Hamilton. 2000. The effectiveness of commercially available sports drinks. *Sports Med.* 29(3):181–209.

Costill, D. L., and B. Saltin. 1974. Factors limiting gastric emptying during rest and exercise. *J. Appl. Physiol.* 37(5):679–83.

Cox, G. R., S. A. Clark, A. J. Cox et al. 2010. Daily training with high carbohydrate availability increases exogenous carbohydrate oxidation during endurance cycling. *J. Appl. Physiol.* 109(1):126–34. doi:10.1152/japplphysiol.00950.2009.

Coyle, E. F. 2004. Fluid and fuel intake during exercise. *J. Sports Sci.* 22(1):39–55. doi:10.1080/0264041031000140545.

Coyle, E. F., J. M. Hagberg, B. F. Hurley, W. H. Martin, A. A. Ehsani, and J. O. Holloszy. 1983. Carbohydrate feeding during prolonged strenuous exercise can delay fatigue. *J. Appl. Physiol.* 55:230–5.

Coyle, E. F., and S. J. Montain. 1992a. Benefits of fluid replacement with carbohydrate during exercise. *Med. Sci. Sports. Exer.* 24(9 Suppl):S324–30.

Coyle, E. F., and S. J. Montain. 1992b. Carbohydrate and fluid ingestion during exercise: Are there trade-offs? *Med. Sci. Sports Exer.* 24:671–8.

Currell, K., S. Conway, and A. E. Jeukendrup. 2009. Carbohydrate ingestion improves performance of a new reliable test of soccer performance. *Int. J. Sport Nutr. Exer. Metab.* 19(1):34–46.

De Ataide e Silva, T., M. E. Di Cavalcanti Alves de Souza, J. F. de Amorim, C. G. Stathis, C. G. Leandro, and A. E. Lima-Silva. 2014. Can carbohydrate mouth rinse improve performance during exercise? A systematic review. *Nutrients* 6:1–10.

De Oliveira, E. P., R. C. Burini, and A. Jeukendrup. 2014. Gastrointestinal complaints during exercise: Prevalence, etiology, and nutritional recommendations. *Sports Med.* 44(Suppl 1):S79–85. doi:10.1007/s40279-014-0153-2.

Dougherty, K. A., L. B. Baker, M. Chow, and W. L. Kenney. 2006. Two percent dehydration impairs and six percent carbohydrate drink improves boys basketball skills. *Med. Sci. Sports. Exer.* 38(9):1650–8. doi:10.1249/01.mss.0000227640.60736.8e.

Edwards, A. M., M. E. Mann, M. J. Marfell-Jones, D. M. Rankin, T. D. Noakes, and D. P. Shillington. 2007. Influence of moderate dehydration on soccer performance: Physiological responses to 45 min of outdoor match-play and the immediate subsequent performance of sport-specific and mental concentration tests. *Br. J. Sports Med.* 41(6):385–91. doi:10.1136/bjsm.2006.033860.

Evans, G. H., S. M. Shirreffs, and R. J. Maughan. 2009a. Postexercise rehydration in man: The effects of carbohydrate content and osmolality of drinks ingested ad libitum. *Appl. Physiol. Nutr. Metab.* 34(4):785–93. doi:10.1139/H09-065.

Evans, G. H., S. M. Shirreffs, and R. J. Maughan. 2009b. Postexercise rehydration in man: The effects of osmolality and carbohydrate content of ingested drinks. *Nutrition* 25(9):905–13. doi:10.1016/j.nut.2008.12.014.

Evans, G. H., S. M. Shirreffs, and R. J. Maughan. 2011. The effects of repeated ingestion of high and low glucose-electrolyte solutions on gastric emptying and blood 2H2O concentration after an overnight fast. *Br. J. Nutr.* 106(11):1732–9. doi:10.1017/S0007114511002169.

Fordtran, J. S. 1975. Stimulation of active and passive sodium absorption by sugars in the human jejunum. *J. Clin. Invest.* 55(4):728–37. doi:10.1172/JCI107983.

Fordtran, J. S., and B. Saltin. 1967. Gastric emptying and intestinal absorption during prolonged severe exercise. *J. Appl. Physiol.* 23(3):331–5.

Garth, A. K., and L. M. Burke. 2013. What do athletes drink during competitive sporting activities? *Sports Med.* 43(7):539–64. doi:10.1007/s40279-013-0028-y.

Gisolfi, C. V. 1994. Use of electrolytes in fluid replacement solutions: What have we learned from intestinal absorption studies? In *Institute of Medicine: Fluid Replacement and Heat Stress*, ed. B.M. Marriott, 11–22. Washington, DC: National Academy Press.

Gisolfi, C. V., G. P. Lambert, and R. W. Summers. 2001. Intestinal fluid absorption during exercise: Role of sport drink osmolality and [Na$^+$]. *Med. Sci. Sports. Exer.* 33(6):907–15.

Gisolfi, C. V., R. D. Summers, H. P. Schedl, and T. L. Bleiler. 1995. Effect of sodium concentration in a carbohydrate-electrolyte solution on intestinal absorption. *Med. Sci. Sports. Exer.* 27(10):1414–20.

Gisolfi, C. V., R. W. Summers, G. P. Lambert, and T. Xia. 1998. Effect of beverage osmolality on intestinal fluid absorption during exercise. *J. Appl. Physiol.* 85(5):1941–8.

Gonzalez-Alonso, J., C. L. Heaps, and E. F. Coyle. 1992. Rehydration after exercise with common beverages and water. *Int. J. Sports Med.* 13(5):399–406.

Holway, F. E., and L. L. Spriet. 2011. Sport-specific nutrition: Practical strategies for team sports. *J. Sports Sci.* 29(Suppl 1):S115–25. doi:10.1080/02640414.2011.605459.

Jeukendrup, A. 2014. A step towards personalized sports nutrition: Carbohydrate intake during exercise. *Sports Med.* 44(Suppl 1):S25–33. doi:10.1007/s40279-014-0148-z.

Jeukendrup, A., F. Brouns, A. J. Wagenmakers, and W. H. Saris. 1997. Carbohydrate-electrolyte feedings improve 1 h time trial cycling performance. *Int. J. Sports Med.* 18(2):125–9. doi:10.1055/s-2007-972607.

Jeukendrup, A. E. 2004. Carbohydrate intake during exercise and performance. *Nutrition* 20(7–8):669–77. doi:10.1016/j.nut.2004.04.017.

Jeukendrup, A. E. 2010. Carbohydrate and exercise performance: The role of multiple transportable carbohydrates. *Curr. Opin. Clin. Nutr. Metab. Care.* 13(4):452–7. doi:10.1097/MCO.0b013e328339de9f.

Jeukendrup, A. E. 2013. Oral carbohydrate rinse: placebo or beneficial? *Curr. Sports Med. Rep.* 12:222–227.

Jeukendrup, A. E., and E. S. Chambers. 2010. Oral carbohydrate sensing and exercise performance. *Curr. Opin. Clin. Nutr. Metab. Care.* 13(4):447–51. doi:10.1097/MCO.0b013e328339de83.

Jeukendrup, A. E., and L. Moseley. 2010. Multiple transportable carbohydrates enhance gastric emptying and fluid delivery. *Scand. J. Med. Sci. Sports* 20(1):112–21. doi:10.1111/j.1600-0838.2008.00862.x.

Kamijo, Y., S. Ikegawa, Y. Okada et al. 2012. Enhanced renal Na$^+$ reabsorption by carbohydrate in beverages during restitution from thermal and exercise-induced dehydration in men. *Am. J. Physiol. Regul. Integr. Comp. Physiol.* 303(8):R824–33. doi:10.1152/ajpregu.00588.2011.

Karelis, A. D., J. W. Smith, D. H. Passe, and F. Peronnet. 2010. Carbohydrate administration and exercise performance: What are the potential mechanisms involved? *Sports Med.* 40(9):747–63. doi:10.2165/11533080-000000000-00000.

Kovacs, E. M., R. M. Schmahl, J. M. Senden, and F. Brouns. 2002. Effect of high and low rates of fluid intake on post-exercise rehydration. *Int. J. Sport Nutr. Exer. Metab.* 12(1):14–23.

Lambert, G. P., J. Lang, A. Bull, J. Eckerson, S. Lanspa, and J. O'Brien. 2008. Fluid tolerance while running: Effect of repeated trials. *Int. J. Sports Med.* 29(11):878–82. doi:10.1055/s-2008-1038620.

Lambert, G. P., X. Shi, and R. Murray. 2012. The gastrointestinal system. In *ACSM's Advanced Exercise Physiology*, eds. P. A. Farrell, M. J. Joyner, and V. J. Caiozzo, 348–62. Philadelphia, PA: Lippincott Williams & Wilkins.

Leiper, J. B. 2001. Gastric emptying and intestinal absorption of fluids, carbohydrates, and electrolytes. In *Sports Drinks: Basic Science and Practical Aspects*, eds. R. J. Maughan and R. Murray, 89–128. Boca Raton, FL: CRC Press LLC.

Maughan, R. J., and J. B. Leiper. 1995. Sodium intake and post-exercise rehydration in man. *Eur J. Appl. Physiol. Occup. Physiol.* 71(4):311–19.

Maughan, R. J., J. B. Leiper, and S. M. Shirreffs. 1997. Factors influencing the restoration of fluid and electrolyte balance after exercise in the heat. *Br. J. Sports Med.* 31(3):175–82.

Maughan, R. J., J. H. Owen, S. M. Shirreffs, and J. B. Leiper. 1994. Post-exercise rehydration in man: Effects of electrolyte addition to ingested fluids. *Eur J. Appl. Physiol. Occup Physiol.* 69(3):209–15.

Maughan, R. J., and S. M. Shirreffs. 2008. Development of individual hydration strategies for athletes. *Int. J. Sport Nutr. Exer. Metab.* 18(5):457–72.

Maughan, R. J., and S. M. Shirreffs. 2010. Development of hydration strategies to optimize performance for athletes in high-intensity sports and in sports with repeated intense efforts. *Scand. J. Med. Sci. Sports* 20(Suppl 2):59–69. doi:10.1111/j.1600-0838.2010.01191.x.

McConell, G. K., C. M. Burge, S. L. Skinner, and M. Hargreaves. 1997. Influence of ingested fluid volume on physiological responses during prolonged exercise. *Acta Physiol. Scand.* 160(2):149–56.

Minehan, M. R., M. D. Riley, and L. M. Burke. 2002. Effect of flavor and awareness of kilojoule content of drinks on preference and fluid balance in team sports. *Int. J. Sport Nutr. Exer. Metab.* 12(1):81–92.

Montain, S. J., and E. F. Coyle. 1992. Influence of graded dehydration on hyperthermia and cardiovascular drift during exercise. *J. Appl. Physiol.* 73(4):1340–50.

Mujika, I., and L. M. Burke. 2010. Nutrition in team sports. *Ann. Nutr. Metab.* 57(Suppl 2): 26–35. doi:10.1159/000322700.

Murray, R. 1987. The effects of consuming carbohydrate-electrolyte beverages on gastric emptying and fluid absorption during and following exercise. *Sports Med.* 4(5):322–51.

Murray, R., W. Bartoli, J. Stofan, M. Horn, and D. Eddy. 1999. A comparison of the gastric emptying characteristics of selected sports drinks. *Int. J. Sport Nutr.* 9(3):263–74.

Murray, R., and J. R. Stofan. 2001. Formulating carbohydrate-electrolyte drinks for optimal efficacy. In *Sports Drinks: Basic Science and Practical Aspects*, eds. R.J. Maughan and R. Murray, 197–223. Boca Raton, FL: CRC Press.

Neufer, P. D., A. J. Young, and M. N. Sawka. 1989. Gastric emptying during exercise: Effects of heat stress and hypohydration. *Eur J. Appl. Physiol. Occup. Physiol.* 58(4):433–9.

Nose, H., G. W. Mack, X. R. Shi, and E. R. Nadel. 1988. Role of osmolality and plasma volume during rehydration in humans. *J. Appl. Physiol.* 65(1):325–31.

Osterberg, K. L., S. E. Pallardy, R. J. Johnson, and C. A. Horswill. 2010. Carbohydrate exerts a mild influence on fluid retention following exercise-induced dehydration. *J. Appl. Physiol.* 108(2):245–50. doi:10.1152/japplphysiol.91275.2008.

Passe, D. H. 2001. Physiological and psychological determinants of fluid intake. In *Sports Drinks: Basic Science and Practical Aspects*, eds. R.J. Maughan and R. Murray, 45–87. Boca Raton, FL: CRC Press.

Passe, D. H., M. Horn, and R. Murray. 2000. Impact of beverage acceptability on fluid intake during exercise. *Appetite* 35(3):219–29. doi:10.1006/appe.2000.0352.

Passe, D. H., M. Horn, J. Stofan, and R. Murray. 2004. Palatability and voluntary intake of sports beverages, diluted orange juice, and water during exercise. *Int. J. Sport Nutr. Exer. Metab.* 14(3):272–84.

Pfeiffer, B., T. Stellingwerff, A. B. Hodgson et al. 2012. Nutritional intake and gastrointestinal problems during competitive endurance events. *Med. Sci. Sports. Exer.* 44(2):344–51. doi:10.1249/MSS.0b013e31822dc809.

Pfeiffer, B., T. Stellingwerff, E. Zaltas, and A. E. Jeukendrup. 2010a. CHO oxidation from a CHO gel compared with a drink during exercise. *Med. Sci. Sports. Exer.* 42(11):2038–45. doi:10.1249/MSS.0b013e3181e0efe6.

Pfeiffer, B., T. Stellingwerff, E. Zaltas, and A. E. Jeukendrup. 2010b. Oxidation of solid versus liquid CHO sources during exercise. *Med. Sci. Sports. Exer.* 42(11):2030–7. doi:10.1249/MSS.0b013e3181e0efc9.

Phillips, S. M., J. Sproule, and A. P. Turner. 2011. Carbohydrate ingestion during team games exercise: Current knowledge and areas for future investigation. *Sports Med.* 41(7): 559–85. doi:10.2165/11589150-000000000-00000.

Phillips, S. M., A. P. Turner, M. F. Sanderson, and J. Sproule. 2012. Carbohydrate gel ingestion significantly improves the intermittent endurance capacity, but not sprint performance, of adolescent team games players during a simulated team games protocol. *Eur. J. Appl. Physiol.* 112(3):1133–41. doi:10.1007/s00421-011-2067-0.

Pliner, P. 1982. The effects of mere exposure on liking for edible substances. *Appetite* 3(3):283–90.

Rehrer, N. J., E. Beckers, F. Brouns, F. ten Hoor, and W. H. Saris. 1989. Exercise and training effects on gastric emptying of carbohydrate beverages. *Med. Sci. Sports. Exer.* 21(5):540–9.

Rehrer, N. J., E. J. Beckers, F. Brouns, F. ten Hoor, and W. H. M. Saris. 1990. Effects of dehydration on gastric emptying and gastrointestinal distress while running. *Med. Sci. Sports Exer.* 20:790–5.

Rehrer, N. J., A. J. Wagenmakers, E. J. Beckers et al. 1992. Gastric emptying, absorption, and carbohydrate oxidation during prolonged exercise. *J. Appl. Physiol.* 72(2):468–75.

Rivera-Brown, A. M., R. Gutierrez, J. C. Gutierrez, W. R. Frontera, and O. Bar-Or. 1999. Drink composition, voluntary drinking, and fluid balance in exercising, trained, heat-acclimatized boys. *J. Appl. Physiol.* 86(1):78–84.

Rolls, B. J. 1994. Palatability and fluid intake. In *Institute of Medicine: Fluid Replacement and Heat Stress*, ed. B.M. Marriott, 161–168. Washington, DC: National Academy Press.

Romijn, J. A., E. F. Coyle, L. S. Sidossis et al. 1993. Regulation of endogenous fat and carbohydrate metabolism in relation to exercise intensity and duration. *Am J Physiol.* 265(3 Pt 1):E380–91.

Rothstein, H., E. F. Adolph, and J. H. Wills. 1947. Voluntary dehydration. In *Physiology of Man in the Desert*, ed. E.R. Adolph, 254–70. New York: Interscience Publishers.

Rozin, P., and T. A. Vollmecke. 1986. Food likes and dislikes. *Annu. Rev. Nutr.* 6:433–56. doi:10.1146/annurev.nu.06.070186.002245.

Russell, M., and M. Kingsley. 2014. The efficacy of acute nutritional interventions on soccer skill performance. *Sports Med.* 44(7):957–70. doi:10.1007/s40279-014-0184-8.

Sawka, M. N., L. M. Burke, E. R. Eichner, R. J. Maughan, S. J. Montain, and N. S. Stachenfeld. 2007. American College of Sports Medicine position stand: Exercise and fluid replacement. *Med. Sci. Sports. Exer.* 39(2):377–90. doi:10.1249/mss.0b013e31802ca597.

Sawka, M. N., A. J. Young, R. P. Francesconi, S. R. Muza, and K. B. Pandolf. 1985. Thermoregulatory and blood responses during exercise at graded hypohydration levels. *J. Appl. Physiol.* 59(5):1394–401.

Shi, X., W. Bartoli, M. Horn, and R. Murray. 2000. Gastric emptying of cold beverages in humans: Effect of transportable carbohydrates. *Int. J. Sport Nutr. Exer. Metab.* 10(4):394–403.

Shi, X., M. K. Horn, K. L. Osterberg et al. 2004. Gastrointestinal discomfort during intermittent high-intensity exercise: Effect of carbohydrate-electrolyte beverage. *Int. J. Sport Nutr. Exer. Metab.* 14(6):673–83.

Shi, X., and D. H. Passe. 2010. Water and solute absorption from carbohydrate-electrolyte solutions in the human proximal small intestine: A review and statistical analysis. *Int. J. Sport Nutr. Exer. Metab.* 20(5):427–42.

Shi, X., R. W. Summers, H. P. Schedl, R. T. Chang, G. P. Lambert, and C. V. Gisolfi. 1994. Effects of solution osmolality on absorption of select fluid replacement solutions in human duodenojejunum. *J. Appl. Physiol.* 77(3):1178–84.

Shi, X., R. W. Summers, H. P. Schedl, S. W. Flanagan, R. Chang, and C. V. Gisolfi. 1995. Effects of carbohydrate type and concentration and solution osmolality on water absorption. *Med. Sci. Sports. Exer.* 27(12):1607–15.

Shirreffs, S. M. 2003. The optimal sports drink. *Sportmedizin und Sporttraumatologie* 51:25–29.

Shirreffs, S. M., L. F. Aragon-Vargas, M. Keil, T. D. Love, and S. Phillips. 2007. Rehydration after exercise in the heat: A comparison of 4 commonly used drinks. *Int. J. Sport Nutr. Exer. Metab.* 17(3):244–58.

Shirreffs, S. M., and M. N. Sawka. 2011. Fluid and electrolyte needs for training, competition, and recovery. *J. Sports Sci.* 29(Suppl 1):S39–46. doi:10.1080/02640414.2011.614269.

Smith, J. W., D. D. Pascoe, D. H. Passe et al. 2013. Curvilinear dose-response relationship of carbohydrate (0-120 g.h^{-1} and performance. *Med. Sci. Sports. Exer.* 45(2):336–41. doi:10.1249/MSS.0b013e31827205d1.

Smith, J. W., J. J. Zachwieja, F. Peronnet et al. 2010. Fuel selection and cycling endurance performance with ingestion of [13C]glucose: Evidence for a carbohydrate dose response. *J. Appl. Physiol.* 108(6):1520–9. doi:10.1152/japplphysiol.91394.2008.

Sohar, E., J. Kaly, and R. Adar. 1962. The prevention of voluntary dehydration. In *Symposium on Environmental Physiology and Psychology in Arid Conditions. Proceedings of the Lucknow Symposium*, 129–35. Paris, France: United Nations Educational Scientific and Cultural Organization.

Stellingwerff, T., and G. R. Cox. 2014. Systematic review: Carbohydrate supplementation on exercise performance or capacity of varying durations. *Appl. Physiol. Nutr. Metab.* 39(9):998–1011. doi:10.1139/apnm-2014-0027.

Szlyk, P. C., I. V. Sils, R. P. Francesconi, R. W. Hubbard, and L. E. Armstrong. 1989. Effects of water temperature and flavoring on voluntary dehydration in men. *Physiol. Behav.* 45(3):639–47.

Takamata, A., G. W. Mack, C. M. Gillen, and E. R. Nadel. 1994. Sodium appetite, thirst, and body fluid regulation in humans during rehydration without sodium replacement. *Am. J. Physiol.* 266(5 Pt 2):R1493–502.

Vandenbogaerde, T. J., and W. G. Hopkins. 2011. Effects of acute carbohydrate supplementation on endurance performance: A meta-analysis. *Sports Med.* 41(9):773–92. doi:10.2165/11590520-000000000-00000.

Van Nieuwenhoven, M. A., B. E. Vriens, R. J. Brummer, and F. Brouns. 2000. Effect of dehydration on gastrointestinal function at rest and during exercise in humans. *Eur. J. Appl. Physiol.* 83(6):578–84. doi:10.1007/s004210000305.

Wemple, R. D., T. S. Morocco, and G. W. Mack. 1997. Influence of sodium replacement on fluid ingestion following exercise-induced dehydration. *Int. J. Sport Nutr.* 7(2):104–16.

Wilk, B., and O. Bar-Or. 1996. Effect of drink flavor and NaCl on voluntary drinking and hydration in boys exercising in the heat. *J. Appl. Physiol.* 80(4):1112–17.

Wong, S. H., C. Williams, M. Simpson, and T. Ogaki. 1998. Influence of fluid intake pattern on short-term recovery from prolonged, submaximal running and subsequent exercise capacity. *J. Sports Sci.* 16(2):143–52. doi:10.1080/026404198366858.

20 Need of Other Elements

Luis Fernando Aragón-Vargas

CONTENTS

20.1 INTRODUCTION

Athletes and physically active people consume many ingredients with purported ergogenic effects, a few of which have already been well documented, while most others have not. A good example is dietary nitrate, mostly in the form of beetroot juice (Hoon et al. 2013; Jones 2014). Some of these components may interact negatively with the major, conventional ingredients in sports drinks; only those that have been studied in association with the rehydration properties of beverages will be discussed in this chapter. Because the main ingredients in rehydration beverages have already been discussed in Chapters 18 and 19, and some others have already been reviewed extensively, emphasis will be placed on some recent work on other elements as they relate to post-exercise rehydration. And since potential ingredients will be presented from the perspective of the renal paradox—the fact that fluid intake by dehydrated humans results in a significant urine output and a compromised rehydration—the chapter begins with a presentation of this concept.

20.2   THE RENAL PARADOX AND THE POTENTIAL ROLE OF POTASSIUM

When it comes to post-exercise rehydration, the renal system does not function quite as expected. From a homeostasis perspective, one would expect the kidneys to retain water as long as the body is in negative fluid balance, but when humans drink fluids, urine production follows. A large overload is not necessary for urine output to exceed *balance* (Pérez-Idárraga and Aragón-Vargas 2014). This phenomenon is apparent with water intake, but even when reasonable concentrations of sodium are used in the drinks, there is a considerable water excretion in the presence of hypohydration. This may be called the *renal paradox*: high urine output after fluid intake undermines the very goal of drinking in a hypohydrated state. One of the major goals in the formulation of a rehydration beverage is to achieve better fluid retention by recurring to sodium, potassium, or even protein, while other resources have been explored with different degrees of success, such as moderate hydration protocols (Kovacs et al. 2002; Mayol-Soto and Aragón-Vargas 2010; Pérez-Idárraga and Aragón-Vargas 2011a).

As an example of this renal paradox, Jimenez et al. (2002) had participants dehydrated by exercise or passive heat and then rehydrated with a homemade beverage containing 5% glucose and 40 mEq/L of NaCl. This sodium content, together with the level of dehydration (2.7% of body mass [BM]) and the ingested volume (92% of sweat losses), would be expected to favor fluid retention. Nevertheless, rehydration resulted in 400–500 mL of urine loss over 3 h of recovery, out of about 1800 mL of fluid intake. Urine output was significantly higher than under the no rehydration conditions ($p < .05$).

As discussed in this book (Chapter 19), sodium has been used in sports drinks as an electrolyte to promote fluid retention of physically active people who dehydrate from sweating. Sodium promotes acute plasma volume expansion during exercise, which is desirable to maintain performance. It also reduces diuresis resulting in lower water losses than when drinking pure water. It is possible, however, that aggressive rehydration with large fluid volumes causes the mechanical signal (baroreceptors) to be stronger than the osmotic and chemical signals which, during hypohydration, would normally help fluid retention by preventing diuresis. In other words, the human body is able to detect when it is hypohydrated and shifts into fluid conservation mode, but if it gets information about an acute plasma volume expansion, this is incorrectly interpreted as being hyperhydrated and diuresis is increased. By the time the error is detected, the body is hypohydrated again. In support of this hypothesis, the previously mentioned study by Jimenez et al. (2002) showed a clear association between acute changes in plasma volume and urine output.

One strategy to solve the problem of inducing excess diuresis would be to identify osmolytes which may promote water movement to the intracellular space, thereby preventing acute plasma volume expansion and helping the body keep the ingested water. In a discussion of the mechanisms of cell volume regulation, Pasantes-Morales et al. (2006) described volume regulation substances (p. 56): "… the organic osmolyte pool involved in volume regulation is formed by a heterogeneous group of small molecules, including aminoacids (taurine, glutamate, glycine, alanine), polyalcohols

(myoinositol, sorbitol), and other compounds such as creatine, phosphocreatine. (...)".
They also mentioned the two major electrolytes that could be involved: potassium
and chloride. Only some of those substances have been experimentally tested; the
possible success of any particular osmolyte would depend on the ability to acutely
and transiently increase its intracellular content.

In a research report not published in a peer-reviewed journal (Aragón-Vargas and
Shirreffs 2014), the authors attempted to design a sports drink for effective post-
exercise rehydration, using potassium and other osmolytes at considerably higher
concentrations than normally used, while sodium was included in small amounts
(about one-third of a conventional sports drink). The rationale was that the particular composition of the experimental drinks would promote intracellular hydration, thus promoting fluid retention, as explained in the previous paragraph. The
drinks evaluated in the study were a commercially available flavored water with
added creatine (A), a high-electrolyte commercially available sports drink (B), a
conventional sports drink with added potassium and creatine (C), a 6% carbohydrate, low-sodium, high-potassium drink (D), and Evian® bottled water as a standard for comparison purposes (see Table 20.1). In a repeated-measures design, ten
healthy, young volunteers did one familiarization trial and then five experimental trials assigned for five consecutive weeks. They arrived in the laboratory in a
fasted, euhydrated state and were dehydrated by intermittent exercise in the heat to
approximately 2% of BM loss. After dehydration and some rest, they began a 1 h
rehydration period with one of the five beverages allocated according to a balanced
randomized design; they drank a volume equivalent to 150% of BM loss. At the end
of rehydration, they were monitored for a total of 4 h, taking blood samples and collecting all urine every 60 min.

Initial conditions were similar among drinks for pre-exercise BM, dehydration level, time to dehydration, mean sweat rate, sweat loss, and fluid intake volume
($p > .05$). The study, however, failed to find a significant difference in total urine
output among trials ($p = .120$), even though acute plasma volume changes tended

TABLE 20.1
Drink Composition

	Drink				
	A	B	C	D	E
Carbohydrate solution (%)	1	6	6	6	0
Na$^+$ (mmol/L)	6.5	35	18	7	0
K$^+$ (mmol/L)	4.5	10	50	50	0
Creatine monohydrate (g/L)	3.3	0	3.3	0	0

Source: Aragón-Vargas, L. F. and Shirreffs, S. M., Rehydration following exercise in the heat: A look at
alternative formulations and the role of plasma volume expansion in excess diuresis, http://
kerwa.ucr.ac.cr/handle/10669/11115, 2014. Used with permission. This work is licensed under
a Creative Commons Attribution 4.0 International License.

to follow the expected pattern. This has also happened in other studies: in one experiment using drinks with 0%, 2%, 5%, and 10% glucose concentrations (Evans et al. 2009), plasma volume changes were negative with the 10% glucose solution at the 10 and 60 min time points, but this did not result in lower urine output.

More recently, a study attempted to achieve better fluid retention by using high-potassium beverages (Pérez-Idárraga and Aragón-Vargas 2014). Because the study included fresh coconut water as one of the experimental drinks, it is further described in Section 20.5.1, but is mentioned here to demonstrate the effect of potassium concentrations in beverages on diuresis. Urine output and fluid retention were compared after rehydration with plain water, fresh coconut water (naturally high in potassium), a conventional sports drink, or a specially formulated, high-potassium drink. Cumulative urine output was significantly higher after rehydrating with water than after the sports drink or the specially formulated drink ($p < .05$), but not than after fresh coconut water ($p = .147$). Fluid retention was only significantly higher for the sports drink compared to plain water ($p = .013$).

The possibility that potassium or other important intracellular molecules in a rehydration beverage may help shift fluids away from the extracellular space, preventing rapid plasma volume increases and hence limiting excess diuresis, has not been confirmed, but it warrants further study. Another possibility for limiting excess diuresis would be to use drinks that are absorbed more slowly, as in the protein studies that follow.

20.3   MAJOR ELEMENTS

20.3.1   PROTEIN

Nutritional products with protein or combinations of amino acids represent an important share of all supplements used by athletes. Because a major goal of protein ingestion is to support or enhance post-exercise muscle recovery by promoting a positive nitrogen balance at a critical time, amino acids or protein have been used in rehydration beverages. It is important to understand to what extent these ingredients contribute to or deter from the rehydration effectiveness of the drinks.

In a discussion about intestinal fluid absorption, Schedl et al. (1994) suggested that because amino acids use sodium-coupled transport systems that are independent from glucose transporters, they could enhance sodium and water absorption beyond the use of carbohydrate and sodium alone. While this possibility has been explored, mainly for the formulation of oral rehydration solutions (ORS) to treat diarrhea, success has been limited if the goal is to improve the results obtained from multiple carbohydrate solutions. This and other hydration-related effects of amino acids have been recently reviewed by Baker and Jeukendrup (2014).

On the other hand, rehydration beverages with protein have been shown to favor fluid retention. Several earlier studies showed the effectiveness of skimmed milk (discussed in detail in Section 20.5.3), but the research design did not allow to study the effect of protein *per se*. In a parallel line of research, a 2006 study by Seifert et al. looked at fluid retention after rehydration with a 6% carbohydrate plus 1.5% protein beverage (including 53 mg sodium and 18 mg potassium), compared with a conventional sports drink (6% carbohydrate, 46 mg sodium, and 12.5 mg potassium) and plain water (Seifert et al. 2006). Each participant ingested a volume equivalent to 100% of weight loss (~1700 mL in 20 min) after exercise-induced dehydration to 2.5% BM. Water showed the least fluid retention (about 53%) of all drinks, and a greater fluid retention was attained with the beverage with protein (88% ± 4.7%) in comparison with the conventional sports drink (75% ± 14.6%). The authors attributed this latter effect to the additional protein, but because the drinks were not isoenergetic and osmolalities were different, other explanations could not be ruled out. More recently, the effort has focused on teasing out the specific role of protein by comparing solutions otherwise matched in electrolyte and energy content.

James et al. (2011) compared two isoenergetic drinks, formulated with the same electrolyte and fat content but differing only in total carbohydrate and protein. One contained 65 g/L of carbohydrate, while the other had 40 g/L of carbohydrate and 25 g/L of milk protein, and they were ingested in a volume equivalent to 150% of sweat loss after exercise-induced dehydration to about 2% BM. All measures of hydration (total urine output, fluid retention, and net fluid balance) were better with the beverage with carbohydrate and protein than the drink with only carbohydrate after 4 h of monitoring. Later, in a very similar study design, James et al. (2013) confirmed these results but using two different carbohydrate and milk protein beverages instead of one (drink A: 40 g/L carbohydrate + 20 g/L milk protein, and drink B: 20 g/L carbohydrate + 40 g/L milk protein), which were better than a 60 g/L carbohydrate beverage but not different between them. In both studies, the authors attributed the milk protein advantage to a possible slower gastric emptying (not measured). While any fluid in the stomach and intestines at a particular time is not contributing to fluid balance, a somewhat slower gastric emptying and intestinal absorption would be a welcome effect to the extent that it might be preventing excess diuresis.

Interestingly, no beneficial effects on rehydration have been found from adding whey protein to carbohydrate–electrolyte drinks, matching energy density and electrolyte content. James et al. (2012) used a very similar protocol to those described in the previous paragraph to compare a 65 g/L carbohydrate solution with a 50 g/L carbohydrate and 15 g/L whey protein isolate solution and found almost identical urine output, fluid retention, and net fluid balance with both beverages. Hobson and James (2015) compared a 62.2 g/L carbohydrate–electrolyte solution (CES) to the same solution with added whey protein (20.4 g/L; in this case, the drinks were not isoenergetic). They obtained similar urine output, fluid retention, and net fluid balance. In another non-isoenergetic comparison, the addition of 20 g/L whey protein isolate to a mineral water was not more effective for rehydration than mineral water alone (James et al. 2014). For additional details on the effects of protein on hydration, the reader is referred to the review by Baker and Jeukendrup (2014).

In summary, CESs with milk protein have been shown to provide better postexercise rehydration than conventional CESs. Furthermore, the addition of whey protein to sports drinks has been shown not to hinder rehydration. Therefore, whenever protein intake may be of interest in addition to fluid replacement during exercise recovery, rehydration beverages properly formulated with milk or whey protein may be used. Practical recommendations and potential contraindications will likely be developed as these newer formulations are more widely used.

20.3.2 GLYCEROL

Research on glycerol as a hydrating agent started in 1987, with a study on human subjects at rest (Riedesel et al. 1987). When glycerol is ingested with water, it promotes hyperhydration by creating an osmotic load and favoring water reabsorption (Freund et al. 1995). By limiting free water clearance, fluid retention is improved; this effect is believed to be independent of hormonal responses (Nelson and Robergs 2007). Glycerol-induced hyperhydration is, naturally, transient, as glycerol will be metabolized by the liver and excreted by the kidneys, but because its metabolic clearance from the blood is slow, hyperhydration can be maintained for a few hours (Nelson and Robergs 2007; Riedesel et al. 1987). The use of glycerol is, however, currently banned by the World Anti-Doping Agency (WADA) because it is considered a masking agent (World Anti-Doping Agency 2014). Glycerol intake levels necessary for enhancing hydration are much higher than the dose shown to give positive results in anti-doping tests (Van Rosendal and Coombes 2012).

Several extensive reviews have addressed the use of glycerol for hydration purposes (Goulet et al. 2007; Nelson and Robergs 2007; Van Rosendal and Coombes 2012; Van Rosendal et al. 2009), but they focus on hyperhydration, as does most of the available research. Few studies have looked at glycerol as a potential ingredient in rehydration beverages for post-exercise utilization.

Briefly, most hyperhydration studies have found an increase in total body water with glycerol. A meta-analysis (Goulet et al. 2007) based on 14 comparisons and a total of 99 subjects found an increased body water of 7.7 ± 2.8 mL/kg BM ($p < .01$); their pooled effect size was 1.64 ± 0.80 ($p < .01$). Almost all studies have been performed comparing plain water and water plus glycerol. While the hyperhydration results are rather clear, the cardiovascular or thermoregulatory benefits are less consistent (Baker and Jeukendrup 2014; Goulet et al. 2007; Van Rosendal et al. 2009). General recommendations for hyperhydration prior to exercise are to ingest 1 g of glycerol/kg BM, with approximately 20 mL of water/kg BM (Baker and Jeukendrup 2014), or 1.2 g of glycerol/kg BM with 26 mL of water/kg BM (Van Rosendal et al. 2009).

The first study to look at glycerol as an ingredient in a rehydration beverage was performed by Scheett et al. (2001). Eight males dehydrated to ~3% BM by exercising in the heat, and rehydrated with water or water plus glycerol (1 g glycerol/kg BM) in a volume equivalent to 100% of sweat losses, over 3 h. The focus of the manuscript was on performance during a subsequent task to exhaustion, which was improved in the glycerol trial. The authors reported no significant difference in total urine output between control and glycerol ($p > .05$), but unfortunately the numbers they presented are internally inconsistent. They did report, however, a significant difference in percent change in body weight at the end of the rehydration protocol, for the control ($-0.73\% \pm 0.09\%$ BM) and glycerol ($-0.50\% \pm 0.08\%$ BM) trials ($p < .05$).

The only other study using a conventional post-exercise rehydration protocol was by Kavouras et al. (2006). Eight highly trained male cyclists completed three separate exercise trials that induced 4% dehydration. Subjects were given no fluid during exercise but were then rehydrated within 80 min with water (placebo) or water plus glycerol equivalent to 3% BM. Ninety minutes later they performed an exercise test to exhaustion. Time to exhaustion was improved in the glycerol trial, but no fluid retention, net fluid balance, or urine output data were reported.

Both abovementioned studies reported improved performance (greater time of exercise to exhaustion) and a greater plasma volume expansion with glycerol; the latter was presented in the original studies as the only clear hydration advantage with glycerol use. Even with limited evidence, guidelines for glycerol use as a rehydrating agent post-exercise have been provided (Van Rosendal et al. 2010). However, plasma volume restoration or expansion is a transient result, which may be beneficial only if subsequent exercise is undertaken in thermally adverse conditions. Given that no data are yet available supporting better fluid retention after exercise-induced dehydration, glycerol use is not warranted as an ingredient in a post-exercise rehydration beverage.

Apart from being prohibited for athletes under the jurisdiction of WADA, glycerol has a few contraindications. Some side effects have been reported, such as nausea, diarrhea, vomiting, and headache, but reviewers agree that these are uncommon, considering the large number of studies without any adverse reports (Baker and Jeukendrup 2014; Van Rosendal et al. 2009); they are apparently associated with higher doses or concentrations of glycerol and should not be a problem with more diluted use (Baker and Jeukendrup 2014; Kavouras et al. 2006; Nelson and Robergs 2007; Van Rosendal et al. 2009). Nevertheless, glycerol ingestion is not recommended for pregnant women or individuals with migraine, headache, or liver disorders, with diabetes, or with renal or cardiovascular disease (Van Rosendal et al. 2009).

20.3.3 Caffeine

Caffeine is widely used as an ergogenic aid and is often present in the regular diet of athletes and sedentary individuals. There are multiple studies and several important literature reviews addressing the ergogenic properties of caffeine (Armstrong 2002; Burke 2008; Doherty and Smith 2005; Kovacs et al. 1998; Spriet 1995) and its effect on hydration (Armstrong et al. 2007; Maughan and Griffin 2003). This section will only discuss the acute rehydration characteristics, with emphasis on the role of caffeine on post-exercise rehydration.

As explained by Armstrong et al. (2007) in their review paper, caffeine increases renal glomerular filtration and inhibits reabsorption of sodium, resulting in higher water and sodium excretion by the kidneys. In addition to this diuretic effect, other physiological responses to caffeine may impair performance during exercise in the heat: sweat rate may be increased because of the stimulation of the sympathetic nervous system, which would compound dehydration, and core temperature may be increased because of a higher resting metabolic rate. Based on their analysis of the available evidence at the time, the authors of the review proposed that caffeine consumption does not result in water–electrolyte imbalances, hyperthermia, or reduced exercise–heat tolerance. Their conclusions followed the same direction of Maughan and Griffin (2003), who concluded *tentatively* that single caffeine doses found in common beverages have little or no diuretic action and that habitual caffeine users develop tolerance to the effects of caffeine, although they acknowledged that a dose greater than 250 mg of caffeine has an acute diuretic action. Baker and Jeukendrup (2014) concluded, in a more recent review, that moderate caffeine intake in the order of 450 mg ($\approx$6.4 mg/kg for a 70 kg individual) has no chronic negative effects on hydration, heat tolerance, or risk of heat illness. The issues warrant further analysis and discussion.

The emphasis of the review by Armstrong et al. (2007) was on chronic hydration and daily living. Interestingly, Maughan and Griffin (2003) made a similar emphasis. This perspective makes sense as caffeine is ingested with a wide variety of beverages in the regular diet, and not only as an ingredient of hydrating beverages. The question they pose is: does the regular intake of moderate doses of caffeine impair hydration status or thermoregulation? The answer seems to be no. But that is different from the question normally posed for other potential ingredients of a drink intended to replace fluids lost through sweat. If the intent is to formulate a beverage to improve exercise performance or enhance immediate post-exercise recovery, the acute effects of caffeine on diuresis, core temperature, and sweating during exercise in the heat, as well as fluid retention during post-exercise rehydration, must be analyzed.

A few experiments are presented here in detail. First, Falk et al. (1990) asked seven trained males, not habitual users of caffeine, to exercise to exhaustion in thermoneutral conditions, once after caffeine ingestion and once after the intake of a placebo in 100 mL of an artificially sweetened drink. Total caffeine intake was 7.5 mg/kg of body weight, all before the exercise. The authors stated that *ad libitum* water consumption was encouraged during exercise and was not different among conditions, but no ingested volume was reported. This study concluded that the amount of caffeine ingested did not significantly affect water deficit, sweat loss, or thermoregulation, but acknowledged that during exercise performed under a greater heat stress, thermoregulation might be hindered.

Later, another study (Wemple et al. 1997) compared caffeinated versus non-caffeinated sports drinks at rest and during exercise. This is a carefully designed study, although it only tested six highly active subjects (4 males and 2 females) who participated in four randomly assigned, counter-balanced trials, two at rest (4 h) and two involving exercise (1 h rest, 3 h of cycling at 60% VO_2max followed by a maximal performance test). All trials were performed in the heat; participants drank a conventional sports drink, 8 mL/kg of body weight ($\approx$560 mL for a 70 kg individual) at the beginning of each trial, and then 3 mL/kg of body weight ($\approx$210 mL) every 20 min, beginning after 60 min from the start of the trial. One rest trial and one exercise trial were performed with caffeine in the drinks (490–680 mg, the equivalent of 8.7 mg/kg); the other exercise and rest trials were performed without caffeine but drinking the same fluid volume. The authors reported lower urine flow rate during exercise than at rest (as expected), and a higher urine flow rate at rest with the caffeinated drink than with the placebo, but there was no effect of caffeine on urine flow during exercise. The reductions in plasma volume, and the increase in sweat rate, heart rate, and core temperature were consistent during the exercise trials, but the effects of caffeine were not significant. They concluded that the caffeine dose provided in this volume of sports drink increases urine production at rest but not during prolonged cycling in the heat.

Diuresis during exercise is undesirable, not only logistically, but because it exacerbates exercise-induced dehydration. To better understand the effects of caffeine on thermoregulation and fluid–electrolyte losses during prolonged exercise in the heat, Del Coso et al. (2009) performed a double-blind, placebo-controlled, randomized experiment with seven endurance-trained, heat-acclimatized males, each serving as his own control. Each participant cycled for 2 h at a moderate intensity in a hot-dry environment on six different occasions. Three trials included no caffeine: without fluid replacement, drinking water to replace 97% of fluid loss, or drinking a carbohydrate–electrolyte beverage to replace 97% of fluid loss. The other three were identical to the aforementioned three, but included the ingestion of 6 mg caffeine/kg of body weight 45 min before the exercise. Fluid intake was ≈2.4 L, distributed evenly over each trial; drinks were at room temperature. Overall, urine flow was increased by 28% in the caffeine conditions (pooled data), while sweat losses of sodium, potassium, and chloride were increased by about 14% ($p < .05$). Caffeine did not alter heat production, forearm skin blood flow, or sweat rate. While rehydration during exercise increased sweat loss, the combination of caffeine ingestion with rehydration had no effect on sweat loss. The combination of water intake and caffeine ingestion increased exercise urine production compared with water alone (664 ± 350 mL vs. 464 ± 278 mL, $p < .05$), but the CES with caffeine resulted in a similar urine output compared with CES alone (422 ± 303 mL vs. 486 ± 380 mL, $p > .05$). The authors highlighted the fact that because diuresis during exercise in the heat is relatively small compared to sweat losses, net fluid balance was not different between caffeine and no-caffeine trials despite higher diuresis in the former. They also pointed out that caffeine is diuretic even during exercise, but suggested that this effect may be counteracted when it is used in combination with carbohydrate–electrolyte drinks.

Published studies on the acute effects of caffeine during post-exercise rehydration are scarce. A seminal study (Gonzalez-Alonso et al. 1992) compared water, a 6% CES, and a diet cola during a 2-h rehydration period after exercise-induced dehydration to 2.5% BM. Ten physically active subjects exercised at a self-selected intensity between 60% and 80% VO_2max in the heat (32°C, 40% relative humidity) until they reached the desired dehydration, on four separate occasions. With the exception of a no-fluid trial used only for monitoring, the rehydration protocol involved drinking one of the three beverages in two boluses. The diet cola contained 128 mg/L of caffeine, for an average intake of 250 mg. Fluid retention at the end of monitoring was significantly lower with the diet cola ($54\% \pm 5\%$) than with water ($64\% \pm 5\%$) or CES ($69\% \pm 5\%$) ($p < .05\%$). The authors attributed the lower rehydration obtained with diet cola to the diuretic effect of caffeine, showing a statistically significant difference in urine output between diet cola and the carbohydrate–electrolyte drink (710 ± 100 mL vs. 480 ± 90 mL, respectively, $p < .05$). However, because of the study objectives, the beverages were different in other ingredients besides caffeine, and the difference between diet cola and water was not significant. No clear conclusions may be reached from this paper regarding post-exercise fluid recovery and a possible diuretic effect of caffeine.

The other study (Brouns et al. 1998) related to the acute effects of caffeine during post-exercise rehydration used voluntary fluid intake. Eight well-trained cyclists participated in three exercise tests, at least four days apart, in a randomized cross-over design. They cycled in a warm environmental chamber to a median dehydration of 3.21% BM. *Ad libitum* rehydration took place over 2 h with one of the beverages: mineral water, carbohydrate–electrolyte drink, and caffeinated soft drink. Participants were then monitored for four additional hours. Fluid intake was greater for the CES than for the mineral water (median = 2.86 and 2.15 kg, respectively), but intake of the caffeinated drink was not different from the others (median = 2.77 kg, for about 379 mg of caffeine). Total urine output after 4 h of the recovery period was similar for all drinks, and fluid restoration was not significantly different either ($p > .1$). The authors concluded that a diuretic effect of caffeine at this level of consumption was ruled out, although they pointed out that acute magnesium and calcium excretion was potentiated with the caffeinated beverage.

In summary, the available evidence suggests that there is no chronic negative effect of moderate caffeine intake on hydration or thermoregulation. There is also some evidence to support the notion that ingestion of caffeine before or during exercise has no biologically significant acute effect on diuresis (especially when combined with carbohydrate–electrolyte drinks), core temperature, or sweating. In the case of the acute effects of caffeine on post-exercise rehydration, two aspects must be considered. First, there would be no rationale for using caffeine during recovery, with the exception of personal preferences. Second, there is a dearth of information on caffeine related to rapid rehydration and recovery after training and competition. More well-designed experiments are warranted before sound recommendations can be made.

20.3.4 ALCOHOL

With the exception of beer, alcohol (ethanol) is rarely ingested as post-exercise rehydration, because of its proven diuretic effect (Eggleton 1942; Murray 1932). Furthermore, alcohol should be avoided before or during exercise, as it will impair motor control and performance, particularly at higher doses (Lecoultre and Schutz 2009; Shirreffs and Maughan 2006). Recently, a high dose of ethanol has been shown to impair myofibrillar protein synthesis during recovery from exercise (Parr et al. 2014). In the particular case of beer, there is a widespread belief that, because of the presence of some electrolytes and due to the lower alcohol concentrations (in comparison with other alcoholic beverages such as vodka, rum, whiskey, or even wine), beer should be a good choice for rehydration. The possibility of rehydrating with beer is appealing to many; there is, however, some strong evidence that beer is not a good choice if the goal is to replace the fluids lost during exercise.

A widely cited study by Shirreffs and Maughan (1997) used especially formulated beers with 0%, 1%, 2%, and 4% ethanol to test the effects of alcohol consumption on restoration of fluid balance after exercise-induced dehydration. They found non-significant differences between the 0% beer and beer with 1% or 2%. Even with the 4% beverage, no significant differences were found in urine output or net fluid balance, although other variables showed small differences. The authors concluded that only at the higher concentration, there was a small detrimental effect of alcohol on fluid retention. This publication has been used to support the claim that beer is a good rehydration beverage, but there is an important problem: apparently, most beers available for consumption contain more than 4% alcohol. According to some marketing data from early 2014 ("Top 10 Best Beers in the World" 2014), the top-ten list of beers with the highest volume consumption in the world include eight which have >4% alcohol by volume (ABV) and two with exactly 4% ABV. Out of another list of more than one thousand of the major world beers, only 62 had less than 4% ABV ("Find the Alcohol Content of Your Favorite Beer" 2014).

More recently, other scientists have experimentally addressed the potential qualities and complications of rehydrating with alcoholic drinks after exercise. Irwin et al. (2013) studied 16 healthy males who were first dehydrated to approximately 2.5% BM by exercising in thermally adverse conditions, in order to assess the impairment in cognitive functions resulting from mild or moderate dehydration combined with moderate alcohol consumption, compared with the consumption of alcohol under fully rehydrated conditions. Because of the study objectives, alcohol was administered with a low volume of fluid and only after a period of rehydration with water and electrolytes or no fluid intake, so the study was not strictly about rehydration with alcoholic drinks. The results, nevertheless, showed a deterioration of choice reaction time, executive function, and response inhibition after alcohol consumption. Perhaps the most important finding was that the negative effects were more pronounced when alcohol intake occurred in a dehydrated condition, which suggests that alcohol intake may be more detrimental to cognitive function when rehydration is needed.

Desbrow et al. (2013) studied rehydration with light beer (2.3% ABV) or regular beer (4.8% ABV), alone or with the addition of 25 mEq L^{-1} Na$^+$. Their purpose was to evaluate the effects of different alcohol concentrations on fluid retention, but also to explore the potential benefits of adding enough sodium to beer, in an attempt to offset the reported increased diuresis that takes place when using regular strength beer (interestingly, this effect was only mentioned anecdotally in the rationale, although it was empirically confirmed in the study). Seven males exercised in a thermoneutral environment but wearing evaporation-restrictive clothing, until dehydrated to ~2% BM. After exercise, they were rehydrated with a volume equivalent to 150% of the total fluid volume lost, or about 2.5 L of beer, resulting in an average ethanol intake close to 60 g with the light beer and 120 g with the regular beer. Following rehydration, urine output was monitored for 4 h. Net fluid balance was negative for all drinks at the end of monitoring, but there were some differences among drinks at the 3 and 4 h time points. The addition of sodium to light beer did not result in a significant improvement of total urine output ($p = .25$) or net fluid balance ($p = .26$). Adding sodium to regular strength beer showed no benefit either (total urine $p = .75$; net fluid balance $p = .77$). Only the combination of low-alcohol and extra sodium was able to improve net fluid balance and total urine output, compared to both full-strength beers. Based on their results, the authors suggested that "beer, irrespective of ingredient profile, is an undesirable postexercise fluid" (p. 598) but, acknowledging the ingestion of high volumes of beer by some individuals after exercise, recommended an emphasis on lower alcohol content to minimize harm during recovery.

Flores-Salamanca and Aragón-Vargas (2014) compared fluid retention, blood alcohol concentration, balance, and reaction time in 11 healthy young men who rehydrated with beer (4.6% ABV), low-alcohol beer (0.5% ABV, often called no-alcohol beer by the industry), or water, after exercising in the heat. Participants dehydrated to ~2% BM and then ingested a volume equivalent to 100% of weight loss of the randomly assigned beverage for each day; this resulted in the regular beer providing almost the same amount of ethanol (60.1 ± 6.5 g) as the light beer in the study by Desbrow et al. (2013), due to the combination of a lower volume and a higher alcohol concentration. Total urine output results are summarized in Figure 20.1. Fluid retention was just about one half with regular beer when compared to low-alcohol beer. Urine output was significantly higher ($p < .05$) after only 30 min of monitoring. Net fluid balance was lower after rehydrating with beer; at the end of monitoring, it was not different from the end of dehydration. In other words, 3 h after ingesting ~1.7 L of regular beer, participants were as dehydrated as they were before drinking.

Figure 20.2 shows the blood alcohol content over time. For the low-alcohol beer and water, it was not possible to register an increase at any time point; with regular beer, however, it reached an average of 0.857 g/L, with a 95% confidence interval of 0.752–0.963 g/L. This is well above legal driving limits in many countries (0.50 g/L). It is to be expected, then, that rehydrating with beer resulted in slower reaction time and impaired balance: reaction time was longer for the beer condition (0.314 ± 0.039 s, mean ± SD) than the low-alcohol beer (0.294 ± 0.034 s, $p = .009$),

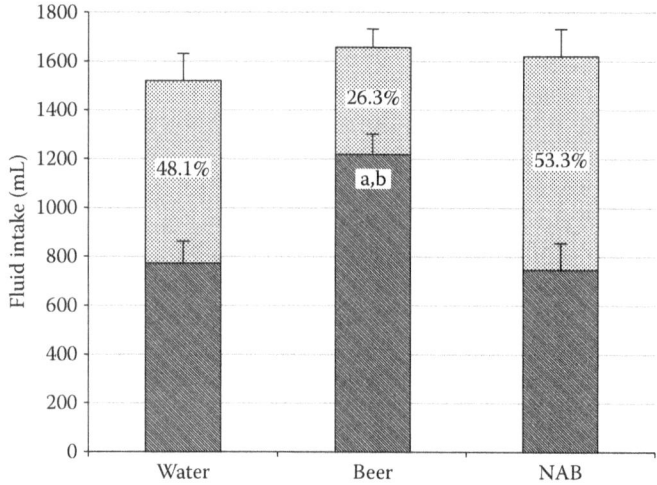

FIGURE 20.1   Fluid intake, total urine output, and retention. NAB: no-alcohol beer (in reality it contained 0.5% ABV). Columns are mean values plus standard error of the mean. F = 8.63, $p = .002$. Lower portion of each column represents urine output, and upper portion is the volume that was retained. (a) Urine volume different from water, $p = .043$. (b) Urine volume different from NAB, $p = .007$. (From Flores-Salamanca, R. and Aragón-Vargas, L.F., Post-exercise rehydration with beer impairs fluid retention, reaction time, and balance, http://kerwa.ucr.ac.cr/handle/10669/9237, 2014. Used with permission. This work is licensed under a Creative Commons Attribution 4.0 International License.)

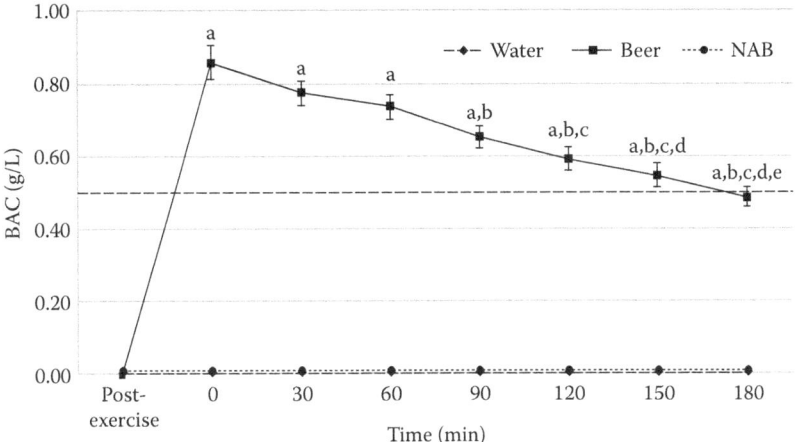

FIGURE 20.2 Blood alcohol content. Points are mean values; bars represent standard error of the mean. Interaction F = 214.1, p = 4.4 × 10^{-8}. Condition main effect F = 442.3, p = 1.3 × 10^{-9}. Time main effect F = 214.1, p = 4.4 × 10^{-8}. a, Different from post-exercise (p < .05). b, Different from 0 min (p < .05). c, Different from 30 min (p < .05). d, Different from 60 min (p < .05). e, Different from 90 min (p < .05). (From Flores-Salamanca, R. and Aragón-Vargas, L.F., Post-exercise rehydration with beer impairs fluid retention, reaction time, and balance, http://kerwa.ucr.ac.cr/handle/10669/9237, 2014. Used with permission. This work is licensed under a Creative Commons Attribution 4.0 International License.)

but not different from water (0.293 ± 0.049 s, p = .077). Balance was significantly impaired for the beer condition from the end of rehydration up to 90 min of follow up, compared with both water and low-alcohol beer.

In summary, recent data show a clear detrimental effect of alcohol on post-exercise rehydration. The presence of alcohol in a rehydration drink is not compatible with the goal of replacing lost fluids quickly and effectively, while it has a negative effect on other aspects of recovery and motor control. This applies to regular beer, despite the presence of a small amount of electrolytes.

20.4 SECONDARY INGREDIENTS

20.4.1 CREATINE

Among all nutritional supplements, creatine is one of the most widely utilized by athletes; it is also well studied by scientists. Various oral loading and maintenance protocols have been shown to increase intramuscular creatine, with a concomitant improvement in sports performance in tasks such as jumping, cycling, and sprinting, where short, high-intensity, limited-recovery repetitive bouts are required (Bemben and Lamont 2005; Mujika and Burke 2010). The potential use of this ingredient in rehydration beverages has received little attention, possibly because of concerns that creatine supplementation may impair thermoregulation or hydration status, mostly from anecdotal reports. Most studies on creatine and rehydration have focused on the consequences of chronic creatine use for conventional purposes (improved anaerobic performance). A recent exhaustive review of the literature on fluid replacement beverages omits the topic of creatine altogether (Baker and Jeukendrup 2014).

Several reviews have been published on chronic creatine use and hydration (Dalbo et al. 2008; Ganio et al. 2007), but the most comprehensive review was performed by Lopez et al. (2009). These reviews agree that chronic creatine supplementation leads to a rapid increase in BM, mostly due to an increase in intracellular water. Many of the studies reviewed showed increases in total body water and intracellular water, while no evidence was found of an impaired thermoregulation. Ganio et al. (2007) argued that several studies showed no change or even an advantageous core temperature in creatine-supplemented subjects exercising in the heat. Lopez et al. (2009) concluded from their meta-analysis that there was no substantial evidence at that point to show that chronic creatine supplementation would hinder thermoregulation or fluid balance.

As mentioned in the previous paragraphs, chronic supplementation with creatine results in increased intracellular water and total body water. No studies have been published addressing the acute hydration effects of creatine in a rehydration beverage. The one report that looked at this issue found no fluid retention advantage when creatine was combined with flavored water or a conventional sports drink with added potassium, in comparison with plain water (Aragón-Vargas and Shirreffs 2014). The time course for the entrance of creatine into muscle cells may be too slow to act as an osmolyte and benefit acute fluid retention. At this point, there is no evidence to support the inclusion of creatine in rehydration beverages; at the same time, there is no basis for contraindicating its chronic use because of thermoregulatory or fluid balance concerns.

20.4.2 Artificial Preservatives

Artificial preservatives are often included in bottled, ready-to-drink beverages to avoid spoilage and therefore extend the shelf life of the product. They act mainly by preventing growth of bacteria, molds, fungi, and yeasts, and they represent a less costly alternative to processes such as pasteurization combined with hot or cold filling industrial procedures. There are, however, two disadvantages of artificial preservatives. First, artificial ingredients are not well accepted in many cultures, particularly in association with a beverage used for hydration, as the drink is meant to be ingested in large volumes, and it should share the healthy image associated with exercise. Second, some preservatives such as sodium benzoate and potassium sorbate may negatively modify the palatability characteristics of drinks, which is undesirable as it may decrease voluntary intake (Passe et al. 2004). Specifically, the presence of sodium benzoate at 0.03% has been reported to reduce voluntary fluid intake by 10.8% during 30 min of moderate intensity exercise in thermoneutral conditions (Passe et al. 1997), even though it had no measurable effects on sensory variables.

A study reported originally in abstract form (Rivera-Brown et al. 2007) but available also as a pre-print (Rivera-Brown et al. 2014) sheds light on the role of artificial preservatives in hydration beverages. Palatability and voluntary intake of

TABLE 20.2
Composition of Commercial Beverages Offered in the Four Experimental Conditions (Per 100 mL)

Beverage	CHO (g)	Na+ (mg)	K+ (mg)	Preservatives	Vitamins	Osmolality (mOsm/L)
Water	0	0	0	None	None	0
CES	6	46	12.5	None	None	325–380
CESP1	6	46	12.5	Sodium hexametaphosphate	None	325–380
CESP2	8	31	21	Potassium sorbate sodium benzoate (<0.06%)	B12 (0.25 μg) B3 (0.75 mg) B6 (0.10 mg)	381

Source: Rivera-Brown, A. et al., Palatability and voluntary intake of three commercially available sports drinks and unflavored water during prolonged exercise in hot and humid conditions, http://www. kerwa.ucr.ac.cr/handle/10669/11105, 2014. Used with permission. This work is licensed under a Creative Commons Attribution 4.0 International License.

four commercially available beverages were compared, in 36 athletes (males and females, mean age 19.5 years), during prolonged exercise in a warm and humid environment (WBGT $= 30.1 \pm 1.1°C$). The beverages varied in carbohydrate and electrolyte composition and preservative content (see Table 20.2). On separate days, athletes completed four 90-min sessions, running or race walking outdoors at 80%–85% of age-predicted maximum heart rate. Each time, they ingested in a randomized order and double-blinded design one of four beverages: bottled water, 6% carbohydrate–electrolyte (6% CES), 6% CES + preservatives solution (CESP1), 8% CES + preservatives + B vitamins solution (CESP2) (see Table 20.2).

Beverages were served cold (~9°C) in squeeze bottles and subjects drank as desired as they ran around a marked, 420-m area, while fluid intake was carefully monitored. Palatability was measured during a 1-min exercise break at 15-min intervals: overall acceptance, liking of flavor and liking of sweetness were measured using 9-point hedonic category scales. Perceived intensity of thirst, sweetness, saltiness, tartness, thirst quenching, palatability, and flavor strength were measured using visual analog 10-point scales. This scale was also used to rate perception of exercise difficulty, how hot/overheated subjects felt, and the question "Can you drink a lot of this beverage?"

Overall, males covered 18.0 ± 2.0 km, while the females completed 13.1 ± 1.7 km in each trial, with no significant differences found among conditions (beverages) in terms of environmental variables, sweat rates, or exercise intensity (see Table 20.3). In other words, any dissimilarity in voluntary fluid intake or fluid balance among conditions could be attributed to the qualities of the beverages used in the study. However, the amounts consumed of the four drinks were not different ($W = 17.0 \pm 4.8$; CES $= 16.9 \pm 5.4$; CESP1 $= 17.8 \pm 5.4$; CESP2 $= 17.5 \pm 5.2$ mL/kg, $p > .05$) (see Figure 20.3 and Table 20.4).

TABLE 20.3

Environmental Variables, Intensity, and Distance Covered under Each of the Four Conditions

	W	CES	CESP1	CESP2
Wet bulb temperature (°C)	26.5 ± 0.5	26.4 ± 0.6	26.5 ± 0.7	26.5 ± 0.5
Dry bulb temperature (°C)	31.5 ± 2.1	31.2 ± 2.2	31.5 ± 1.4	31.6 ± 2.2
Globe temperature (°C)	42.2 ± 3.5	41.5 ± 4.1	41.4 ± 4.2	42.4 ± 3.9
WBGT index (°C)	30.1 ± 0.9	29.9 ± 1.2	30.0 ± 1.3	30.2 ± 1.1
Relative humidity (%)	66.7 ± 5.8	67.6 ± 6.8	67.9 ± 5.9	66.2 ± 6.7
Heart rate (beats min⁻¹)	163.3 ± 5.0	163.9 ± 5.9	163.8 ± 5.9	163.7 ± 7.0
% Maximum heart rate	81.8 ± 2.4	82.0 ± 2.9	81.9 ± 3.0	81.9 ± 3.3
Distance covered (km)	15.6 ± 3.2	15.5 ± 3.2	15.6 ± 3.1	15.5 ± 3.1

Source: Rivera-Brown, A. et al., Palatability and voluntary intake of three commercially available sports drinks and unflavored water during prolonged exercise in hot and humid conditions, http://www.kerwa.ucr.ac.cr/handle/10669/11105, 2014. Used with permission. This work is licensed under a Creative Commons Attribution 4.0 International License.

Note: No significant differences were found between conditions, $p > .05$.

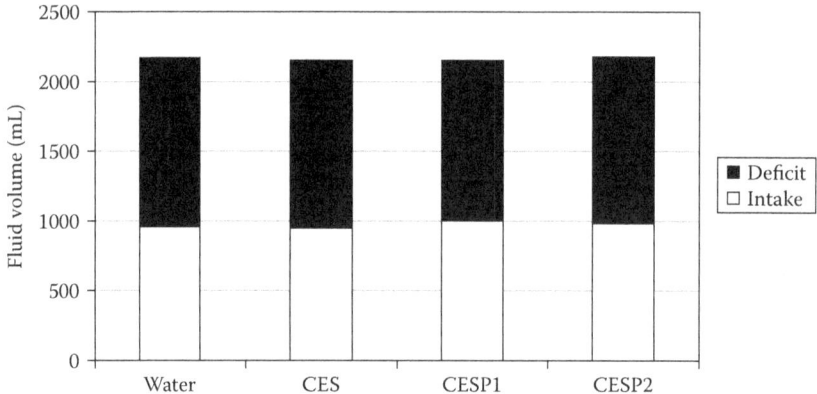

FIGURE 20.3 Fluid balance under the four conditions. (Adapted from Rivera-Brown, A. et al., Palatability and voluntary intake of three commercially available sports drinks and unflavored water during prolonged exercise in hot and humid conditions, http://www.kerwa. ucr.ac.cr/handle/10669/11105, 2014. Used with permission. This work is licensed under a Creative Commons Attribution 4.0 International License.)

As may be seen on Table 20.4, sweat losses and voluntary fluid intake were remarkably similar for the different beverages, resulting in nearly identical dehydration. In addition, fluid intake was insufficient to match sweat rates.

Careful analysis of sensory data shows that insufficient intake was not related to palatability, as the overall acceptance was close to 8.0 on a 9-point scale for all

TABLE 20.4
Body Fluid Balance

	Water	CES	CESP1	CESP2
Body weight pre (kg)	56.3 ± 7.6	56.4 ± 7.6	56.3 ± 7.6	56.6 ± 7.9
Body weight post (kg)	55.1 ± 7.1	55.2 ± 7.1	55.1 ± 7.1	55.4 ± 7.5
Δ in body weight (kg)	1.2 ± 0.6	1.2 ± 0.7	1.2 ± 0.6	1.2 ± 0.6
Fluid intake (mL kg^{-1})	17.0 ± 4.8	16.9 ± 5.4	17.8 ± 5.4	17.5 ± 5.3
Sweat loss (mL kg^{-1})	37.2 ± 8.7	37.1 ± 9.0	36.9 ± 8.9	37.6 ± 8.8
Sweat rate (L $\cdot$ h^{-1})	1.41 ± 0.45	1.42 ± 0.47	1.41 ± 0.48	1.44 ± 0.48
Fluid loss (mL kg^{-1})	38.3 ± 8.7	38.0 ± 9.0	38.0 ± 9.2	38.5 ± 8.7
Dehydration (% BW)	2.1 ± 1.0	2.1 ± 1.0	2.0 ± 0.9	2.1 ± 0.9
Rehydration (%)	46.9 ± 17.2	47.1 ± 18.7	48.9 ± 16.6	47.1 ± 16.9

Source: Rivera-Brown, A. et al., Palatability and voluntary intake of three commercially available sports drinks and unflavored water during prolonged exercise in hot and humid conditions, http://www.kerwa.ucr.ac.cr/handle/10669/11105, 2014. Used with permission. This work is licensed under a Creative Commons Attribution 4.0 International License.

Note: No significant differences between conditions, $p > .05$.

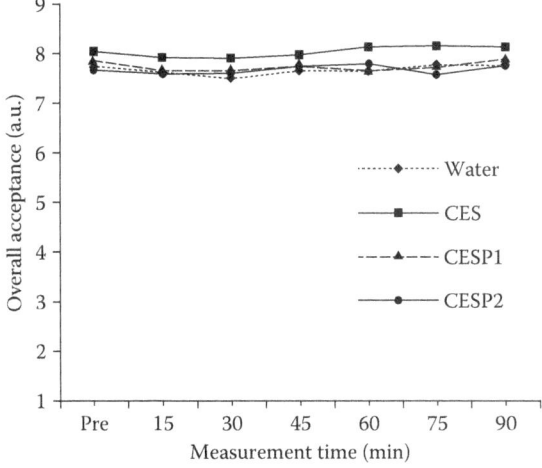

FIGURE 20.4 Overall acceptance of the drinks. (From Rivera-Brown, A. et al., Palatability and voluntary intake of three commercially available sports drinks and unflavored water during prolonged exercise in hot and humid conditions, http://www.kerwa.ucr.ac.cr/handle/10669/11105, 2014. Used with permission. This work is licensed under a Creative Commons Attribution 4.0 International License.)

four beverages, and it did not decline over time (Figure 20.4). There were no significant differences among beverages in terms of liking of flavor, thirst quenching, or declared willingness to drink a substantial volume; on the other hand, the scales were able to detect the expected differences between water and the other three drinks regarding flavor strength, sweetness, saltiness, and tartness. From this study, it may be concluded that the reluctance of young male and female athletes to drink while running or race walking in hot and humid conditions did not seem to be related to the presence of artificial preservatives in these beverages.

In summary, the few available data suggest that while artificial preservatives such as sodium benzoate may have a negative impact on voluntary fluid intake during moderate intensity exercise in warm conditions, the effect disappears when heat-acclimated humans exercise at a moderate-to-high intensity in hot and humid conditions.

20.5   NATURAL DRINKS

Independently from the balance of the evidence in favor or against the presence of different artificial ingredients in beverages at any particular point in time, natural drinks appeal to an important segment of the population. These include fruit juices and diluted drinks, mineral waters, infusions, and even milk. Scientific studies are limited because the composition of any particular natural drink varies widely, depending not only on preparation and freshness but also, in the case of fruit, on ripeness and site of cultivation.

It is also necessary to recognize the weight of what could be called *the cultural ingredient*: acceptability of a natural drink varies widely, depending on cultural aspects. It has long been recognized that voluntary fluid intake is strongly influenced by factors such as flavor, temperature, and texture, making them key elements in the formulation of beverages (Baker and Jeukendrup 2014). In the case of natural drinks, those elements are often not manipulated, yet remain an important component of the effectiveness of the beverages, as the best rehydration fluid in the world will not be optimal unless a sufficient amount is ingested.

Each drink in this section is discussed as it is normally consumed, presenting its acceptability and composition but not necessarily focusing on single ingredients. Natural drinks are included in this chapter because they comprise an important segment of all rehydrating beverages.

20.5.1   COCONUT WATER

Perhaps the most popular natural drink, coconut water has been studied for decades as a rehydration fluid. A few papers published during or immediately after World War II pointed out its high biological value (Picado Twight 1942), and some medical uses such as child feeding or its administration to dehydrated patients orally or intravenously with very few allergic reactions (Eiseman 1954; Soto et al. 1942). Other papers have reported its utilization in medical emergencies, also for rehydration (Campbell-Falck et al. 2000, Kuberski et al. 1979).

Coconut water is currently packaged and sold in many countries. Typically, commercial processing and the addition of preservatives degrades the flavor, producing an inferior product. The Food and Agricultural Organization (FAO) has been granted a patent for a process that allows manufacturers to bottle coconut water, theoretically preserving its microbiological and organoleptic properties (Rolle 2014).

Fresh coconut water is a crystal clear, biologically sterile fluid found in green coconuts, about 6–9 months old. One green coconut may have about 500–750 mL of fluid. Many people in the tropics see fresh coconut water as the natural choice for hydration. Studies have shown that coconut water is subject to variability in its carbohydrate and electrolyte content, according to place of growth, plant species, and maturation (age upon harvesting) (Child and Nathanael 1950; Vigliar et al. 2006). On average, there are 2–5 g of carbohydrate in 100 mL of fresh coconut water, compared with 6 g in a conventional sports drink. Sodium content (about 4 mEq/L) is low for rehydration purposes, about one-fifth the content in a conventional sports drink. Potassium content (about 50 mEq/L) is much higher than normally found in sports drinks (about 3 mEq/L). Pérez-Idárraga and Aragón-Vargas (2014) report 31 mEq/L of chloride. Major carbohydrates are glucose, sucrose, and fructose, in a ratio of about 50:35:15. Coconut water also contains inulin, a fructose-containing carbohydrate that has been shown to boost calcium absorption (Coxam 2007), although coconut water has not been tested in this regard. Osmolality has been reported to be between 282 and 452 mOsm/kg water (Pérez-Idárraga and Aragón-Vargas 2014; Vigliar et al. 2006).

The first experimental study looking at the effectiveness of coconut water for post-exercise rehydration was published as an abstract (Aragón-Vargas and Madriz-Davila 2000). Nineteen teenage boys ran intermittently in the heat until they were dehydrated to 2.3% BM. Each one rehydrated with water, fresh coconut water, or a sports drink on three different occasions, one week apart, in random order. After 3 h of monitoring, their net fluid balance was about 300 g greater ($p < .05$) with both the coconut water and the sports drink than with plain water.

Later, Saat et al. (2002) failed to find any difference in fluid retention between water and fresh coconut water. Recent results have also been inconsistent, with a majority of the studies showing fresh coconut water to be more effective than plain water, and another showing no difference. Ismail et al. (2007), after dehydrating subjects to 3% BM and rehydrating them with a volume equivalent to 120% of sweat losses of plain water, fresh coconut water, sports drink, or sodium-enriched fresh coconut water, obtained a better fluid retention with coconut water (65.1% ± 1.7%) than with plain water (58.9% ± 1.8%). Pérez-Idárraga and Aragón-Vargas (2011b) exercised 11 young participants in the heat to 1.84% ± 0.2% BM dehydration and had them drink within an hour plain water, fresh coconut water, or a sports drink in a volume equivalent to 120% BM loss. Fluid retention was similar for the latter two beverages (71%), but both were better than water (56%, $p < .001$). Another recent paper by Pérez-Idárraga and Aragón-Vargas (2014) was unable to confirm the advantage of fresh coconut water compared with plain water. Twelve physically active participants exercised in the heat to about 2% dehydration and then ingested 120% of weight loss of one of four beverages randomly assigned to each trial: plain water, fresh coconut water, a conventional sports drink, or a high-potassium drink. All beverages were well tolerated and scored high in palatability, but fluid retention was similar between coconut water (62.5% ± 15.4%) and plain water (51.3% ± 12.6%). It is possible that statistics did not reach significance due to the higher number of comparisons (4 instead of 3) in this study.

The studies of the previous two paragraphs refer to post-exercise rehydration with fresh coconut water. A recent paper by Laitano et al. (2014) evaluated the effects of prior ingestion of a commercial coconut water on fluid retention and exercise capacity in the heat. They provided 10 mL/kg BM of a flavored drink, plain water, or commercially available coconut water to eight subjects before performing an exercise capacity trial on a cycle ergometer in the heat. Time to exhaustion was about 24% longer with the commercial coconut water ($p < .05$), and urine output was significantly reduced by 25% when compared with plain water, despite being collected over a longer period (urine collection took place after the end of the exercise capacity trial). Another recent paper evaluated commercially available bottled coconut water and coconut water from concentrate, but failed to provide the beverage composition of the drinks and presented internally inconsistent results (Kalman et al. 2012).

In summary, available research on post-exercise rehydration with fresh coconut water shows either better fluid retention or no difference in comparison with plain water. For that reason and because of its good palatability and tolerance when ingested in large volumes, coconut water is deemed suitable as a rehydration beverage.

20.5.2 JAMAICA (ROSELLE) FLOWER INFUSION

Many herbal infusions are ingested—not necessarily for exercise hydration purposes—in different regions of the world, and some have been studied for their rehydration effectiveness, such as rooibos tea (Utter et al. 2010), regular tea (Chang et al. 2010), and lemon tea (Wong and Chen 2011), but they have been found to be no more effective than plain water. An interesting beverage in this category is Roselle flower infusion, a beverage ingested in large volumes as a refreshing drink in Mexico and other parts of the world. Roselle flower or Jamaica flower (*Hibiscus sabdariffa*) infusion has purported medicinal properties and is popularly thought to be a diuretic; this would make it a bad selection for post-exercise rehydration. A study by Mayol-Soto and Aragón-Vargas (2002) compared Roselle infusion with plain water and a conventional sports drink, with the purpose of verifying whether Roselle tea would induce diuresis beyond the normal effect of a high water volume load. Sixteen young men exercised in the heat and lost approximately 2.3% of their BM. They ingested one of the three beverages on each opportunity, in random order, in a volume equivalent to 150% of sweat lost. While rehydration was almost identical with the three drinks, urine composition and output dynamics were different with each beverage. Because urine output was not higher with Roselle infusion than with plain water, the authors concluded that Roselle tea did not show a diuretic effect 3 h after consumption of ~2.4 L in exercise-dehydrated subjects. Therefore, there seems to be no basis for contraindicating Roselle infusion as a post-exercise rehydration beverage.

20.5.3 MILK AND CHOCOLATE MILK

Milk is not necessarily considered a natural drink, because of the high processing normally involved in pasteurization, bottling, and distribution. It is, however, widely consumed by humans, particularly in regions where lactose intolerance (a characteristic strongly associated with race [De Vrese et al. 2001]) is not prevalent. Several

TABLE 20.5

Comparison of a Conventional Sports Drink with Different Types of Milk and Chocolate Milk

	Gatorade®	Whole Milk	2% Milk	Skim Milk	Chocolate Milk
Energy (kcal)	52	155	125	83	213
Energy (kJ)	218	649	525	347	892
Protein (g)	0	8	8	8	8
Fat (g)	0	8.3	5	0.3	5
CHO (g)	15	12	12	12	34
Sodium (mg)	115	126	129	133	150
Potassium (mg)	31	391	398	431	422
Calcium (mg)	0	300	300	300	248

Source: Translated from Aragón-Vargas, L.F. La leche... ¿Bebida deportiva? http://www.kerwa.ucr.ac.cr/ handle/10669/444, 2009. Used with permission. This work is licensed under a Creative Commons Attribution 4.0 International License.

Data are reported for 250 mL, as obtained from nutritional information labels from Dos Pinos® Dairy Products (San José, Costa Rica), with the exception of sodium and potassium contents which are adapted from Roy, B.D., *J. Int. Soc. Sports Nutr.*, 5, 15, 2008.

reviews have discussed the qualities of milk in association with sports performance, protein synthesis during recovery, and post-exercise rehydration (Aragón-Vargas 2009; James 2012; Roy 2008); this section will focus on the hydration aspect.

Before looking at milk and its potential use as a rehydration beverage, a quick analysis of its composition is warranted (see Table 20.5). Milk is not very different from a conventional sports drink in terms of carbohydrate and sodium content, but it has a higher energy density and it contains fat and protein, together with a higher concentration of potassium and calcium. Probably the most important differences among milk types, for rehydration purposes, are total energy, and fat and carbohydrate content. Skim milk is the most commonly tested dairy product in association with exercise and hydration.

Shirreffs et al. (2007) compared the effectiveness of low fat milk (with or without added sodium) with a conventional sports drink and water. Eleven subjects exercised intermittently in the heat until they reached a dehydration of ~1.8% BM. They replaced 150% of the sweat losses (~2 L) on four different opportunities, one with each beverage, in a cross over design; the total volume was ingested in 1 h. Subjects were monitored for 4 h after rehydration was completed. Partial urine output was remarkably different among the drinks, because water and the sports drink resulted in the typical pattern of a marked increase in urine volume 1 and 2 h after ingestion, but the two milk products did not. Furthermore, total urine output was lower with milk (611 ± 207 mL; mean ± SD) and with milk + sodium (550 ± 141 mL), compared with water (1184 ± 321 mL) or the sports drink (1205 ± 142 mL) ($p < .001$). The authors concluded that milk can be an effective beverage for post-exercise rehydration, as long as the individual has no lactose intolerance.

Watson et al. published a study in 2008, comparing the effects of milk and a carbohydrate-electrolyte drink on the restoration of fluid balance and exercise capacity. They dehydrated seven males to ~2% BM with an intermittent exercise protocol in the heat, and rehydrated them with 150% of their sweat losses with a sports drink or skim milk, following them up for 3 h. No comparison with water was attempted in this case. Exercise capacity results are beyond the scope of this chapter, but were almost identical between drinks. While net fluid balance was not different between beverages at the end of monitoring ($p < .051$), it was positive for the milk ($p = .02$ vs. baseline) but not different from zero for the sports drink ($p = .796$). Cumulative urine output was not different at the end of follow-up ($p = .056$), but it was lower for milk at the 2-h time point ($p < .05$). Another rehydration measure, percentage of the drink retained at the end of the recovery period, was greater with the milk than with the carbohydrate-electrolyte drink ($p = .045$). The authors pointed out the ability of skim milk to match a sports drink specifically designed for rehydration and performance, but warned individuals who are lactose intolerant against its use.

A recent study comparing water, a sports drink, and skim milk in children and adolescents confirmed earlier findings with adults (Volterman et al. 2014). Thirty-eight heat-acclimated boys and girls, 7–17 years of age, completed a fixed-time, moderate-intensity exercise protocol in the heat, on three separate opportunities, in a randomized, repeated-measures cross-over design. They dehydrated to ~1.3% BM, and ingested each of the three beverages in a volume equivalent to 100% of BM loss. The authors made several important comparisons by gender and age group, which are beyond the scope of this chapter; most importantly, the rehydration measures showed a better fraction of beverage retention for the skim milk than the sports drink, and both were better than water ($p < .05$). Body fluid balance was negative for all three drinks at the end of the 2 h of monitoring, but it was significantly less negative for the skim milk than for water ($p < .05$), and cumulative urine output was lower for skim milk than both sports drink and water ($p < .001$). Beverage acceptance was measured on a scale from 1 to 9, with 9 being the worst score, and participants reported enjoying the taste of the sports drink (2.3 ± 1.5) better than water (3.7 ± 1.7) and skim milk (3.4 ± 1.5) ($p < .001$). The authors concluded that skim milk was found to be more effective than water and the selected sports drink at replacing fluid losses that occur during exercise in the heat, but mentioned the limitation of skim milk for practical uses since it was not as palatable as the sports drink.

Because of the important role played by voluntary fluid intake in rehydration during and after exercise in the heat. Mateos Román and Aragón-Vargas (2011) completed a study with 31 male soccer players, 10–14 years of age, who rehydrated *ad libitum* with water and partially skimmed milk (session A), or water and partially skimmed chocolate milk (session B) in a randomized fashion while exercising in the heat. The boys exercised in the heat in a controlled environment chamber at a moderate intensity, for a total of 120 min. Each dairy drink was always presented simultaneously with a bottle of water to avoid producing a floor effect, that is, that boys would drink more of a beverage they didn't like simply because they were hot and thirsty; beverage temperature was ~16°C.

The boys arrived to sessions A and B in practically identical conditions, and recorded the same exercise intensity and thermal stress. Sweat rates were very similar (about 460 mL/h), and fluid balance was positive and the same for both sessions at the end of exercise ($\approx 0.76\%$ BM). Boys drank the same volume of water, milk, and chocolate milk but, in general terms, milk and chocolate milk scored better for palatability than water. Chocolate milk was assigned higher scores for sweetness (7.6, $p < .001$), flavor intensity (5.8, $p < .001$), liking (7.9, $p < .001$), and overall acceptance (8.7, $p < .001$) when compared with milk and water; these two were not significantly different. In terms of overall acceptance, though, chocolate milk showed a tendency for declining scores over time, while milk remained constant and water increased over time. The authors concluded that when presented simultaneously with water, both partially skimmed milk and chocolate milk were effective in preventing voluntary dehydration in boys exercising in the heat. Palatability scores were favorable and GI symptoms were not clinically relevant.

To summarize, there is good experimental evidence to support skim milk as an effective beverage for rehydration after exercise in the heat, both in children, adolescents, and adults. Apparently, not only does the sodium content promote fluid retention, but the total energy content in the form of carbohydrate and protein may be inducing a slower gastric emptying, delaying fluid delivery to the bloodstream and causing a smaller plasma volume expansion with the concomitant lower diuresis (Roy 2008). As far as chocolate milk is concerned, more experimental evidence is needed, particularly in the light of its high palatability. Rehydration with milk products is contraindicated for lactose-intolerant individuals.

20.5.4 MINERAL WATER

Once it has been understood that the presence of electrolytes (minerals) in beverages favors rehydration, it naturally follows that mineral water should be a good choice for this purpose. However, naturally occurring minerals in melted ice and snow filtered through the mountains and obtained from natural springs are usually present in very low concentrations. In fact, many commercially available waters boast of their low mineral content, a quality called *oligominerale*. In spite of their common use, even by professional teams (Shirreffs et al. 2005), few studies have experimentally evaluated the efficacy of mineral waters for post-exercise rehydration.

Shirreffs et al. (2007) compared the hydration effectiveness of four commonly used drinks after exercise in the heat in eight subjects who dehydrated to about 2% BM. In a repeated-measures design, each participant ingested one of four beverages in randomized order: a sports drink with 6% carbohydrate, 23 mEq/L sodium, 6 mEq/L potassium and 17 mEq/L chloride; mineral water A with no carbohydrate, and virtually no sodium or potassium, and 3 mEq/L chloride; mineral water B with no carbohydrate, 1 mEq/L sodium, 0 mEq/L potassium, and 5 mEq/L chloride; and a homemade beverage mixing four parts of apple juice and six parts of carbonated mineral water (Apfelschörle), providing 6.7% carbohydrate, 8 mEq/L sodium, 30 mEq potassium, and 1 mEq/L chloride. Participants drank a volume equivalent to 150% of sweat losses in 1 h, and were monitored for 4 h. Only the sports drink was able to maintain euhydration by the end of monitoring, suggesting that the mineral content in the waters used was too low to have a positive impact on net fluid balance. The mineral water used by Real Madrid and reported by Shirreffs et al. in 2005 had similarly low concentrations of sodium (2.3 mEq/L), potassium (0.3 mEq/L) and chloride (2.2 mEq/L), and would not be expected to produce better results.

20.6   OTHER INGREDIENTS

Very little work has been published on the rehydration qualities of other potential ingredients such as magnesium (Brilla et al. 2003), calcium (Brancaccio et al. 2012), oxygen, taurine (Janeke et al. 2003), betaine (Sayed and Downing 2011), and vitamins (Snell et al. 2010), even though they have all otherwise been studied in association with sports performance. Some of the experiments have not yet been performed in humans, and most of them have not studied these ingredients independently. None of them seem promising as possible elements in rehydration beverages.

There is some information on glutamine, which has been tested as a potential ingredient in ORS for the treatment of diarrhea. Schedl et al. had proposed in 1994 that a glutamine-glucose ORS had the potential to maximize water, electrolyte, and energy transport in the small intestine. Unfortunately, glutamine has not proven to be more effective when compared with a standard World Health Organization ORS (Gutiérrez et al. 2007, Ribeiro Júnior et al. 1994). A recently published experiment, focusing on exercise performance (Hoffman et al. 2010), is probably the only exercise-related study on glutamine and hydration; hormone concentrations associated with fluid balance were reported, but no data were presented regarding fluid retention, net fluid balance, or urine output. Available data on glutamine are not enough to support its use for post-exercise rehydration.

20.7   SUMMARY

A number of natural drinks and ingredients used in hydration beverages have been discussed, focusing on their effectiveness for post-exercise rehydration. Because of excess diuresis, complete, rapid, and sustained restoration of fluid balance remains a challenge. The most effective ingredient to promote fluid retention is sodium, although the possible merits of milk protein, creatine, and additional potassium were presented.

Cultural preferences may lead some individuals to rehydrate with Roselle infusion, mineral water, or coconut water; doing so has been shown not to deter from the goal of restoring body fluid. Skimmed milk has been shown to be more effective than conventional sports drinks and water and may be a good rehydration choice in the absence of lactose intolerance. Furthermore, sports drinks with added milk or whey protein, which may be desirable to provide amino acids and help recovery, have been shown not to impair rehydration. The use of caffeinated beverages, although favored by many consumers and shown not to impair chronic hydration, remains more of an open question for quick post-exercise rehydration. Evidence was provided to show why regular beer is contraindicated when effective post-exercise rehydration is desired, together with evidence about the serious limitations associated with the use of glycerol.

REFERENCES

Aragón-Vargas, L. F. October 2009. La leche... ¿Bebida deportiva?. http://www.kerwa.ucr. ac.cr/handle/10669/444.

Aragón-Vargas, L. F., and K. Madriz-Davila. 2000. Incomplete warm-climate, post-exercise rehydration with water, coconut water, or a sports drink. *Med. Sci. Sports Exer.* 32(5):s238.

Aragón-Vargas, L. F., and S. M Shirreffs. August 2014. Rehydration following exercise in the heat: A look at alternative formulations and the role of plasma volume expansion in excess diuresis. http://kerwa.ucr.ac.cr/handle/10669/11115.

Armstrong, L. E. 2002. Caffeine, body fluid-electrolyte balance, and exercise performance. *Int. J. Sport Nutr. Exer. Metab.* 12(2):189–206.

Armstrong, L. E., D. J. Casa, C. M. Maresh, and M. S. Ganio. 2007. Caffeine, fluid-electrolyte balance, temperature regulation, and exercise-heat tolerance. *Exer. Sport Sci. Rev.* 35(3):135–40.

Baker, L. B., and A. E. Jeukendrup. 2014. Optimal composition of fluid-replacement beverages. *Compr. Physiol.* 4(2):575–620. doi:10.1002/cphy.c130014.

Bemben, M. G., and H. S. Lamont. 2005. Creatine supplementation and exercise performance: Recent findings. *Sports Med.* 35(2):107–25.

Brancaccio, P., F. M. Limongelli, I. Paolillo, A. D'Aponte, V. Donnarumma, and L. Rastrelli. 2012. Supplementation of Acqua Lete® (Bicarbonate Calcic Mineral Water) improves hydration status in athletes after short term anaerobic exercise. *J. Int. Soc. Sports Nutr.* 9(1):35. doi:10.1186/1550-2783-9-35.

Brilla, L. R., M. S. Giroux, A. Taylor, and K. M. Knutzen. 2003. Magnesium-creatine supplementation effects on body water. *Metabolism* 52(9):1136–40.

Brouns, F., E. M. Kovacs, and J. M. Senden. 1998. The effect of different rehydration drinks on post-exercise electrolyte excretion in trained athletes. *Int. J. Sports Med.* 19(1):56–60.

Burke, L. M. 2008. Caffeine and sports performance. *Appl. Physiol. Nutr. Metab.* 33(6):1319–34. doi:10.1139/H08-130.

Campbell-Falck, D., T. Thomas, T. M. Falck, N. Tutuo, and K. Clem. 2000. The intravenous use of coconut water. *Am. J. Emerg. Med.* 18(1):108–11.

Chang, C.-Q., Y.-B. Chen, Z.-M. Chen, and L.-T. Zhang. 2010. Effects of a carbohydrate-electrolyte beverage on blood viscosity after dehydration in healthy adults. *Chinese Med. J.* 123(22):3220–5.

Child, R, and W. Nathanael. 1950. Changes in the sugar composition of coconut water during maturation and germination. *J. Sci. Food Agric.* 1:326–9.

Coxam, V. 2007. Current data with inulin-type fructans and calcium, targeting bone health in adults. *J. Nutr.* 137(11 Suppl):2527S–33S.

Dalbo, V. J., M. D. Roberts, J. R. Stout, and C. M. Kerksick. 2008. Putting to rest the myth of creatine supplementation leading to muscle cramps and dehydration. *Br. J. Sports Med.* 42(7): 567–73. doi:10.1136/bjsm.2007.042473.

Del Coso, J., E. Estevez, and R. Mora-Rodriguez. 2009. Caffeine during exercise in the heat: Thermoregulation and fluid-electrolyte balance. *Med. Sci. Sports Exer.* 41(1):164–73.

Desbrow, B., D. Murray, and M. Leveritt. 2013. Beer as a sports drink? Manipulating beer's ingredients to replace lost fluid. *Int. J. Sport Nutr. Exer. Metab.* 23(6):593–600.

De Vrese, M., A. Stegelmann, B. Richter, S. Fenselau, C. Laue, and J. Schrezenmeir. 2001. Probiotics-compensation for lactase insufficiency. *Am. J. Clin. Nutr.* 73(2 Suppl):421S–9S.

Doherty, M., and P. M. Smith. 2005. Effects of caffeine ingestion on rating of perceived exertion during and after exercise: A meta-analysis. *Scand. J. Med. Sci. Sports* 15(2):69–78.

Eggleton, M. G. 1942. The diuretic action of alcohol in man. *J. Physiol.* 101(2):172–91.

Eiseman, B. 1954. Intravenous infusion of coconut water. *AMA Arch. Surg.* 68:167–78.

Evans, G. H., S. M. Shirreffs, and R. J. Maughan. 2009. Acute effects of ingesting glucose solutions on blood and plasma volume. *Br. J. Nutr.* 101(10):1503–8.

Falk, B., R. Burstein, J. Rosenblum, Y. Shapiro, E. Zylber-Katz, and N. Bashan. 1990. Effects of caffeine ingestion on body fluid balance and thermoregulation during exercise. *Can. J. Physiol. Pharmacol.* 68(7):889–92.

Find the alcohol content of your favorite beer. 2014. Accessed August 22. http://www. alcoholcontents.com/beer/.

Flores-Salamanca, R., and L. F. Aragón-Vargas. 2014. Postexercise rehydration with beer impairs fluid retention, reaction time, and balance. *Appl. Physiol. Nutr. Metab.* 39:1175–81. doi:10.1139/apnm-2013-0576. http://kerwa.ucr.ac.cr/handle/10669/9237.

Freund, B. J., S. J. Montain, A. J. Young et al. 1995. Glycerol hyperhydration: Hormonal, renal, and vascular fluid responses. *J. Appl. Physiol.* 79(6):2069–77.

Ganio, M. S., D. J. Casa, L. E. Armstrong, and C. M. Maresh. 2007. Evidence-based approach to lingering hydration questions. *Clin. Sports Med.* 26(1):1–16.

Gonzalez-Alonso, J., C. L. Heaps, and E. F. Coyle. 1992. Rehydration after exercise with common beverages and water. *Int. J. Sports Med.* 13(5):399–406.

Goulet, E. D. B., M. Aubertin-Leheudre, G. E. Plante, and I. J. Dionne. 2007. A meta-analysis of the effects of glycerol-induced hyperhydration on fluid retention and endurance performance. *Int. J. Sport Nutr. Exer. Metab.* 17(4):391–410.

Gutiérrez, C., S. Villa, F. R. Mota, and J. J. Calva. 2007. Does an L-glutamine-containing, glucose-free, oral rehydration solution reduce stool output and time to rehydrate in children with acute diarrhoea? A double-blind randomized clinical trial. *J. Health Pop. Nutr.* 25(3):278–84.

Hobson, R., and L. James. 2015. The addition of whey protein to a carbohydrate-electrolyte drink does not influence post-exercise rehydration. *J. Sports Sci.* 33(1):77–84. doi:10. 1080/02640414.2014.925570.

Hoffman, J. R., N. A. Ratamess, J. Kang et al. 2010. Examination of the efficacy of acute L-alanyl-L-glutamine ingestion during hydration stress in endurance exercise. *J. Int. Soc. Sports Nutr.* 7:8. doi:10.1186/1550-2783-7-8.

Hoon, M. W., N. A. Johnson, P. G. Chapman, and L. M. Burke. 2013. The effect of nitrate supplementation on exercise performance in healthy individuals: A systematic review and meta-analysis. *Int. J. Sport Nutr. Exer. Metab.* 23(5):522–32.

Irwin, C., M. Leveritt, D. Shum, and B. Desbrow. 2013. The effects of dehydration, moderate alcohol consumption, and rehydration on cognitive functions. *Alcohol* 47(3):203–13.

Ismail, I., R. Singh, and R. G. Sirisinghe. 2007. Rehydration with sodium-enriched coconut water after exercise-induced dehydration. *Southeast Asian J. Trop. Med. Public Health* 38(4):769–85.

James, L. 2012. Milk protein and the restoration of fluid balance after exercise. *Med. Sport Sci.* 59:120–6.

James, L. J., D. Clayton, and G. H. Evans. 2011. Effect of milk protein addition to a carbohydrate-electrolyte rehydration solution ingested after exercise in the heat. *Br. J. Nutr.* 105(3):393–9.

James, L. J., G. H. Evans, J. Madin et al. 2013. Effect of varying the concentrations of carbohydrate and milk protein in rehydration solutions ingested after exercise in the heat. *Br. J. Nutr.* 110(7): 285–91. doi:10.1017/S0007114513000536.

James, L. J., R. Gingell, and G. H. Evans. 2012. Whey protein addition to a carbohydrate-electrolyte rehydration solution ingested after exercise in the heat. *J. Athl. Train.* 47(1):61–6.

James, L. J., L. Mattin, P. Aldiss, R. Adebishi, and R. M. Hobson. 2014. Effect of whey protein isolate on rehydration after exercise. *Amino Acids* 46(5):1217–24. doi:10.1007/ s00726-014-1680-8.

Janeke, G., W. Siefken, S. Carstensen et al. 2003. Role of taurine accumulation in keratinocyte hydration. *J. Invest. Dermatol.* 121(2):354–61. doi:10.1046/j.1523-1747.2003.12366.x.

Jimenez, C., N. Koulmann, I. Mischler et al. 2002. Plasma compartment filling after exercise or heat exposure. *Med. Sci. Sports Exer.* 34(10):1624–31.

Jones, A. M. 2014. Influence of dietary nitrate on the physiological determinants of exercise performance: A critical review. *Appl. Physiol. Nutr. Metab.* 39(9):1019–28. doi:10.1139/apnm-2014-0036.

Kalman, D. S., S. Feldman, D. R. Krieger, and R. J. Bloomer. 2012. Comparison of coconut water and a carbohydrate-electrolyte sport drink on measures of hydration and physical performance in exercise-trained men. *J. Int. Soc. Sports Nutr.* 9(1):1.

Kavouras, S. A., L. E. Armstrong, C. M. Maresh et al. 2006. Rehydration with glycerol: Endocrine, cardiovascular, and thermoregulatory responses during exercise in the heat. *J. Appl. Physiol.* 100(2):442–50. doi:10.1152/japplphysiol.00187.2005.

Kovacs, E. M., R. M. Schmahl, J. M. Senden, and F. Brouns. 2002. Effect of high and low rates of fluid intake on post-exercise rehydration. *Int. J. Sport Nutr. Exer. Metab.* 12(1):14–23.

Kovacs, E. M. R., J. H. C. H. Stegen, and F. Brouns. 1998. Effect of caffeinated drinks on substrate metabolism, caffeine excretion, and performance. *J. Appl. Physiol.* 85(2):709–15.

Kuberski T, Roberts A, Lineham B, Bryden N, and Teburae M. 1979. Coconut water as a rehydration fluid. *N. Z. Med. J.* 641(90):98–100.

Laitano, O., S. J. Trangmar, D. de Melo Marins, E. Soares Menezes, and G. da Silva Reis. 2014. Improved exercise capacity in the heat followed by coconut water consumption. *Motriz: Rev. Educação Física* 20(1):107–11. doi:10.1590/S1980-65742014000100016.

Lecoultre, V., and Y. Schutz. 2009. Effect of a small dose of alcohol on the endurance performance of trained cyclists. *Alcohol* 44(3):278–83. doi:10.1093/alcalc/agn108.

Lopez, R. M., D. J. Casa, B. P. McDermott, M. S. Ganio, L. E. Armstrong, and C. M. Maresh. 2009. Does creatine supplementation hinder exercise heat tolerance or hydration status? A systematic review with meta-analyses. *J. Athl. Train.* 44(2):215–23. doi:10.4085/1062-6050-44.2.215.

Mateos Román, A., and L. F. Aragón-Vargas. September 2011. Voluntary fluid intake with milk or chocolate milk in boys exercising in the heat. http://www.kerwa.ucr.ac.cr/handle/10669/528.

Maughan, R. J., and J. Griffin. 2003. Caffeine ingestion and fluid balance: A review. *J. Hum. Nutr. Diet.* 16(6):411–20.

Mayol-Soto, M. L., and L. F. Aragón-Vargas. 2002. Rehidratación post-ejercicio con diferentes tipos de bebidas: Agua pura, bebida deportiva, y agua de Jamaica. *Pensar en Movimiento: Revista de Ciencias Del Ejercicio Y La Salud* 2(1):41–54.

Mayol-Soto, M. L., and L. F. Aragón-Vargas. 2010. Estrategias de rehidratación post-ejercicio: tasa de ingesta de líquido y tipo de bebida. *Pensar En Movimiento: Revista de Ciencias Del Ejercicio Y La Salud* 7(1):1. doi:10.15517/pensarmov.v7i1.372.

Mujika, I., and L. M. Burke. 2010. Nutrition in team sports. *Ann. Nutr. Metab.* 57(Suppl 2):26–35. doi:10.1159/000322700.

Murray, M. M. 1932. The diuretic action of alcohol and its relation to pituitrin. *The Journal of Physiology* 76(3):379–86.

Nelson, J. L., and R. A. Robergs. 2007. Exploring the potential ergogenic effects of glycerol hyperhydration. *Sports Med.* 37(11):981–1000.

Parr, E. B., D. M. Camera, J. L. Areta et al. 2014. Alcohol ingestion impairs maximal post-exercise rates of myofibrillar protein synthesis following a single bout of concurrent training. *PloS One* 9(2):e88384. doi:10.1371/journal.pone.0088384.

Pasantes-Morales, H., R. A. Lezama, G. Ramos-Mandujano, and K. L. Tuz. 2006. Mechanisms of cell volume regulation in hypo-osmolality. *Am. J. Med.* 119(7 Suppl 1):S4–11.

Passe, D., M. Horn, and R. Murray. 1997. The effects of beverage carbonation on sensory responses and voluntary fluid intake following exercise. *Int. J. Sport Nutr.* 7:286–97.

Passe, D. H., M. Horn, J. Stofan, and R. Murray. 2004. Palatability and voluntary intake of sports beverages, diluted orange juice, and water during exercise. *Int. J. Sport Nutr. Exer. Metab.* 14(3):272–84.

Pérez-Idárraga, A., and L. F. Aragón-Vargas. 2011a. Rehidratación posejercicio: la forma de distribuir la ingesta de un volumen constante de líquido no altera su conservación. *Pensar En Movimiento: Revista de Ciencias Del Ejercicio Y La Salud* 9(1):12–21.

Pérez-Idarraga, A., and L. F. Aragón-Vargas. 2011b. Rehidratación post-ejercicio con agua de coco: igual o más efectiva que una bebida deportiva? *MH Salud* 8(1):1–17.

Pérez-Idárraga, A., and L. F. Aragón-Vargas. 2014. Postexercise rehydration: Potassium-rich drinks versus water and a sports drink. *Appl. Physiol. Nutr. Metab.* doi:10.1139/apnm-2013-0434.

Picado Twight, C. 1942. El Agua de coco como medio de cultivo. *Boletín de La Oficina Sanitaria Panamericana.* 21(10):960–965.

Ribeiro Júnior, H., T. Ribeiro, A. Mattos et al. 1994. Treatment of acute diarrhea with oral rehydration solutions containing glutamine. *J. Am. Coll. Nutr.* 13(3):251–5.

Riedesel, M. L., D. Y. Allen, G. T. Peake, and K. Al-Qattan. 1987. Hyperhydration with glycerol solutions. *J. Appl. Physiol.* 63(6):2262–8.

Rivera-Brown, A., L. F. Aragón-Vargas, Y. Cabrera-Dávila, and L. E. Berríos. August 2014. Palatability and voluntary intake of three commercially available sports drinks and unflavored water during prolonged exercise in hot and humid conditions. http://www.kerwa.ucr.ac.cr/handle/10669/11105.

Rivera-Brown, A. M., L. F. Aragón-Vargas, Y. Cabrera-Dávila, and L. E. Berríos. 2007. Voluntary intake of sports drinks and water in males and females running in hot environment. *Med. Sci. Sports Exer.* 39(Supplement):S315. doi:10.1249/01.mss.0000274223.49529.eb.

Rolle, R. 2014. Good practice for the small-scale production of bottled coconut water. Accessed September 4. http://www.fao.org/docrep/010/a1418e/a1418e00.htm.

Roy, B. D. 2008. Milk: The new sports drink? A review. *J. Int. Soc. Sports Nutr.* 5:15.

Saat, M., R. Singh, R. G. Sirisinghe, and M. Nawawi. 2002. Rehydration after exercise with fresh young coconut water, carbohydrate-electrolyte beverage and plain water. *J. Physiol. Anthropol. Appl. Human Sci.* 21(2):93–104.

Sayed, M. A. M., and J. Downing. 2011. The effects of water replacement by oral rehydration fluids with or without betaine supplementation on performance, acid-base balance, and water retention of heat-stressed broiler chickens. *Poultry Science* 90(1):157–67. doi:10.3382/ps.2009-00594.

Schedl, H. P., R. J. Maughan, and C. V. Gisolfi. 1994. Intestinal absorption during rest and exercise: Implications for formulating an oral rehydration solution (ORS). Proceedings of a roundtable discussion, April 21–22, 1993. *Med. Sci. Sports Exer.* 26(3):267–80.

Scheett, T. P., M. J. Webster, and K. D. Wagoner. 2001. Effectiveness of glycerol as a rehydrating agent. *Int. J. Sport Nutr. Exer. Metab.* 11(1):63–71.

Seifert, J., J. Harmon, and P. DeClercq. 2006. Protein added to a sports drink improves fluid retention. *Int. J. Sport Nutr. Exer. Metab.* 16(4):420–9.

Shirreffs, S. M., L. F. Aragon-Vargas, M. Chamorro, R. J. Maughan, L. Serratosa, and J. J. Zachwieja. 2005. The sweating response of elite professional soccer players to training in the heat. *Int. J. Sports Med.* 26:90–5.

Shirreffs, S. M., L. F. Aragon-Vargas, M. Keil, T. D. Love, and S. Phillips. 2007. Rehydration after exercise in the heat: A comparison of 4 commonly used drinks. *Int. J. Sport Nutr. Exer. Metab.* 17(3):244–58.

Shirreffs, S. M., and R. J. Maughan. 1997. Restoration of fluid balance after exercise-induced dehydration: Effects of alcohol consumption. *J. Appl. Physiol.* 83(4):1152–8.

Shirreffs, S. M., and R. J. Maughan. 2006. The effect of alcohol on athletic performance. *Curr. Sports Med. Rep.* 5(4):192–6.

Shirreffs, S. M., P. Watson, and R. J. Maughan. 2007. Milk as an effective post-exercise rehydration drink. *Br. J. Nutr.* 98(1):173–80.

Snell, P. G., R. Ward, C. Kandaswami, and S. J. Stohs. 2010. Comparative effects of selected non-caffeinated rehydration sports drinks on short-term performance following moderate dehydration. *J. Int. Soc. Sports Nutr.* 7:28.

Soto, E, Fernandez, E, and Calderin, O. 1942. Coconut water: A clinical and experimental study. *Am. J. Dis. Child* 64:977–96.

Spriet, L. L. 1995. Caffeine and performance. *Int. J. Sport Nutr.* 5:S84–99.

Top 10 best beers in the World. 2014. *List Crux.* Accessed August 21. http://listcrux.com/top-10-best-beers-in-the-world/.

Utter, A. C., J. C. Quindry, G. P. Emerenziani, and J. S. Valiente. 2010. Effects of rooibos tea, bottled water, and a carbohydrate beverage on blood and urinary measures of hydration after acute dehydration. *Res. Sports Med.* 18(2):85–96.

Van Rosendal, S. P., and J. S. Coombes. 2012. Glycerol use in hyperhydration and rehydration: Scientific update. *Med. Sport Sci.* 59:104–12.

Van Rosendal, S. P., M. A. Osborne, R. G. Fassett, and J. S. Coombes. 2009. Physiological and performance effects of glycerol hyperhydration and rehydration. *Nutr. Rev.* 67(12):690–705. doi:10.1111/j.1753-4887.2009.00254.x.

Van Rosendal, S. P., M. A. Osborne, R. G. Fassett, and J. S. Coombes. 2010. Guidelines for glycerol use in hyperhydration and rehydration associated with exercise. *Sports Med.* 40(2):113–29.

Vigliar, R., V. L. Sdepanian, and U. Fagundes-Neto. 2006. Biochemical Profile of coconut water from coconut palms planted in an Inland Region. *J. Pediatr. (Rio J.)* 82(4):308–12.

Volterman, K. A., J. Obeid, B. Wilk, and B. W. Timmons. 2014. Effect of milk consumption on rehydration in youth following exercise in the heat. *Appl. Physiol. Nutr. Metab.* doi:10.1139/apnm-2014-0047.

Watson, P., T. D. Love, R. J. Maughan, and S. M. Shirreffs. 2008. A comparison of the effects of milk and a carbohydrate-electrolyte drink on the restoration of fluid balance and exercise capacity in a hot, humid environment. *Eur. J. Appl. Physiol.* 104(4):633–42.

Wemple, R. D., D. R. Lamb, and K. H. McKeever. 1997. Caffeine vs caffeine-free sports drinks: Effects on urine production at rest and during prolonged exercise. *Int. J. Sports Med.* 18(1):40–6. doi:10.1055/s-2007-972593.

Wong, S. H., and Y. Chen. 2011. Effect of a carbohydrate-electrolyte beverage, lemon tea, or water on rehydration during short-term recovery from exercise. *Int. J. Sport Nutr. Exer. Metab.* 21(4):300–10.

World Anti-Doping Agency. 2014. Welcome to the List | Wada Prohibited List. *English.* Accessed September 18. http://list.wada-ama.org/.

Index

Note: Locator followed by '*f*' and '*t*' denotes figure and table in the text